MW01259936

Fundamentals of Linear State Space Systems

Fundamentals of Linear State Space Systems

John S. Bay

Virginia Polytechnic Institute and State University

Boston Burr Ridge, IL Dubuque, IA Madison, WI New York San Francisco St. Louis
Bangkok Bogotá Caracas Lisbon London Madrid
Mexico City Milan New Delhi Seoul Singapore Sydney Taipei Toronto

WCB/McGraw-Hill 

A Division of The McGraw-Hill Companies

FUNDAMENTALS OF LINEAR STATE SPACE SYSTEMS

This book is printed on acid-free paper.

1 2 3 4 5 6 7 8 9 0 DOC/DOC 9 3 2 1 0 9 8

ISBN 0-256-24639-4

Vice president and editorial director: *Kevin T. Kane*
Publisher: *Thomas Casson*
Executive editor: *Elizabeth A. Jones*
Developmental editor: *Bradley Kosirog*
Marketing manager: *John T. Wannemacher*
Senior project manager: *Mary Conzachi*
Senior production supervisor: *Heather D. Burbridge*
Freelance design coordinator: *JoAnne Schopler*
Cover designer: *James E. Shaw*
Supplement coordinator: *Marc Mattson*
Printer: *R. R. Donnelley & Sons Company*

Library of Congress Cataloging-in-Publication Data

Bay, John S.
Fundamentals of linear state space systems / John. S. Bay.
p. cm.
Includes index.
ISBN 0-256-24639-4
1. Linear systems. 2. State-space methods. I. Title.
QA402.B39 1999
003'.74—dc21

98-25177

http://www.mhhe.com

Preface

This book wrote itself over the period of August 1990 to December 1997. It is a result of my teaching the graduate course *Linear System Theory* at Virginia Polytechnic Institute and State University. This is a first course in linear systems taught in the Bradley Department of Electrical and Computer Engineering.

Target Audience

The book is intended to be a comprehensive treatment of the use of linear state space system theory in engineering problems. It is targeted at seniors and first-year graduate students, although much of the material will be accessible to students with only an understanding of basic signals and systems principles. It is intended to gather into a single volume the linear algebra, vector space, and state space theories now used in many engineering texts, but which are often covered in separate courses and separate departments. The book will have appeal to students in all engineering departments.

Whereas many texts introduce state space theory, it is often presented as a supplement to frequency-domain material, such as after classical methods in control systems or after transfer functions in signals and systems texts. Such texts often forsake the mathematical basics necessary for true understanding of state space modeling and analysis. Rather than use frequency-domain analysis as a prelude to state space, this text uses the more natural and meaningful foundation of vector spaces and linear algebra. Thus, state space analysis can be understood from the mathematical foundations of its own domain, rather than as a counterpart to frequency-domain methods. This text would be ideal in a course dedicated to time-domain analysis (both continuous and discrete). It would also be an appropriate text for a school that treats state variable analysis as a stand-alone course, independent of a student's interest or preparation in control systems. It is written in such a way that it can be *read*; it is not merely a collection of linear algebraic facts arranged in an optimal manner.

Content and Organization

The text is organized into two parts. Part 1 begins with a review of linear algebra and vector spaces, both from a geometric viewpoint. This is done in a manner that complements the material presented in a student's mathematics courses, which sometimes leave the student confused by the distinction between linear algebra and matrix theory. It is assumed that students know some matrix theory (determinants, inverses, gaussian elimination, etc.), but not necessarily linear algebra on any abstract level. Furthermore, it exploits the engineering student's training in spatial relationships, facilitating intuitive understanding. By addressing the engineering student, we can focus on the practical matters, geometry, applications, and implementation issues of linear systems, thus maintaining a student's engineering context throughout. This mathematical introduction is rigorous enough to stand on its own, but not so encumbered by proofs that engineering relevance is sacrificed. While graduate students with sufficient mathematical background might skip it, even a student with a good understanding of vector spaces might benefit from the geometric perspective offered.

As part of the discussion of the mathematical preliminaries, linear algebraic systems are treated. Topics such as subspaces, orthogonal projections, basis changes, inner products, and linear transformations are critical to true understanding of the state space, so it is important that they be covered in some detail. Again, these methods are used to study the geometry of physical systems sometimes neglected in engineering texts. A student without knowledge of such issues would otherwise miss the underlying meaning of such common concepts of eigenvalues and eigenvectors, simultaneous equations, Fourier analysis, and similarity transformations.

Only after these algebraic topics are covered are linear differential methods introduced in Part 2. It is then that we cover the topics that are often given in controls texts as the linear system "basics." The latter part of the book contains control system applications and principles. For all of these latter chapters of the book, a familiarity with s-domain and ω-domain analysis is useful, but a deep understanding of classical control or signal processing is not required.

Both continuous-time and discrete-time systems are discussed throughout, although z-domain material is minimized. Because certain developments in state space systems are more easily understood in one domain or the other, this parallel presentation gives us the flexibility to introduce examples from either domain at our convenience. For example, controllability tests are particularly easy to derive in discrete-time, so that is where they should be first introduced.

It is inescapable that computer-aided engineering (CAE) is an integral component of linear system usage. There are now dozens of books dedicated to the use of MATLAB® for linear system and control system design. Recognizing the importance of the computer but wishing to avoid a book tied too closely to computer simulations, we will make use of margin notes wherever a MATLAB command might do the numerical work of the accompanying discussion, denoted by a superscript M on the applicable term, e.g., *rank*^{M}. For example, we use the margin to indicate the availability of certain MATLAB commands, functions, and toolboxes, but we do not assume that MATLAB programming is a required component of the course, nor does the book instruct on its usage. The margin notes refer the reader to Appendix B, which contains summaries for the commands most relevant to the topic at hand. In addition, the end-of-chapter exercises include some computer-based problems, but these problems will not necessarily be tailored to MATLAB.

Most of the material contained in this book can be covered in a single three semester-hour course. If the course is indeed an undergraduate or first-year graduate course, then Chapter 11 might not be covered in that time span. It is recommended that even students with extensive mathematical preparation not omit the early chapters, because the geometric perspective established in these chapters is maintained in the latter chapters. Furthermore, the applications-oriented examples denoted by italics in the table of contents are concentrated in Part 1. In most of the examples thereafter, it will usually be assumed that the physical application has been modeled in state space form. However, if little mathematical review is necessary, then Chapter 11, *Introduction to Optimal Control and Estimation*, can be used to tie together the contents of the previous ten chapters, and make the course more of a *control* than a *systems* course.

John S. Bay
Blacksburg, Virginia
1998

Acknowledgments

I would first like to thank my family and friends for their forbearance during the writing of this book. This includes my students and colleagues at Virginia Tech, who were all eager consultants. I am also grateful to the peer reviewers of the manuscript, many of whose suggestions I have incorporated within. I thank Ms. Melanie George and Mr. Jae Park for assisting me in the preparation of the final manuscript.

Next I would like to thank the teachers and thinkers in my past who have inspired me.

> Mr. Gary Franz, who admitted that he does math while sitting on the beach.
>
> Ms. Gail Colman, who was really very good at what she did.
>
> Mrs. Virginia McWhorter, who showed me that the academic life was a lot of fun.
>
> Professor Bogdan Baishanski, who in a very short time made me understand a lifetime of math that I never previously understood.
>
> Professor Charles Bostian, who has served me in many roles.
>
> Most of all, Professor Hooshang Hemami, a gentleman after whom I model my professional conduct.

Finally, I thank all those too many to name who supported my career and research pursuits, ultimately giving me this opportunity. Some of these people I have known for only a very short time, but they all made lasting impacts.

For my wife, Judy,
And to my mother, Mildred,
And to the memory of my father, John

Table of Contents

Listings in *italics* denote applications-oriented examples.

Part I

Mathematical Introduction to State Space

1

Models of Linear Systems

Linear systems are usually mathematically described in one of two domains: time-domain and frequency-domain. The frequency-domain approach (s- or ω-domain) usually results in a system representation in the form of a *transfer function*. Transfer functions represent the ratio of a system's frequency-domain output to the frequency-domain input, assuming that the initial conditions on the system are zero. Such descriptions are the subject of many texts in *signals and systems*.

In time-domain, the system's representation retains the form of a differential equation. However, as any student of engineering will attest, differential equations can be difficult to analyze. The mathematics gets more burdensome as the order of the equations increases, and the combination of several differential equations into one single system can be difficult.

In this chapter, we will introduce a time-domain representation of systems that alleviates some of the problems of working with single, high-order differential equations. We will describe a system with *state variables*, which collectively can be thought of as a vector. Using the language of vector analysis, we will demonstrate that state variables provide a convenient time-domain representation that is essentially the same for systems of all order. Furthermore, state variable descriptions do not assume zero initial conditions, and allow for the analysis and design of system characteristics that are not possible with frequency-domain representations. We will begin with some elementary definitions and a review of mathematical concepts. We will give a number of examples of state variable descriptions and introduce several of their important properties.

1.1 Linear Systems and State Equations

To define what we mean by a *linear system*, we will categorize the types of systems encountered in nature. First, a *system* is simply the mathematical description of a relationship between externally supplied quantities (i.e., those coming from outside the system) and the dependent quantities that result from the action or effect on those external quantities. We use the term "input" or u to refer to the independent quantity, although we indeed may have no control over it at all. It merely represents an excitation for the system. The response of the system will be referred to as the output y. These input and output signals may be constant, defined as functions over continuous-time or discrete-time, and may be either deterministic or stochastic. The system that relates the two may be defined in many ways, so for the time being, we depict it as in Figure 1.1, simply a block that performs some mathematical operation.

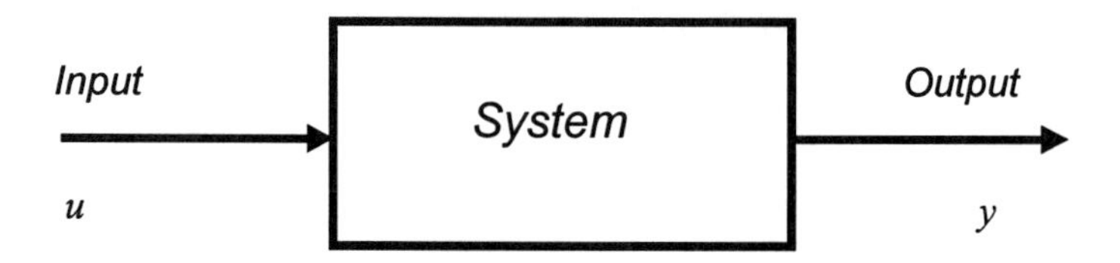

Figure 1.1 Elementary representation of a system acting on an input and producing an output.

1.1.1 Definitions and Review

In this section, we present some definitions for systems that will be useful in subsequent chapters. It is expected that the reader already has some familiarity and practice with these concepts from signals and systems studies.

> ***Memory:*** A system with *memory* is one whose output depends on itself from an earlier point in time. A system whose output depends only on the current time and the current input is *memoryless*. (1.1)

Systems with memory most often occur as differential equations (continuous-time), or as difference equations (discrete-time) because closed-form solutions of such systems require integration (or summation) of a quantity over past time. Systems with hysteresis are also examples of systems with memory because the portion of curve on which they operate depends on the past state and the direction of change of the input. For our purposes, we will have systems we call *algebraic*, which are memoryless, and *differential* or *difference*, which represent differential equations or difference equations. Furthermore, our treatment of algebraic systems will serve as a tool for the more detailed discussion of differential systems in the latter chapters.

> ***Causality:*** A system is said to be *causal* if the value of the output at time t_0 depends on the values of the input and output for all t up to time t_0 but no further, i.e., only for $t \le t_0$. (1.2)

Systems that are *not* causal are sometimes called *anticipatory*, because they violate the seemingly impossible condition that they can anticipate future values of a signal, predicting it at some future time for use at the current time. Anticipatory systems are often used in data filtering and image processing applications, wherein an entire data set is first acquired, then processed in batch mode. In such situations, the "next data" is already available for processing at any given time.

It is known that for a system to be causal, its transfer function (if it has one) must be proper. That is, the degree of its numerator polynomial must be no greater than its denominator polynomial. This is true in both continuous-time systems (*s*-domain) and discrete-time systems (*z*-domain).

> ***Time Invariance:*** Given an initial time t_0, the output of a system will in general depend on the current time as well as this initial time, $y = y(t, t_0)$. A *time-invariant system* is one whose output depends only on the difference between the initial time and the current time, $y = y(t - t_0)$. Otherwise, the system is *time-varying*. (1.3)

Time-varying systems are typically systems in which time appears as an explicit variable in the differential, difference, or algebraic equation that describes the system. Thus, a time-invariant differential equation must, by necessity, be one with constant coefficients. Time-varying systems have outputs that depend, in a sense, on the actual "clock" time at which they were "turned on." Time-invariant systems have outputs that depend on time only to the extent that they depend on how long it has been since they were "turned on." Thus if the input were shifted in time, the output would be simply shifted in time as well. Time-varying equations are very difficult to solve, rivaling *nonlinear* equations.

To define linearity, we consider the action of the system to be represented by the symbol *S*, i.e., using our previous notation, $y = S(u)$. If we consider two inputs, u_1 and u_2, and a scaling factor, *a*, we introduce the definition:

> ***Linearity:*** A *linear system* is one that satisfies *homogeneity* and *additivity*. A *homogeneous system* is one for which $S(au) = aS(u)$ for all a and u, and an *additive system* is one for which $S(u_1 + u_2) = S(u_1) + S(u_2)$ for all u_1 and u_2. (1.4)

Linear systems are thus systems for which the principle of superposition holds. We will later consider so-called multivariable systems, which have more than one input and more than one output. When such systems are linear, the effect of each input can be considered independently of one another. In systems with memory, the term *linear* refers to systems that are linear in *all* of the variables on which they depend. Therefore, for example, a linear n^{th} order differential equation is one whose n^{th} derivative depends in a linear way on each of the lower derivatives, and also in a linear way on the forcing function, if any.

Nonlinear systems are notoriously difficult to analyze and solve, partly because they exist in such an infinite variety of forms, preventing any cohesive theory for analysis.

In the next section, we will review the process by which models of linear systems are derived, followed by some examples for practical physical systems.

1.1.2 Physical System Modeling

The underlying motivation for all the analysis tools presented in this book is the understanding of physical systems. Whether the system is mechanical, electrical, or chemical, a mathematical description must be written in a unified way so that a single theory of stability, control, or analysis can be applied to the model. This is often the first task of an engineer in any design problem. In this section, we will introduce linear modeling principles for electrical, mechanical, and some fluid systems, and we will attempt to illustrate the unity in such models.

Physical Variables

We start by categorizing the physical quantities of interest to us. The first quantities available in any problem specification are the *constants*. These are, of course, constant numerical values specifying the dimensions, ranges, amounts, and other physical attributes of the masses in the system. These are often available as known quantities, but are sometimes unknown or poorly known and are subject to a process known as *system identification*. System identification is also useful when the physical attributes of a system are not constant, but vary with time. For example, the weight of a vehicle may change as its fuel is consumed, and the resistance of a resistor may change with temperature. In this chapter and for most of this book, we will not consider time-varying quantities in much detail.

The second class of number to be considered is the *variables*, which are of interest to us because they do usually vary with time. Some variables, i.e., those considered *inputs*, are known a priori, while others (the outputs) are to be determined. We separate these into two broad categories: *flow* variables and *potential* variables. Flow variables are quantities that must be measured through the cross-section of the medium through which they are transmitted. The easiest flow variable to imagine is a fluid (nonviscous, incompressible). The flow of such a fluid through a conduit must be measured by breaking the pipe and "counting" the amount (mass) of fluid passing through the cross-section.* For electrical systems, the analogous flow variable can be considered to be the current. Because current is defined as the amount of charge per unit area flowing across a cross-section per unit time, we can equally well consider charge to be a flow variable. In mechanical systems, force is a flow variable. Although it may not conform to the fluid analogy, it is nevertheless a quantity that must be measured by breaking a connection and inserting a measurement device.

The second type of variable is the potential variable. Potential variables are physical quantities that must be measured at two locations; the value of the measurement is the relative difference between the locations. Pressure, voltage, and displacement (position or velocity) are all potential variables because their definitions all require a reference location. Although we speak of a voltage appearing at a particular location in a circuit, it is always understood that this measurement was taken relative to another point.

Physical Laws

For many simple systems, there are only a few basic physical laws that must be obeyed in order to generate a sufficient set of equations that describe a system. If we consider only the basic necessities (i.e., using finite-dimensional, lumped-parameter, Newtonian dynamics rather than relativistic mechanics), we can categorize these into two types of laws: *mass* conservation and *circuit* conservation laws. Mass conservation laws are defined on nodes, and circuit conservation laws are defined on closed paths. A *node* is an interconnection point between two or more conduits transmitting a flow variable. In mechanical systems, nodes are associated with masses so that applied forces are shown to be acting on something. These circuit laws are integrally related to the two types of variables above: flow and potential.

Mass conservation laws take the basic form:

$$\sum \begin{pmatrix} \text{all flow variables} \\ \text{entering a node} \end{pmatrix} = \sum \begin{pmatrix} \text{net equivalent flow} \\ \text{into node} \end{pmatrix} \tag{1.5}$$

For an electrical network, this type of law translates to Kirchoff's current law (KCL), which states that the sum of all currents entering a node must equal zero, $\sum i_i = 0$. For a mechanical system, the mass conservation law takes the form of Newton's law: $\sum F_i = ma$. Note that in Newton's law, the sum of flow variables need not equal zero but must be proportional to the net acceleration of the object on which the forces act. In electrical and fluid systems, the net

* In reality, clever fluid flow measurement systems have been devised that can measure flow variables without interrupting them, e.g., ultrasonic flowmeters and inductive ammeters.

equivalent flow is zero, because it is impossible for net charges to accumulate indefinitely in a wire connection, just as it would be for fluid molecules to accumulate in a conduit junction.

Circuit conservation laws take the following form:

$$\sum\begin{pmatrix}\text{signed changes in a potential}\\ \text{variable around a closed path}\end{pmatrix}=0 \tag{1.6}$$

Such laws enforce the intuitive notion that if a potential variable is measured at one location relative to a fixed reference, and if relative changes are added as components are traversed in a closed path, then the potential measured at the original location upon returning should not have changed. Thus, Kirchoff's voltage law (KVL) specifies that around any closed path in a network, $\sum v_i = 0$, being careful to include the appropriate algebraic signs in the terms. In mechanical systems, circuit conservation allows us to measure what we consider absolute position by summing a sequence of relative displacements (although in truth all positions are relative to something). For fluids, the sum of pressure drops and rises throughout a closed network of pipes and components must equal zero.

These laws go a long way toward generating the equations that describe physical systems. We are ignoring a great many physical quantities and processes, such as deformation, compressibility, and distributed parameters, that usually provide a more complete and accurate model of a system. Usually, though, it is best to attempt a simple model that will suffice until its deficiencies can be discovered later.

Constitutive Relationships

The physical laws above are not by themselves sufficient to write complete equations. Flow variables and potential variables are not unrelated, but their relationship depends on the physical device being considered. Aside from sources, which provide, e.g., input forces, voltages, currents, and flows to systems, we also have *components*, which we assume to be lumped, i.e., their effects are modeled as being concentrated at a single location as opposed to being distributed over space. Each type of component has associated with it a *constitutive relationship* that relates the flow variable through it and the potential variable across it.

Electrical Components

The three most basic linear components common in electrical networks are the resistor (R, measured in ohms, Ω), capacitor (C, measured in farads, F), and inductor (L, measured in henrys, H). These are pictured in Figure 1.2.

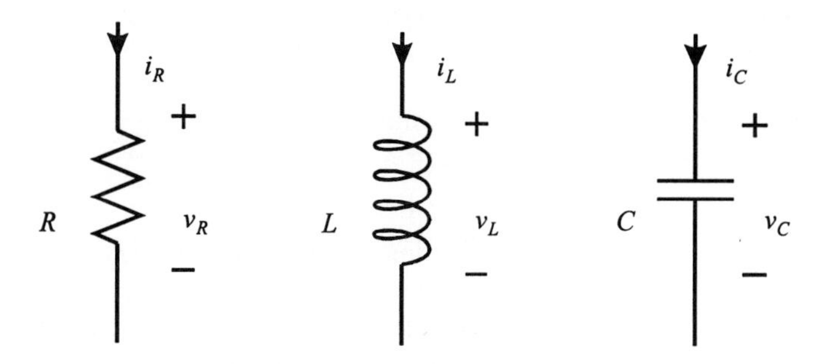

Figure 1.2 Electrical components: resistor, inductor, and capacitor (left to right).

For each component, we define reference directions for the flow variable (indicated by an arrow) and the potential variable (indicated by $+/-$ signs), so that the appropriate algebraic sign can be defined for the component.

For these three components, the constitutive relationships are:

$$v_r = i_R R \qquad v_L = L\frac{di_L}{dt} \qquad i_c = C\frac{dv_C}{dt} \tag{1.7}$$

At any such component, these relationships can be used to substitute a flow variable for a potential variable or vice versa. Note that with the differential relationships, an integration is necessary when expressing the reverse relationship.

Mechanical Components

For mechanical systems, the fundamental components are the mass, damper, and spring. These are pictured in Figure 1.3 below.

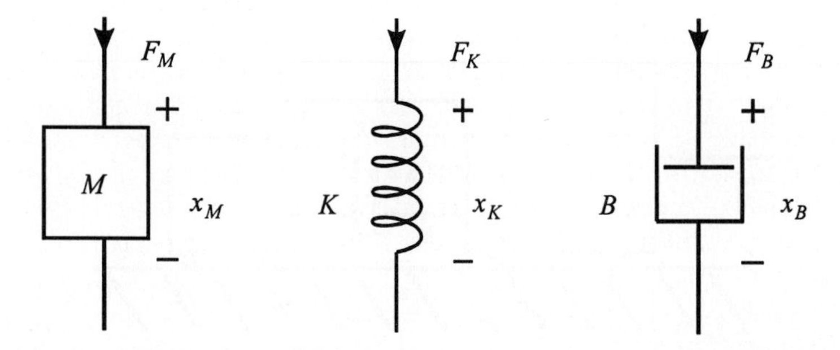

Figure 1.3 Mechanical components: mass, spring, and damper (left to right).

In the figure, a displacement is shown which indicates that one side of the component is displaced relative to the other side, except for the mass, which is displaced relative to a fixed reference defined elsewhere. The constitutive equations for these elements are

$$F_M = M\frac{d^2x_M}{dt^2} \qquad F_K = Kx_K \qquad F_B = B\frac{dx_B}{dt} \tag{1.8}$$

where F is the force applied to (or resulting from) the component. It should be noted that these equations do not appear entirely analogous to those of the electrical system (1.7). Most notably, these equations have a second derivative, and are all expressions of the flow variables in terms of the potential variable. This is partly based on convention so that the requisite computations are more convenient, and partly natural, since electrical and mechanical quantities are perceived differently by the systems on which they act. Nevertheless, the mathematical analogies remain. For fluid systems, the analogy becomes even weaker, with tanks and valves being the primary components. These components have somewhat more complex, sometimes nonlinear constitutive relationships. In truth, though, all constitutive relationships become nonlinear when the limits of their capacity are approached.

Example 1.1: Mechanical System Equations

Derive the equations of motion for the system of two masses shown in Figure 1.4. In the system, the two masses are connected by a spring with Hooke's law constant K and a damper with damping constant B, both initially unstretched and stationary. Their positions on the horizontal plane are measured as $x_1(t)$ and $x_2(t)$ from points on the masses such that they are initially equal. In addition, an externally applied force F pushes on the first mass.

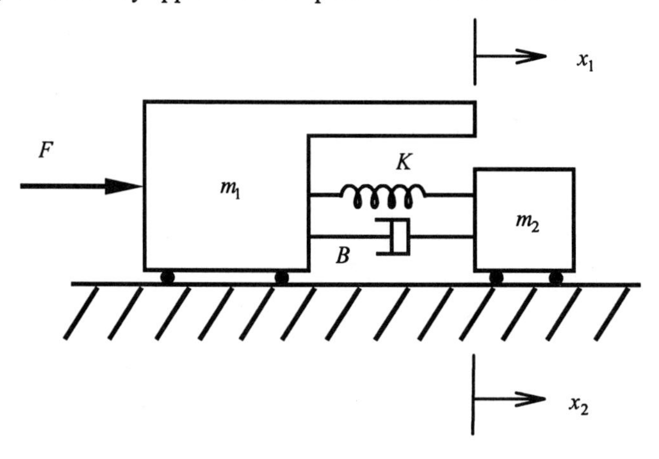

Figure 1.4 Two-mass system containing a spring and damper, and a forcing function.

Solution:

As mass 1 moves in the direction of positive x_1, the spring will compress and react with a force $K(x_1 - x_2)$ against the motion. Likewise, the damper will resist motion with viscous friction force $B(\dot{x}_1 - \dot{x}_2)$. The free-body diagram of the system in Figure 1.5 shows the two masses and all of the forces acting on them.

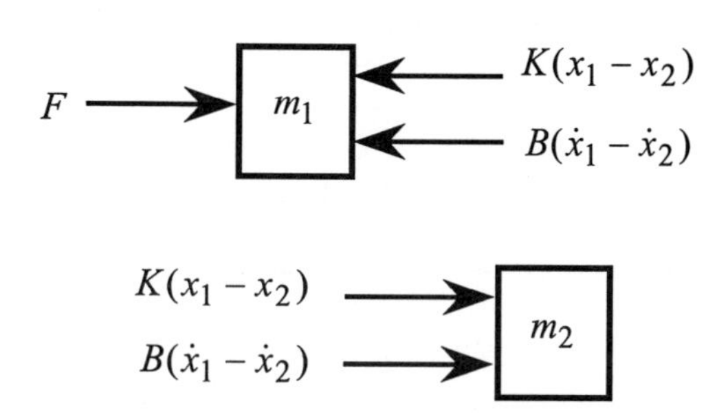

Figure 1.5 Free-body diagrams showing the masses in Figure 1.4 and the forces that act on them.

Applying Newton's law, $ma = \sum F$, we get

$$m_1\ddot{x}_1 = F - K(x_1 - x_2) - B(\dot{x}_1 - \dot{x}_2) \tag{1.9}$$

For the second mass, only the spring and damper provide forces, which are equal and opposite to the forces seen in (1.9). Therefore,

$$m_2\ddot{x}_2 = K(x_1 - x_2) + B(\dot{x}_1 - \dot{x}_2) \tag{1.10}$$

Rearranging these equations to a more convenient form,

$$\begin{aligned} m_1\ddot{x}_1 + B(\dot{x}_1 - \dot{x}_2) + K(x_1 - x_2) &= F \\ m_2\ddot{x}_2 + B(\dot{x}_2 - \dot{x}_1) + K(x_2 - x_1) &= 0 \end{aligned} \tag{1.11}$$

These simple linear equations will be used for further examples later.

Example 1.2: Electrical System Equations

For the circuit shown in Figure 1.6, derive differential equations in terms of the capacitor voltage $v_c(t)$ and the inductor current $i(t)$.

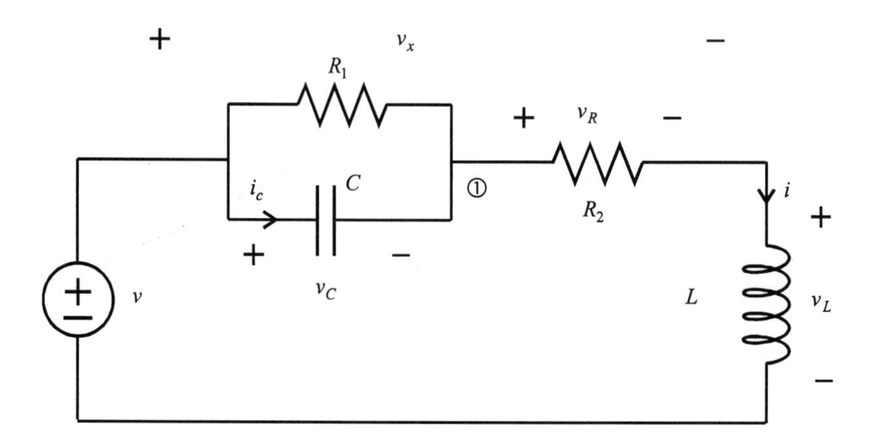

Figure 1.6 Electric circuit example with variables $v_c(t)$ and $i(t)$.

Solution:

In this example, the sum of the three currents entering node 1 must be zero. While Ohm's law $v = iR$ may be used for the resistor, the current through the capacitor is given by $i_c = C\,dv_c/dt$. Thus,

$$-\frac{v_c}{R_1} - C\frac{dv_c}{dt} + i = 0 \tag{1.12}$$

where care has been taken to use only the desired variables $v_c(t)$ and $i(t)$. For a second relationship, we note that the sum of all voltages around the main loop of the circuit must be zero. The voltage across the inductor is given by $v = L\,di/dt$, so

$$-v + v_c + R_2 i + L\frac{di}{dt} = 0 \tag{1.13}$$

where $v(t)$ is the forcing function provided by the voltage source in the circuit. Rewriting (1.12) and (1.13) in a more convenient form,

$$C\frac{dv_c}{dt} + \frac{1}{R_1}v_c - i = 0 \qquad L\frac{di}{dt} + R_2 i + v_c = v \tag{1.14}$$

1.1.3 State Equations

The sets of equations derived in (1.11) and (1.14) are coupled, in the sense that the variables in one appear in the other. This implies that they must be solved simultaneously, or else they must be combined into a single, larger-order differential equation by taking derivatives and substituting one into the other. The standard methods for solving differential equations are then applied.

However, such a process can be tedious, and the methods employed vary in complexity as the order of the differential equation increases. Instead, we prefer to write the dynamic equations of physical systems as *state equations*. State equations are simply collections of first-order differential equations that together represent exactly the same information as the original larger differential equation. Of course, with an nth-order differential equation, we will need n first-order equations. However, the variables used to write these n first-order equations are not unique. These so-called "state variables" may be chosen for convenience, as one set of state variables may result in mathematical expressions that make the solution or other characteristic of the system more apparent.

In a strict sense, the collection of state variables at any given time is known as the *state* of the system, and the set of all values that can be taken on by the state is known as the *state space*. The state of the system represents complete information of the system, such that if we know the state at time t_0, it is possible to compute the state at all future times. We will model the state spaces for linear systems as linear vector spaces, which we begin to discuss in the next chapter.

State Variables

Consider a nth-order linear, time-invariant differential equation:

$$\frac{d^n x(t)}{dt^n} + a_{n-1}\frac{d^{n-1}x(t)}{dt^{n-1}} + \cdots + a_1\frac{dx(t)}{dt} + a_0 x(t) = b_0 u(t) \tag{1.15}$$

The most straightforward method for choosing n state variables to represent this system is to let the state variables be equal to $x(t)$ and its first $(n-1)$ derivatives. Thus, if the state variables are denoted by ξ, then

$$\begin{aligned}
\xi_1(t) &= x(t)\\
\xi_2(t) &= \frac{dx(t)}{dt}\\
&\vdots\\
\xi_n(t) &= \frac{d^{n-1}x(t)}{dt^{n-1}}
\end{aligned}$$

These definitions of state variables are also called *phase variables*. The n differential equations resulting from these definitions become

$$\begin{aligned}
\dot{\xi}_1(t) &= \xi_2(t) \\
\dot{\xi}_2(t) &= \xi_3(t) \\
&\vdots \\
\dot{\xi}_{n-1}(t) &= \xi_n(t) \\
\dot{\xi}_n(t) &= -a_0\xi_1(t) - a_1\xi_2(t) - \cdots - a_{n-2}\xi_{n-1} - a_{n-1}\xi_n + b_0 u(t)
\end{aligned} \tag{1.16}$$

We will find it convenient to express such a system of equations in vector-matrix form:

$$\begin{bmatrix} \dot{\xi}_1 \\ \dot{\xi}_2 \\ \vdots \\ \dot{\xi}_n \end{bmatrix} = \begin{bmatrix} 0 & 1 & 0 & \cdots & 0 \\ 0 & 0 & 1 & \ddots & 0 \\ 0 & 0 & 0 & \ddots & 0 \\ 0 & 0 & \cdots & 0 & 1 \\ -a_0 & -a_1 & -a_2 & \cdots & -a_{n-1} \end{bmatrix} \begin{bmatrix} \xi_1 \\ \xi_2 \\ \vdots \\ \xi_n \end{bmatrix} + \begin{bmatrix} 0 \\ \vdots \\ 0 \\ b_0 \end{bmatrix} u(t) \tag{1.17}$$

If in addition, one of the state variables, say $\xi_1(t)$, is designated as the "output" of interest, denoted $y(t)$, then we can also write the so-called "output equation" in vector-matrix form as well:

$$y(t) = \begin{bmatrix} 1 & 0 & \cdots & 0 \end{bmatrix} \begin{bmatrix} \xi_1 \\ \xi_2 \\ \vdots \\ \xi_n \end{bmatrix} \tag{1.18}$$

More generally, we may designate as the output a weighted sum of the state variables and sometimes also a sum of state variables and *input* variables. Defining $x(t) \triangleq \begin{bmatrix} \xi_1 & \cdots & \xi_n \end{bmatrix}^T$, the two equations (1.17) and (1.18) are together written[M] as

$$\begin{aligned}
\dot{x} &= Ax + bu \\
y &= cx + du
\end{aligned} \tag{1.19}$$

```
ss(a,b,c,d)
ssdata(sys)
```

where the matrices A, b, c, and d are the corresponding matrices[M] in (1.17) and (1.18). These will be referred to as the *state matrix* (A), the *input matrix* (b), the *output matrix* (c), and the *feedthrough matrix* (d) (so named because it is the gain through which inputs feed directly into outputs). These equations are expressed for a single input, single output (SISO) system. For multi-input, multioutput (MIMO) or multivariable systems, the equations in (1.19) can be written exactly the same except that the input matrix is a capital B, the output matrix is a capital C, and the feedthrough matrix is a capital D. These changes indicate that they are matrices rather than simple columns (b), rows (c), or scalars (d). The equations in (1.19) will become quite familiar, as they are the format used for studying the properties of linear systems of equations throughout this book.

Of course, if the original Equation (1.15) were time-varying, then the coefficients might be functions of time, i.e., $a_i(t)$ and $b_i(t)$. In that case, (1.19) might contain $A(t)$, $b(t)$, $c(t)$ and $d(t)$.

Example 1.3: State Variables for the Mechanical System Example

Write a state variable expression for the differential equations already derived for the mechanical system of Example 1.1, using force F as the input and the difference $x_2 - x_1$ as the output.

Solution:

In the mechanical system, we derived two separate equations, each being second order (1.11). To generate state equations, we will introduce the variables $\xi_1 = x_1$, $\xi_2 = \dot{x}_1$, $\xi_3 = x_2$, and $\xi_4 = \dot{x}_2$. Then, by inspection of (1.11), the state equations are:

$$\begin{bmatrix} \dot{\xi}_1 \\ \dot{\xi}_2 \\ \dot{\xi}_3 \\ \dot{\xi}_4 \end{bmatrix} = \begin{bmatrix} 0 & 1 & 0 & 0 \\ -K/m_1 & -B/m_1 & K/m_1 & B/m_1 \\ 0 & 0 & 0 & 1 \\ K/m_2 & B/m_2 & -K/m_2 & -B/m_2 \end{bmatrix} \begin{bmatrix} \xi_1 \\ \xi_2 \\ \xi_3 \\ \xi_4 \end{bmatrix} + \begin{bmatrix} 0 \\ 1/m_1 \\ 0 \\ 0 \end{bmatrix} F(t)$$

$$y(t) = \begin{bmatrix} -1 & 0 & 1 & 0 \end{bmatrix} \begin{bmatrix} \xi_1 \\ \xi_2 \\ \xi_3 \\ \xi_4 \end{bmatrix} \tag{1.20}$$

Example 1.4: State Variables for the Electrical System Example

Find a state variable expression for the electrical system equations in (1.14) from Example 1.2. As the system output, use the voltage across the inductor.

Solution:

In this system, the two equations in (1.14) are each first order. The total system is then second order. Using v_c and $i(t)$ as the two state variables, i.e., $\xi_1(t) = v_c(t)$ and $\xi_2(t) = i(t)$, we can immediately write the equations:

$$\begin{bmatrix} \dot{\xi}_1 \\ \dot{\xi}_1 \end{bmatrix} = \begin{bmatrix} -\frac{1}{CR_1} & \frac{1}{C} \\ -\frac{1}{L} & -\frac{R_2}{L} \end{bmatrix} \begin{bmatrix} \xi_1 \\ \xi_2 \end{bmatrix} + \begin{bmatrix} 0 \\ \frac{1}{L} \end{bmatrix} v(t)$$

$$y(t) = \begin{bmatrix} -1 & -R_2 \end{bmatrix} \begin{bmatrix} \xi_1 \\ \xi_2 \end{bmatrix} + (1)v(t) \tag{1.21}$$

The output equation in this result comes from the KVL equation $v_L = v - v_c - iR_2$.

Alternative State Variables

If we had attempted to use the definitions of state variables in (1.16) to write state equations for the more general differential equation:

$$\begin{aligned} \frac{d^n x(t)}{dt^n} + a_{n-1} \frac{d^{n-1}x(t)}{dt^{n-1}} + \cdots + a_1 \frac{dx(t)}{dt} + a_0 x(t) \\ = b_n \frac{d^n u(t)}{dt^n} + \cdots + b_0 u(t) \end{aligned} \tag{1.22}$$

we would have required derivatives of $u(t)$ in the state equations. According to the standard format of (1.19), this is not allowed. Instead, there are a number of commonly used formulations for state variables for equations such as (1.22). These are best represented in the simulation diagrams on the following pages.

In Figure 1.7, the state variables shown are similar to those found in (1.16), in the sense that $n-1$ of them are simply derivatives of the previous ones. The state equations that describe this diagram are given immediately following Figure 1.7, in Equation (1.25). We see that in this form, the feedback coefficients a_i appear in only the final state equation. Having feedforward connections with b_i coefficients allows for derivatives of $u(t)$ in (1.22) without appearing in (1.25) below, which is the state variable equivalent of (1.22). Note that in this form each state variable is assigned to be the output of an integrator, just as with the phase variables discussed above.

A second common choice of state variables can be generated from the following manipulations. Suppose that in an attempt to solve the second-order equation of the form:

$$x''(t)+a_1x'(t)+a_0x(t)=b_2u''(t)+b_1u'(t)+b_0u(t) \tag{1.23}$$

both sides of the equation are integrated twice with respect to the time variable. This would result in:

$$\int_{-\infty}^{t}\int_{-\infty}^{s}\left[x''(\sigma)+a_1x'(\sigma)+a_0(\sigma)x(\sigma)\right]d\sigma\, ds$$
$$=\int_{-\infty}^{t}\int_{-\infty}^{s}\left[b_2u''(\sigma)+b_1u'(\sigma)+b_0(\sigma)u(\sigma)\right]d\sigma\, ds$$

giving

$$x(t)+\int_{-\infty}^{t}a_1x(s)\,ds+\int_{-\infty}^{t}\int_{-\infty}^{s}a_0x(\sigma)\,d\sigma\, ds$$
$$=b_2u(t)+\int_{-\infty}^{t}b_1u(s)\,ds+\int_{-\infty}^{t}\int_{-\infty}^{s}b_0u(\sigma)\,d\sigma\, ds$$

or

$$\begin{aligned}x(t)&=b_2u(t)+\int_{-\infty}^{t}\left[b_1u(s)-a_1x(s)\right]ds+\int_{-\infty}^{t}\int_{-\infty}^{s}\left[b_0u(\sigma)-a_0x(\sigma)\right]d\sigma\, ds\\&=b_2u(t)+\int_{-\infty}^{t}\left\{\left[b_1u(s)-a_1x(s)\right]+\int_{-\infty}^{s}\left[b_0u(\sigma)-a_0x(\sigma)\right]d\sigma\right\}ds\end{aligned} \tag{1.24}$$

For higher-order systems, this process continues until an equation of the form of (1.24) is derived. From the form of (1.24), the simulation diagram shown in Figure 1.8 can be drawn, with the associated state equations appearing in Equation (1.26). Note that in this formulation, the integrators in the network are not connected end-to-end. Thus the state variables are not simply derivatives of one another as are phase variables. Instead, the state equations are written as in (1.26). The state variables are, however, still defined as the outputs of the integrators. This is commonly done, but is not necessary. Additional examples will be shown in Chapter 9.

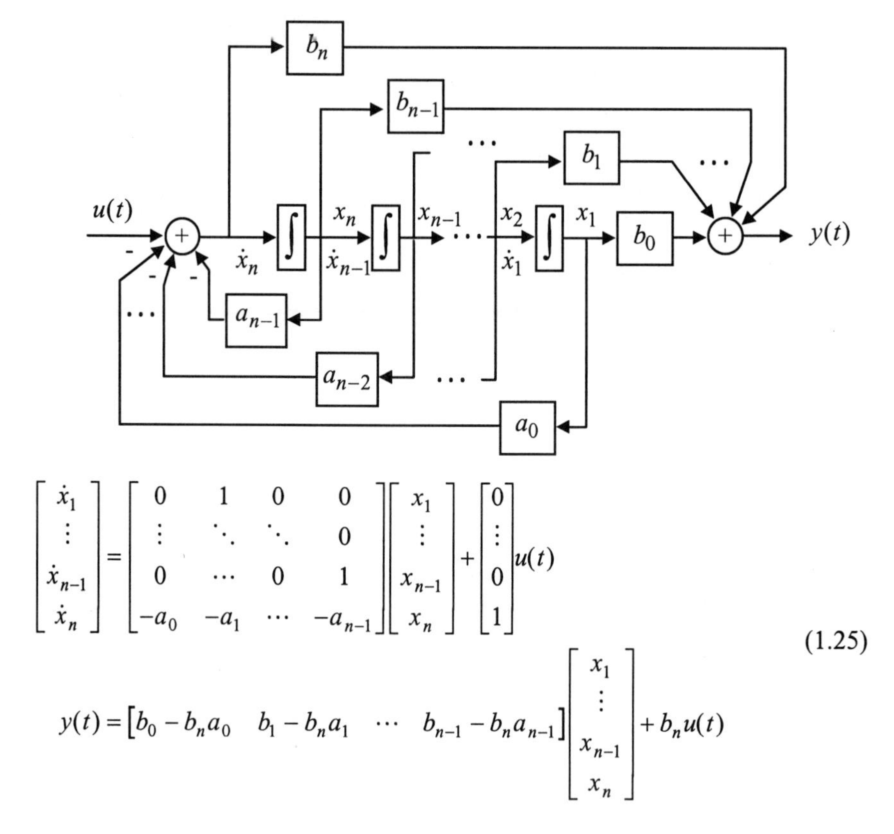

$$\begin{bmatrix} \dot{x}_1 \\ \vdots \\ \dot{x}_{n-1} \\ \dot{x}_n \end{bmatrix} = \begin{bmatrix} 0 & 1 & 0 & 0 \\ \vdots & \ddots & \ddots & 0 \\ 0 & \cdots & 0 & 1 \\ -a_0 & -a_1 & \cdots & -a_{n-1} \end{bmatrix} \begin{bmatrix} x_1 \\ \vdots \\ x_{n-1} \\ x_n \end{bmatrix} + \begin{bmatrix} 0 \\ \vdots \\ 0 \\ 1 \end{bmatrix} u(t)$$

$$y(t) = \begin{bmatrix} b_0 - b_n a_0 & b_1 - b_n a_1 & \cdots & b_{n-1} - b_n a_{n-1} \end{bmatrix} \begin{bmatrix} x_1 \\ \vdots \\ x_{n-1} \\ x_n \end{bmatrix} + b_n u(t) \tag{1.25}$$

Figure 1.7 Simulation diagram and state equations for phase variable definitions of state variables.

These special choices are examples of different ways in which state variables can be assigned to a particular system. They have some convenient properties that we will examine in later chapters. Such special forms of equations are known as *canonical* forms.

It should be noted that there are an infinite number of other ways in which to derive definitions for state variables, a fact that will soon become readily apparent. It is also important to realize that in our simple physical system examples, the variables we choose are physically measurable or are otherwise meaningful quantities, such as voltages or displacements. Often, variables will be selected purely for the effect they have on the structure of the state equations, not for the physical meaning they represent. State variables need not be physically meaningful. This is one of the primary advantages to the state variable technique.

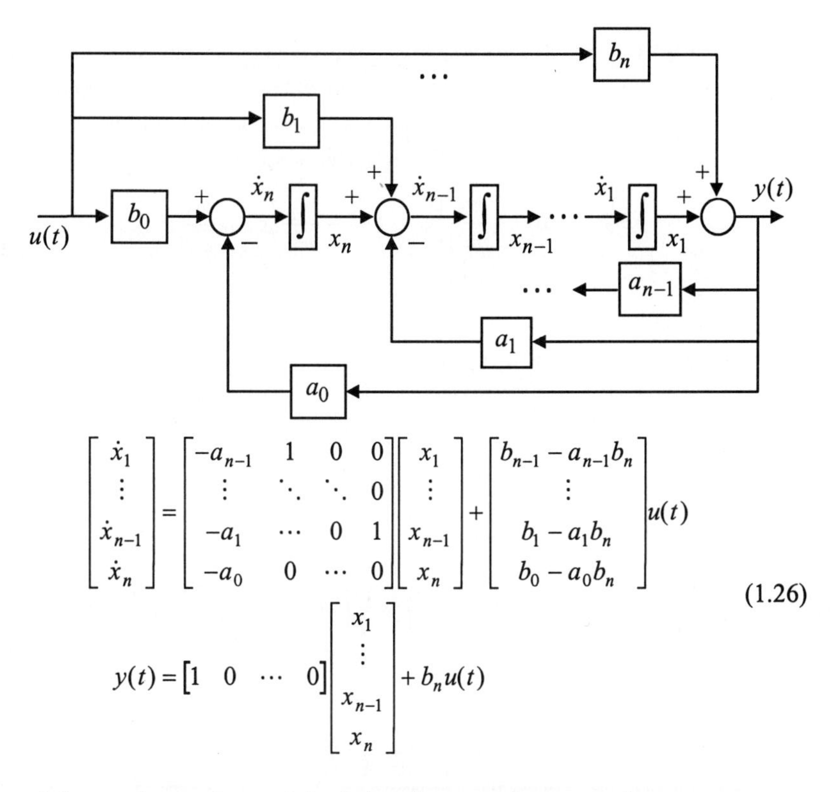

$$
\begin{bmatrix} \dot{x}_1 \\ \vdots \\ \dot{x}_{n-1} \\ \dot{x}_n \end{bmatrix} = \begin{bmatrix} -a_{n-1} & 1 & 0 & 0 \\ \vdots & \ddots & \ddots & 0 \\ -a_1 & \cdots & 0 & 1 \\ -a_0 & 0 & \cdots & 0 \end{bmatrix} \begin{bmatrix} x_1 \\ \vdots \\ x_{n-1} \\ x_n \end{bmatrix} + \begin{bmatrix} b_{n-1} - a_{n-1}b_n \\ \vdots \\ b_1 - a_1 b_n \\ b_0 - a_0 b_n \end{bmatrix} u(t)
$$

$$
y(t) = \begin{bmatrix} 1 & 0 & \cdots & 0 \end{bmatrix} \begin{bmatrix} x_1 \\ \vdots \\ x_{n-1} \\ x_n \end{bmatrix} + b_n u(t) \tag{1.26}
$$

Figure 1.8 A second type of simulation diagram and state equations.

Changing State Variables

In the state variable description for the mechanical system, we might be inclined to write equations not for the independent variables, $x_1(t)$ and $x_2(t)$, but rather for the variables $x_1(t)$ and $(x_1(t) - x_2(t))$. Perhaps we make this choice because the device is being used as a seismograph and the motion of the second mass relative to the first is more important than the absolute position. For this system, the equations in (1.11) are relatively simple to rewrite with this change of variables. If we introduce $\hat{x}_1(t) = x_1(t)$ and $\hat{x}_2(t) = x_1(t) - x_2(t)$, then (1.11) becomes:

$$
\begin{aligned}
m_1 \ddot{\hat{x}}_1 + B\dot{\hat{x}}_2 + K\hat{x}_2 &= F \\
m_2\left(\ddot{\hat{x}}_1 - \ddot{\hat{x}}_2\right) - B\dot{\hat{x}}_2 - K\hat{x}_2 &= 0
\end{aligned} \tag{1.27}
$$

In order to get each equation to contain second derivatives in only one of the variables, we can solve the first equation of (1.27) for $\ddot{\hat{x}}_1$ and substitute it into the second equation. Performing this operation and simplifying, we obtain

$$m_1\ddot{\hat{x}}_1 + B\dot{\hat{x}}_2 + K\hat{x}_2 = F$$

$$m_2\ddot{\hat{x}}_2 + B\left(1+\frac{m_2}{m_1}\right)\dot{\hat{x}}_2 + K\left(1+\frac{m_2}{m_1}\right)\hat{x}_2 = \frac{m_2}{m_1}F$$

Now using new state variables $\hat{\xi}_1 = \hat{x}_1$, $\hat{\xi}_2 = \dot{\hat{x}}_1$, $\hat{\xi}_3 = \hat{x}_2$ and $\hat{\xi}_4 = \dot{\hat{x}}_2$, the new state equations are readily apparent:

$$\begin{bmatrix} \dot{\hat{\xi}}_1 \\ \dot{\hat{\xi}}_2 \\ \dot{\hat{\xi}}_3 \\ \dot{\hat{\xi}}_4 \end{bmatrix} = \begin{bmatrix} 0 & 1 & 0 & 0 \\ 0 & 0 & -\frac{K}{m_1} & -\frac{B}{m_1} \\ 0 & 0 & 0 & 1 \\ 0 & 0 & -K\left(\frac{1}{m_1}+\frac{1}{m_2}\right) & -B\left(\frac{1}{m_1}+\frac{1}{m_2}\right) \end{bmatrix} \begin{bmatrix} \hat{\xi}_1 \\ \hat{\xi}_2 \\ \hat{\xi}_3 \\ \hat{\xi}_4 \end{bmatrix} + \begin{bmatrix} 0 \\ \frac{1}{m_1} \\ 0 \\ \frac{1}{m_1} \end{bmatrix} F$$

$$y = \begin{bmatrix} 0 & 0 & -1 & 0 \end{bmatrix} \begin{bmatrix} \hat{\xi}_1 \\ \hat{\xi}_2 \\ \hat{\xi}_3 \\ \hat{\xi}_4 \end{bmatrix}$$

Here again we have let the output variable be $y(t) = x_2 - x_1$ $(= -\hat{x}_2(t) = -\hat{\xi}_3)$. Notice that the state equations above are significantly different from (1.20), although they are both valid equations representing the same physical system.

Clearly, redefining all the state variables and rederiving or manipulating the equations can be time consuming. To demonstrate an easier method that uses the vector-matrix notation we have adopted, we turn to the electrical system example (Example 1.2). Suppose we wish to write new state equations, where the relevant state variables are v_R and v_x (see the circuit diagram in Figure 1.6). Because we can write the new state variables as weighted sums of the old state variables

$$\begin{aligned} v_R &= R_2 i \\ v_x &= v_c + R_2 i \end{aligned} \tag{1.28}$$

we can use vector-matrix notation immediately to write

$$\begin{bmatrix} v_R \\ v_x \end{bmatrix} = \begin{bmatrix} 0 & R_2 \\ 1 & R_2 \end{bmatrix} \begin{bmatrix} v_c \\ i \end{bmatrix} \tag{1.29}$$

Alternatively, using the inverse relationship,

$$\begin{bmatrix} v_c \\ i \end{bmatrix} = \begin{bmatrix} 0 & R_2 \\ 1 & R_2 \end{bmatrix}^{-1} \begin{bmatrix} v_R \\ v_x \end{bmatrix} = \frac{\begin{bmatrix} R_2 & -R_2 \\ -1 & 0 \end{bmatrix}}{-R_2} \begin{bmatrix} v_R \\ v_x \end{bmatrix} = \begin{bmatrix} -1 & 1 \\ 1/R_2 & 0 \end{bmatrix} \begin{bmatrix} v_R \\ v_x \end{bmatrix} \tag{1.30}$$

If we use the notation $\mathrm{X} = \begin{bmatrix} v_C & i \end{bmatrix}^{\mathrm{T}} \triangleq \begin{bmatrix} \xi_1 & \xi_2 \end{bmatrix}^{\mathrm{T}}$ for the state vector already introduced in (1.21) and $\hat{\mathrm{X}} = \begin{bmatrix} v_R & v_x \end{bmatrix}^{\mathrm{T}} \triangleq \begin{bmatrix} \hat{\xi}_1 & \hat{\xi}_2 \end{bmatrix}^{\mathrm{T}}$ for the new state vector, then we can write (1.30) symbolically as

$$\mathrm{X} = M\hat{\mathrm{X}} \tag{1.31}$$

where matrix M is defined in (1.30). Likewise, the symbolic form for (1.21) is as given in (1.19), i.e.,

$$\begin{aligned} \dot{\mathrm{X}} &= A\mathrm{X} + bv \\ y &= c\mathrm{X} + dv \end{aligned} \tag{1.32}$$

Equation (1.31) directly implies that $\dot{\mathrm{X}} = M\dot{\hat{\mathrm{X}}}$, so substituting into (1.32),

$$\begin{aligned} M\dot{\hat{\mathrm{X}}} &= AM\hat{\mathrm{X}} + bv \\ y &= cM\hat{\mathrm{X}} + dv \end{aligned} \tag{1.33}$$

or

$$\begin{aligned} \dot{\hat{\mathrm{X}}} &= M^{-1}AM\hat{\mathrm{X}} + M^{-1}bv \triangleq \hat{A}\hat{\mathrm{X}} + \hat{b}v \\ y &= cM\hat{\mathrm{X}} + dv \triangleq \hat{c}\hat{\mathrm{X}} + \hat{d}v \end{aligned} \tag{1.34}$$

where

$$\hat{A} \triangleq M^{-1}AM = \begin{bmatrix} 0 & R_2 \\ 1 & R_2 \end{bmatrix} \begin{bmatrix} -\frac{1}{CR_1} & \frac{1}{C} \\ -\frac{1}{L} & -\frac{R_2}{L} \end{bmatrix} \begin{bmatrix} -1 & 1 \\ \frac{1}{R_2} & 0 \end{bmatrix}$$

$$= \begin{bmatrix} 0 & -\frac{R_2}{L} \\ \frac{1}{CR_1} + \frac{1}{CR_2} & -\frac{1}{CR_1} - \frac{R_2}{L} \end{bmatrix}$$

$$\hat{b} \triangleq M^{-1}b = \begin{bmatrix} 0 & R_2 \\ 1 & R_2 \end{bmatrix} \begin{bmatrix} 0 \\ 1/L \end{bmatrix} = \begin{bmatrix} \frac{R_2}{L} \\ \frac{R_2}{L} \end{bmatrix}$$

$$\hat{c} \triangleq cM = \begin{bmatrix} -1 & -R_2 \end{bmatrix} \begin{bmatrix} -1 & 1 \\ \frac{1}{R_2} & 0 \end{bmatrix} = \begin{bmatrix} 0 & -1 \end{bmatrix} \tag{1.35}$$

$$\hat{d} \triangleq d = 1$$

This gives a new system of state equations that are equivalent to those in (1.21) but appear quite different. Notice that the feedthrough matrix, d, is the same in each case.

The procedure we have just introduced will be considered from a different perspective in later chapters. The important points here are:

- n^{th}-order linear differential equations (or several coupled systems constituting an nth-order system) can be written as n first-order state equations.

- State equations are entirely equivalent to the original differential equations but are not unique.

- State equations can be changed with matrix-vector operations, resulting in a new form for each of the four system matrices.

In future chapters, we will see what conveniences can be realized with different choices of state vector definition and what insight might be gained into the original physical system by using the state variable notation. There will be certain characteristics of differential equations and control systems that are more apparent in state space (time-domain) than in transfer function form (frequency-domain). In order to understand these properties, some details of linear vector spaces and linear algebras will be necessary.

1.1.4 Discrete-Time Systems

Throughout this book we will, on occasion, use discrete-time systems as examples or for other illustrative purposes. Perhaps more so than in frequency-domain, state space methods for discrete-time systems are very similar to those used in continuous-time systems. For example, most of the first part of the book on vector spaces, linear operators, and functions of matrices is indifferent to the time-domain, because the equations being considered are independent of time. In latter chapters, some concepts, such as controllability and pole placement are common to continuous and discrete-time, while other concepts, including stability and advanced controller design, can vary significantly in discrete-time. The basic concepts remain parallel, but the matrix equations may look different.

We do not give a detailed treatment here of digital filtering or z-domain methods, or of continuous filtering and s-domain methods. Nevertheless, it is useful to review the terminology.

Discrete-time systems may be inherently discrete, as, for example, in the equations that describe the balance of a bank account that undergoes withdrawals, deposits, and interest postings at regular intervals. Alternatively, they may be discretizations of continuous-time systems. A discretization is a conversion of a continuous-time equation into discrete-time. These discretizations may be performed in a number of different ways, e.g., by using integrator equivalence, pole-zero mapping, or hold equivalence [6]. In either case, we end up with a *difference* equation rather than a differential equation. Difference equations are expressed in terms of time delays rather than derivatives. If, for example, the sampling period of a discrete-time system is T, then a simple difference equation might appear as

$$x(kT+2T)+a_1x(kT+T)+a_0x(kT)=b_0u(kT) \tag{1.36}$$

where k is an integer. In such a situation, we will sometimes simplify the notation by dropping the (constant) sampling time T and using a subscript rather than an argument in parentheses, i.e.,

$$x_{k+2}+a_1x_{k+1}+a_0x_k=b_0u_k \tag{1.37}$$

With this difference equation, we may define a state vector in any of the ways discussed above for continuous-time systems. For example, if we let $\mathrm{X}\stackrel{\Delta}{=}\begin{bmatrix} x_k & x_{k+1}\end{bmatrix}^{\mathrm{T}}$, then we will obtain the discrete-time state equation

$$\begin{aligned} \mathrm{X}_{k+1} &= \begin{bmatrix} 0 & 1 \\ -a_0 & -a_1 \end{bmatrix}\mathrm{X}_k + \begin{bmatrix} 0 \\ b_0 \end{bmatrix} u_k \\ &\stackrel{\Delta}{=} A_d\mathrm{X}_k + b_d u_k \end{aligned} \tag{1.38}$$

where the notation A_d and b_d are introduced to distinguish these discrete-time system matrices from their continuous-time counterparts. An output equation can be similarly defined.

Furthermore, if an arbitrary difference equation is specified:

$$x_{k+n} + a_{n-1}x_{k+(n-1)} + \cdots + a_1 x_{k+1} + a_0 x_k = b_n u_{k+n} + \cdots + b_0 u_k \tag{1.39}$$

then the simulation diagrams of Figures 1.7 and 1.8 (and all other possibilities) may also be used, with the exception that the integrators are replaced with unit-time delays, i.e., delay blocks representing one sample period, *T*, in length. When such a system is time-varying, the coefficients a_i and b_i above may be replaced with $a_i(k)$ and $b_i(k)$. (The inclusion of this extra time argument is the motivation for the subscript notation; writing $a_i(k, kT-T)$ would be cumbersome.)

The following examples illustrate a number of ways a difference equation can be used to model physical systems. The first two examples illustrate the derivation of difference equations by direct discrete-time modeling, and in the third example, a method is given for approximating a differential equation by a discrete-time system. Another method for representing discrete-time systems as approximations of continuous-time systems is given in Chapter 6.

Example 1.5: Direct Difference Equation Modeling of a Savings Account

The simplest model of an interest-bearing savings account provides an example of a first-order difference equation. Derive the difference equation model for the balance of an account earning *i*% per year, compounded monthly. Assume that interest is computed on the previous month's balance and that the account owner may make any number of deposits and withdrawals during that month.

Solution:

The balance, after compounding, in month *k* is denoted $x(k)$. Then $x(k)$ will be equal to the previous month's balance plus the interest for month *k* and the net total of the owner's deposits and withdrawals, which we will denote by $u(k)$. Then we have the difference equation

$$\begin{aligned} x(k) &= x(k-1) + \tfrac{i}{12}x(k-1) + u(k-1) \\ &= \left(1 + \tfrac{i}{12}\right)x(k-1) + u(k-1) \end{aligned}$$

This is a first-order difference equation and, hence, is already in linear state variable form. Note that time can be arbitrarily shifted to reflect that $k = 0$ is the time origin and that $x(1)$ is the first balance that must be computed. Because the system is time-invariant, this is merely a notational convenience:

$$x(k+1) = \left(1 + \frac{i}{12}\right)x(k) + u(k)$$

Example 1.6: A Difference Equation for a Predator-Prey System

A tropical fish enthusiast buys a new fish tank and starts his collection with $P_p(0)$ piranhas and $P_g(0)$ guppies. He buys an ample supply of guppy food (which the piranhas will not eat) and expects the piranhas to eat the guppies. He samples the populations of the two species each day, $P_p(d)$ and $P_g(d)$, and finds, as he expected, that they change. Generate a linear difference equation model for these "population dynamics."

Solution:

Several assumptions must be made to derive a suitable model. First, we will assume that the birthrate of the piranhas is directly proportional to their food supply, which of course is the population of guppies. Further, we assume that because piranhas are large, their death rate is proportional to overcrowding, i.e., to the level of their own population. Therefore, we can write the relationship

$$\begin{aligned} P_p(d+1) &= P_p(d) + k_1 P_g(d) - k_2 P_p(d) \\ &= (1-k_2)P_p(d) + k_1 P_g(d) \end{aligned} \tag{1.40}$$

where k_1 and k_2 are constants of proportionality. Of course, $P_p(0)$ and $P_g(0)$ are the initial conditions.

Now assume that the birthrate of the guppies is proportional to their food supply, which is the input of guppy food, $u(d)$. The death rate of the guppies will be proportional to the population of the piranhas. Therefore,

$$P_g(d+1) = P_g(d) + u(d) - k_3 P_p(d) \tag{1.41}$$

Together, these two equations can be combined into the state space difference equation model

$$\begin{bmatrix} P_p(d+1) \\ P_g(d+1) \end{bmatrix} = \begin{bmatrix} 1-k_2 & k_1 \\ -k_3 & 1 \end{bmatrix} \begin{bmatrix} P_p(d) \\ P_g(d) \end{bmatrix} + \begin{bmatrix} 0 \\ 1 \end{bmatrix} u(d) \tag{1.42}$$

Example 1.7: Discretized Differential Equations for a Harmonic Oscillator
By approximating the derivative as a finite difference, find a discrete-time version of the forced harmonic oscillator equations

$$\ddot{x}(t)+\omega^2 x(t)=u(t) \tag{1.43}$$

Solution:

At time $t=kT$, where T is a sampling time, the approximation of the first derivative is

$$\dot{x}(kT)\approx\frac{x(kT+T)-x(kT)}{T}$$

This implies that the approximation for the second derivative would be:

$$\begin{aligned}\ddot{x}(kT)&\approx\frac{\dot{x}(kT+T)-\dot{x}(kT)}{T}\\ &=\frac{\dfrac{x(kT+2T)-x(kT+T)}{T}-\left(\dfrac{x(kT+T)-x(kT)}{T}\right)}{T}\\ &=\frac{x(kT+2T)-2x(kT+T)+x(kT)}{T^2}\end{aligned}$$

Substituting these approximations into the original Equation (1.43), we get

$$\frac{x(kT+2T)-2x(kT+T)+x(kT)}{T^2}+\omega^2 x(kT)=u(kT)$$

or simplifying and dropping the T 's in the time arguments,

$$x(k+2)-2x(k+1)+\left(\omega^2T^2+1\right)x(k)=T^2u(k) \tag{1.44}$$

This can be seen as a discrete-time equation of the form (1.39).

If we now choose state variables as $x_1(k)=x(k)$ and $x_2(k)=x(k+1)$, then the discrete-time state space description of the system becomes

$$\begin{bmatrix}x_1(k+1)\\ x_2(k+2)\end{bmatrix}=\begin{bmatrix}0 & 1\\ -(\omega^2T^2+1) & 2\end{bmatrix}\begin{bmatrix}x_1(k)\\ x_2(k)\end{bmatrix}+\begin{bmatrix}0\\ T^2\end{bmatrix}u(k) \tag{1.45}$$

If we now wish to draw a simulation diagram for this system that is similar to the one in Figure 1.7, we must remember that instead of integrators, we must use the unit delay operator, z^{-1}. Because (1.45) has the same form as (1.39), which itself is similar in form to (1.22), the simulation diagram of Figure 1.9 is obtained.

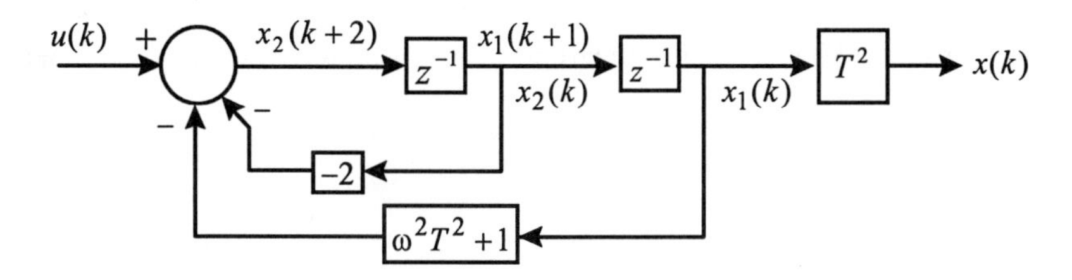

Figure 1.9 Simulation diagram for a discrete-time state space system. The unit delay operator is signified by the symbol z^{-1}.

1.1.5 Relationship to Transfer Functions

It may be reasonably asked at this point how the state variable description, which is in time-domain, relates to the transfer function representation of systems that is usually introduced in lower level signals and systems courses. The most fundamental answer to this question is that transfer functions are defined for systems with zero initial conditions only. State variable descriptions are therefore more appropriate when transients due to nonzero initial conditions are important.

However there are many other distinctions as well. The mathematical procedures for solving time- and frequency-domain systems differ greatly. From a control systems point of view, there exist entirely different design tools in time and frequency-domains, and certain aspects of system behavior are more obvious in one domain or the other. Design criteria can be specified in one domain that are not easily expressed in the other. Other concepts that will be introduced in later chapters include controllability, observability, and optimal compensator design, concepts that are all straightforward in time-domain but not necessarily in frequency-domain.

For the time being, though, we will give a flavor of the intrinsic relationship between the two domains. Consider the form of general SISO state equations:

$$\begin{aligned} \dot{x} &= Ax + bu \\ y &= cx + du \end{aligned} \tag{1.46}$$

If we assume that the initial condition is $x(t_0)=0$ for this system, Equations (1.46) can be LaPlace transformed simply as

$$\begin{aligned} sX(s) &= AX(s)+bU(s) \\ Y(s) &= cX(s)+dU(s) \end{aligned} \tag{1.47}$$

Simplifying the first of these equations,

$$sX(s)-AX(s)=(sI-A)X(s)=bU(s)$$

or

$$X(s)=(sI-A)^{-1}bU(s) \tag{1.48}$$

Substituting (1.48) into the output equation in (1.47), we get

$$Y(s)=\left[c(sI-A)^{-1}b+d\right]U(s)$$

or, for a SISO system in which $Y(s)$ and $U(s)$ are scalar quantities and the division operation exists,

$$\frac{Y(s)}{U(s)}=\left[c(sI-A)^{-1}b+d\right]\stackrel{\Delta}{=}P(s) \tag{1.49}$$

giving the transfer function, which is a ratio of polynomials.

It is worth noting that many textbooks pay little attention to the feedthrough term of the state equations, i.e., d (D). This is because its effect can be "factored out" such that it represents an algebraic relationship between the input and output, rather than a differential one. It occurs only in nonstrictly proper transfer functions wherein the order of the numerator is the same as the order of the denominator. By using polynomial division, such a transfer function can always be rewritten as the sum of a constant factor and a strictly proper transfer function. For example, consider the transfer function

$$P(s)=\frac{2s+4}{s+1}=2+\frac{2}{s+1} \tag{1.50}$$

Converting this system to state space representation will always give $d=2$ as the feedthrough term, regardless of the choice of variables. Recall that if the transfer function is *im*proper, then positive powers of s would divide out of such a fraction, and a noncausal system would be apparent.

Equation (1.49) shows how it is possible to obtain a transfer function from the matrices given in a state variable representation.[M] To get a state variable representation from a transfer function,[M] we usually return to the original differential equation or use the simulation diagrams in Figures 1.7 and 1.8. This subject is complicated by controllability and observability issues introduced in Chapters 8 and 9.

```
tf2ss(num,den)
ss2tf(a,b,c,d)
```

We have asserted that no matter which (valid) state variables we choose, the state equations represent the original system exactly. One may than ask if a change of variables is performed as in (1.31), how is the transfer function (1.49) altered? Suppose in the equation for the transfer function in terms of system matrices in (1.49), we substitute the "transformed" matrices from (1.35). Then we obtain

$$\begin{aligned}
\hat{P}(s) &= \hat{c}\left(sI - \hat{A}\right)^{-1}\hat{b} + \hat{d} \\
&= cM\left(sI - M^{-1}AM\right)^{-1}M^{-1}b + d \\
&= cM\left(sM^{-1}M - M^{-1}AM\right)^{-1}M^{-1}b + d \\
&= cMM^{-1}(sI - A)^{-1}MM^{-1}b + d \\
&= c(sI - A)^{-1}b + d \\
&= P(s)
\end{aligned} \tag{1.51}$$

So we have shown that the same transfer function results regardless of the choice of state equations. Note that this same result holds for discrete-time systems in exactly the same form; simply replace the s operator in this section with the z operator.

1.2 Linearization of Nonlinear Equations

It is an unfortunate fact that most physical systems encountered in practice are not linear. It is almost always the case that when one encounters a linear model for a physical system, it is an idealized or simplified version of a more accurate but much more complicated nonlinear model. In order to create a linear model from a nonlinear system, we introduce a linearization method based on the Taylor series expansion of a function.

1.2.1 Linearizing Functions

Recall that the Taylor series expansion expresses a general function $f(x)$ as the infinite series

$$f(x)=\sum_{n=0}^{\infty}\frac{1}{n!}\frac{df(x)}{dx}\bigg|_{x=x_0}(x-x_0)^n \tag{1.52}$$

This series is said to be expanded about the point $x = x_0$. The point x_0 is interchangeably referred to as the *bias point, operating point*, or, depending on some stability conditions discussed in Chapter 7, the *equilibrium point.*

Any function for which such a series converges is said to be *analytic.* Writing out the first few terms of such a series,

$$f(x)=f(x_0)+\frac{df(x)}{dx}\bigg|_{x=x_0}(x-x_0)+\frac{1}{2}\frac{d^2f(x)}{dx^2}\bigg|_{x=x_0}(x-x_0)^2+\ldots \tag{1.53}$$

For functions that are relatively smooth, the magnitudes of the terms in this series decrease as higher order derivatives are introduced, so an approximation of a function can be achieved by selecting only the low-order terms. Choosing only the first two, for example, we obtain

$$\begin{aligned} f(x) &\approx f(x_0)+f'(x_0)(x-x_0) \\ &= f'(x_0)x+\left[f(x_0)-f'(x_0)x_0\right] \end{aligned} \tag{1.54}$$

It can be seen from (1.54) that by keeping only the first two terms of the Taylor series, an equation of a line results. The approximation of (1.54) will be referred to as the *linearization* of $f(x)$. This linearization is illustrated in Figure 1.10. In the figure, the original curve $f(x)$ appears as the wavy line. The point of expansion, $x = x_0$, is the point at which the curve is approximated as a straight line by (1.54). For most curves, it is important that the straight-line approximation not be used if the value of x strays too far from x_0. However if x remains close to x_0, the Taylor series approximation is sometimes a very good one for practical purposes. The accuracy and so-called "linear region" depend, of course, on the particular function $f(x)$.

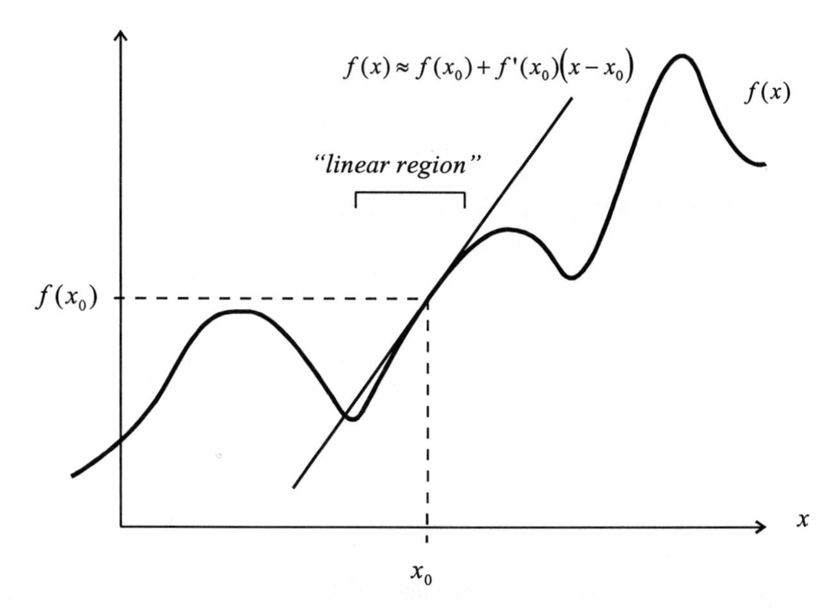

Figure 1.10 Taylor series linearization of a curve.

When the function *f* depends on several variables, such as $x_1, x_2, \ldots, x_n$, not only must all the partial derivatives of the individual variables x_i be used in (1.52), but all their cross-derivatives as well. That is, the Taylor series may be expressed as

$$
\begin{aligned}
f(x_1, x_2, \ldots x_n) &= f(x_{10}, x_{20}, \ldots x_{n0}) \\
&\quad + \left((x_1 - x_{10}) \frac{\partial}{\partial x_1} + \cdots + (x_n - x_{n0}) \frac{\partial}{\partial x_n} \right) f(x_1, x_2, \ldots, x_n) \bigg|_{x_i = x_{i0}} \\
&\quad + \frac{1}{2!} \left((x_1 - x_{10}) \frac{\partial}{\partial x_1} + \cdots + (x_n - x_{n0}) \frac{\partial}{\partial x_n} \right)^2 f(x_1, x_2, \ldots x_n) \bigg|_{x_i = x_{i0}} \\
&\quad + \cdots
\end{aligned}
$$

Of course, for a linear approximation, only the first two terms of this series need be retained.

It is also quite common that several functions of several variables need to be linearized, such as

$$\begin{array}{c} f_1(x_1,x_2,\ldots x_n) \\ f_2(x_1,x_2,\ldots x_n) \\ \vdots \\ f_m(x_1,x_2,\ldots x_n) \end{array} \tag{1.55}$$

In this case, each function $f_j(x_1,x_2,\ldots,x_n)$, $j=1,\ldots,m$ can be expanded into a Taylor series and thus linearized separately. Alternatively, we can use *matrix-vector notation* and rewrite (1.55) as

$$\boldsymbol{f}(\boldsymbol{x}) = \begin{bmatrix} f_1(x_1,x_2,\ldots x_n) \\ f_2(x_1,x_2,\ldots x_n) \\ \vdots \\ f_m(x_1,x_2,\ldots x_n) \end{bmatrix} \tag{1.56}$$

where $\boldsymbol{f} = \begin{bmatrix} f_1 & f_2 & \cdots & f_m \end{bmatrix}^{\mathrm{T}}$ and $\boldsymbol{x} = \begin{bmatrix} x_1 & x_2 & \cdots & x_n \end{bmatrix}^{\mathrm{T}}$. Using this notation, the linearized version of the nonlinear functions are (by taking the first two terms of the Taylor series):

$$\boldsymbol{f}(\boldsymbol{x}) \approx \boldsymbol{f}(\boldsymbol{x}_0) + \left.\frac{\partial \boldsymbol{f}(\boldsymbol{x})}{\partial \boldsymbol{x}}\right|_{x_0} (\boldsymbol{x} - \boldsymbol{x}_0) \tag{1.57}$$

In this expression, the derivative of $\boldsymbol{f}(\boldsymbol{x})$ is a derivative of an $m \times 1$ vector with respect to an $n \times 1$ vector, resulting in an $m \times n$ matrix whose $(i,j)^{th}$ element is $\partial f_i / \partial x_j$. See Appendix A.

1.2.2 Linearizing Differential Equations

The Taylor series linearization process can be performed on differential equations as well as on functions. When each term of a nonlinear differential equation is linearized in terms of the variables on which it depends, we say that the entire equation has been *linearized*. When this has been done, all the linear analysis tools presented here and elsewhere may be applied to the equation(s), remembering that this linear equation is just an approximation of the original system. There are some situations, such as in chaotic systems, wherein linear approximations are not very good approximators of the true solution of the system, and still more, such as when the nonlinear equations are not analytic (e.g., systems with static friction), when the linearization does not apply at all. In many physical systems, though, the two-term Taylor series approximation provides a reasonably accurate representation of the system, usually good enough that linear controllers can be applied, provided that the approximation is taken sufficiently near the operating point.

Example 1.8: Linearization of a Differential Equation for an Inverted Pendulum

The equation of motion can be derived for the model of an inverted pendulum on a cart shown in Figure 1.11. In the model, $\theta(t)$ is the angle of the pendulum clockwise with respect to the vertical, $x(t)$ is the horizontal position of the cart relative to some arbitrary fixed reference location, 2ℓ is the length of the pendulum, M and m are the masses of the cart and the pendulum, respectively, and I is the moment of inertia of the pendulum about its center of gravity. F is a force applied to the body of the cart.

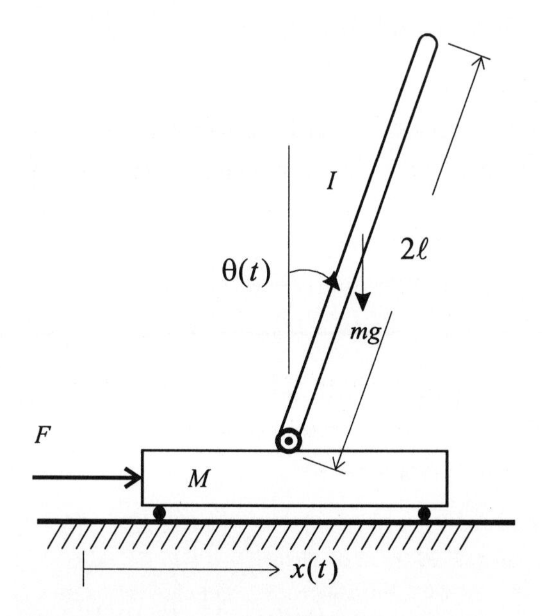

Figure 1.11 Physical model of an inverted pendulum on a cart.

It can be shown that the two coupled differential equations that describe the motion of this system are:

$$\begin{aligned}(m+M)\ddot{x}+m\ell\ddot{\theta}\cos\theta-m\ell\dot{\theta}^2\sin\theta&=F\\ \left(I+m\ell^2\right)\ddot{\theta}+m\ell\ddot{x}\cos\theta-mg\ell\sin\theta&=0\end{aligned} \tag{1.58}$$

Linearize these two equations using the first two terms of the Taylor series.

Solution:

Although only $x(t)$ and $\theta(t)$ are considered "coordinate variables," when linearizing we consider their respective derivatives to also constitute independent variables. Likewise, the input force F is a variable (although the equations already happen to be linear in terms of F). We therefore are linearizing *two* equations, each in terms of the *seven independent* variables $\theta(t)$, $\dot{\theta}(t)$, $\ddot{\theta}(t)$, $x(t)$, $\dot{x}(t)$, $\ddot{x}(t)$, and F.

To begin, we select an operating point. For convenience, we choose the "zero" position of $\theta_0(t)=\dot{\theta}_0(t)=\ddot{\theta}_0(t)=0$, $x_0(t)=\dot{x}_0(t)=\ddot{x}_0(t)=0$, and $F_0=0$. We should note that these values are not all independent of one another. If we arbitrarily select the first six values, then substitution into (1.58) allows us to determine $F_0=0$. If, for example, the cart and pendulum were moving up a wall, then a different F_0 (nonzero) would have to be chosen in order to counteract gravity. The equations in (1.58) must hold at all times.

We will apply (1.54) to the four terms in the first equation of (1.58) and leave the details of the second equation as an exercise.

For $(m+M)\ddot{x}$,

$$\begin{aligned}(m+M)\ddot{x} &\approx (m+M)\ddot{x}_0+(m+M)(\ddot{x}-\ddot{x}_0)\\ &=0+(m+M)\ddot{x}\\ &=(m+M)\ddot{x}\end{aligned}$$

which of course returns the original expression since it was linear to begin with.

For $m\ell\ddot{\theta}\cos\theta$,

$$\begin{aligned}m\ell\ddot{\theta}\cos\theta &\approx m\ell\ddot{\theta}_0\cos\theta_0+\frac{\partial}{\partial\theta}\left(m\ell\ddot{\theta}\cos\theta\right)\Big|_{\substack{\theta_0\\ \ddot{\theta}_0}}(\theta-\theta_0)\\ &\quad+\frac{\partial}{\partial\ddot{\theta}}\left(m\ell\ddot{\theta}\cos\theta\right)\Big|_{\substack{\theta_0\\ \ddot{\theta}_0}}\left(\ddot{\theta}-\ddot{\theta}_0\right)\\ &=0-m\ell\ddot{\theta}_0\sin\theta_0(\theta)+m\ell\cos\theta_0\left(\ddot{\theta}\right)\\ &=m\ell\,\ddot{\theta}\end{aligned}$$

For $m\ell\dot{\theta}^2\sin\theta$,

$$
\begin{aligned}
m\ell\dot{\theta}^2 \sin\theta &\approx m\ell\dot{\theta}_0{}^2 \sin\theta_0 + \frac{\partial}{\partial\theta}\left(m\ell\dot{\theta}^2 \sin\theta\right)\Big|_{\substack{\theta_0 \\ \dot{\theta}_0}} (\theta - \theta_0) \\
&\quad + \frac{\partial}{\partial\dot{\theta}}\left(m\ell\dot{\theta}^2 \sin\theta\right)\Big|_{\substack{\theta_0 \\ \dot{\theta}_0}} (\dot{\theta} - \dot{\theta}_0) \\
&= 0 + m\ell\dot{\theta}_0{}^2 \cos\theta_0(\theta) + 2m\ell\dot{\theta}_0 \sin\theta_0(\dot{\theta}) \\
&= 0
\end{aligned}
$$

It may at first seem somewhat surprising that this term should disappear entirely. However, recalling that the equilibrium point we have chosen has $\dot{\theta}_0 = 0$, the "linear" region in this problem will include only small velocities. Because the velocity in this term appears only as the square, and we are assuming that second order and higher terms are negligible, this entire term should be negligible near the equilibrium.

As for the final term, *F*, it is clear that this is already linear, and we will leave it as is. The "linearized" version of the first of Equations (1.58) is therefore:

$$
(m + M)\ddot{x} + m\ell\ddot{\theta} = F \tag{1.59}
$$

Note that this equation has two second-order derivatives in it, which are the highest derivatives of each variable $\theta(t)$ and $x(t)$. The second equation will linearize to

$$
(I + m\ell^2)\ddot{\theta} + m\ell\ddot{x} - mg\ell\theta = 0 \tag{1.60}
$$

It also has two second-order derivatives. In order to construct a state space representation for such a system, in both Equation (1.59) and (1.60) we will have solve for one such second derivative and substitute it into the other, so that each equation contains only one of the highest derivatives of any of the state variables. This procedure and the subsequent formulation of the state equations is left as Exercise 1.12.

The Choice of Operating Point

One should be careful in applying these methods with the goal of arriving at an equation of the form

$$
\dot{\boldsymbol{x}} = \boldsymbol{f}(\boldsymbol{x}_0) + \frac{\partial \boldsymbol{f}(\boldsymbol{x})}{\partial \boldsymbol{x}}\Big|_{\boldsymbol{x}_0} (\boldsymbol{x} - \boldsymbol{x}_0) \tag{1.61}
$$

If in (1.61) we have

$$f(x_0) - \left.\frac{\partial f(x)}{\partial x}\right|_{x_0} (x_0) \neq 0$$

then (1.61) is *not linear*! It has a constant offset and will therefore violate the homogeneity and additivity conditions required for linearity. Equation (1.61) will indeed look like a straight line on the $\dot{x}$ versus x plane, but it will not pass through the origin of this plane. Such a system is called *affine*.

In order to make the system linear, one must choose an appropriate operating value for the variable x. In particular, choose as the operating point the equilibrium value x_e such that $f(x_e) = 0$. Then to eliminate the constant term in (1.61), we must make a change of variable $\xi \stackrel{\Delta}{=} x - x_e$. Then, $x = \xi + x_e$ and $\dot{x} = \dot{\xi}$. Substituting these values into (1.61), we get

$$\dot{\xi} = f(x_e) + \left.\frac{\partial f(x)}{\partial x}\right|_{x_e} (\xi) = \left.\frac{\partial f(x)}{\partial x}\right|_{x_e} \xi \tag{1.62}$$

This equation is then truly *linear*. If after solving for ξ, we desire to know the value of x, then the change of variables may be reversed.

1.3 Summary

In this chapter we have introduced the state space notation to simple models of physical systems. Our purpose has been to establish terms and definitions that are both common to and different from the transfer function descriptions with which the reader may be more familiar. Using the state variable technique, our main points have been as follows:

- State variables are simply regarded as a set of variables in which the behavior of a system may be mathematically modeled.
- Mathematical models for use in state variable systems may be derived, simplified, linearized, and discretized just as any other mathematical model is produced. It is only the choice of state variables that is new so far.
- State variables and state variable equations together constitute the same information about a physical system as do the original differential equations derived from familiar conservation and constitutive laws.

- State variables may be changed to represent either other physical variables or variables that have no discernible physical significance at all. This changes the structure of the matrix elements in the state equations but does not alter the fact that the equations describe the same physical system equally well.

We have tried to stress the need for matrix and vector analysis and linear algebra in the discussions of this chapter. In the next chapter, we will begin to study these mathematical tools. They will later lead us to see the benefits and drawbacks of the state variable techniques.

1.4 Problems

1.1 A system has an input $u(t)$ and an output $y(t)$, which are related by the information provided below. Classify each system as linear or nonlinear and time-invariant or time-varying.

a) $y(t)=0$ for all t.

b) $y(t)=a$, $a \neq 0$, for all t.

c) $y(t)=-3u(t)+2$.

d)

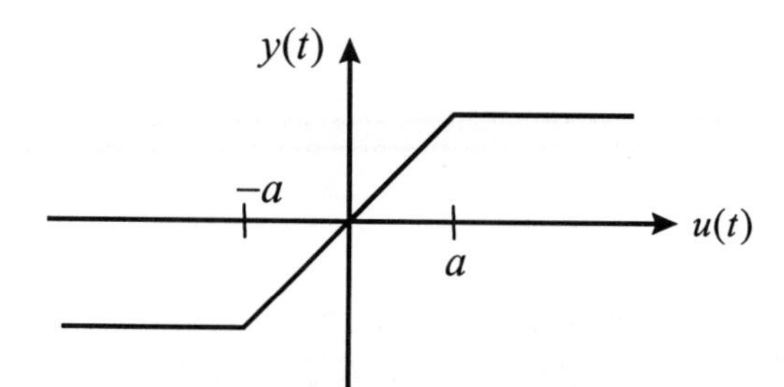

e)

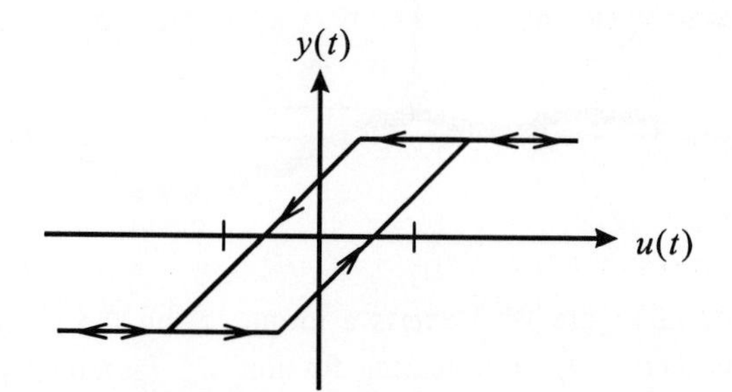

f) $\ddot{y}(t)+3e^{-t}\dot{y}(t)+y(t)=u(t)$.

g) $\ddot{y}(t)+(1-y^2(t))\dot{y}(t)+\omega^2 y(t)=u(t)$, $\omega = \text{constant} \neq 0$.

h) $y(t) = u(t-3)$.

i) $\ddot{y}(t) + u(t)y(t) = 0$.

j) $y(t) = \int_0^t e^{-\tau} u(t-\tau) d\tau$.

k) $y(k+2) = -0.4y(k+1) - y(k) + 3u(k+2) - u(k)$.

l) $y(k) = \sum_{i=0}^{k} \sin(iT) e^{-iT} u(k-i)$, $T = \text{constant} > 0$.

1.2 Figure P1.2 shows a model commonly used for automobile suspension analysis. In it, the uneven ground specifies the position of the wheel's contact point. The wheel itself is not shown, as its mass is considered negligible compared to the mass of the rest of the car. Write a differential equation and a state variable description for this system, considering the height of the car, $x(t)$, to be the output, and the road height, $y(t)$, to be the input.

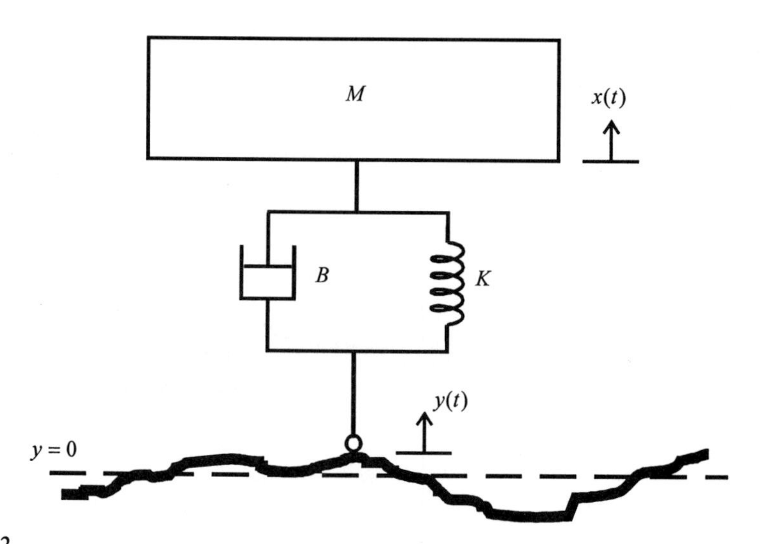

P1.2

1.3 The motor in Figure P1.3 exerts a torque τ on the shaft of a motor, which has inertia J_m and bearing friction B_m (assumed viscous). This motor is attached to a load inertia J_L, which itself has a viscous friction B_L. The motor coupling is slightly flexible and is modeled as a torsional spring K. Write the equations of motion and a set of state equations for the system, taking $\tau(t)$ as the input, and $\theta_L(t)$ as the output.

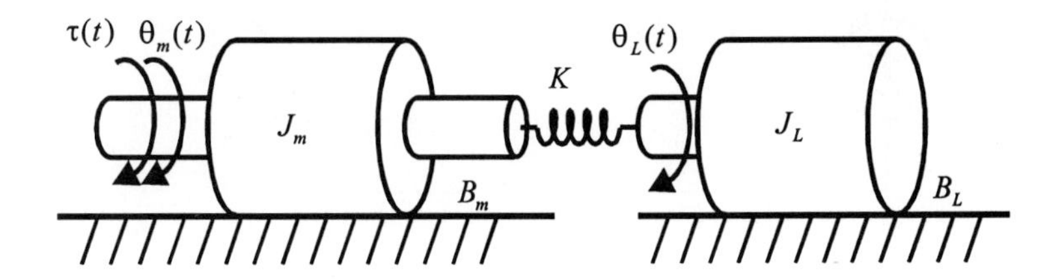

P1.3

1.4 For the mechanical system in Figure P1.4 with a spring, friction forces, and an external force, write state equations using the following sets of variables:

a) State variables $\xi_1 = x_1$, $\xi_2 = \dot{x}_1$, $\xi_3 = x_2$, and $\xi_4 = \dot{x}_2$; and output variable $y = x_2$.

b) State variables $\xi_1 = x_1$, $\xi_2 = \dot{x}_1$, $\xi_3 = x_2 - x_1$, and $\xi_4 = \dot{x}_2 - \dot{x}_1$; and output variable $y = x_2 - x_1$.

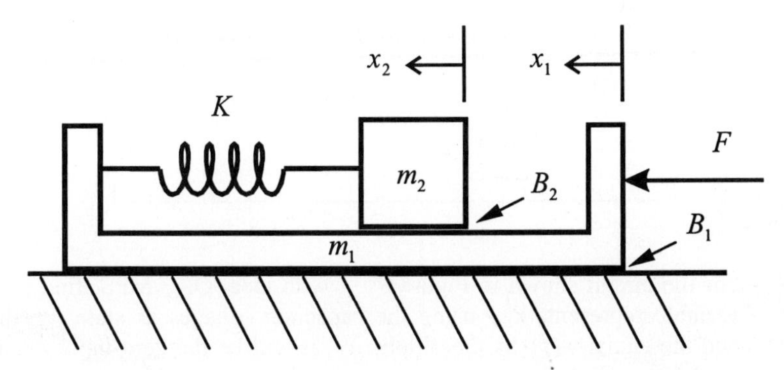

P1.4

1.5 For the circuit shown in Figure P1.5, choose appropriate state variables and write state equations, taking as output the voltage across the resistor.

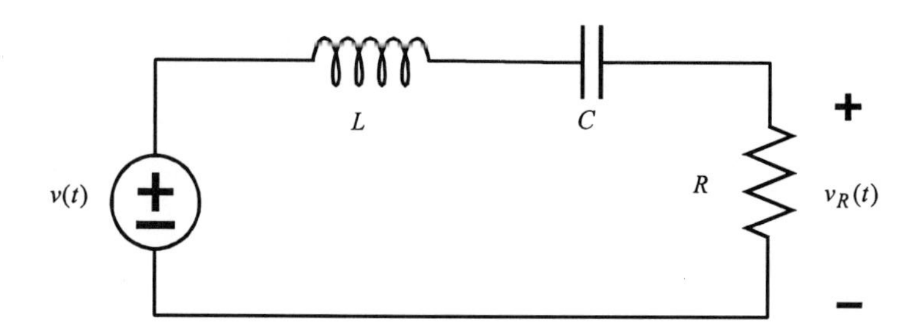

P1.5

1.6 For the circuit shown in Figure P1.6, write a single second-order differential equation in terms of the voltage $v_x(t)$ and the input $x(t)$. Then write state equations using state variables $v_C(t)$ and $i_L(t)$, where the output of the system is considered to be $v_x(t)$, and the input is $x(t)$.

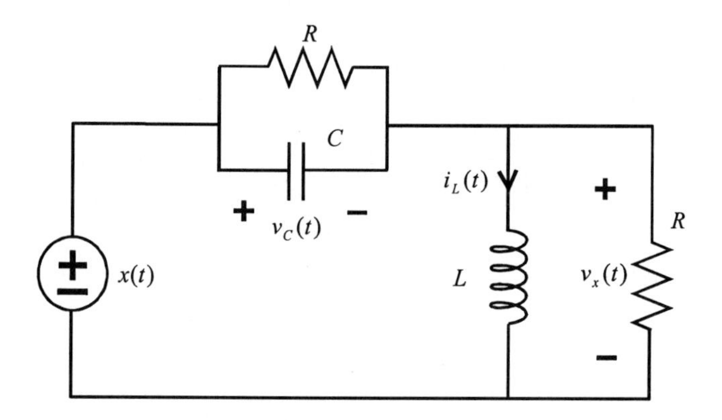

P1.6

1.7 For the circuit shown in Figure P1.7, with ideal OP-AMPS, find a state variable representation, using the capacitor voltages as state variables, and the signal $x(t)$ as the output. What will be the zero-input solution for $x(t)$?

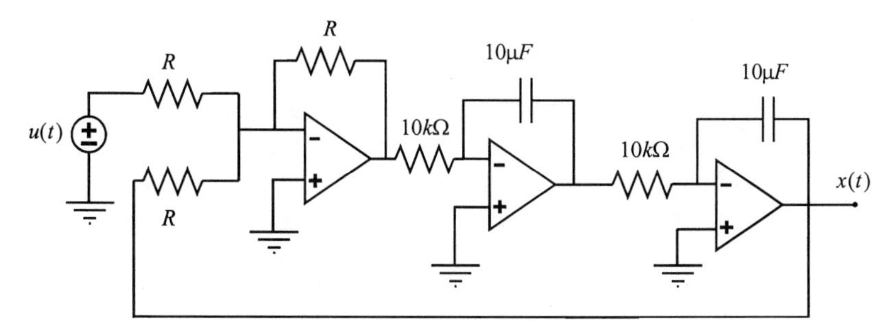

P1.7

1.8 Write two differential equations that describe the behavior of the circuit of Example 1.2, using $v_L(t)$ and $v_C(t)$ as the state variables.

1.9 Given the state variable description of the system in terms of $x_i(t)$ below, change the state variables and write new state equations for the variables $\xi_1(t) = 3x_1(t) + 2x_2(t)$, and $\xi_2(t) = 7x_1(t) + 5x_2(t)$.

$$\begin{bmatrix} \dot{x}_1(t) \\ \dot{x}_2(t) \end{bmatrix} = \begin{bmatrix} 11 & 10 \\ -21 & -18 \end{bmatrix} \begin{bmatrix} x_1(t) \\ x_2(t) \end{bmatrix} + \begin{bmatrix} 0 \\ 1 \end{bmatrix} u(t)$$

$$y(t) = \begin{bmatrix} 2 & 2 \end{bmatrix} \begin{bmatrix} x_1(t) \\ x_2(t) \end{bmatrix} + u(t)$$

1.10 For the state variable description of the system in terms of $x_i(t)$,

$$\begin{bmatrix} \dot{x}_1(t) \\ \dot{x}_2(t) \\ \dot{x}_3(t) \end{bmatrix} = \begin{bmatrix} 18 & 9 & 13 \\ 50 & 23 & 35 \\ -65 & -31 & -46 \end{bmatrix} \begin{bmatrix} x_1(t) \\ x_2(t) \\ x_3(t) \end{bmatrix} + \begin{bmatrix} -1 \\ 0 \\ 1 \end{bmatrix} u(t)$$

$$y(t) = \begin{bmatrix} 5 & -5 & 5 \end{bmatrix} \begin{bmatrix} x_1(t) \\ x_2(t) \\ x_3(t) \end{bmatrix}$$

change the state variables and write new state equations for variables

$$\xi_1(t) = -4x_1(t) - 2x_2(t) - 3x_3(t)$$

$$\xi_2(t) = 15x_1(t) + 7x_2(t) + 10x_3(t)$$

and

$$\xi_3(t) = -5x_1(t) - 2x_2(t) - 3x_3(t)$$

1.11 The robot shown in Figure P1.11 has the differential equations of motion given. Symbols $m_1, m_2, I_1, I_2, \ell_1$, and g are constant parameters, representing the characteristics of the rigid body links. Quantities θ_1 and d_2 are the coordinate variables and are functions of time. The inputs are τ_1 and τ_2. Linearize the two equations about the operating point $\theta_1 = \dot{\theta}_1 = \ddot{\theta}_1 = 0$, $d_2 = 3$, and $\dot{d}_2 = \ddot{d}_2 = 0$.

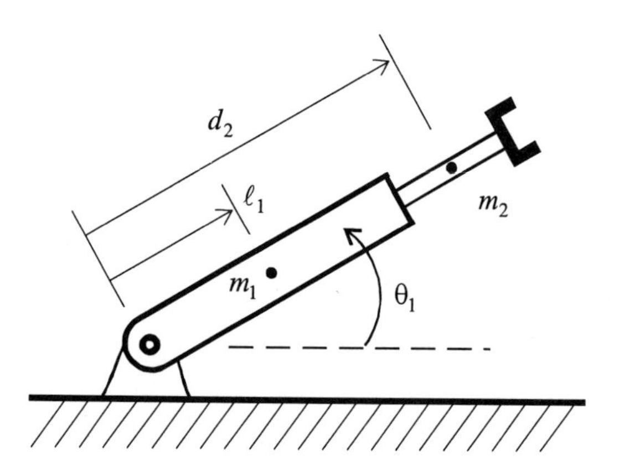

P1.11

$$(m_1\ell_1^2 + I_1 + I_2 + m_2 d_2^2)\ddot{\theta}_1 + 2m_2 d_2 \dot{\theta}_1 \dot{d}_2 + (m_1\ell_1 + m_2 d_2) g \cos\theta_1 = \tau_1$$

$$m_2 \ddot{d}_2 - m_2 d_2 \dot{\theta}_1^2 + m_2 g \sin\theta_1 = \tau_2$$

1.12 For the system in Example 1.8, the nonlinear equations of motion are given in (1.58). Show that (1.60) is the linear approximation of the second of these equations, and combine the two linearized equations (1.59) and (1.60) into state space form.

1.13 A permanent-magnet DC motor with a connected inertial and friction load is depicted in Figure P1.13. The motor armature is driven with voltage V, and the motor turns through angle θ with torque τ. The armature has resistance R and inductance L, and the armature current is denoted i. The mechanical load is an inertia J, and the bearings have viscous friction coefficient f. The motor produces a back-emf of $e = k_b\dot{\theta}$ and a torque of $\tau = k_a i$, where k_a and k_b are constants. Determine a set of describing equations for the electrical and mechanical components of the system. Then express these equations as a set of linear state equations, using the voltage V as input and the angle θ as output.

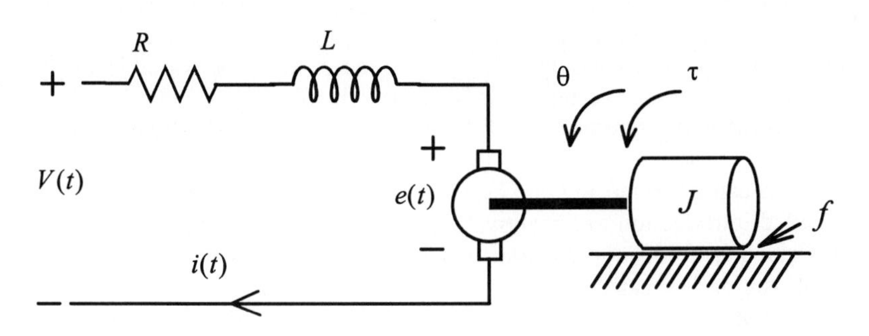

P1.13

1.5 References and Further Reading

The formulation of the dynamic equations for physical systems can be found in any physics text. For information specific to mechanical systems, texts in statics and dynamics can be consulted, and for circuit analysis, there are a great many elementary texts, including [1], which focuses on state variable analysis of electrical networks. A reference that ties them together and introduces the unified modeling terminology of "through" variables and "across" variables is [4]. Other good introductions to the state space representation for physical systems can be found in [2], [7], [8], and [10]. In particular, [10] gives a very detailed introduction to linear system terminology and definitions. For systems described in frequency-domain, which we do not treat in much depth in this book, the student can consult [3] and [7].

Additional state variable models can be found in [5] and [8], both of which provide numerous examples from systems that engineering students do not traditionally encounter, such as genetics, populations, economics, arms races, air pollution, and predator-prey systems.

Further information on nonlinear systems and linearization is given in [9].

[1] Belevitch, V., *Classical Network Theory*, Holden-Day, 1968.

[2] Brogan, William L., *Modern Control Theory*, 3rd edition, Prentice-Hall, 1991.

[3] Callier, Frank M., and Charles A. Desoer, *Multivariable Feedback Systems*, Springer-Verlag, 1982.

[4] Cannon, Robert H. Jr., *Dynamics of Physical Systems*, McGraw-Hill, 1967.

[5] Casti, John L., *Linear Dynamical Systems*, Academic Press, 1987.

[6] Franklin, Gene, and J. David Powell, *Digital Control of Dynamic Systems*, Addison-Wesley, 1981.

[7] Kailath, Thomas, *Linear Systems*, Prentice-Hall, 1980.

[8] Luenberger, David G., *Introduction to Dynamic Systems*, John Wiley & Sons, 1979.

[9] Slotine, Jean-Jacques, and Weiping Li, *Applied Nonlinear Control*, Prentice-Hall, 1991.

[10] Zadeh, Lotfi A. and Charles A. Desoer, *Linear System Theory: The State Space Approach*, McGraw-Hill, 1963.

2

Vectors and Vector Spaces

There are several conceptual levels to the understanding of vector spaces and linear algebra. First, there is the mechanical interpretation of the term *vector* as it is often taught in physics and mechanics courses. This implies a magnitude and direction, usually with clear physical meaning, such as the magnitude and direction of the velocity of a particle. This is often an idea restricted to the two and three dimensions of physical space perceptible to humans. In this case, the familiar operations of dot product and cross product have physical implications, and can be easily understood through the mental imagery of projections and right-hand rules.

Then there is the idea of the vector as an n-tuple of numbers, usually arranged as a column, such as often presented in lower-level mathematics classes. Such columns are given the same properties of magnitude and direction but are not constrained to three dimensions. Further, they are subject to more generalized operators in the form of a matrix. Matrix-vector multiplication can also be treated very mechanically, as when students are first taught to find solutions to simultaneous algebraic equations through gaussian elimination or by using matrix inverses.[M]

`inv(A)`

These simplified approaches to vector spaces and linear algebra are valuable tools that serve their purposes well. In this chapter we will be concerned with more generalized treatments of linear spaces. As such, we wish to present the properties of vector spaces with somewhat more abstraction, so that the applications of the theory can be applied to broad classes of problems. However, because the simplified interpretations are familiar, and they are certainly consistent with the generalized concept of a linear space, we will not discard them. Instead, we will use simple mechanical examples of vector spaces to launch the discussions of more general settings. In this way, one can transfer existing intuition into the abstract domain for which intuition is often difficult.

2.1 Vectors

To begin with, we appeal to the intuitive sense of a vector, i.e., as a magnitude and direction, as in the two-dimensional case shown below.

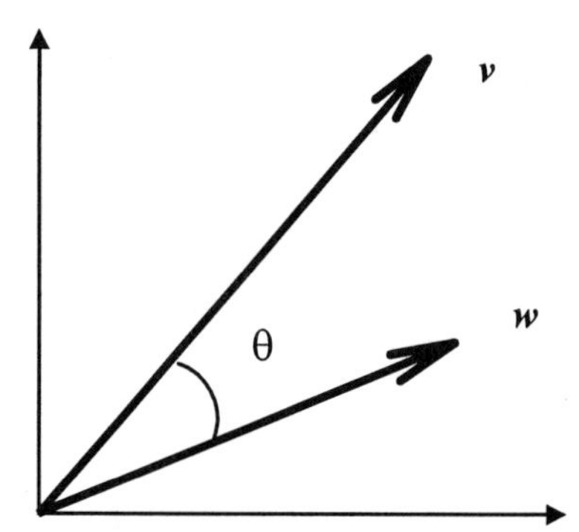

Figure 2.1 The concept of a vector with magnitude and direction. Such vectors can be easily pictured in two and three dimensions only.

In Figure 2.1, vectors $\boldsymbol{v}$ and $\boldsymbol{w}$ are depicted as arrows, the length of which indicates their magnitude, with directions implied by the existence of the reference arrows lying horizontally and vertically. It is clear that the two vectors have an angle between them (θ), but in order to uniquely fix them in "space," they should also be oriented with respect to these references. The discussion of these reference directions is as important as the vectors themselves.

2.1.1 Familiar Euclidean Vectors

We begin the discussion of the most basic properties of such vectors with a series of definitions, which we first present in familiar terms but which may be generalized later. We give these as a reminder of some of the simpler operations we perform on vectors that we visualize as "arrows" in space.

`dot(x,y)`

> ***Inner (dot) product***[M]***:*** The inner product of two vectors, $\boldsymbol{v}$ and $\boldsymbol{w}$, is denoted as $\langle \boldsymbol{v}, \boldsymbol{w} \rangle$. Although there may exist many definitions for the computation of an inner product, when vectors are interpreted as columns of n real numbers, the inner product is customarily computed as:

$$\langle \boldsymbol{v}, \boldsymbol{w} \rangle = \boldsymbol{v}^T \boldsymbol{w} = \boldsymbol{w}^T \boldsymbol{v} = \sum_{i=1}^{n} v_i w_i \tag{2.1}$$

where v_i and w_i are the individual elements of the vectors. In terms of the magnitudes of the vectors (defined below), this is also sometimes given the definition:

$$\langle v, w \rangle = \|v\| \, \|w\| \cos\theta \tag{2.2}$$

This gives a geometric relationship between the size and direction of the vectors and their inner product.

Norm (magnitude): The norm[M] of a vector $\boldsymbol{v}$, physically interpreted as its magnitude, can be "induced" from this definition of the inner product above as:

norm(x)

$$\|v\| = \langle v, v \rangle^{1/2} = \sqrt{\sum_{i=1}^{n} v_i^2} \tag{2.3}$$

As we will see, there may be many different norms. Each different inner product may produce a different norm. When applied to n-tuples, the most common is the euclidean, as given in (2.3) above.

Outer (tensor) product: The outer product of two vectors $\boldsymbol{v}$ and $\boldsymbol{w}$ is defined as:

$$v\rangle\langle w = vw^{\mathrm{T}} = -wv^{\mathrm{T}} \tag{2.4}$$

We will not use this definition much in this text.

Cross (vector) product[M]: This product produces a vector quantity from two vectors $\boldsymbol{v}$ and $\boldsymbol{w}$.

cross(x,y)

$$\|v \times w\| = \|v\| \, \|w\| \sin\theta \tag{2.5}$$

where the resulting vector has a new direction that is perpendicular to the plane of the two vectors $\boldsymbol{v}$ and $\boldsymbol{w}$ and that is generated by the so-called "right-hand rule." As such, this operation is generally used in three dimensions only. Again, it will have limited usefulness for us.

2.2 Vector Spaces

Given these familiar definitions, we turn now to more abstract definitions of vectors. Our concept of vectors will retain the properties of inner products and norms as defined above, but we will consider vectors consisting of quantities that appear to be very different from n-tuples of numbers, in perhaps infinite dimensions. Nevertheless, analogies to the intuitive concepts above will survive.

2.2.1 Fields

First, we cannot introduce the idea of a vector space without first giving the definition of a *field*. Conceptually, a field is the set from which we select elements that we call *scalars*. As the name implies, scalars will be used to scale vectors.

> ***Field:*** A field consists of a set of two or more elements, or members, which must include the following:
>
> 1. There must exist in the field a unique element called 0 (zero). For $0 \in F$ and any other element $a \in F$, $0(a) = 0$ and $a + 0 = a$.
> 2. There must exist in the field another unique element called 1 (one). For $1 \in F$ and $a \in F$, $1(a) = a(1) = (a/1) = a$.
> 3. For every $a \in F$, there is a unique element called its negative, $-a \in F$, such that $a + (-a) = 0$.
>
> A field must also provide definitions of the operations of addition, multiplication, and division. There is considerable flexibility in the definition of these operations, but in any case the following properties must hold:
>
> 1. If $a \in F$ and $b \in F$, then $(a + b) = (b + a) \in F$. That is, the sum of any two vectors in a field is also in the same field. This is known as *closure under addition.*
> 2. If $a \in F$ and $b \in F$, then $(ab) = (ba) \in F$. That is, their product remains in the field. This is known as *closure under multiplication.*
> 3. If $a \in F$ and $b \in F$, and if $b \neq 0$, then $a/b \in F$.

Finally, for the addition and multiplication operations defined, the usual associative, commutative, and distributive laws apply. (2.6)

Example 2.1: Candidate Fields

Determine which of the following sets of elements constitute fields, using elementary arithmetic notions of addition, multiplication, and division (as we are familiar with them).*

1. The set of real numbers $\{0, 1\}$
2. The set of all real matrices of the form $\begin{bmatrix} x & -y \\ y & x \end{bmatrix}$ where x and y are real numbers
3. The set of all polynomials in s
4. The set of all real numbers
5. The set of all integers

Solutions:

1. No. This cannot be a field, because $1+1=2$, and the element 2 is not a member of the set.
2. Yes. This set has the identity matrix and the zero matrix, and inverse matrices of the set have the same form. However, this set has the special property of being commutative under multiplication, which is not true of matrices in general. Therefore, the set of all 2×2 matrices is *not* a field.
3. No. The inverse of a polynomial is not usually a polynomial.
4. Yes. This is the most common field that we encounter.
5. No. Like polynomials, the inverse of an integer is not usually an integer.

Now we can proceed to define vector spaces. The definition of a vector space is dependent on the field over which we specify it. We therefore must refer to *vector spaces over fields*.

---

* Such arithmetic operations need not generally follow our familiar usage. For instance, in Example 1, the answer will vary if the set is interpreted as binary digits with binary operators rather than the two real numbers 0 and 1.

Linear Vector Space: A *linear vector space* $\mathbf{X}$ is a collection of elements called *vectors*, defined over a field F. Among these vectors must be included:

1. A vector $\mathbf{0} \in \mathbf{X}$ such that $\boldsymbol{x} + \mathbf{0} = \mathbf{0} + \boldsymbol{x} = \boldsymbol{x}$.
2. For every vector $\boldsymbol{x} \in \mathbf{X}$, there must be a unique vector $\boldsymbol{y} \in \mathbf{X}$ such that $\boldsymbol{x} + \boldsymbol{y} = \mathbf{0}$. This condition is equivalent to the existence of a negative element for each vector in the vector space, so that $\boldsymbol{y} = -\boldsymbol{x}$.

As with a field, operations on these elements must satisfy certain requirements. (In the rules that follow, the symbols $\boldsymbol{x}$, $\boldsymbol{y}$, and $\boldsymbol{z}$ are elements of the space $\mathbf{X}$, and symbols a and b are elements of the field F.) The requirements are:

1. *Closure* under addition: If $\boldsymbol{x} + \boldsymbol{y} = \boldsymbol{v}$, then $\boldsymbol{v} \in \mathbf{X}$.
2. *Commutativity* of addition: $\boldsymbol{x} + \boldsymbol{y} = \boldsymbol{y} + \boldsymbol{x}$.
3. *Associativity* of addition: $(\boldsymbol{x} + \boldsymbol{y}) + \boldsymbol{z} = \boldsymbol{x} + (\boldsymbol{y} + \boldsymbol{z})$.
4. *Closure* under scalar multiplication: For every $\boldsymbol{x} \in \mathbf{X}$ and scalar $a \in F$, the product $a\boldsymbol{x}$ gives another vector $\boldsymbol{y} \in \mathbf{X}$. Scalar a may be the *unit* scalar, so that $a\boldsymbol{x} = 1 \cdot \boldsymbol{x} = \boldsymbol{x} \cdot 1 = \boldsymbol{x}$.
5. *Associativity* of scalar multiplication: For any scalars $a, b \in F$, and for any vector $\boldsymbol{x} \in \mathbf{X}$, $a(b\boldsymbol{x}) = (ab)\boldsymbol{x}$.
6. *Distributivity* of scalar multiplication over vector addition:

$$
\begin{aligned}
(a+b)\boldsymbol{x} &= a\boldsymbol{x} + b\boldsymbol{x} \\
a(\boldsymbol{x}+\boldsymbol{y}) &= a\boldsymbol{x} + a\boldsymbol{y}
\end{aligned}
\tag{2.7}
$$

Example 2.2: Candidate Vector Spaces

Determine whether the following sets constitute vector spaces when defined over the associated fields.

1. The set of all n-tuples of scalars from any field F, defined over F. For example, the set of n-tuples $\Re^n$ over the field of reals $\Re$, or the set of complex n-tuples $\mathbf{C}^n$ over the field of complex numbers $\mathbf{C}$.
2. The set of complex numbers over the reals.

3. The set of real numbers over the complex numbers.
4. The set of all $m \times n$ matrices, over the reals. The same set over complex numbers.
5. The set of all piecewise continuous functions of time, over the reals.
6. The set of all polynomials in s of order less than n, with real coefficients, over $\Re$.
7. The set of all symmetric matrices over the real numbers.
8. The set of all nonsingular matrices.
9. The set of all solutions to a particular linear, constant-coefficient, finite-dimensional homogeneous differential equation.
10. The set of all solutions to a particular linear, constant-coefficient, finite-dimensional *non*-homogeneous differential equation.

Solutions:

1. Yes. These are the so-called "euclidean spaces," which are most familiar. The intuitive sense of vectors as necessary for simple mechanics fits within this category.
2. Yes. Note that a complex number, when multiplied by a real number, is still a complex number.
3. No. A real number, when multiplied by a complex number, is not generally real. This violates the condition of closure under scalar multiplication.
4. Yes in both cases. Remember that such matrices are not generally fields, as they have no unique inverse, but they do, as a collection, form a space.
5. Yes.
6. Yes. Again, such a set would not form a field, but it does form a vector space. There is no requirement for division in a vector space.
7. Yes. The sum of symmetric matrices is again a symmetric matrix.
8. No. One can easily find two nonsingular matrices that add to form a singular matrix.
9. Yes. This is an important property of solutions to differential equations.
10. No. If one were to add the particular solution of a nonhomogeneous differential equation to itself, the result would not be a solution to the same differential equation, and thus the set would not be closed under addition. Such a set is not closed under scalar multiplication either.

With the above definitions of linear vector spaces and their vectors, we are ready for some definitions that relate vectors to each other. An important concept is the *independence* of vectors.

2.2.2 Linear Dependence and Independence

It may be clear from the definitions and examples above that a given vector space will have an infinite number of vectors in it. This is because of the closure rules, which specify that any multiple or sum of vectors in a vector space must also be in the space. However, this fact does not imply that we are doomed to manipulating large numbers of vectors whenever we use vector spaces as descriptive tools. Instead, we are often as interested in the directions of the vectors as we are in the magnitudes, and we can gather vectors together that share a common set of directions.

In future sections, we will learn to decompose vectors into their exact directional components. These components will have to be independent of one another so that we can create categories of vectors that are in some way similar. This will lead us to the concepts of bases and dimension. For now, we start with the concepts of *linear dependence* and *independence*.

> ***Linear Dependence:*** Consider a set of n vectors $\{\boldsymbol{x}_1, \boldsymbol{x}_2, \ldots, \boldsymbol{x}_n\} \subset \mathbf{X}$. Such a set is said to be *linearly dependent* if there exists a set of scalars $\{a_i\}$, $i = 1, \ldots, n$, *not all of which are zero*, such that
>
> $$a_1 \boldsymbol{x}_1 + a_2 \boldsymbol{x}_2 + \cdots + a_n \boldsymbol{x}_n = \sum_{i=1}^{n} a_i \boldsymbol{x}_i = 0 \tag{2.8}$$
>
> The sum on the left side in this equation is referred to as a *linear combination.*

> ***Linear Independence:*** If the linear combination shown above, $\sum_{i=1}^{n} a_i \boldsymbol{x}_i = 0$, requires that *all* of the coefficients $\{a_i\}$, $i = 1, \ldots, n$ be zero, then the set of vectors $\{\boldsymbol{x}_i\}$ is linearly independent. (2.9)

If we consider the vectors $\{\boldsymbol{x}_i\}$ to be columns of numbers, we can use a more compact notation. Stacking such vectors side-by-side, we can form a matrix X as follows:

$$X = \begin{bmatrix} x_1 & \vdots & x_2 & \vdots & \cdots & \vdots & x_3 \end{bmatrix}$$

Now considering the set of scalars a_i to be similarly stacked into a column called a (this does not mean they constitute a vector), then the condition for linear independence of the vectors $\{x_i\}$ is that the equation $Xa = \mathbf{0}$ has only the trivial solution $a = 0$.

Linear Independence of Functions

When the vectors themselves are functions, such as $x_i(t)$ for $i = 1, \ldots, n$, then we can define linear independence on intervals of t. That is, if we can find scalars a_i, *not all zero*, such that the linear combination

$$a_1 x_1(t) + \cdots + a_n x_n(t) = 0$$

for all $t \in [t_0, t_1]$, then the functions $\{x_1(t), \ldots, x_n(t)\}$ are linearly dependent in that interval. Otherwise they are independent. Independence outside that interval, or for that matter, in a subinterval, is not guaranteed. Furthermore, if $X(t) = [x_1(t) \quad \cdots \quad x_n(t)]^{\mathrm{T}}$ is an $n \times 1$ vector of functions, then we can define the *Gram matrix* (or *grammian*) of $X(t)$ as the $n \times n$ matrix:

$$G_X(t_1, t_2) = \int_{t_1}^{t_2} X(t) X^*(t) \, dt$$

It can be shown (see Problem 2.26) that the functions $x_i(t)$ are linearly independent if and only if the *Gram determinant* is nonzero, i.e., $|G_X(t_1, t_2)| \neq 0$.

Example 2.3: Linear Dependence of Vectors

Consider the set of three vectors from the space of real n-tuples defined over the field of reals:

$$x_1 = \begin{bmatrix} 2 \\ -1 \\ 0 \end{bmatrix} \quad x_2 = \begin{bmatrix} 1 \\ 3 \\ 4 \end{bmatrix} \quad x_3 = \begin{bmatrix} 0 \\ 7 \\ 8 \end{bmatrix}$$

This is a linearly dependent set because we can choose the set of a-coefficients as $a_1 = -1$, $a_2 = 2$, and $a_3 = -1$. Clearly not all (indeed, none) of these scalars is zero, yet

$$\sum_{i=1}^{3} a_i \boldsymbol{x}_i = -\boldsymbol{x}_1 + 2\boldsymbol{x}_2 - \boldsymbol{x}_3 = \boldsymbol{0}$$

det(A)

Some readers may be familiar with the test from matrix theory that allows one to determine the linear independence of a collection of vectors from the determinant[M] formed by the matrix X as we constructed it above. That is, the dependence of the vectors in the previous example could be ascertained by the test

$$\det(X) = \begin{vmatrix} 2 & 1 & 0 \\ -1 & 3 & 7 \\ 0 & 4 & 8 \end{vmatrix} = 0$$

If this determinant were nonzero, then the vectors would have been found linearly independent.

However, some caution should be used when relying on this test, because the number of components in the vector may not be equal to the number of vectors in the set under test, so that a nonsquare matrix would result. Nonsquare matrices have no determinant defined for them.* We will also be examining vector spaces that are not such simple n-tuples defined over the real numbers. They therefore do not form such a neat determinant. Furthermore, this determinant test does not reveal the underlying geometry that is revealed by applying the definition of linear dependence. The following example illustrates the concept of linear independence with a different kind of vector space.

Example 2.4: Vectors of Rational Polynomials

The set $R(s)$ of all rational polynomial functions in s is a vector space over the field of real numbers $\Re$. It is also known to be a vector space over the field of rational polynomials themselves. Consider two such vectors of the space of *ordered pairs* of such rational polynomials:

---

* In such a situation, one can examine all the square submatrices that can be formed from subsets of the rows (or columns) of the nonsquare matrix. See the problems at the end of the chapter for examples.

$$x_1 = \begin{bmatrix} \frac{1}{s+1} \\ \frac{1}{s+2} \end{bmatrix} \qquad x_2 = \begin{bmatrix} \frac{s+2}{(s+1)(s+3)} \\ \frac{1}{s+3} \end{bmatrix}$$

If the chosen field is the set of all real numbers, then this set of two vectors is found to be linearly independent. One can verify that if a_1 and a_2 are real numbers, then setting

$$\mathbf{0} = \begin{bmatrix} 0 \\ 0 \end{bmatrix} = a_1 x_1 + a_2 x_2 = a_1 \begin{bmatrix} \frac{1}{s+1} \\ \frac{1}{s+2} \end{bmatrix} + a_2 \begin{bmatrix} \frac{s+2}{(s+1)(s+3)} \\ \frac{1}{s+3} \end{bmatrix}$$

will imply that $a_1 = a_2 = 0$. See Problem 2.3 at the end of this chapter.

If instead the field is chosen as the set of rational polynomials, then we have an entirely different result. Through careful inspection of the vectors, it can be found that if the scalars are chosen as

$$a_1 = 1 \qquad \text{and} \qquad a_2 = -\frac{s+3}{s+2}$$

then

$$\sum_{i=1}^{2} a_i x_i = \begin{bmatrix} \frac{1}{s+1} \\ \frac{1}{s+2} \end{bmatrix} + \begin{bmatrix} -\frac{1}{s+1} \\ -\frac{1}{s+2} \end{bmatrix} = \begin{bmatrix} 0 \\ 0 \end{bmatrix}$$

The vectors are now seen to be linearly dependent. Note that this is not a matter of the *same* two vectors being dependent where they once were independent. We prefer to think of them as entirely *different vectors*, because they were selected from entirely different spaces, as specified over two different fields.

This particular example also raises another issue. When testing for linear independence, we should check that equality to zero of the linear combinations is true *identically*. The vectors shown are themselves functions of the variable s. Given a set of linearly independent vectors, there may therefore be isolated values for s that might produce the condition $\sum a_i \boldsymbol{x}_i = 0$ even though the vectors are independent. We should not hastily mistake this for linear dependence, as this condition is not true identically, i.e., for *all s*.

Geometrically, the concept of linear dependence is a familiar one. If we think of two-dimensional euclidean spaces that we can picture as a plane, then any two collinear vectors will be linearly dependent. To show this, we can always scale one vector to have the negative magnitude of the other. This scaled sum would of course equal zero. Two linearly independent vectors will have to be noncollinear. Scaling with nonzero scalars and adding them will never produce zero, because scaling them cannot change their directions.

In three dimensions, what we think of as a volume, we can have three linearly independent vectors if they are noncoplanar. This is because if we add two scaled vectors, then the result will of course lie in the plane that is formed by the vectors. If the third vector is not in that same plane, then no amount of nonzero scaling will put it there. Thus there is no possibility of scaling it so that it is the negative sum of the first two. This geometric interpretation results in a pair of lemmas that are intuitively obvious and can be easily proven.

> LEMMA: If we have a set of linearly dependent vectors, and we add another vector to this set, the resulting set will also have to be linearly dependent. (2.10)

This is obvious from the geometric description we have given. If the original set of vectors is dependent, it adds to zero with at least one nonzero scaling coefficient. We can then include the new vector into the linear combination with a zero coefficient, as in Figure 2.2. At least one of the other scalars is already known to be nonzero, and the new sum will remain zero. Thus, the augmented set is still dependent.

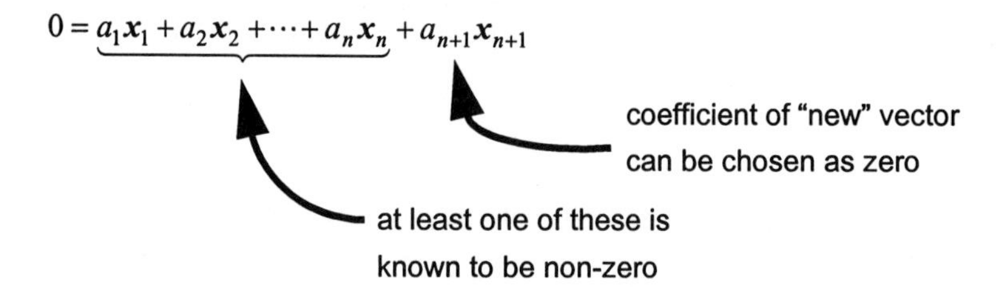

Figure 2.2 Manipulations necessary to show linear dependence of an augmented set of already dependent vectors.

The other lemma is:

LEMMA: If a given set of vectors is linearly dependent, then one of the vectors can be written as a linear combination of the other vectors. (2.11)

This lemma is sometimes given as the definition of a linearly dependent set of vectors. In fact, it can be derived as a result of our definition.

PROOF: If the set $\{x_i\}$ is linearly dependent, then $\sum_{i=1}^{n} a_i x_i = 0$ with at least one a coefficient not equal to zero. Suppose that $a_j \neq 0$. Then without risking a divide-by-zero, we can perform the following operation:

$$x_j = \frac{-a_1 x_1 - a_2 x_2 - \cdots - a_{j-1} x_{j-1} - a_{j+1} x_{j+1} - \cdots - a_n x_n}{a_j}$$

$$= -\frac{a_1}{a_j} x_1 - \frac{a_2}{a_j} x_2 - \cdots - \frac{a_{j-1}}{a_j} x_{j-1} - \frac{a_{j+1}}{a_j} x_{j+1} - \cdots - \frac{a_n}{a_j} x_n \quad (2.12)$$

By defining $b_i \stackrel{\Delta}{=} -a_i / a_j$,

$$x_j = b_1 x_1 + b_2 x_2 + \cdots + b_n x_n \quad (2.13)$$

thus proving that one vector can be written as a linear combination of the others.

Already, our intuitive sense is that we can have (at most) as many independent vectors in a set as we have "dimensions" in the space. This is strictly true, although we will need to rigorously show this after discussion of the mathematical notion of *dimension*.

Dimension: The *dimension* of a linear vector space is the largest possible number of linearly *in*dependent vectors that can be taken from that space. (2.14)

2.2.3 Bases

If we are working within a particular vector space and we select the maximum number of linearly independent vectors, the set we will have created is known as a *basis*. Officially, we have the definition:

Basis: A set of linearly independent vectors in vector space **X** is a *basis* of **X** if and only if every vector in **X** can be written as a *unique* linear combination of vectors from this set. (2.15)

One must be careful to note that in an n-dimensional vector space, there must be exactly n vectors in any basis set, but there are an infinite number of such sets that qualify as a basis. The main qualification on these vectors is that they be linearly independent and that there be a maximal number of them, equal to the dimension of the space. The uniqueness condition in the definition above results from the fact that if the basis set is linearly independent, then each vector contains some unique "direction" that none of the others contain. In geometric terms, we usually think of the basis as being the set of coordinate axes. Although coordinate axes, for our convenience, are usually orthonormal, i.e., mutually orthogonal and of unit length, basis vectors need not be.

THEOREM: In an n-dimensional linear vector space, *any* set of n linearly *in*dependent vectors qualifies as a basis. (2.16)

PROOF: This statement implies that a vector $\boldsymbol{x}$ should be described uniquely by any n linearly independent vectors, say

$$\{\boldsymbol{e}_i\} = \{\boldsymbol{e}_1, \boldsymbol{e}_2, \ldots, \boldsymbol{e}_n\}$$

That is, for every vector $\boldsymbol{x}$,

$$\begin{aligned} \boldsymbol{x} &= \text{linear combination of } \boldsymbol{e}_i\text{'s} \\ &= \sum_{i=1}^{n} \alpha_i \boldsymbol{e}_i \end{aligned} \tag{2.17}$$

Because the space is n-dimensional, the set of $n+1$ vectors $\{\boldsymbol{x}, \boldsymbol{e}_1, \boldsymbol{e}_2, \ldots, \boldsymbol{e}_n\}$ must be linearly dependent. Therefore, there exists a set of scalars $\{\alpha_i\}$, not all of which are zero, such that

$$\alpha_0 \boldsymbol{x} + \alpha_1 \boldsymbol{e}_1 + \cdots + \alpha_n \boldsymbol{e}_n = \boldsymbol{0} \tag{2.18}$$

Suppose $\alpha_0 = 0$. If this were the case and if the set $\{\boldsymbol{e}_i\}$ is, as specified, linearly independent, then we must have $\alpha_1 = \alpha_2 = \cdots = \alpha_n = 0$. To avoid this trivial situation, we assume that $\alpha_0 \neq 0$. This would allow us to write the equation:

$$\begin{aligned} x &= -\frac{\alpha_1}{\alpha_0} e_1 - \frac{\alpha_2}{\alpha_0} e_2 - \cdots - \frac{\alpha_n}{\alpha_0} e_n \\ &\equiv \beta_1 e_1 + \beta_2 e_2 + \cdots + \beta_n e_n \\ &= \sum_{i=1}^{n} \beta_i e_i \end{aligned} \tag{2.19}$$

So we have written vector $\boldsymbol{x}$ as a linear combination of the $\boldsymbol{e}_i$'s.

Now we must show that this expression is *unique*. To do this, suppose there were *another* set of scalars $\{\bar{\beta}_i\}$ such that $\boldsymbol{x} = \sum_{i=1}^{n} \bar{\beta}_i \boldsymbol{e}_i$. We already have $\boldsymbol{x} = \sum_{i=1}^{n} \beta_i \boldsymbol{e}_i$, so

$$\begin{aligned} x - x &= \sum_{i=1}^{n} \bar{\beta}_i e_i - \sum_{i=1}^{n} \beta_i e_i \\ &= \sum_{i=1}^{n} (\bar{\beta}_i - \beta_i) e_i \\ &= 0 \end{aligned} \tag{2.20}$$

But the set $\{\boldsymbol{e}_i\}$ is known to be a basis; therefore, for the above equality to hold, we must have $\bar{\beta}_i - \beta_i = 0$ for all $i = 1, \ldots, n$. So $\bar{\beta}_i = \beta_i$ and the uniqueness of the representation is proven.

Once the basis $\{\boldsymbol{e}_i\}$ is chosen, the set of numbers $\{\beta_i\}$ is called the *representation* of x in $\{\boldsymbol{e}_i\}$. We can then refer to vector $\boldsymbol{x}$ by this representation, which is simply an n-tuple of numbers $\beta = [\beta_1 \;\; \beta_2 \;\; \cdots \;\; \beta_n]^T$. This is a very important by-product of the theorem, which implies that any finite dimensional vector space is *isomorphic** to the space of n-tuples. This allows us to always use the n-tuple operations such as dot-products and the induced norm on arbitrary spaces. It also allows us to rely on the intuitive picture of a vector space as a set of coordinate axes. For these reasons, it is often much more convenient to use the representation of a vector than to use the vector itself. However, it should be remembered that the same vector will have different representations in different bases.

There is weaker terminology to describe the expansion of a vector in terms of vectors from a particular set.

> ***Span:*** A set of vectors X is *spanned* by a set of vectors $\{x_i\}$ if every $x \in X$ can be written as a linear combination of the x_i's. Equivalently, the x_i's *span* X. The notation is
>
> $$X = \mathrm{sp}\{x_i\}\,. \tag{2.21}$$

Note that in this terminology, the set $\{x_i\}$ is not necessarily a basis. It may be linearly dependent. For example, five noncollinear vectors in two dimensions suffice to span the two dimensional euclidean space, but as a set they are not a basis. We sometimes use this definition when defining subspaces, which will be discussed later.

We will see after the following examples how one can take a single vector and convert its representation from one given basis to another.

Example 2.5: Spaces and Their Bases

1. Let **X** be the space of all vectors written $x = \begin{bmatrix} x_1 & x_2 & \cdots & x_n \end{bmatrix}^T$ such that $x_1 = x_2 = \cdots = x_n$. One legal basis for this space is the single vector $v = \begin{bmatrix} 1 & 1 & \cdots & 1 \end{bmatrix}^T$. This is therefore a one-dimensional space, despite it consisting of vectors containing n "components."
2. Consider the space of all polynomials in s of degree less than 4, with real coefficients, defined over the field of reals. One basis for this space is the set

$$e_1 = 1 \qquad e_2 = s \qquad e_3 = s^2 \qquad e_4 = s^3$$

In this basis, the vector

---

* Spaces are said to be isomorphic if there exists a bijective mapping (i.e., one-to-one and onto) from one to the other.

$$x = 3s^3 + 2s^2 - 2s + 10$$

can obviously be expanded as

$$x = 3e_4 + 2e_3 + (-2)e_2 + 10e_1$$

So the vector x has the representation $x = \begin{bmatrix} 10 & -2 & 2 & 3 \end{bmatrix}^T$ in the $\{e_i\}$ basis. We can write another basis for this same space:

$$\{f_i\} = \{f_1, f_2, f_3, f_4\} = \{s^3 - s^2, s^2 - s, s - 1, 1\}$$

In this different basis, the same vector has a different representation:

$$x = 3(s^3 - s^2) + 5(s^2 - s) + 3(s - 1) + 13$$

or

$$x = \begin{bmatrix} 3 \\ 5 \\ 3 \\ 13 \end{bmatrix}$$

This can be verified by direct expansion of this expression.

Example 2.6: Common Infinite-Dimensional Spaces

While at first the idea of infinite-dimensional spaces may seem unfamiliar, most engineers have had considerable experience with function spaces of infinite dimension. For the following spaces, give one or more examples of a valid basis set.

1. The space of all functions of time that are analytic in the interval $a < t < b$.
2. The space of all bounded periodic functions of period T with at most a finite number of finite discontinuities and a finite number of extrema.

Solution:

The first function space can be described with power series expansions because of the use of the key word *analytic*. Therefore, for example, we can write a Taylor series expansion

$$f(t) = \sum_{n=0}^{\infty} \frac{1}{n!} \left. \frac{df(t)}{dt} \right|_{t=t_0} (t - t_0)^n \tag{2.22}$$

which suggests the bases $e_i(t) = \{(t - t_0)^i\}$ for $i = 0, 1, \cdots$, or $e_i(t) = \{1, (t - t_0)^1, (t - t_0)^2, \cdots\}$. There are, as usual, an infinite number of such valid bases (consider the MacLaurin series).

The second set of functions is given by the conditions that guarantee a convergent Fourier series. The Fourier expansion of any such function can be written either as

$$f(t) = \frac{a_0}{2} + \sum_{n=1}^{\infty} \left[a_n \cos\left(\tfrac{2n\pi}{T} t\right) + b_n \sin\left(\tfrac{2n\pi}{T} t\right) \right] \tag{2.23}$$

or

$$f(t) = \frac{1}{2} \sum_{n=-\infty}^{+\infty} c_n e^{j(2n\pi/T)t} \tag{2.24}$$

for appropriately calculated coefficients a_0, a_n, b_n, and c_n (where $j = \sqrt{-1}$). Therefore, two suitable bases for these function spaces are

$$e_i = \left\{ 1, \cos\left(\tfrac{2\pi}{T}\right), \sin\left(\tfrac{2\pi}{T}\right), \cos\left(\tfrac{4\pi}{T}\right), \sin\left(\tfrac{4\pi}{T}\right), \cdots \right\} \tag{2.25}$$

or

$$f_i = \left\{ 1, e^{j(2\pi/T)t}, e^{j(4\pi/T)t}, \cdots \right\} \tag{2.26}$$

2.2.4 Change of Basis

Suppose a vector $\boldsymbol{x}$ has been given in a basis $\{\boldsymbol{v}_j\}$, $j = 1, \ldots, n$. We are then given a different basis, consisting of vectors $\{\hat{\boldsymbol{v}}_i\}$, $i = 1, \ldots, n$, coming from the same space. Then the same vector $\boldsymbol{x}$ can be expanded into the two bases with different representations:

$$\boldsymbol{x} = \sum_{j=1}^{n} x_j \boldsymbol{v}_j = \sum_{i=1}^{n} \hat{x}_i \hat{\boldsymbol{v}}_i \tag{2.27}$$

But since the basis $\{v_j\}$ consists of vectors in a common space, these vectors themselves can be expanded in the new basis $\{\hat{v}_i\}$ as:

$$v_j = \sum_{i=1}^{n} b_{ij}\hat{v}_i \tag{2.28}$$

By gathering the vectors on the right side of this equation side-by-side into a matrix, this expression can also be expressed as

$$v_j = \begin{bmatrix} \hat{v}_1 & \hat{v}_2 & \cdots & \hat{v}_n \end{bmatrix} \begin{bmatrix} b_{1j} \\ b_{2j} \\ \vdots \\ b_{nj} \end{bmatrix}$$

or, when all j-values are gathered in a single notation,

$$\begin{aligned} \begin{bmatrix} v_1 & v_2 & \cdots & v_n \end{bmatrix} &= \begin{bmatrix} \hat{v}_1 & \hat{v}_2 & \cdots & \hat{v}_n \end{bmatrix} \begin{bmatrix} b_{11} & b_{12} & \cdots & b_{1n} \\ b_{21} & b_{22} & \cdots & b_{2n} \\ \vdots & \vdots & \ddots & \vdots \\ b_{n1} & b_{n2} & \cdots & b_{nn} \end{bmatrix} \\ &\triangleq \begin{bmatrix} \hat{v}_1 & \hat{v}_2 & \cdots & \hat{v}_n \end{bmatrix} B \end{aligned} \tag{2.29}$$

where the matrix B so defined in this equation will hereafter be referred to as the "change of basis matrix."

Substituting the expression (2.28) into Equation (2.27),

$$\sum_{j=1}^{n} x_j \left[\sum_{i=1}^{n} b_{ij}\hat{v}_i \right] = \sum_{i=1}^{n} \hat{x}_i \hat{v}_i \tag{2.30}$$

By changing the order of summation on the left side of this equality, moving both terms to the left of the equal sign, and factoring,

$$\sum_{i=1}^{n} \left(\sum_{j=1}^{n} b_{ij} x_j - \hat{x}_i \right) \hat{v}_i = 0 \tag{2.31}$$

For this to be true given that $\{\hat{v}_i\}$ is an independent set, we must have each constant in the above expansion being zero, implying

$$\hat{x}_i = \sum_{j=1}^{n} b_{ij} x_j \tag{2.32}$$

This is how we get the components of a vector in a new basis from the components of the old basis. The coefficients b_{ij} in the expansion come from our knowledge of how the two bases are related, as determined from Equation (2.29). Notice that this relationship can be written in vector-matrix form by expressing the two vectors as n-tuples, $\boldsymbol{x}$ and $\hat{\boldsymbol{x}}$. Denoting by B the $n \times n$ matrix with coefficient b_{ij} in the ith row and j^{th} column, we can write

$$\hat{\boldsymbol{x}} = B\boldsymbol{x} \tag{2.33}$$

Example 2.7: Change of Basis

Consider the space $\Re^2$ and the two bases:

$$\{\boldsymbol{e}_1, \boldsymbol{e}_2\} = \left\{\begin{bmatrix}1\\0\end{bmatrix}, \begin{bmatrix}0\\1\end{bmatrix}\right\} \quad \text{and} \quad \{\hat{\boldsymbol{e}}_1, \hat{\boldsymbol{e}}_2\} = \left\{\begin{bmatrix}1\\2\end{bmatrix}, \begin{bmatrix}-1\\0\end{bmatrix}\right\}$$

Let a vector $\boldsymbol{x}$ be represented by $\boldsymbol{x} = \begin{bmatrix}2 & 2\end{bmatrix}^{\mathrm{T}}$ in the $\{\boldsymbol{e}_j\}$ basis. To find the representation in the $\{\hat{\boldsymbol{e}}_i\}$ basis, we must first write down the relationship between the two basis sets. We do this by expressing the vectors $\boldsymbol{e}_j$ in terms of vectors $\hat{\boldsymbol{e}}_i$:

$$\boldsymbol{e}_1 = 0\hat{\boldsymbol{e}}_1 + (-1)\hat{\boldsymbol{e}}_2 = \begin{bmatrix}\hat{\boldsymbol{e}}_1 & \hat{\boldsymbol{e}}_2\end{bmatrix}\begin{bmatrix}0\\-1\end{bmatrix}$$

$$\boldsymbol{e}_2 = \tfrac{1}{2}\hat{\boldsymbol{e}}_1 + \tfrac{1}{2}\hat{\boldsymbol{e}}_2 = \begin{bmatrix}\hat{\boldsymbol{e}}_1 & \hat{\boldsymbol{e}}_2\end{bmatrix}\begin{bmatrix}\tfrac{1}{2}\\\tfrac{1}{2}\end{bmatrix}$$

From this expansion we can extract the matrix

$$B = \begin{bmatrix}0 & \tfrac{1}{2}\\-1 & \tfrac{1}{2}\end{bmatrix}$$

giving, in the $\{\hat{e}_i\}$ basis,

$$\hat{x} = Bx = \begin{bmatrix} 0 & \frac{1}{2} \\ -1 & \frac{1}{2} \end{bmatrix} \begin{bmatrix} 2 \\ 2 \end{bmatrix} = \begin{bmatrix} 1 \\ -1 \end{bmatrix}$$

REMARK: The reader might verify that the expansion of the $\hat{e}_i$ vectors in terms of the e_j vectors is considerably easier to write by inspection:

$$\hat{e}_1 = 1e_1 + 2e_2 = \begin{bmatrix} e_1 & e_2 \end{bmatrix} \begin{bmatrix} 1 \\ 2 \end{bmatrix}$$

$$\hat{e}_2 = (-1)e_1 + 0e_2 = \begin{bmatrix} e_1 & e_2 \end{bmatrix} \begin{bmatrix} -1 \\ 0 \end{bmatrix}$$

Therefore, the inverse of the change-of-basis matrix is more apparent:

$$B^{-1} = \begin{bmatrix} 1 & -1 \\ 2 & 0 \end{bmatrix}$$

Note that the columns of this matrix are simply equal to the columns of coefficients of the basis vectors $\{\hat{e}_i\}$ because the $\{e_j\}$ basis happens to be the same as the standard basis, but this is not always the case. We can display the relationship between the two basis sets and the vector x as in Figure 2.3.

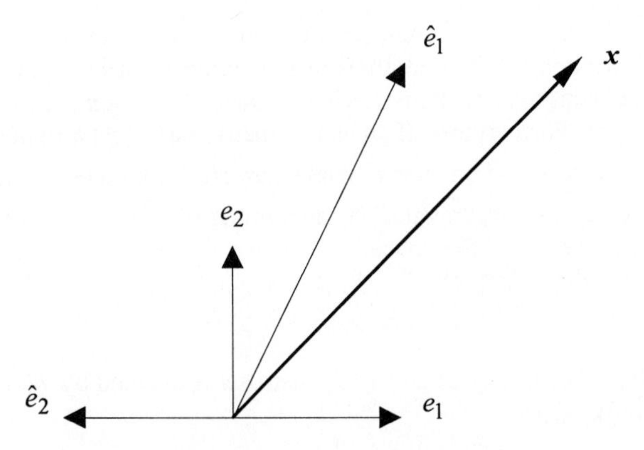

Figure 2.3 Relative positions of vectors in Example 2.7. It can be seen that $x = 2e_1 + 2e_2$ and that $x = 1 \cdot \hat{e}_1 - 1 \cdot \hat{e}_2$.

2.2.5 Rank and Degeneracy

rank(A)

In matrix theory, the *rank*[M] of a matrix A, denoted $r(A)$, is defined as the size of the largest nonzero determinant that can be formed by taking square subsets of the rows and columns of A. Given our knowledge of bases and representations, we can construct matrices from collections of vectors by considering each column to be the representation of a different vector (in the same basis):

$$A = \begin{bmatrix} \mathbf{x}_1 & \mathbf{x}_2 & \cdots & \mathbf{x}_n \end{bmatrix} = \left[\begin{array}{c|c|c|c} x_{11} & x_{12} & & x_{1n} \\ x_{21} & x_{22} & \cdots & x_{2n} \\ \vdots & \vdots & & \vdots \\ x_{m1} & x_{m2} & & x_{mn} \end{array}\right] \tag{2.34}$$

With this interpretation of a matrix, the concept of rank can be reconsidered.

> ***Rank:*** The *rank* of a matrix A, $r(A)$, is the maximum number of linearly independent columns in the matrix. This will be the same number as the maximum number of linearly independent rows in the matrix. That is, the *row* and *column* rank of a matrix are identical. (2.35)

Note that this definition for rank is consistent with the test for linear independence of a collection of vectors. Therefore, it is apparent also that the rank is the dimension of the space formed by the columns of the matrix (the *range* space, see Chapter 3). If the collection of columns forms a matrix with a nonzero determinant, then the set of vectors is linearly independent. Of course, this test only works with a number of vectors equal to the number of coefficients in the representation (which is equal to the dimension of the space). If the number of vectors is different from the dimension of their space, then all square subsets of the rows and columns of the matrix must be separately tested with the determinant test. If matrix A is of dimension $m \times n$, then of course, $r(A) \le \min(m, n)$. Furthermore, if an $m \times n$ matrix has $r(A) = \min(m, n)$, it is said to be of "full rank." If an $m \times n$ matrix has $r(A) < \min(m, n)$, it is said to be "rank deficient," or "degenerate." If the number of vectors considered is the same as the dimension of the space, so that the matrix A is square, then the matrix is also called "singular." A square matrix that is of *full* rank is "nonsingular."

> ***Nullity***: The *nullity* of an $m \times n$ matrix A is denoted by $q(A)$ and is defined as

$$q(A) = n - r(A) \tag{2.36}$$

NOTE: This is also sometimes referred to as the *degeneracy*, but this is not accurate terminology. It is possible for a full-rank matrix (i.e., not degenerate) to have a nonzero nullity, even though its degeneracy is zero. For example, consider a 3×5 matrix with rank 3. The degeneracy of this matrix is zero $[rank = \min(m, n)]$, but its nullity is nonzero $[q(A) = n - r(A) = 5 - 3 = 2]$. For a *square* matrix, the degeneracy and nullity are equivalent.

In addition, there are some useful properties for the rank and nullity of products of matrices, given by the following theorem.

THEOREM: If matrix A is $m \times n$ and matrix B is $n \times p$, and we form the $m \times p$ matrix product $AB = C$, then the following properties hold:

$$\begin{gathered} r(A) + r(B) - n \le r(C) \le \min(r(A), r(B)) \\ q(C) \le q(A) + q(B) \end{gathered} \tag{2.37}$$

[The second line can be derived from the first, using the definition of nullity (2.36)]. Furthermore, if D is an $m \times m$ nonsingular matrix and E is an $n \times n$ nonsingular matrix, then

$$r(DA) = r(A) \qquad \text{and} \qquad r(AE) = r(A) \tag{2.38}$$

2.2.6 Inner Products

In the beginning of this chapter we presented the definitions of some inner products and norms that are commonly used in euclidean n-space, $\Re^n$. Now that more general vector spaces have been introduced, we are prepared for more general definitions of these operators on vectors.

Inner Product: An *inner product* is an operation on two vectors, producing a scalar result. The inner product of two vectors $\boldsymbol{x}$ and $\boldsymbol{y}$ is denoted $\langle \boldsymbol{x}, \boldsymbol{y} \rangle$ and must have the following properties:

1. $\langle \boldsymbol{x}, \boldsymbol{y} \rangle = \overline{\langle \boldsymbol{y}, \boldsymbol{x} \rangle}$ (where the overbar indicates complex conjugate). (2.39)
2. $\langle \boldsymbol{x}, \alpha \boldsymbol{y}_1 + \beta \boldsymbol{y}_2 \rangle = \alpha \langle \boldsymbol{x}, \boldsymbol{y}_1 \rangle + \beta \langle \boldsymbol{x}, \boldsymbol{y}_2 \rangle$. (2.40)

3. $\langle x, x \rangle \geq 0$ for all x, and $\langle x, x \rangle = 0$ if and only if $x = 0$.
(2.41)

THEOREM: The following properties are true for inner products and can be derived from the three properties in the definition of the inner product:

4. $\langle x, \alpha y \rangle = \alpha \langle x, y \rangle$. (2.42)

5. $\langle \alpha x, y \rangle = \bar{\alpha} \langle x, y \rangle$. (2.43)

Vector spaces with inner products defined for them are referred to as *inner product spaces*. One can verify that the inner products given at the beginning of this chapter for $\Re^n$ are valid inner products as defined above. Note, however, that if the vectors are defined over a complex field, then the conjugate operator as shown in the conditions must be observed. For example, the inner product of two complex-valued vectors can be written $\langle x, y \rangle = x^* y$, where x^* is the complex-conjugate transpose.

Example 2.8: Inner Product for a Function Space

Consider the space of complex-valued functions of time defined on the interval $t \in [a, b]$, consisting of vectors of the form $x(t)$. An inner product for this space might be:

$$\langle x, y \rangle = \int_a^b x^*(t) y(t)\, dt \tag{2.44}$$

2.2.7 Norms

A norm generalizes our concept of "length" or "magnitude" of a vector.

norm(x)

Norm: A *norm*[M] is a function of a single vector that produces a scalar result. For a vector x, it is denoted $\|x\|$ and must satisfy the following rules:

1. $\|x\| = 0$ if and only if $x = 0$. (2.45)

2. $\|\alpha x\| = |\alpha| \|x\|$ for any scalar α. (2.46)

3. $\|x + y\| \leq \|x\| + \|y\|$ ("triangle inequality"). (2.47)

4. $|\langle x, y\rangle| \le \|x\| \cdot \|y\|$ ("Cauchy-Schwarz inequality"). (2.48)

Vector spaces with such norms defined for them are referred to as *normed linear spaces*.

Norms are sometimes induced from the given inner product. For example, the euclidean norm given earlier in this chapter is induced by whatever inner product is defined:

$$\|x\| = \langle x, x\rangle^{1/2} \tag{2.49}$$

A *unit vector* is one that has unit length, i.e., whose norm is 1. Of course, any vector can be converted to a unit vector ("normalized") by dividing it by its own norm. This will not change its direction, only scale its magnitude.

Example 2.9: Norms

One can use the properties listed above to verify that each of the following is a valid norm for vectors (considered as n-tuples, $x = [x_1 \quad x_2 \quad \cdots \quad x_n]^{\mathrm{T}}$). These are known as the "ℓ_k norms":

$$\|x\|_k = \sqrt[k]{\sum_{i=1}^{n} x_i^{\,k}} \tag{2.50}$$

of which the following common norms are special cases:

1. ℓ_1 norm: $$\|x\|_1 = \sum_{i=1}^{n} |x_i| \tag{2.51}$$

2. ℓ_2 norm: $$\|x\|_2 = \sqrt{\sum_{i=1}^{n} |x_i|^2} \tag{2.52}$$

   (i.e., the euclidean norm). If the vector is a signal vector, $x(t)$, the ℓ_2 norm becomes:

3. $$\|x(t)\|_2 = \sqrt{\int_{-\infty}^{\infty} x^*(t) \cdot x(t)\, dt} \tag{2.53}$$

4. ℓ_∞ norm: $$\|x\|_\infty = \max_i |x_i| \tag{2.54}$$

2.2.8 Some Other Terms

Until now we have referred to spaces consisting of vectors of n-tuples as *euclidean n-spaces*. This is actually a general term that refers to any real finite-dimensional linear space that has an inner product defined for it. A similar space consisting of complex vectors with an inner product satisfying the rules given above is known as a *unitary space*. In this section we will give definitions for other terms, some of which are not critical for our purposes, but which are nevertheless mathematically very important and quite common.

> ***Metric:*** A *metric* is a function of two vectors and gives a scalar measure of the distance between two vectors. For two vectors $\boldsymbol{x}$ and $\boldsymbol{y}$, it is denoted as $\rho(\boldsymbol{x},\boldsymbol{y})$ and is defined as:
>
> $$\rho(\boldsymbol{x},\boldsymbol{y}) = \|\boldsymbol{x}-\boldsymbol{y}\| \tag{2.55}$$
>
> A space with such a defined metric is referred to as *metric space*.

Sometimes we wish to define the distance between vectors independently of the lengths of the vectors involved. Such a concept is available in the form of the *angle* between two vectors.

> ***Angle:*** The *angle* between two vectors is denoted by $\theta(\boldsymbol{x},\boldsymbol{y})$ and satisfies the following equation:
>
> $$\langle \boldsymbol{x},\boldsymbol{y}\rangle = \|\boldsymbol{x}\|\,\|\boldsymbol{y}\|\cos\theta \tag{2.56}$$

In low dimensions (two- and three-dimensional spaces) and using the euclidean norm, this concept of angle is exactly what one might imagine, yet it is entirely valid for higher dimensions and other norms as well. In addition, this concept of angle extends to sets and subspaces[M] (see Section 2.4) of vectors as well, taking as the angle between two sets X and Y the *minimum* of all angles between pairs of vectors $\boldsymbol{x}$ and $\boldsymbol{y}$, where $\boldsymbol{x}\in X$ and $\boldsymbol{y}\in Y$.

`subspace(A,B)`

> ***Hilbert Space:*** A Hilbert space is an inner-product euclidean space with the following additional conditions:
>
> 1. The space is *infinite dimensional.*
> 2. The space is *complete* with respect to the metric $\rho(\boldsymbol{x},\boldsymbol{y}) = \|\boldsymbol{x}-\boldsymbol{y}\|$.

A space is said to be *complete* if all Cauchy sequences converge to an element within the space. A *Cauchy sequence* is a sequence of elements $\{\boldsymbol{x}_n\}$ that satisfies the following criterion: Given any $\varepsilon > 0$, there exists an integer $N(\varepsilon)$ such that $\rho(\boldsymbol{x}_m, \boldsymbol{x}_n) < \varepsilon$ for all $m, n > N(\varepsilon)$ (see [4]). (2.57)

All euclidean spaces are complete, and we will not generally be concerned with the completeness of the spaces we discuss in this book. We will therefore not elaborate on this definition further.

Banach Space: A *Banach space* is a normed linear space that is complete according to the definition of completeness given for Hilbert spaces above. (2.58)

Orthogonality and Orthonormality: Two vectors are said to be *orthogonal* if their inner product is zero, i.e., $\langle \boldsymbol{x}, \boldsymbol{y} \rangle = 0$. A pair of vectors is *orthonormal* if $\langle \boldsymbol{x}, \boldsymbol{y} \rangle = 0$ and, in addition, $\|\boldsymbol{x}\| = \|\boldsymbol{y}\| = 1$. (2.59)

A *set* of vectors $\{\boldsymbol{x}_i\}$ is *orthogonal* if

$$\begin{aligned} \langle \boldsymbol{x}_i, \boldsymbol{x}_j \rangle &= 0 \text{ if } i \neq j \text{ and} \\ \langle \boldsymbol{x}_i, \boldsymbol{x}_j \rangle &\neq 0 \text{ if } i = \text{j} \end{aligned} \tag{2.60}$$

A set of vectors is *orthonormal* if

$$\begin{aligned} \langle \boldsymbol{x}_i, \boldsymbol{x}_j \rangle &= 0 \text{ if } i \neq j \text{ and} \\ \langle \boldsymbol{x}_i, \boldsymbol{x}_j \rangle &= 1 \text{ if } i = \text{j} \end{aligned} \tag{2.61}$$

2.3 *Gram-Schmidt Orthonormalization*

It is computationally convenient to use basis sets that are orthonormal. However, if we are provided with a basis set that is not orthonormal, the Gram-Schmidt orthonormalization procedure provides us with a technique for orthonormalizing[M] it. By this we mean that if we are provided with the nonorthonormal set of basis vectors $\{\boldsymbol{y}_i\}$, $i = 1, \ldots, n$, we generate a new set of

`orth(A)`

vectors $\{v_i\}$, $i = 1,\ldots,n$, which is orthonormal. In addition, each subset of vectors, $\{v_j\}$, $j = 1,\ldots,m$, spans $\{y_j\}$, $j = 1,\ldots,m$, for every $m < n$. We build the orthonormal set by starting with one vector from the original nonorthonormal set, and successively include additional vectors, one at a time, while subtracting from each new vector any component that it shares with the previously collected vectors. Therefore, as we add more vectors to the orthonormal set, each will contain only components along directions that were not included in any previously found vectors.

The technique will be presented as the following algorithm:

1. Let $v_1 = y_1$.

2. Choose v_2 to be y_2 with all components along the previous vector, subtracted out, i.e.,

$$v_2 = y_2 - av_1$$

Here, the coefficient a represents the magnitude of the component of y_2 in the v_1 direction. It is not initially known, so it will be computed. Because the v_i's are orthogonal,

$$\langle v_1, v_2 \rangle = 0 = \langle v_1, y_2 \rangle - a\langle v_1, v_1 \rangle$$

so

$$a = \frac{\langle v_1, y_2 \rangle}{\langle v_1, v_1 \rangle}$$

3. Choose v_3 as y_3 with the components along *both* previous v_i vectors subtracted out:

$$v_3 = y_3 - a_1 v_1 - a_2 v_2 \tag{2.62}$$

Again, coefficients a_1 and a_2, which represent the components of y_3 along the previously computed v_i's, must be calculated with the knowledge that the v_i's are orthogonal. As before, we find the inner product of Equation (2.62) with both v_1 and v_2:

$$\begin{aligned} \langle v_1, v_3 \rangle &= 0 = \langle v_1, y_3 \rangle - a_1 \langle v_1, v_1 \rangle - a_2 \overbrace{\langle v_1, v_2 \rangle}^{0} \\ \langle v_2, v_3 \rangle &= 0 = \langle v_2, y_3 \rangle - a_1 \underbrace{\langle v_2, v_1 \rangle}_{0} - a_2 \langle v_2, v_2 \rangle \end{aligned} \tag{2.63}$$

These two equations can be solved independently of one another to give

$$a_1 = \frac{\langle v_1, y_3 \rangle}{\langle v_1, v_1 \rangle} \qquad \text{and} \qquad a_2 = \frac{\langle v_2, y_3 \rangle}{\langle v_2, v_2 \rangle}$$

4. Continue using this same process to give

$$v_i = y_i - \sum_{k=1}^{i-1} a_k v_k = y_i - \sum_{k=1}^{i-1} \frac{\langle v_k, y_i \rangle}{\langle v_k, v_k \rangle} v_k \tag{2.64}$$

5. Finish by normalizing each of the new basis vectors using:

$$\hat{v}_i = \frac{v_i}{\|v_i\|} \tag{2.65}$$

Alternatively, this normalization can be done at the end of each stage, after which the terms $\langle v_k, v_k \rangle = 1$ in (2.64).

It should be noted that there are a number of variations on this procedure. For example, the normalization step (step 5) could be executed each time a new vector v is computed. This will affect the numerical results in subsequent steps, and make the end result look somewhat different, though still entirely correct. It has the benefit, though, that all terms $\langle v_i, v_i \rangle = 1$ for subsequent iterations. Also, the results will depend on which basis vector is chosen at the start. There is no special reason that we chose y_1 to start the process; another vector would suffice.

If we always use orthonormal basis sets, we can always find the component of vector x along the jth basis vector by performing a simple inner product operation with that j^{th} basis vector. That is, if $\{e_i\}$, $i = 1, \ldots, n$ is an orthonormal basis for space $\mathbf{X}$ and if $x \in \mathbf{X}$, then $x = \sum_{i=1}^{n} x_i e_i$ and

$$x_j = \langle e_j, x \rangle = \sum_{i=1}^{n} x_i \langle e_j, e_i \rangle$$

$$= x_1 \underbrace{\langle e_j, e_1 \rangle}_{0} + x_2 \underbrace{\langle e_j, e_2 \rangle}_{0} + \cdots + x_j \underbrace{\langle e_j, e_j \rangle}_{1} + \cdots + x_n \underbrace{\langle e_j, e_n \rangle}_{0} \tag{2.66}$$

Furthermore, if x is expressed in an orthonormal basis $\{e_i\}$, $i = 1, \ldots, n$, then the norm of x can be found using a form of *Parseval's theorem*.

THEOREM (Parseval's theorem): If a vector in an n-dimensional linear space is given in *any* orthonormal basis by the representation $\{c_i\}$, then the euclidean norm of that vector can be expressed as

$$\|x\| = \left[\sum_{i=1}^{n} c_i^2 \right]^{1/2} \tag{2.67}$$

PROOF: If a vector is given in an orthonormal basis by the expression $x = \sum_{i=1}^{n} x_i e_i$, then

$$\begin{aligned}
\|x\| &= \langle x, x \rangle^{1/2} \\
&= \left\langle \sum_{j=1}^{n} c_j e_j, \sum_{i=1}^{n} c_i e_i \right\rangle^{1/2} \\
&= \left[\sum_{i=1}^{n} c_i \left(\left\langle \sum_{j=1}^{n} c_j e_j, e_i \right\rangle \right) \right]^{1/2} \\
&= \left[\sum_{i=1}^{n} c_i (c_i) \right]^{1/2} = \left[\sum_{i=1}^{n} (c_i)^2 \right]^{1/2}
\end{aligned}$$

Therefore for the vector x above, we can write:

$$\|x\| = \left[\sum_{i=1}^{n} x_i^2 \right]^{1/2} \tag{2.68}$$

Note that this is the formula often given for the norm of any vector, but it may not be true if the vector is not expressed in orthonormal bases. See Example 2.7 (Change of Basis) on page 64. Parseval's theorem is particularly useful when finding the norm of a *function* that has been expanded into an orthonormal set of basis functions (see Example 2.11 on page 79); the theorem applies to infinite-dimensional spaces as well.

2.4 Subspaces and the Projection Theorem

Often, we have a linear vector space that contains a particular portion in which we are interested. For example, although we are sometimes interested in the mechanics of a particle moving in a plane, we cannot forget that that plane is a *subspace* of the larger three-dimensional space of physical motion. In words, a *subspace* is a *subset* of a linear vector space that itself qualifies as a vector space. To make this notion mathematically general, we must give a definition for a subspace:

2.4.1 Subspaces

> ***Subspace:*** A *subspace* **M** of a vector space **X** is a subset of **X** such that if $x, y \in \mathbf{M}$, α and β are scalars, and $z = \alpha x + \beta y$, then $z \in \mathbf{M}$. (2.69)

Strictly, this definition specifies only that the subspace is closed under scalar multiplication and vector addition. Some of the other properties that must be satisfied for the subspace to be considered a space unto itself are automatically satisfied given the fact that vectors in subspace **M** are gathered from the space **X**. Other properties can be inferred from closure. For example, that the subspace by necessity contains a zero vector is implied by the above definition if one chooses $\alpha = \beta = 0$.

Note that space **X** can be considered a subspace of itself. In fact, if the dimension of **M**, $\dim(\mathbf{M})$, is the same as the dimension of **X**, then $\mathbf{X} = \mathbf{M}$. If $\dim(\mathbf{M}) < \dim(\mathbf{X})$, then **M** is called a *proper subspace* of **X**. Because of the closure requirement and the fact that each subspace must contain the zero vector, we can make the following observations on euclidean spaces:

1. All proper subspaces of $\Re^2$ are straight lines that pass through the origin. (In addition, one could consider the trivial set consisting of only the zero vector to be a subspace.)

2. All proper subspaces of $\Re^3$ are either straight lines that pass through the origin, or planes that pass through the origin.

3. All proper subspaces of $\Re^n$ are surfaces of dimension $n-1$ or less, and all must pass through the origin. Such subspaces that we refer to as "surfaces" but that are not two-dimensional, as we usually think of surfaces, are called *hypersurfaces*.

2.4.2 The Projection Theorem

Often, we will encounter a vector in a relatively large-dimensional space, and a subspace of which the vector is not a member. The most easily envisioned example is a vector in a three-dimensional euclidean space and a two-dimensional subspace consisting of a plane. If the vector of interest is not in the plane, we can ask the question: Which vector that *does* lie in the plane is closest to the given vector, which does *not* lie in the plane? We will use the projection theorem to answer this question, which arises in the solution of simultaneous algebraic equations, as encountered in Chapter 3.

> THEOREM: Suppose $\mathbf{U}$ is a proper subspace of $\mathbf{X}$, so that $\dim(\mathbf{U}) < \dim(\mathbf{X})$. Then for every $\boldsymbol{x} \in \mathbf{X}$, there exists a vector $\boldsymbol{u} \in \mathbf{U}$ such that $\langle \boldsymbol{x}-\boldsymbol{u}, \boldsymbol{y} \rangle = 0$ for every $\boldsymbol{y} \in \mathbf{U}$. The vector $\boldsymbol{u}$ is *the orthogonal projection* of $\boldsymbol{x}$ into $\mathbf{U}$. (2.70)

This situation is depicted in Figure 2.4. In the figure, vector $\boldsymbol{x}$ is in the three-dimensional space defined by the coordinate axes shown.

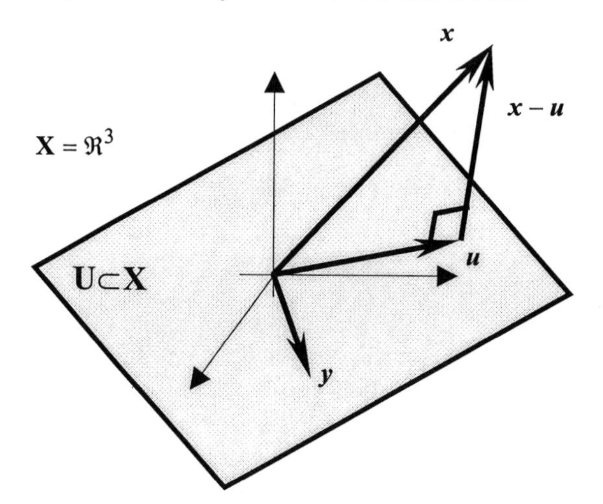

Figure 2.4 Illustration of the orthogonal projection theorem. Vectors $\boldsymbol{u}$ and $\boldsymbol{y}$ lie in the shaded plane. Vector $\boldsymbol{x}$ extends off the plane, and $\boldsymbol{x}-\boldsymbol{u}$ is orthogonal to $\boldsymbol{u}$ and every other vector in the plane.

The shaded plane that passes through the origin is the subspace **U**. Vector $\boldsymbol{x}$ does not lie in the plane of **U**. The vector $\boldsymbol{x}-\boldsymbol{u}$ referred to in the theorem is orthogonal to the plane itself; i.e., it is orthogonal to every vector in the plane, such as the vector $\boldsymbol{y}$. It is this orthogonality condition that enables us to find the projection.

In the following example, we will practice using the projection theorem by considering subspaces that are the *orthogonal complements* of one another. For this we give the definition:

> ***Orthogonal Complement:*** If **W** is a subspace of **X**, the *orthogonal complement* of **W**, denoted $\mathbf{W}^{\perp}$, is also a subspace of **X**, all of whose vectors are orthogonal to the set **W**. (2.71)

Example 2.10: Projection of a Vector

Consider vectors in $\Re^4$: $\boldsymbol{f}=[4\ \ 0\ \ 2\ \ -1]^{\mathrm{T}}$, $\boldsymbol{x}_1=[1\ \ 2\ \ 2\ \ 1]^{\mathrm{T}}$, and $\boldsymbol{x}_2=[0\ \ 0\ \ 1\ \ 1]^{\mathrm{T}}$. Decompose the vector $\boldsymbol{f}$ into a sum $\boldsymbol{f}=\boldsymbol{g}+\boldsymbol{h}$, where $\boldsymbol{g}\in\mathbf{W}=span\{\boldsymbol{x}_1,\boldsymbol{x}_2\}$ and $\boldsymbol{h}$ is in the *orthogonal complement* of **W**.

Solution:

We will use two approaches to solve this problem. In the first, we use the projection theorem without first orthonormalizing the vectors $\boldsymbol{x}_1$ and $\boldsymbol{x}_2$. To directly apply the projection theorem, we propose that the projection of $\boldsymbol{f}$ onto **W** be written in the following form, with as yet undetermined coefficients α_i:

$$P\boldsymbol{f}=\alpha_1\boldsymbol{x}_1+\alpha_2\boldsymbol{x}_2$$

Then by the projection theorem, we must have $(\boldsymbol{f}-P\boldsymbol{f})\perp\mathbf{W}$, so that $\langle\boldsymbol{f}-P\boldsymbol{f},\boldsymbol{x}_1\rangle=0$. But $\langle\boldsymbol{f}-P\boldsymbol{f},\boldsymbol{x}_1\rangle=\langle\boldsymbol{f},\boldsymbol{x}_1\rangle-\langle P\boldsymbol{f},\boldsymbol{x}_1\rangle=0$, so

$$\langle\boldsymbol{f},\boldsymbol{x}_1\rangle=\langle P\boldsymbol{f},\boldsymbol{x}_1\rangle=\alpha_1\langle\boldsymbol{x}_1,\boldsymbol{x}_1\rangle+\alpha_2\langle\boldsymbol{x}_2,\boldsymbol{x}_1\rangle$$

Similarly,

$$\langle\boldsymbol{f},\boldsymbol{x}_2\rangle=\langle P\boldsymbol{f},\boldsymbol{x}_2\rangle=\alpha_1\langle\boldsymbol{x}_1,\boldsymbol{x}_2\rangle+\alpha_2\langle\boldsymbol{x}_2,\boldsymbol{x}_2\rangle$$

Together, these two equations give the pair of simultaneous equations:

$$7 = \alpha_1 10 + \alpha_2 3$$
$$1 = \alpha_1 3 + \alpha_2 2$$

that can be solved (see Chapter 3) to give $\alpha_1 = 1$ and $\alpha_2 = -1$. Therefore,

$$g = Pf = x_1 + (-1) \cdot x_2 = \begin{bmatrix} 1 \\ 2 \\ 1 \\ 0 \end{bmatrix}$$

This being the desired vector in **W**, it is a simple matter to compute:

$$h = f - g = \begin{bmatrix} 3 \\ -2 \\ 1 \\ -1 \end{bmatrix} \in \mathbf{W}^{\perp}$$

As an alternative path to the solution, which avoids the simultaneous equation solving at the expense of the Gram-Schmidt operations, we could first orthonormalize the x_i vectors. Denote the orthonormalized version as $\{v_i\}$ and compute them as described:

$$v_1 = x_1 = \begin{bmatrix} 1 \\ 2 \\ 2 \\ 1 \end{bmatrix}$$

$$v_2 = x_2 - a v_1$$

where

$$a = \frac{\langle v_1, x_2 \rangle}{\langle v_1, v_1 \rangle} = 0.3$$

giving

$$v_2 = \begin{bmatrix} -0.3 \\ -0.6 \\ 0.4 \\ 0.7 \end{bmatrix}$$

Normalizing the two new basis vectors:

$$\hat{v}_1 = \frac{v_1}{\|v_1\|} = \begin{bmatrix} 0.3162 \\ 0.6325 \\ 0.6325 \\ 0.3162 \end{bmatrix} \quad \text{and} \quad \hat{v}_2 = \frac{v_2}{\|v_2\|} = \begin{bmatrix} -0.2860 \\ -0.5721 \\ 0.3814 \\ 0.6674 \end{bmatrix}$$

Then to find the components of f along these directions,

$$f_1 = \langle f, \hat{v}_1 \rangle = 2.2136$$
$$f_2 = \langle f, \hat{v}_2 \rangle = -1.0488$$

giving

$$g = Pf = f_1\hat{v}_1 + f_2\hat{v}_2 = \begin{bmatrix} 1 \\ 2 \\ 1 \\ 0 \end{bmatrix}$$

exactly as before.

Example 2.11: Finite Fourier Series

Let **V** be a linear vector space consisting of all piecewise continuous, real-valued functions of time defined over the interval $t \in [-\pi, \pi]$. A valid inner product for this space is

$$\langle f(t), g(t) \rangle = \frac{1}{\pi} \int_{-\pi}^{\pi} f(t) g(t) \, dt \tag{2.72}$$

Now consider a subspace of this space, $\mathbf{W} \subset \mathbf{V}$ consisting of all functions that can be formed by linear combinations of the functions 1, $\cos t$, $\sin t$, $\cos 2t$, and $\sin 2t$, i.e.,

$$\mathbf{W} = span\{1, \cos t, \sin t, \cos 2t, \sin 2t\}.$$

1. From the inner product given, induce a norm on this space.
2. If it is known that the vectors that span **W** are linearly independent and therefore form a basis for this 5-dimensional subspace, show that the set of vectors $\{1, \cos t, \sin t, \cos 2t, \sin 2t\}$ is an orthogonal basis for **W**. Normalize this set to make it an orthonormal basis.
3. Find the orthogonal projection of the following function, $f(t)$, defined in Figure 2.5, onto the space **W.**

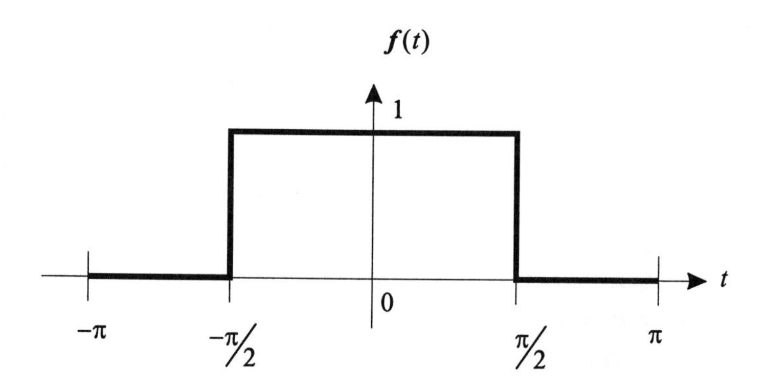

Figure 2.5 Square wave signal for Example 2.11.

Solution:

To induce a norm, we use

$$\|f(t)\| = \langle f(t), f(t)\rangle^{1/2} = \left[\frac{1}{\pi}\int_{-\pi}^{\pi} f^2(t)\,dt\right]^{1/2} \tag{2.73}$$

Use the notation:

$$\{e_1, e_2, e_3, e_4, e_5\} = \{1, \cos t, \sin t, \cos 2t, \sin 2t\}$$

If this set is linearly independent (can you show this?), then it automatically qualifies as a basis. To show that the set is orthogonal, we must demonstrate that condition (2.60) holds. Finding all these inner products,

$$\langle e_1, e_1 \rangle = \frac{1}{\pi} \int_{-\pi}^{\pi} 1\, dt = 2$$

$$\langle e_2, e_2 \rangle = \frac{1}{\pi} \int_{-\pi}^{\pi} \cos^2 t \; dt = 1$$

$$\langle e_3, e_3 \rangle = \frac{1}{\pi} \int_{-\pi}^{\pi} \sin^2 t \; dt = 1$$

$$\langle e_4, e_4 \rangle = \frac{1}{\pi} \int_{-\pi}^{\pi} \cos^2 2t \; dt = 1$$

$$\langle e_5, e_5 \rangle = \frac{1}{\pi} \int_{-\pi}^{\pi} \sin^2 2t \; dt = 1$$

$$\langle e_1, e_2 \rangle = \langle e_2, e_1 \rangle = 0$$

$$\vdots$$

$$\langle e_i, e_j \rangle = 0 \;\text{ for all } i \neq j$$

which indeed satisfies (2.60).

To normalize the set $\{e_i\}$, we simply have to divide each basis vector by its own norm, as defined by the induced norm above. These computations were mostly done while showing orthogonality, and it was found that the only vector with a nonunity norm was e_1, with $\|e_1\| = \sqrt{\langle e_1, e_1 \rangle} = \sqrt{2}$. Therefore an orthonormal set of basis vectors can be written as:

$$\{\hat{e}_i\} = \left\{ \frac{1}{\sqrt{2}},\; \cos t,\; \sin t,\; \cos 2t,\; \sin 2t \right\}$$

To find the orthogonal projection of $f(t)$ onto the space spanned by any orthonormal basis, we have only to find the components of the given vector along those basis vectors, which, according to (2.66), can simply be found via an inner product. Suppose we denote this projected vector (in **W**) as:

$$g(t) = C_1 \hat{e}_1 + C_2 \hat{e}_2 + C_3 \hat{e}_3 + C_4 \hat{e}_4 + C_5 \hat{e}_5$$

Then the undetermined coefficients C_i can be found by (2.66) as

$$C_1 = \langle f(t), \hat{e}_1 \rangle = \left\langle f(t), \frac{1}{\sqrt{2}} \right\rangle = \frac{1}{\pi} \int_{-\pi}^{\pi} \frac{1}{\sqrt{2}} f(t)\, dt = \frac{1}{\pi} \int_{-\pi/2}^{\pi/2} \frac{1}{\sqrt{2}}\, dt = \frac{1}{\sqrt{2}}$$

$$C_2 = \langle f(t), \hat{e}_2 \rangle = \frac{1}{\pi} \int_{-\pi}^{\pi} \cos t \; f(t)\, dt = \frac{1}{\pi} \int_{-\pi/2}^{\pi/2} \cos t \; dt = \frac{2}{\pi}$$

$$C_3 = \langle f(t), \hat{e}_3 \rangle = \frac{1}{\pi} \int_{-\pi}^{\pi} \sin t \; f(t)\, dt = \frac{1}{\pi} \int_{-\pi/2}^{\pi/2} \sin t \; dt = 0$$

$$C_4 = \langle f(t), \hat{e}_4 \rangle = \frac{1}{\pi} \int_{-\pi}^{\pi} \cos 2t \; f(t)\, dt = \frac{1}{\pi} \int_{-\pi/2}^{\pi/2} \cos 2t \; dt = 0$$

$$C_5 = \langle f(t), \hat{e}_5 \rangle = \frac{1}{\pi} \int_{-\pi}^{\pi} \sin 2t \; f(t)\, dt = \frac{1}{\pi} \int_{-\pi/2}^{\pi/2} \sin 2t \; dt = 0$$

so that the projection of $f(t)$ onto **W** is

$$g(t) = \frac{1}{2} + \frac{2}{\pi} \cos t \tag{2.74}$$

Note that this is simply the five-term Fourier series approximation for the given signal $f(t)$. The original function and its projection are shown in Figure 2.6 below.

It is useful at this point to also remember Parseval's theorem. Because the projection $g(t)$ is written in an orthonormal basis, its norm can easily be found from the coefficients in this basis as:

$$\|g(t)\| = \left[\sum_{i=1}^{5} C_i^2 \right]^{1/2} = \left[\frac{1}{2} + \frac{4}{\pi^2} \right]^{1/2} \approx 0.95 \tag{2.75}$$

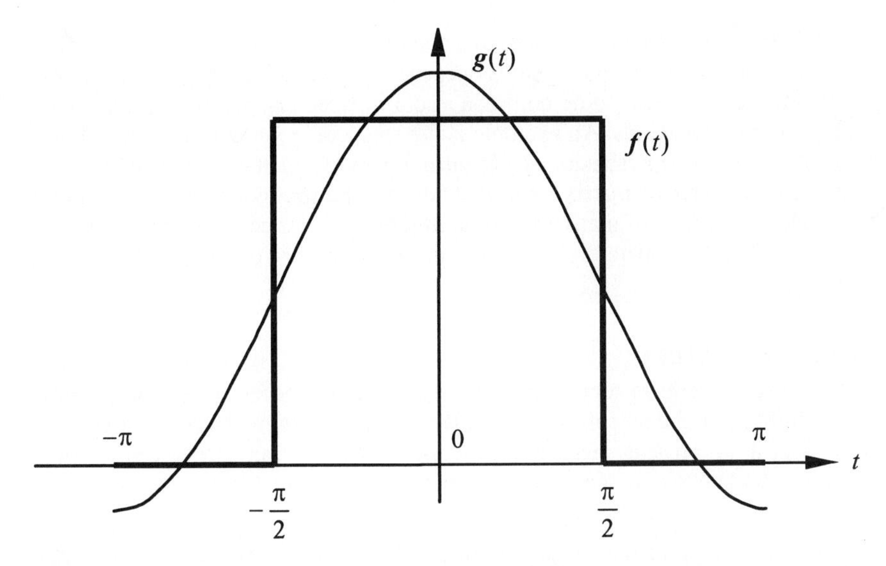

Figure 2.6 Original function $f(t)$ and its projection $g(t)$ onto the subspace **W**.

2.5 Linear Algebras

We will close this chapter by defining a term that is often used to refer to the general mathematical process of manipulating vectors, matrices, and simultaneous equations:

> ***Linear Algebra:*** A *linear algebra* is a linear vector space that has, in addition to all its other requirements for being a linear space, an operation called *multiplication*. This operation satisfies the following rules for all vectors x, y, and z, and scalars α from its field:
>
> 1. Associativity of scalar multiplication, i.e.,
>
> $$\alpha(xy) = (\alpha x)y = x(\alpha y)$$
>
> 2. Associativity of vector multiplication, i.e.,
>
> $$(xy)z = x(yz)$$
>
> 3. Distributivity of vector multiplication,
>
> $$(x + y)z = xz + yz .$$
>
> Note that the vector multiplication need not be commutative in a linear algebra. (2.76)

The euclidean vectors that we have been using as examples throughout this chapter are in fact *not* linear algebras because we have defined no multiplication for them. Nevertheless, some common vector spaces that we will discuss will indeed be linear algebras. An example is the space of $n \times n$ matrices, which we will introduce as representations of linear operators in the next chapter. The familiar procedure for matrix multiplication will be the multiplication operator that makes the *space* of matrices into an *algebra* of matrices. The interpretation of these matrices as operators on vector spaces is the topic of the next chapter.

2.6 Summary

In this chapter we have reviewed the basics of vector spaces and linear algebra. These fundamental concepts are rightfully the domain of mathematics texts, but it is useful to present them here in order that we can stress engineering relevance and the terminology we will find important later. The most important concepts include the following:

- Vectors need not be thought of as "magnitude and direction" pairs, but this is a good way to imagine them in low dimensions.
- All linear finite dimensional vector spaces, such as time functions or rational polynomials, may be treated as vectors of n-tuples because they are isomorphic to n-tuples. We then treat their representations as we would n-tuples, unifying the vector-matrix operations we must perform into one computational set.
- The basis of a space is an important generalization of the "coordinate system" that we are familiar with from basic mechanics and physics. Basis expansions and change-of-basis operations can be performed with vector-matrix arithmetic.
- The rank of a matrix represents the dimension of the space formed by its columns, i.e., it is the number of linearly independent columns the matrix possesses.
- Inner products are used to relate the differences in direction of two vectors, and norms are used to gauge their "size."
- Orthonormal bases are the most convenient kind of basis for computational purposes and can be constructed with the Gram-Schmidt procedure.
- There exists a simple theorem that enables us to find the vector closest to a subspace but not actually in it. This process will be important in engineering approximations wherein we must find the best solution to a problem when an exact solution is unavailable or impractical.

In the next chapter, we will consider linear operators, which will help us solve simultaneous equations, map vectors from one space into another or into itself, and compute other functions performed on vectors.

2.7 *Problems*

2.1 The set **W** of $n \times n$ matrices with real entries is known to be a linear vector space. Determine which of the following sets are subspaces of **W**:

a) The set of $n \times n$ skew-symmetric matrices?

b) The set of $n \times n$ diagonal matrices?

c) The set of $n \times n$ upper-diagonal matrices?

d) The set of $n \times n$ singular matrices?

2.2 Consider the set of all polynomials in s of degree less than or equal to four, with real coefficients, with the conditions:

$$\begin{aligned} p'(0) + p(0) &= 0 \\ p(1) &= 0 \end{aligned}$$

Show that this set is a linear vector space, and find its dimension. Construct a basis for this space.

2.3 For Example 2.4 on page 54 where we tested the linear independence of vectors of rational polynomials, show that indeed, if a_1 and a_2 are real, then they must both be zero.

2.4 Are the following vectors in $\Re^4$ linearly independent over the reals?

$$\begin{bmatrix} 2 \\ 0 \\ -1 \\ 3 \end{bmatrix}, \begin{bmatrix} 1 \\ -3 \\ 4 \\ 0 \end{bmatrix}, \text{ and } \begin{bmatrix} 1 \\ 1 \\ -2 \\ 2 \end{bmatrix}$$

2.5 Which of the following sets of vectors are linearly independent?

a) $\left\{\begin{bmatrix}-1\\-9\\0\end{bmatrix},\begin{bmatrix}1\\3\\0\end{bmatrix},\begin{bmatrix}2\\-2\\1\end{bmatrix}\right\}$ in the space of real 3-tuples over the field of reals.

b) $\left\{\begin{bmatrix}2-i\\-i\end{bmatrix},\begin{bmatrix}1+2i\\-i\end{bmatrix},\begin{bmatrix}-i\\3+4i\end{bmatrix}\right\}$ in the space of complex pairs over the field of reals.

c) $\left\{2s^2+2s-1,\ -2s^2+2s+s,\ s^2-s-5\right\}$ in the space of polynomials over the field of reals.

2.6 Prove or disprove the following claims: if u, v, and w are linearly independent in vector space **V**, then so are

a) u, $u+v$, and $u+v+w$.

b) $u+2v-w$, $u-2v-w$, and $4v$.

c) $u-v$, $v-w$, and $w-u$.

d) $-u+v+w$, $u-v+w$, and $-u+v-w$.

2.7 Determine a basis for the space $\mathbf{W}\subset\Re^4$ spanned by the four vectors

$$\{y_1,y_2,y_3,y_4\}=\left\{\begin{bmatrix}-1\\1\\-5\\7\end{bmatrix},\begin{bmatrix}2\\1\\7\\-8\end{bmatrix},\begin{bmatrix}3\\2\\10\\-11\end{bmatrix},\begin{bmatrix}5\\5\\15\\-15\end{bmatrix}\right\}$$

Is $z=\begin{bmatrix}3\\-1\\13\\-17\end{bmatrix}\in\mathbf{W}$? Is $u=\begin{bmatrix}4\\9\\12\\-8\end{bmatrix}\in\mathbf{W}$? Is $v=\begin{bmatrix}-1\\-1\\-3\\3\end{bmatrix}\in\mathbf{W}$?

2.8 Use the matrix form of the change of basis operation, i.e., Equation (2.28), to perform the change of basis on the vector $x=\begin{bmatrix}10 & -2 & 2 & 3\end{bmatrix}^{\mathrm{T}}$ given in Example 2.5 on page 60, arriving at the new representation shown in that example.

2.9 Prove that the number of linearly independent rows in a matrix is equal to the number of linearly independent columns.

2.10 For the following matrices, find the ranks and degeneracies.

$$\text{a)} \begin{bmatrix} 1 & 2 & -3 \\ -2 & 2 & 2 \\ 4 & 1 & -4 \end{bmatrix} \qquad \text{b)} \begin{bmatrix} 0 & 2 & 0 \\ 2 & 0 & 0 \\ 0 & 2 & 0 \end{bmatrix} \qquad \text{c)} \begin{bmatrix} 1 & 6 & 3 & 2 & -2 \\ 2 & 12 & 4 & 6 & -10 \\ 3 & 18 & 0 & 15 & -15 \end{bmatrix}$$

2.11 Prove or disprove the following statement: If $\{v_1, v_2, v_3, v_4\}$ is a basis for the vector space $\Re^4$ and $\mathbf{W}$ is a subspace of $\Re^4$ then some subset of $\{v_1, v_2, v_3, v_4\}$ forms a basis for $\mathbf{W}$.

2.12 If it is known that a valid basis for the space of polynomials of degree less than or equal to three is $\{1, t, t^2, t^3\}$, show that every such polynomial has a unique representation as

$$p(t) = a_3 t^3 + a_2 t^2 (1-t) + a_1 t (1-t)^2 + a_0 (1-t)^3$$

2.13 Construct an orthogonal basis of the space spanned by the set of vectors

$$\left\{ \begin{bmatrix} -2 \\ -2 \\ -3 \\ 1 \end{bmatrix}, \begin{bmatrix} 1 \\ 0 \\ 1 \\ -1 \end{bmatrix}, \begin{bmatrix} 5 \\ -3 \\ 4 \\ 8 \end{bmatrix} \right\}$$

Repeat for

$$\left\{ \begin{bmatrix} 2 \\ -1 \\ 3 \\ -3 \end{bmatrix}, \begin{bmatrix} -1 \\ -2 \\ -1 \\ -1 \end{bmatrix}, \begin{bmatrix} -14 \\ -3 \\ -19 \\ 11 \end{bmatrix} \right\}$$

2.14 Using the Gram-Schmidt technique, construct an orthonormal basis for the subspace of $\Re^4$ defined by the vectors

$$x_1 = \begin{bmatrix} -1 \\ 4 \\ 1 \\ 1 \end{bmatrix}, \quad x_2 = \begin{bmatrix} 4 \\ -3 \\ 1 \\ 1 \end{bmatrix}, \quad \text{and} \quad x_3 = \begin{bmatrix} 3 \\ -1 \\ -2 \\ 2 \end{bmatrix}$$

2.15 Let $\mathbf{C}^4$ denote the space of all complex 4-tuples, over the complex numbers. Use the "standard" inner product and the Gram-Schmidt method to orthogonalize the vectors

$$x_1 = \begin{bmatrix} 0 \\ 0 \\ -i \\ 1+i \end{bmatrix}, \quad x_2 = \begin{bmatrix} 2-i \\ i \\ -1 \\ 1+i \end{bmatrix}, \quad x_3 = \begin{bmatrix} 1+i \\ 1-i \\ 2 \\ 1-i \end{bmatrix}$$

2.16 Consider the vector $y = \begin{bmatrix} 1 & 2 & 3 \end{bmatrix}^T$. Find the orthogonal projection of y onto the plane $x_1 - 2x_2 + 4x_3 = 0$.

2.17 Let

$$f_1(x) = \begin{cases} 4x & \text{if } 0 \le x \le 0.25 \\ 2 - 4x & \text{if } 0.25 < x \le 0.5 \\ 0 & \text{otherwise} \end{cases}$$

Also let $f_2(x) = f_1(x - 0.25)$ and $f_3(x) = f_1(x - 0.5)$. Let $\mathbf{V}$ be the space of real-valued continuous functions on the interval $x \in [0, 1]$ using the inner product

$$\langle f, g \rangle = \int_0^1 f(x) g(x)\, dx$$

and let $\mathbf{W}$ be the subspace spanned by the three functions f_1, f_2, and f_3. Find the dimension of the subspace $\mathbf{W}$.

Now let

$$g(x) = \begin{cases} 1 & \text{if } 0.5 \le x \le 1 \\ 0 & \text{if } 0 \le x < 0.5 \end{cases}$$

Find the projection of $\boldsymbol{g}$ in **W**.

2.18 Let V be the vector space of all polynomials, with inner product

$$\langle \boldsymbol{f}, \boldsymbol{g} \rangle = \int_0^\infty \boldsymbol{f}(x)\boldsymbol{g}(x)e^{-x}\, dx$$

Let **W** be the subspace of **V** spanned by $\{1, x, x^2\}$.

a) Use Gram-Schmidt to orthogonalize $\{1, x, x^2\}$.

b) Find the orthogonal projection of $\boldsymbol{f}(x) = x^3$ onto **W**.

2.19 The LeGendre polynomials may be defined as the orthogonalization of the basis $\{1, t, t^2, \ldots\}$ for the space of polynomials in t, defined on $t \in [-1, 1]$, using the inner product

$$\langle \boldsymbol{f}, \boldsymbol{g} \rangle = \int_{-1}^{1} \boldsymbol{f}(t)\boldsymbol{g}(t)\, dt$$

Find the first four such LeGendre polynomials.

2.20 Let **V** be the vector space of continuous real-valued functions of t defined on the interval $t \in [0, 3]$. Let **W** be the subspace of **V** consisting of functions that are linear for $0 < t < 1$, linear for $1 < t < 2$, and linear for $2 < t < 3$.

a) Construct a basis for **W**.

b) If an inner product for **W** is given by

$$\langle \boldsymbol{f}, \boldsymbol{g} \rangle = \int_0^3 \boldsymbol{f}(x)\boldsymbol{g}(x)\, dx$$

compute the orthogonal projection of $\boldsymbol{f}(x) = x^2$ onto **W**.

2.21 Let $\mathbf{V}$ be an inner product space. If $x, y \subset \mathbf{V}$, and $\langle x, z \rangle = \langle y, z \rangle$ for every $z \in \mathbf{V}$, prove that $x = y$.

2.22 Use the complex vector $x = [1 + 2i \quad 2 - i]^{\mathrm{T}}$ to demonstrate that, in a complex vector space, $\langle x, y \rangle = x^{\mathrm{T}} y$ is not a valid inner product.

2.23 Find the polynomial $p(t) = t^3 + a_2 t^2 + a_1 t + a_0$ for which the integral

$$J = \int_{-1}^{1} p^2(t)\, dt$$

has the smallest value.

2.24 Let $\mathbf{V}$ be the space of all polynomials with real coefficients, with the inner product

$$\langle f, g \rangle = \int_{-\infty}^{\infty} f(x) g(x) e^{-x^2}\, dx$$

Orthogonalize the set of vectors $\{1, x, x^2, x^3\}$.

2.25 Let $\mathbf{V}$ be the linear space of real-valued continuous functions defined on the interval $t \in [-1, 1]$, with inner product

$$\langle f, g \rangle = \int_{-1}^{1} f(x) g(x)\, dx$$

Let $\mathbf{W}$ be a subspace consisting of polynomials p of degree less than or equal to four that satisfy $p(1) = p(-1) = 0$. Find the projection of the polynomial $p(t) = 1$ in the subspace $\mathbf{W}$.

2.26 Prove that a vector of functions $[x_1(t) \quad \cdots \quad x_n(t)]^{\mathrm{T}}$ is linearly independent if and only if the Gram determinant is nonzero.

2.8 References and Further Reading

The engineering perspective on vectors and vector spaces presented here is comparable to those in [1], [6], and [7]. Further details and a more mathematical treatment of vector spaces and vector-matrix computations can be found in [2], [3], [5], and [8].

[1] Brogan, William L., *Modern Control Theory*, 3rd edition, Prentice-Hall, 1991.

[2] Golub, Gene H., and Charles F. Van Loan, *Matrix Computations*, Johns Hopkins University Press.

[3] Halmos, Paul R., *Finite Dimensional Vector Spaces*, 2nd edition, D. Van Nostrand, 1958.

[4] Kolmogorov, A. N., and S. V. Fomin, *Introductory Real Analysis*, Dover, 1970.

[5] Naylor, A. W., and G. R. Sell, *Linear Operator Theory in Engineering and Science*, Springer-Verlag, 1982.

[6] Shilov, Georgi E., *Linear Algebra*, Dover, 1977.

[7] Strang, Gilbert, *Linear Algebra and Its Applications*, Academic Press, 1980.

[8] Zadeh, Lotfi, A., and Charles A. Desoer, *Linear System Theory: The State Space Approach*, McGraw-Hill, 1963.

3

Linear Operators on Vector Spaces

In the previous chapter, we discussed bases and the methods for changing the basis representation of a vector. This change of basis was performed by multiplying the representation (the n-tuple of components) with an appropriate matrix, which described the relationship between the two sets of basis vectors [see Equation (2.33)]. When considering the change of basis operation, it is important to remember that the vector itself does not change, only the coordinate system in which it is expressed.

In this chapter, we will consider linear operators. Linear operators are functions on the vector space but are fundamentally different from the change of basis, although they will also be expressed in terms of a matrix multiplication. A linear operator, or linear transformation, is a process by which a given vector is transformed into an entirely different vector. As we shall see, linear operators might also implicitly perform a change of basis, they might take a vector in one space and produce a vector in an entirely different space, or they may simply change a vector in a given space into a different vector in the same space.

3.1 Definition of a Linear Operator

A linear operator is simply a rule that associates one vector with a different vector, which may or not be in the same linear vector space. It also must satisfy the properties of linearity. To be exact, we give the following definition:

> ***Linear Operator (Transformation):*** An operator A from linear vector space $\mathbf{X}$ to linear vector space $\mathbf{Y}$, denoted $A\colon \mathbf{X} \rightarrow \mathbf{Y}$, is linear if
>
> $$A(\alpha_1 \boldsymbol{x}_1 + \alpha_2 \boldsymbol{x}_2) = \alpha_1 A\boldsymbol{x}_1 + \alpha_2 A\boldsymbol{x}_2 \tag{3.1}$$
>
> for any $\boldsymbol{x}_1, \boldsymbol{x}_2 \in \mathbf{X}$, and scalars α_1 and α_2.

3.1.1 Range and Null Spaces

A linear operator such as the one in the definition above, $A\colon \mathbf{X} \rightarrow \mathbf{Y}$, will produce a vector $\boldsymbol{y} \in \mathbf{Y}$ and can be written as $A\boldsymbol{x} = \boldsymbol{y}$. The vector $\boldsymbol{y}$ produced by this operation is said to be in the *range* of operator A and is sometimes referred to as the *image* of $\boldsymbol{x}$ in $\mathbf{Y}$ ($\boldsymbol{x}$ being the *preimage* of $\boldsymbol{y}$). The concept of the range of an operator is similar to that of scalar functions, but linear operators in general will have multidimensional ranges. These ranges will themselves be linear vector spaces.

> ***Range Space:*** The range space of an operator $A\colon \mathbf{X} \rightarrow \mathbf{Y}$, denoted R($A$), is the set of all vectors $\boldsymbol{y}_i \in \mathbf{Y}$ such that for every $\boldsymbol{y}_i \in \mathrm{R}(A)$ there exists an $\boldsymbol{x} \in \mathbf{X}$ such that $A\boldsymbol{x} = \boldsymbol{y}_i$.
>
> (3.2)

If the range of operator A is the entire space $\mathbf{Y}$, then A is said to be *onto*, or *surjective*. If A maps elements of $\mathbf{X}$ to *unique* values in $\mathbf{Y}$, i.e., if $\boldsymbol{x}_1 \neq \boldsymbol{x}_2$ implies that $A(\boldsymbol{x}_1) \neq A(\boldsymbol{x}_2)$, then A is said to be *one-to-one*, or *injective*. If operator A is both one-to-one and onto, then it is *bijective*, in which case it is also *invertible*.[M] An operator A is invertible if and only if there exists another operator, denoted $A^{-1}\colon \mathbf{Y} \rightarrow \mathbf{X}$ such that $A^{-1}(A(x)) = x$ and $A(A^{-1}(y)) = y$. Then $A^{-1}A$ is denoted I_{X} and AA^{-1} is denoted I_{Y}, which are the *identity*[M] *operators* in their respective spaces. When A maps a space into itself, as in $A\colon \mathbf{X} \rightarrow \mathbf{X}$, we write simply $AA^{-1} = A^{-1}A = I$.

`inv(A)`

`eye(n)`

We will often need to know which, among all vectors $\boldsymbol{x} \in \mathbf{X}$, will map to the zero vector in $\mathbf{Y}$. For this reason we define the null space[M] of an operator.

`null(A)`

Null Space: The null space of operator A, denoted N(A), is the set of all vectors $\boldsymbol{x}_i \in \mathbf{X}$ such that $A(\boldsymbol{x}_i) = \mathbf{0}$:

$$\mathrm{N}(A) = \{\boldsymbol{x}_i \in \mathbf{X} | A\boldsymbol{x}_i = \mathbf{0}\} \tag{3.3}$$

Example 3.1: The Range and Null Space of a Projection

Consider the projection theorem given in Chapter 2. Show that such a projection $P: \Re^3 \to U$ is a linear operator and find its range and null spaces.

Solution:

Suppose we have two vectors $\boldsymbol{x}_1, \boldsymbol{x}_2 \in \mathbf{X}$ and two real scalars α_1 and α_2. Then let $\boldsymbol{x} = \alpha_1 \boldsymbol{x}_1 + \alpha_2 \boldsymbol{x}_2 \in \mathbf{X}$. The projection theorem says that there exists a vector $\boldsymbol{u} \in \mathbf{U}$ such that if we denote the projection $\boldsymbol{u} = P\boldsymbol{x}$, then $\langle \boldsymbol{x} - P\boldsymbol{x}, \boldsymbol{y} \rangle = 0$ for all $\boldsymbol{y} \in U$. So by the definition of linearity, we have

$$\begin{aligned} \langle \alpha_1 \boldsymbol{x}_1 + \alpha_2 \boldsymbol{x}_2 - P(\alpha_1 \boldsymbol{x}_1 + \alpha_2 \boldsymbol{x}_2), \boldsymbol{y} \rangle &= \langle \alpha_1 \boldsymbol{x}_1 + \alpha_2 \boldsymbol{x}_2, \boldsymbol{y} \rangle - \langle P(\alpha_1 \boldsymbol{x}_1 + \alpha_2 \boldsymbol{x}_2), \boldsymbol{y} \rangle \\ &= \alpha_1 \langle \boldsymbol{x}_1, \boldsymbol{y} \rangle + \alpha_2 \langle \boldsymbol{x}_2, \boldsymbol{y} \rangle - \langle P(\alpha_1 \boldsymbol{x}_1 + \alpha_2 \boldsymbol{x}_2), \boldsymbol{y} \rangle \\ &= 0 \end{aligned}$$

so

$$\langle P(\alpha_1 \boldsymbol{x}_1 + \alpha_2 \boldsymbol{x}_2), \boldsymbol{y} \rangle = \alpha_1 \langle \boldsymbol{x}_1, \boldsymbol{y} \rangle + \alpha_2 \langle \boldsymbol{x}_2, \boldsymbol{y} \rangle \tag{3.4}$$

But considering the projections $P\boldsymbol{x}_1$ and $P\boldsymbol{x}_2$ separately, we have $\langle \boldsymbol{x}_1 - P\boldsymbol{x}_1, \boldsymbol{y} \rangle = 0$ and $\langle \boldsymbol{x}_2 - P\boldsymbol{x}_2, \boldsymbol{y} \rangle = 0$, which give $\langle \boldsymbol{x}_1, \boldsymbol{y} \rangle = \langle P\boldsymbol{x}_1, \boldsymbol{y} \rangle$ and $\langle \boldsymbol{x}_2, \boldsymbol{y} \rangle = \langle P\boldsymbol{x}_2, \boldsymbol{y} \rangle$, or, equivalently by multiplying each term by a scalar, $\alpha_1 \langle \boldsymbol{x}_1, \boldsymbol{y} \rangle = \alpha_1 \langle P\boldsymbol{x}_1, \boldsymbol{y} \rangle$ and $\alpha_2 \langle \boldsymbol{x}_2, \boldsymbol{y} \rangle = \alpha_2 \langle P\boldsymbol{x}_2, \boldsymbol{y} \rangle$. We can substitute these expressions into (3.4) above, to give

$$\begin{aligned} \langle P(\alpha_1 \boldsymbol{x}_1 + \alpha_2 \boldsymbol{x}_2), \boldsymbol{y} \rangle &= \alpha_1 \langle P\boldsymbol{x}_1, \boldsymbol{y} \rangle + \alpha_2 \langle P\boldsymbol{x}_2, \boldsymbol{y} \rangle \\ &= \langle \alpha_1 P\boldsymbol{x}_1 + \alpha_2 P\boldsymbol{x}_2, \boldsymbol{y} \rangle \end{aligned} \tag{3.5}$$

Recalling that this must be true *for all* $\boldsymbol{y} \in U$, then comparing (3.4) and (3.5) gives $P(\alpha_1 \boldsymbol{x}_1 + \alpha_2 \boldsymbol{x}_2) = \alpha_1 P\boldsymbol{x}_1 + \alpha_2 P\boldsymbol{x}_2$ and linearity is proven. The relationships between all these vectors are shown in Figure 3.1.

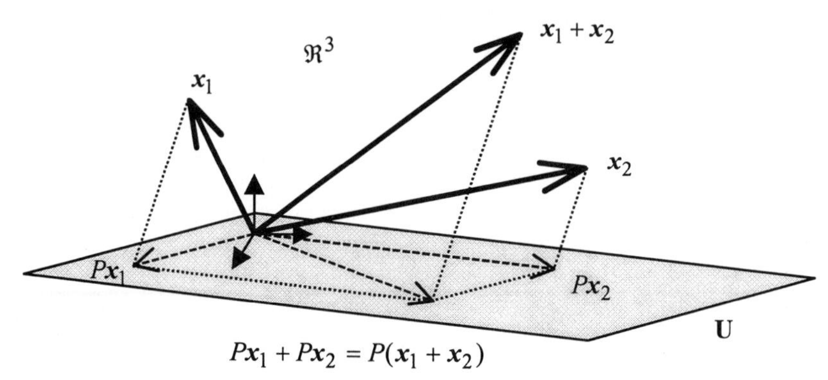

Figure 3.1 Graphical representation of the linearity of the projection operator. The projection of the sum of vectors is the sum of the projections of the vectors.

The range space of operator P, R(P), is obviously the entire subspace **U**. This is apparent from the observation that every vector in the entire subspace will have a preimage that can be constructed by adding to it any vector orthogonal to the plane **U**. One can see that, in fact, every vector in **U** will have infinitely many preimages because there are infinitely many such orthogonal vectors that can be used. The range space therefore has dimension two, the same as the dimension of **U** itself.

The null space N(P) consists of the set of vectors in **X** that project onto the zero vector of **U**. These must therefore be constructed by taking all vectors orthogonal to **U** and adding them to this zero vector, which is the origin of the plane. Suppose, for example, the subspace **U** is described by the equation of a plane $f(x) = 5x_1 + 3x_2 - 2x_3 = 0$. Then the set of all vectors orthogonal to the plane consists of the set of vectors $\alpha \nabla f = \alpha \begin{bmatrix} 5 & 3 & -2 \end{bmatrix}^{\mathrm{T}}$, where α is any scalar and $\nabla(\cdot)$ is the gradient operator described in Appendix A. Therefore, N(P) is the straight line passing through the origin, extending in the direction of the vector $\begin{bmatrix} 5 & 3 & -2 \end{bmatrix}^{\mathrm{T}}$. One may wish to verify this using the projection theorem and any vector (other than the zero vector) that lies in **U**. Being a single line, the null space is therefore of dimension one.

We will next show that any finite-dimensional linear operator can be represented for computational purposes as a matrix, also called A, acting on the representations of vectors. For linear operator A, the range space R(A) is a subspace of **Y** whose dimension is the rank of matrix A. Conversely, the null space N(A) is a subspace of **X**, whose dimension is the nullity of matrix A. It can be shown that not only can any operator be represented by a matrix, but any matrix can be interpreted as a linear operator. This correspondence will prevent any confusion regarding the notation. We can use the same symbol A for both a matrix and an operator without ambiguity because they are essentially one and the same.

3.1.2 Matrix Representations of Linear Operators

For generality, we will discuss the matrix representation of linear operators that might also change the basis of the vector on which it operates. This is obviously necessary if we speak of an operator that operates on vectors in a space of one dimension and returns vectors in a different dimension. Clearly a single basis would not suffice to describe both vectors.

Consider two vectors, one from an n-dimensional space and one from an m-dimensional space: $x \in \mathbf{X}^n$ and $y \in \mathbf{X}^m$. We will derive a matrix representation A for an operator that transforms a vector from one space into a vector in the other, i.e., $A: \mathbf{X}^n \rightarrow \mathbf{X}^m$. Denote arbitrary bases for the two spaces as

$$\{v_j\} = \{v_1, v_2, \ldots, v_n\} \text{ for space } \mathbf{X}^n$$

and

$$\{u_i\} = \{u_1, u_2, \ldots, u_m\} \text{ for space } \mathbf{X}^m$$

By expanding x out as a representation in its basis as $x = \sum_{j=1}^{n} \alpha_j v_j$ and applying the linearity property of the operator A,

$$y = Ax = A\left(\sum_{j=1}^{n} \alpha_j v_j\right) = \sum_{j=1}^{n} \alpha_j A(v_j) \tag{3.6}$$

This simple but important result means that we can determine the effect of A on any vector x by knowing only the effect that A has on the basis vectors in which x is written. Equation (3.6) can be alternatively written in vector-matrix notation as:

$$y = Ax = \left[Av_1 \mid Av_2 \mid \cdots \mid Av_n \right] \begin{bmatrix} \alpha_1 \\ \alpha_2 \\ \vdots \\ \alpha_n \end{bmatrix}$$

Now we note that each vector $A\boldsymbol{v}_j$ is by definition a vector in the range space $\mathbf{X}^m$. Then like any other vector in that space, it can be expanded in the basis defined for that space, $\{\boldsymbol{u}_i\}$. This expansion gives

$$A\boldsymbol{v}_j = \sum_{i=1}^{m} a_{ij}\boldsymbol{u}_i \tag{3.7}$$

for some set of coefficients a_{ij}. We can also expand $\boldsymbol{y}$ out directly in terms of its own basis vectors, using coefficients β_i,

$$\boldsymbol{y} = \sum_{i=1}^{m} \beta_i \boldsymbol{u}_i \tag{3.8}$$

Substituting (3.7) and (3.8) into Equation (3.6), we can write:

$$\boldsymbol{y} = \sum_{j=1}^{n} \alpha_j \left(\sum_{i=1}^{m} a_{ij}\boldsymbol{u}_i \right) = \sum_{i=1}^{m} \beta_i \boldsymbol{u}_i \tag{3.9}$$

By changing the order of summation in this expression, we obtain

$$\sum_{i=1}^{m} \left(\sum_{j=1}^{n} a_{ij}\alpha_j \right) \boldsymbol{u}_i = \sum_{i=1}^{m} \beta_i \boldsymbol{u}_i \tag{3.10}$$

But because of the uniqueness of any vector's expansion in a basis [see Section (2.2.3)], the expansion of $\boldsymbol{y}$ in $\{\boldsymbol{u}_i\}$ must be unique. This implies that

$$\beta_i = \sum_{j=1}^{n} a_{ij}\alpha_j \quad \text{for all } i = 1, \ldots, m \tag{3.11}$$

The reader may notice that the expression (3.11) above is in the form of a matrix-vector multiplication. In fact, this is how we will usually apply this result. If $\alpha = \begin{bmatrix} \alpha_1 & \alpha_2 & \cdots & \alpha_n \end{bmatrix}^{\mathrm{T}}$ is the representation of $\boldsymbol{x} \in \mathbf{X}^n$ in the $\{\boldsymbol{v}_j\}$ basis and $\beta = \begin{bmatrix} \beta_1 & \beta_2 & \cdots & \beta_m \end{bmatrix}^{\mathrm{T}}$ is the representation of $\boldsymbol{y} \in \mathbf{X}^m$ in the $\{\boldsymbol{u}_i\}$ basis, then we can find this representation of $\boldsymbol{y}$ with the matrix multiplication $\beta = A\alpha$, where we use our new matrix representation for operator A:

$$A = \begin{bmatrix} a_{11} & a_{12} & \cdots & a_{1n} \\ a_{21} & a_{22} & \cdots & a_{2n} \\ \vdots & \vdots & \ddots & \vdots \\ a_{m1} & a_{m2} & \cdots & a_{mn} \end{bmatrix} \tag{3.12}$$

That is, the $(i, j)^{th}$ element of the $(m \times n)$ matrix A is the coefficient from (3.7), which describes how the operator A transforms the basis vectors from one space into the basis of the range space. Of course, this implies that the bases of each space must be specified before determining the matrix representation for any linear operator. In general, the following observation can be made from a comparison of (3.7) and (3.12):

> The jth column of the matrix representation of any linear operator $A\!:\!\mathbf{X}^n \rightarrow \mathbf{X}^m$ can be constructed as the representation of the vector that is the result of A acting on the jth basis vector of $\mathbf{X}^n$, where this new vector is expanded into the basis of the range space. That is, the jth column of A is simply $A\boldsymbol{v}_j$, written as a representation in the basis $\{\boldsymbol{u}_i\}$.

This is a very useful property, as we shall see in the subsequent examples. It allows us a convenient way to determine the matrix representation of an arbitrary linear operator. Use of the matrix representation will be necessary for any computational process, such as the ones we will use in future chapters.

The representation of the linear operator as a matrix also has implications for the computation of the range and null spaces. As we will demonstrate in Section 3.3, the linear operation $\boldsymbol{y} = A\boldsymbol{x}$ can be viewed numerically as

$$\begin{aligned} \boldsymbol{y} &= A\boldsymbol{x} \\ &= \begin{bmatrix} a_1 & a_2 & \cdots & a_n \end{bmatrix} \begin{bmatrix} x_1 \\ x_2 \\ \vdots \\ x_n \end{bmatrix} \end{aligned} \tag{3.13}$$

where a_i denotes the ith column of the matrix A and $\boldsymbol{x}_i$ is the ith component of the vector $\boldsymbol{x}$ or of its representation. Therefore, if the range space is the space of all possible values of $A\boldsymbol{x}$, it may be represented as the span of all of the columns of A. This implies that the rank of *matrix A* is equal to the dimension of the range space of *operator A*, i.e., $r(A) = \dim(\mathrm{R}(A))$. Similarly, the null space of *operator A* may be represented as the space of all solutions of the simultaneous linear equations

$$\begin{aligned} 0 &= Ax \\ &= \begin{bmatrix} a_1 & a_2 & \cdots & a_n \end{bmatrix} \begin{bmatrix} x_1 \\ x_2 \\ \vdots \\ x_n \end{bmatrix} \end{aligned} \tag{3.14}$$

Therefore the dimension of the null space of *operator* A is equal to the nullity of *matrix* A, i.e., $q(A) = \dim(\mathrm{N}(A))$. For Example 3.1, $r(P) = 2$ and $q(P) = 1$. See Section 3.3 for a complete discussion the computation of these spaces.

Example 3.2: Rotation Matrices

Consider a linear operator $A: \Re^2 \to \Re^2$ that takes a vector $\boldsymbol{x}$ and rotates it counterclockwise by an angle θ as shown in Figure 3.2. Find the matrix representation of the linear operator that performs such planar rotations on arbitrary vectors in $\Re^2$, and test this matrix by rotating the vector $\boldsymbol{x} = \begin{bmatrix} 1 & 2 \end{bmatrix}^{\mathrm{T}}$ by an angle $\theta = 30°$.

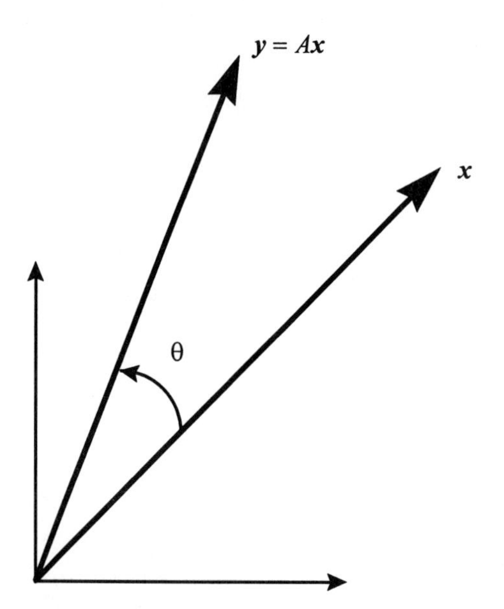

Figure 3.2 Vector $\boldsymbol{x}$ is transformed by operator A into a vector $\boldsymbol{y} = A\boldsymbol{x}$ that is simply a rotated version of $\boldsymbol{x}$ (by angle θ counterclockwise).

Solution:

To find the matrix representation of any operator, we need the basis of each space, and the effect of the operator on those basis vectors. We will use the standard basis vectors as shown in Figure 3.3.

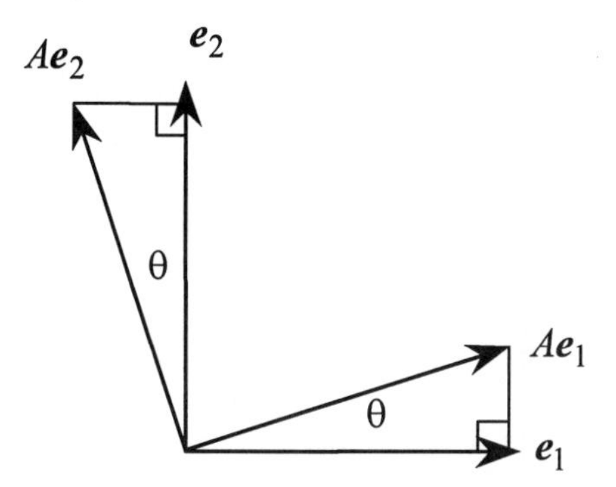

Figure 3.3 The effect of operator A on the basis vectors.

By decomposing the rotated vectors Ae_1 and Ae_2 along the original basis directions e_1 and e_2, we can see using basic trigonometry that

$$\begin{aligned} Ae_1 &= \cos\theta \cdot e_1 + \sin\theta \cdot e_2 = [e_1 \mid e_2]\begin{bmatrix}\cos\theta \\ \sin\theta\end{bmatrix} \\ Ae_2 &= -\sin\theta \cdot e_1 + \cos\theta \cdot e_2 = [e_1 \mid e_2]\begin{bmatrix}-\sin\theta \\ \cos\theta\end{bmatrix} \end{aligned} \tag{3.15}$$

Therefore, the matrix representation of A is

$$A = \begin{bmatrix}\cos\theta & -\sin\theta \\ \sin\theta & \cos\theta\end{bmatrix}$$

Note that each column of this A-matrix is simply the corresponding representation in (3.15) above.

Applying this to the vector $x = [1 \quad 2]^T$, we get

$$Ax = \begin{bmatrix}\cos 30° & -\sin 30° \\ \sin 30° & \cos 30°\end{bmatrix}\begin{bmatrix}1 \\ 2\end{bmatrix} = \begin{bmatrix}-0.134 \\ 2.23\end{bmatrix}$$

This rotation can be verified in Figure 3.4.

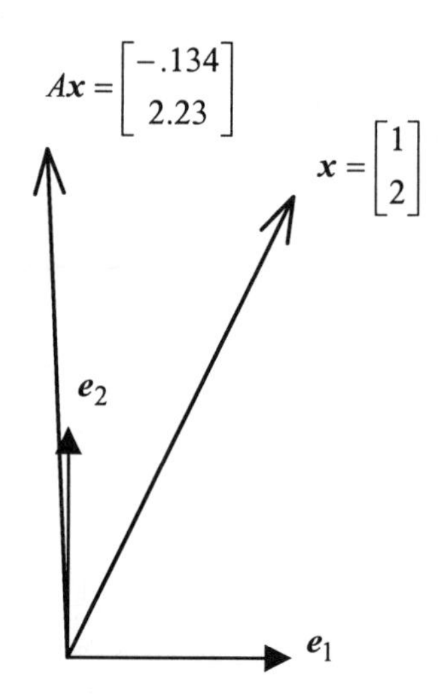

Figure 3.4 Pictorial representation of vector rotation.

Example 3.3: Composite Rotations: Roll, Pitch, and Yaw

Above we considered rotations in a plane. We can also consider rotations in three dimensions. However, in three dimensions, we must remember that rotations about three distinct axes are possible. The matrices that describe these rotations depend on which axis was used as the axis of rotation. The three possibilities, rotation about e_1, e_2, or e_3, are depicted in Figure 3.5.

To achieve an arbitrary three-dimensional rotation, these rotations about coordinate axes may be applied in sequence, but such sequences would not be unique. If a three-dimensional orthonormal frame is arbitrarily oriented with respect to another three-dimensional orthonormal frame, there is generally an infinite number of different axis rotation steps that can be taken to arrive at one frame from the other.

Find the change of basis matrices for the three individual rotations and compute the composite rotation matrix that rotates a coordinate system by θ_R around e_1, θ_P around e_2, and θ_Y around e_3, in that order.

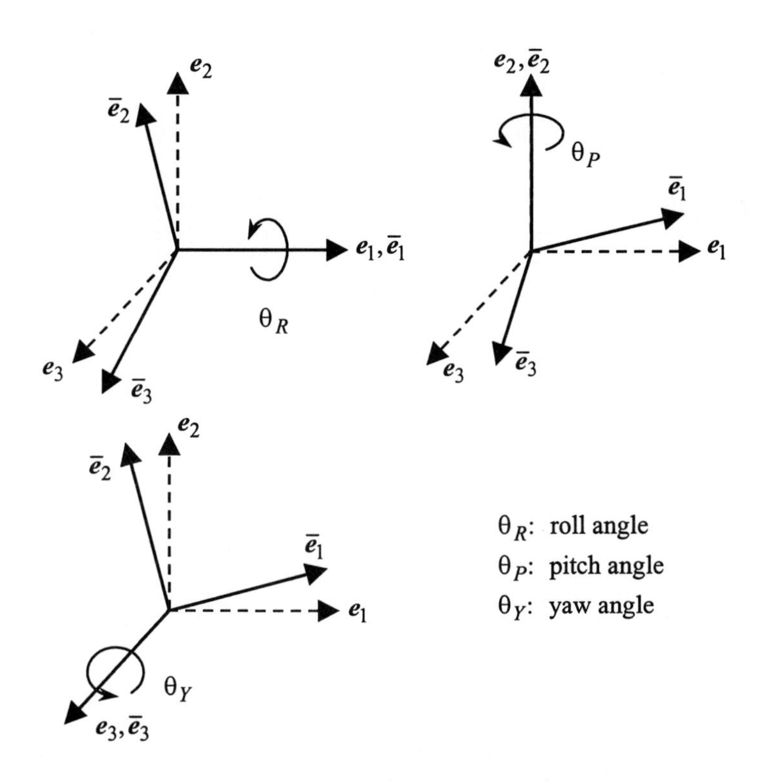

Figure 3.5 Axis rotations for pitch, roll, and yaw.

Solution:

The individual matrices may be computed by viewing the rotations in the plane in which they occur and decomposing the rotated axes in that plane, just as we did with the planar rotation example in Figure 3.3. With those same planar decompositions, the change of basis matrix, which can also be considered a rotation operator on a vector by Equation (2.33), for each rotation is:

$$R_R = \begin{bmatrix} 1 & 0 & 0 \\ 0 & \cos\theta_R & -\sin\theta_R \\ 0 & \sin\theta_R & \cos\theta_R \end{bmatrix} \qquad R_P = \begin{bmatrix} \cos\theta_P & 0 & \sin\theta_P \\ 0 & 1 & 0 \\ -\sin\theta_P & 0 & \cos\theta_P \end{bmatrix}$$

and

$$R_Y = \begin{bmatrix} \cos\theta_P & \sin\theta_P & 0 \\ \sin\theta_P & \cos\theta_P & 0 \\ 0 & 0 & 1 \end{bmatrix}$$

If a roll rotation is applied to a vector $\boldsymbol{x}$ first, followed by a pitch rotation, then a yaw rotation, the composite rotation can be written as the product

$$\begin{aligned} R &= R_Y R_P R_R \\ &= \begin{bmatrix} \cos\theta_Y \cos\theta_P & \cos\theta_Y \sin\theta_P \sin\theta_R - \sin\theta_Y \cos\theta_R & \cdots \\ \sin\theta_Y \cos\theta_P & \sin\theta_Y \sin\theta_P \sin\theta_R + \cos\theta_Y \cos\theta_R & \cdots \\ -\sin\theta_P & \cos\theta_P \sin\theta_R & \end{bmatrix} \\ &\qquad \begin{matrix} \cos\theta_Y \sin\theta_P \cos\theta_R + \sin\theta_Y \sin\theta_R \\ \sin\theta_Y \sin\theta_P \cos\theta_R - \cos\theta_Y \sin\theta_R \\ \cos\theta_P \cos\theta_R \end{matrix} \Bigg] \end{aligned} \tag{3.16}$$

Such rotations are useful for free-floating bodies, such as aircraft, missiles, spacecraft, and submarines, whose orientation is best described with respect to inertial coordinate frames. Rotations about body-centered coordinate axes are somewhat different (see a robotics text such as [5]).

Example 3.4: Matrix Representation of the Projection Operator

Let A be the orthogonal projection operator that projects vectors from $\Re^3$ into $\Re^2$. Suppose $\Re^2$ is given as the $\{e_1, e_2\}$-plane, where the basis of $\Re^3$ is $\{e_1, e_2, e_3\}$. This situation is pictured in Figure 3.6. Find the matrix representation of the orthogonal projection operator $P: \Re^3 \rightarrow \Re^2$.

Solution:

To find the matrix representation of P, we again determine its effect on the basis vectors, two of which are reused in the new space $\Re^2$. In this case, we find that $Pe_1 = e_1$, $Pe_2 = e_2$, and $Pe_3 = \mathbf{0}$. Therefore the matrix of the projection operator is

$$P = \begin{bmatrix} 1 & 0 & 0 \\ 0 & 1 & 0 \end{bmatrix}$$

Of course this matrix must be (2×3) because the transformation takes vectors from three dimensions into two dimensions.

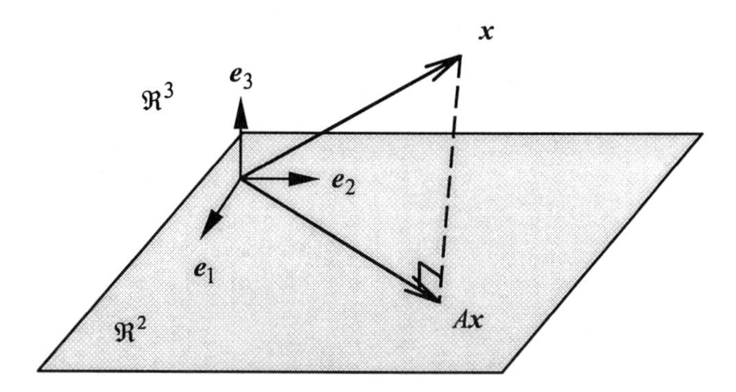

Figure 3.6 Orthogonal projection operation.

Example 3.5: Projection onto an Arbitrary Plane

Give the formula for the orthogonal projection of a vector $\boldsymbol{x}$ onto the plane that is orthogonal to a given vector $\boldsymbol{n}$, as depicted in Figure 3.7 (unlike the example above, such a plane will, in general, not contain any of the coordinate axes).

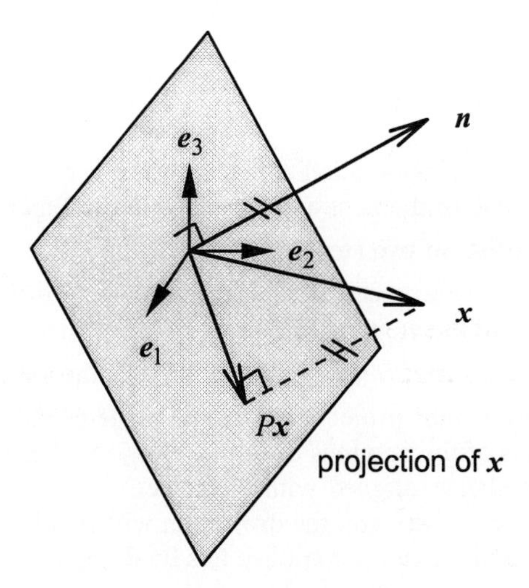

Figure 3.7 Orthogonal projection onto an arbitrary plane.

Solution:

To project $\boldsymbol{x}$ onto the plane normal to $\boldsymbol{n}$, we need to remove from $\boldsymbol{x}$ any of its components that lie along $\boldsymbol{n}$. To find such a component, we use an inner product. We then simply subtract this component. We must remember to normalize the vector $\boldsymbol{n}$ first: $\hat{\boldsymbol{n}} \triangleq \boldsymbol{n}/\|\boldsymbol{n}\|$. Then the length of the component of $\boldsymbol{x}$ along $\boldsymbol{n}$ will be $\hat{\boldsymbol{n}} \cdot \boldsymbol{x} = \hat{\boldsymbol{n}}^{\mathrm{T}} \boldsymbol{x}$. Therefore, the projection is

$$P\boldsymbol{x} = \boldsymbol{x} - \hat{\boldsymbol{n}}(\hat{\boldsymbol{n}}^{\mathrm{T}} \boldsymbol{x}) \tag{3.17}$$

or

$$P\boldsymbol{x} = \left(I - \hat{\boldsymbol{n}}\hat{\boldsymbol{n}}^{\mathrm{T}}\right)\boldsymbol{x} \tag{3.18}$$

Example 3.6: Computer Graphics and Object Rotation

Consider a three-dimensional wire frame object that is to be drawn on a computer screen as it would be viewed by a person whose eye position is located by the vector $\boldsymbol{v}$. This eye position is subject to change, so a program is desired that can rotate the graphical representation of the object according to the viewer's location. Consider the object itself to be specified by a set of vectors $\boldsymbol{p}_i$, $i = 1, \cdots, n$ that indicate the locations of the vertices. Assume that a graphics library is available that can draw lines between any pair of vertices provided, so that the user need only recompute the locations of these vertices in order to redraw the object. Formulate a set of operators that can be used to redraw the object.

Solution:

The same operator will be applied to each vertex vector, so we can consider working with any one of them, say $\boldsymbol{p}$. As depicted in Figure 3.8, the necessary transformation consists of two steps:

1. Rotation of the vertex vectors such that the eye position vector coincides with the z-axis of the viewing screen
2. Projection of the vertex vector onto the plane of the viewing screen

We could perform the projection onto the plane of the viewing screen first by using the results of the previous example. However, because the plane of the viewing screen is always aligned with the screen's z-axis, it will be simpler to first rotate the object. Afterward, the projection will simply be a projection onto a coordinate axis, and we can accomplish this by dropping its z-component.

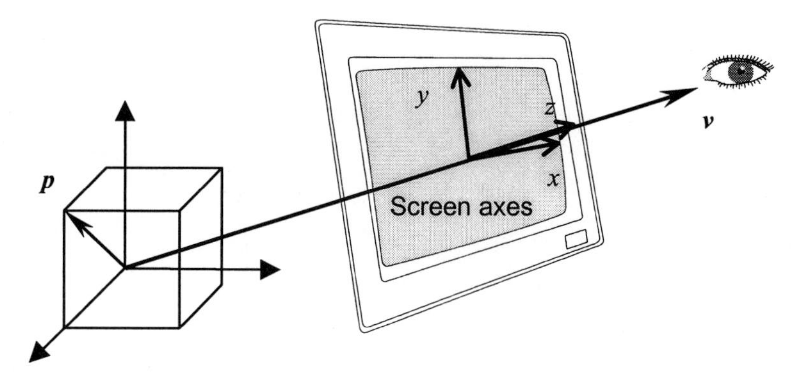

Figure 3.8 The rotation and projection of an object onto a viewing screen normal to eye position vector $\boldsymbol{v}$.

To do the rotation, we will find a rotation matrix that aligns $\boldsymbol{v}$ with the *z*-axis of the screen. From Figure 3.9 it is apparent that this operation can be accomplished in two steps: first a rotation by angle θ about the *z*-axis, followed by a rotation by angle ϕ about the *x*-axis. This is equivalent to a rotation using matrices R_Y and R_R with angles θ and ϕ, respectively. The composite rotation matrix is then

$$\begin{aligned} R &= \begin{bmatrix} 1 & 0 & 0 \\ 0 & \cos\phi & -\sin\phi \\ 0 & \sin\phi & \cos\phi \end{bmatrix} \begin{bmatrix} \cos\theta & -\sin\theta & 0 \\ \sin\theta & \cos\theta & 0 \\ 0 & 0 & 1 \end{bmatrix} \\ &= \begin{bmatrix} \cos\theta & -\sin\theta & 0 \\ \cos\phi\sin\theta & \cos\phi\cos\theta & -\sin\phi \\ \sin\phi\sin\theta & \sin\phi\cos\theta & 1 \end{bmatrix} \end{aligned} \tag{3.19}$$

The angles θ and ϕ can be found from

$$\theta = \tan^{-1}\frac{\hat{v}_1}{\hat{v}_2} \tag{3.20}$$

and

$$\phi = \tan^{-1}\frac{\sqrt{\hat{v}_1^2 + \hat{v}_1^2}}{\hat{v}_3} \tag{3.21}$$

where $\hat{v}_1$, $\hat{v}_2$, and $\hat{v}_3$ are the components of $\hat{\boldsymbol{v}} \stackrel{\Delta}{=} \boldsymbol{v}/\|\hat{v}\|$.

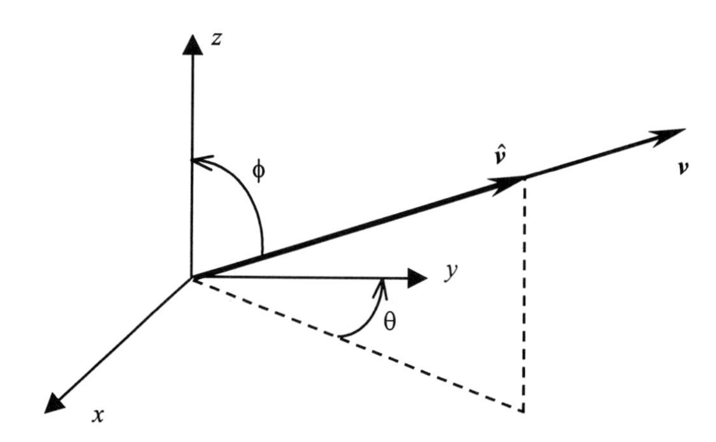

Figure 3.9 Rotation of the eye position to align with the screen z-axis.

After this alignment, the projection can be accomplished with the matrix operator

$$P = \begin{bmatrix} 1 & 0 & 0 \\ 0 & 1 & 0 \end{bmatrix} \tag{3.22}$$

but this is equivalent to dropping the z-component of the rotated vector.

In summary, each vertex should be processed with the operators given in Equations (3.19) and (3.22):

$$(\mathbf{x}_i)_{\text{new}} = PR\mathbf{x}_i \tag{3.23}$$

after which the new two-dimensional vectors $(\mathbf{x}_i)_{\text{new}}$ are drawn and connected in screen coordinates.

3.1.3 Linear Operators in Different Bases

In the next chapter, we will be particularly concerned with transformations from a space into itself, such as $A: \Re^n \to \Re^n$. Of course, this will give a square $(n \times n)$ matrix, and it might simultaneously accomplish a change of basis. Let $\mathbf{X}_v$ denote such an n-dimensional space, equipped with basis $\{v\}$. We say that the matrix A that transforms $\mathbf{X}_v \to \mathbf{X}_v$ is a *transformation* in the basis $\{v\}$. This is for now a departure from our practice of referring to *vectors* in a basis, but since the representation of a vector will change as its basis does, then it should be expected that the representation of an operator on that vector should change with the basis as well. The operator that transforms vectors expressed in a different basis $\{\hat{v}\}$ is therefore denoted by $\hat{A}$. Using this notation, we say that $y = Ax$ and $\hat{y} = \hat{A}\hat{x}$, where, clearly, $\hat{y}$ and $\hat{x}$ are vectors represented in basis $\{\hat{v}\}$.

The question of interest now is how we transform a matrix representation of an operator from one basis into another. We will do this using the change of basis matrices developed in the previous chapter. Using the change of basis matrix B as developed in Equation (2.33), we can write $\hat{y} = By$ and $\hat{x} = Bx$. Using these changes of basis, we have

$$\begin{aligned}\hat{y} &= \hat{A}\hat{x} \\ By &= \hat{A}Bx\end{aligned}$$

so

$$y = B^{-1}\hat{A}Bx$$

B^{-1} must exist because it is an $(n \times n)$ change of basis matrix and of full rank. Comparing this to the expression $y = Ax$, we see that

$$A = B^{-1}\hat{A}B \tag{3.24}$$

Equivalently, if we denote $M = B^{-1}$, then $A = M\hat{A}M^{-1}$, or

$$\hat{A} = M^{-1}AM \tag{3.25}$$

This change of basis for a matrix representation of a linear operator is known as a *similarity transformation*.[M] It is said that A is *similar to* $\hat{A}$ and vice versa. It can be recognized that this transformation is exactly the process used in Chapter 1 when we first introduced state variables in different coordinates. What we were actually doing was changing the basis of the space of solutions to the differential equations.

`ss2ss(sys,T)`

Example 3.7: Matrix Operators and Change of Basis

Consider the linear vector space $\mathbf{X}$ of polynomials in s of degree less than four, with real coefficients, over the field of reals. The operator $A: \mathbf{X} \to \mathbf{X}$ that takes such vectors $v(s)$ and returns the new vectors $w(s) = v''(s) + 2v'(s) + 3v(s)$ (where the prime denotes differentiation in s) is a linear operator (as is differentiation in general). In this example, we shall find the matrix representation of this operator in two bases,

$$\{e_i\} = \{s^3, s^2, s, 1\} \equiv \{e_1(s), e_2(s), e_3(s), e_4(s)\}$$

and

$$\{\bar{e}_i\} = \{s^3 - s^2, s^2 - s, s - 1, 1\} \equiv \{\bar{e}_1(s), \bar{e}_2(s), \bar{e}_3(s), \bar{e}_4(s)\}$$

as well as the transformation between the two.

Solution:

First we will determine the effect of A on the basis vectors $\{e_i\}$:

$$\begin{aligned} Ae_1 &= e_1''(s) + 2e_1'(s) + 3e_1(s) \\ &= 6s + 6s^2 + 3s^3 \\ &= \begin{bmatrix} e_1 & e_2 & e_3 & e_4 \end{bmatrix} \begin{bmatrix} 3 \\ 6 \\ 6 \\ 0 \end{bmatrix} \end{aligned}$$

$$\begin{aligned} Ae_2 &= e_2''(s) + 2e_2'(s) + 3e_2(s) \\ &= 2 + 4s + 3s^2 \\ &= \begin{bmatrix} e_1 & e_2 & e_3 & e_4 \end{bmatrix} \begin{bmatrix} 0 \\ 3 \\ 4 \\ 2 \end{bmatrix} \end{aligned}$$

$$\begin{aligned} Ae_3 &= e_3''(s) + 2e_3'(s) + 3e_3(s) \\ &= 2 + 3s \\ &= \begin{bmatrix} e_1 & e_2 & e_3 & e_4 \end{bmatrix} \begin{bmatrix} 0 \\ 0 \\ 3 \\ 2 \end{bmatrix} \end{aligned}$$

$$\begin{aligned} Ae_4 &= e_4''(s) + 2e_4'(s) + 3e_4(s) \\ &= 3 \\ &= \begin{bmatrix} e_1 & e_2 & e_3 & e_4 \end{bmatrix} \begin{bmatrix} 0 \\ 0 \\ 0 \\ 3 \end{bmatrix} \end{aligned} \tag{3.26}$$

Taking the four columns of coefficients in these four representations, we obtain the matrix

$$A = \begin{bmatrix} 3 & 0 & 0 & 0 \\ 6 & 3 & 0 & 0 \\ 6 & 4 & 3 & 0 \\ 0 & 2 & 2 & 3 \end{bmatrix} \tag{3.27}$$

We can check this result by applying this transformation to the vector $v(s) = s^2 + 1$, whose representation in $\{e_i\}$ is $v = \begin{bmatrix} 0 & 1 & 0 & 1 \end{bmatrix}^T$. Multiplying this representation of w by the matrix A results in

$$Av = \begin{bmatrix} 3 & 0 & 0 & 0 \\ 6 & 3 & 0 & 0 \\ 6 & 4 & 3 & 0 \\ 0 & 2 & 2 & 3 \end{bmatrix} \begin{bmatrix} 0 \\ 1 \\ 0 \\ 1 \end{bmatrix} = \begin{bmatrix} 0 \\ 3 \\ 4 \\ 5 \end{bmatrix} \tag{3.28}$$

Applying the transformation by direct differentiation gives

$$w(s) = Av(s) = v''(s) + 2v'(s) + 3v(s) = 3s^2 + 4s + 5 \tag{3.29}$$

which is clearly the same vector as computed in (3.28) above.

How does this operator appear in the different basis $\{\bar{e}_i\}$? We have two methods to determine the answer. We can either derive the matrix transformation directly in this basis, or we can perform a similarity transformation on the matrix A that was computed above in (3.27), as in (3.24). First, if we find the change of basis matrix between the two bases, we must compute the expansion

$$e_j = \sum_{i=1}^{n} b_{ij} \bar{e}_i$$

However, it is apparent by inspection that it is easier to compute the *inverse* relationship:

$$\bar{e}_1 = e_1 - e_2 = \begin{bmatrix} e_1 & e_2 & e_3 & e_4 \end{bmatrix} \begin{bmatrix} 1 \\ -1 \\ 0 \\ 0 \end{bmatrix}$$

$$\bar{e}_2 = e_2 - e_3 = \begin{bmatrix} e_1 & e_2 & e_3 & e_4 \end{bmatrix} \begin{bmatrix} 0 \\ 1 \\ -1 \\ 0 \end{bmatrix}$$

$$\bar{e}_3 = e_3 - e_4 = \begin{bmatrix} e_1 & e_2 & e_3 & e_4 \end{bmatrix} \begin{bmatrix} 0 \\ 0 \\ 1 \\ -1 \end{bmatrix}$$

$$\bar{e}_4 = e_4 = \begin{bmatrix} e_1 & e_2 & e_3 & e_4 \end{bmatrix} \begin{bmatrix} 0 \\ 0 \\ 0 \\ 1 \end{bmatrix}$$

Because this is the inverse relationship, the matrix we arrive at is the inverse matrix, $M = B^{-1}$. By gathering the coefficient columns, this matrix is

$$M = \begin{bmatrix} 1 & 0 & 0 & 0 \\ -1 & 1 & 0 & 0 \\ 0 & -1 & 1 & 0 \\ 0 & 0 & -1 & 1 \end{bmatrix}$$

from which we can compute

$$B = M^{-1} = \begin{bmatrix} 1 & 0 & 0 & 0 \\ 1 & 1 & 0 & 0 \\ 1 & 1 & 1 & 0 \\ 1 & 1 & 1 & 1 \end{bmatrix}$$

Now using the formula $A = B^{-1}\overline{A}B$ from Equation (3.24), we find that

$$\overline{A} = BAB^{-1} = \begin{bmatrix} 3 & 0 & 0 & 0 \\ 6 & 3 & 0 & 0 \\ 8 & 4 & 3 & 0 \\ 6 & 4 & 2 & 3 \end{bmatrix}$$

Now to check this result by finding $\overline{A}$ in another way, we can find the effect of the original operator on vectors that have already been expressed in the $\{\overline{e}_i\}$ basis. First, we should find the expression for $\overline{v}(s)$, i.e., the same vector $\boldsymbol{v}$, but in the "bar" basis:

$$\overline{v} = Bv = \begin{bmatrix} 1 & 0 & 0 & 0 \\ 1 & 1 & 0 & 0 \\ 1 & 1 & 1 & 0 \\ 1 & 1 & 1 & 1 \end{bmatrix} \begin{bmatrix} 0 \\ 1 \\ 0 \\ 1 \end{bmatrix} = \begin{bmatrix} 0 \\ 1 \\ 1 \\ 2 \end{bmatrix}$$

which is simply the matrix-vector notation for

$$1 \cdot \overline{e}_2(s) + 1 \cdot \overline{e}_3(s) + 2 \cdot \overline{e}_4(s) = \left(s^2 - s\right) + \left(s - 1\right) + 2 = s^2 + 1 = \boldsymbol{v}(s)$$

Now applying the matrix operation in the $\{\overline{e}_i\}$ basis,

$$\overline{A}\overline{v} = \begin{bmatrix} 3 & 0 & 0 & 0 \\ 6 & 3 & 0 & 0 \\ 8 & 4 & 3 & 0 \\ 6 & 4 & 2 & 3 \end{bmatrix} \begin{bmatrix} 0 \\ 1 \\ 1 \\ 2 \end{bmatrix} = \begin{bmatrix} 0 \\ 3 \\ 7 \\ 12 \end{bmatrix}$$

Writing this representation out explicitly in terms of the "barred" basis vectors,

$$\overline{A}\overline{v}(s) = 3\overline{e}_2 + 7\overline{e}_3 + 12\overline{e}_4 = 3(s^2 - s) + 7(s-1) + 12(1) = 3s^2 + 4s + 5$$

which of course is exactly as we expect: the same vector we computed in Equation (3.28). In other terms, we have demonstrated the commutativity of the Figure 3.10. So-called "commutative diagrams" show how different paths, each representing an operation or transformation taken in order, achieve the same result. To get from one vector to another, one performs the operation indicated next to the arrow, in the direction indicated. If a path must be traversed in reverse ("against" the arrow), then the inverse transformation is used. One can use this diagram to illustrate that $\overline{A} = BAB^{-1}$.

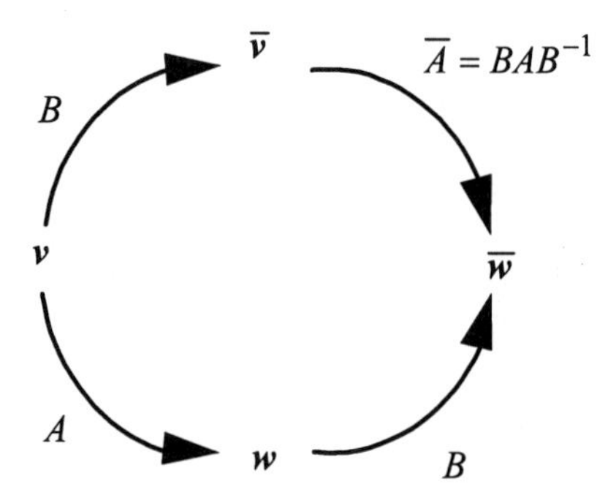

Figure 3.10 Commutative diagram showing different paths to the transformed vector $\overline{\boldsymbol{w}}$.

3.2 Operators as Spaces

It can be shown that the set of all linear operators from one space into another, or from a space into itself, is itself a linear vector space. Because we have been representing linear operators as matrices, the space of operators is, in fact, a space of matrices. It is also an *algebra*, as defined in the previous chapter, because it has a multiplication operation, which we have been using all along whenever we multiplied two compatible matrices.

Because the set of operators is a space, it must satisfy all the other requirements of linear vector spaces, as presented in Chapter 2. For example, operators have an addition operation, which takes the form of the familiar element-by-element matrix addition, where

$$A_1\boldsymbol{x} + A_2\boldsymbol{x} = (A_1 + A_2)\boldsymbol{x}$$

3.2.1 Operator Norms

Operators also have norms. An operator norm,[M] like a vector norm, is intuitively a measure of how big the operator is, and how "big" it is depends on what it does when it operates on a vector. In this sense, a norm describes an operator's maximal "magnification" properties. With this interpretation, we state a general norm for an operator (i.e., a matrix) as

`norm(A)`

$$\|A\| = \sup_{x \neq 0} \frac{\|Ax\|}{\|x\|} = \sup_{\|x\|=1} \|Ax\| \tag{3.30}$$

where *sup* ("supremum") indicates the least upper bound. Of course, as we have seen, there are many ways to define the norm $\|x\|$ of x. Consequently, there are many different operator norms. However, all operator norms must satisfy the following properties:

1. $\|Ax\| \leq \|A\| \cdot \|x\|$ for all vectors x (3.31)
2. $\|A_1 + A_2\| \leq \|A_1\| + \|A_2\|$ (3.32)
3. $\|A_1 A_2\| \leq \|A_1\| \cdot \|A_2\|$ (3.33)
4. $\|\alpha A\| = |\alpha| \|A\|$ (3.34)

Corresponding to the vector norm definitions given in Chapter 2, the following matrix norms may be used:

1. $\|A\|_1 = \max_j \left(\sum_{i=1}^{n} |a_{ij}| \right)$ (3.35)

2. $\|A\|_2 = \left[\max_{\|x\|=1} \{ \bar{x}^{\mathrm{T}} \bar{A}^{T} A x \} \right]^{1/2}$ (3.36)

   where the overbar denotes complex conjugate. This is the euclidean matrix norm. Note that this definition does not indicate how this norm is actually computed. For that we will appeal to the definitions of eigenvalues and singular values in the next chapter.

3. $\|A\|_\infty = \max_i \left(\sum_{j=1}^{n} |a_{ij}| \right)$ (3.37)

4. $\|A\|_F = \left[\mathrm{tr}\left(\bar{A}^{\mathrm{T}} A \right) \right]^{1/2} = \left[\sum_{i=1}^{m} \sum_{j=1}^{n} |a_{ij}|^2 \right]^{1/2}$ (3.38)

trace(A)

where tr(·) denotes the trace[M] of the matrix argument, which is the sum of the elements on the diagonal. This is called the *Frobenius norm*.

Regardless of which norm we choose, we say that the linear operator A is *bounded* if there exists a finite constant κ such that $\|Ax\| \le \kappa \|x\|$ for all vectors $\boldsymbol{x}$ in the space.

3.2.2 Adjoint Operators

'(single apostrophe denotes transpose)

We have used the transpose[M] of several matrices throughout the definitions of such quantities as inner products and norms. The transpose, however, is a special kind of operator, which satisfies the following definition:

Adjoint Operator: The *adjoint* of linear operator A is denoted A^* and must satisfy the relationship

$$\langle Ax, y \rangle = \langle x, A^* y \rangle \tag{3.39}$$

for all vectors $\boldsymbol{x}$ and $\boldsymbol{y}$.

The operator A^* may itself be represented by a matrix. For matrix representations of finite-dimensional operators, A^* is equal to the complex conjugate transpose of matrix A, $A^* = \overline{A}^{\mathrm{T}}$ *in an orthonormal basis*. If A is real, therefore, the representation of A^*, in an orthonormal basis, is simply equal to the matrix A^{T}.

Sometimes the adjoint of an operator (matrix) is the same as the operator (matrix) itself, in which case it is called a *hermitian* operator (matrix), and the operator is referred to as *self-adjoint*. Note that if the basis in which an operator is represented is not orthonormal, then a symmetric matrix representation does *not* imply a self-adjoint operator. If a hermitian matrix is actually real, then it is *symmetric*, and $A = A^{\mathrm{T}}$.

3.3 Simultaneous Linear Equations

The solution of systems of simultaneous linear algebraic equations is an important application of spaces and operator theory. Undoubtedly the reader has had some exposure to systems of simultaneous linear equations before, and has some experience solving such systems using methods such as gaussian elimination or Cramer's rule. It is not our purpose to discuss such numerical computations but rather to consider the solution of simultaneous algebraic equations from a vector space viewpoint. This will provide the geometric insight needed to explain certain phenomena observed in engineering, such as the existence of multiple solutions to a problem or the nonexistence of solutions. At the end of this section, we will present examples that will lead to important conclusions about properties of systems that are not discussed until later in the book, e.g., controllability, observability, and estimation.

To begin, consider a set of m equations in terms of n unknown quantities, $x_1, \ldots, x_n$:

$$\begin{aligned} a_{11}x_1 + a_{12}x_2 + \cdots + a_{1n}x_n &= y_1 \\ a_{21}x_1 + a_{22}x_2 + \cdots + a_{2n}x_n &= y_2 \\ &\vdots \\ a_{m1}x_1 + a_{m2}x_2 + \cdots + a_{mn}x_n &= y_m \end{aligned} \tag{3.40}$$

By writing this set of equations in the form

$$\begin{bmatrix} a_{11} & a_{12} & \cdots & a_{1n} \\ a_{21} & a_{22} & \cdots & a_{2n} \\ \vdots & \vdots & \ddots & \vdots \\ a_{m1} & a_{m2} & \cdots & a_{mn} \end{bmatrix} \begin{bmatrix} x_1 \\ x_2 \\ \vdots \\ x_n \end{bmatrix} = \begin{bmatrix} y_1 \\ y_2 \\ \vdots \\ y_m \end{bmatrix}$$

we can express this relationship as $Ax = y$, where $x \in \Re^n$, and $y \in \Re^m$, and the matrix A is considered to be a linear operator $A: \Re^n \to \Re^m$.

As we have said, we wish to investigate the circumstances of when, if, and how many solutions exist for this system of equations. By interpreting the system as a linear operator, we can rephrase the question, "Is the given vector $\boldsymbol{y}$ in the range space R(A)?" We can use yet another notation to clarify this interpretation of the problem. Suppose the ith column of the matrix A is considered to be a vector $\boldsymbol{a}_i$ from the space of m-tuples $\Re^m$. Then, the above notation becomes

$$Ax = \begin{bmatrix} \\ \boldsymbol{a}_1 \\ \\ \end{bmatrix} x_1 + \begin{bmatrix} \\ \boldsymbol{a}_2 \\ \\ \end{bmatrix} x_2 + \cdots + \begin{bmatrix} \\ \boldsymbol{a}_n \\ \\ \end{bmatrix} x_n = \begin{bmatrix} \\ \boldsymbol{y} \\ \\ \end{bmatrix}$$

For equality to hold, then, the vector $\boldsymbol{y}$ would have to be a linear combination of the columns of A. If this is the case, then $\boldsymbol{y}$ will be linearly dependent on the columns of A. The problem thus reduces to a test of linear independence of vectors.

We can perform this test of independence by forming the new matrix $W = [A \mid \boldsymbol{y}]$ by appending[M] $\boldsymbol{y}$ onto A as its rightmost column. If indeed $\boldsymbol{y}$ is

```
:(colon denotes
matrix append)
```

dependent on the columns of A, then we should have $r(A) = r(W)$, because y will not have added any new "directions" to the space already spanned by the columns of A. Then, of course, $y \in R(A)$. Otherwise, i.e., if y were linearly independent of the columns of A, then $r(W)$ would be equal to $r(A)+1$. In summary, then, we can say that

$r(A) = r(W)$ if *at least* one solution exists to $Ax = y$

$r(A) \neq r(W)$ if there are no solutions to $Ax = y$

3.3.1 Multiple and Nonexistent Solutions

Note that in the first statement above, we have emphasized the fact that if $r(A) = r(W)$, then there is not only a solution, but there may be more than one solution. In fact, if $r(A) = r(W)$, there is either a unique solution or else there are an infinite number of solutions. This situation depends on the relationship between $r(A) \left[= r(W) \right]$ and the dimension of the space $\Re^n$, which of course is n.

First, if $r(A) = n$, then the solution to the system $Ax = y$ will be *unique*. This is because there are exactly n columns in the matrix A, and if $r(A) = n$, then all the columns are linearly independent. This set of n columns therefore forms a basis for $\Re^n$. Knowing that $y \in R(A)$, then we know from Chapter 2 that the representation of y in this basis is unique. The representation will be exactly the set of coefficients $\{x_i\}$ that we are seeking.

If, on the other hand, $r(A) < n$, then there are fewer than n columns of A that are linearly independent, i.e., the columns of A form a linearly dependent set. The set $\{a_i\}$ still spans the subspace in which y lies, because $r(A) = r(W)$, but because there are too many a_i's to form a basis, the representation of y in the set $\{a_i\}$ is not unique.

If the solution is not unique, one may ask why the set of possible solutions is infinite and not finite. To answer this question, recall that $q(A) = n - r(A)$. If we have $r(A) < n$, then our result is that $q(A) > 0$. The significance of this is that the nullity is the same as the dimension of the null space of a matrix. Therefore, $q(A) > 0$ implies that the matrix (operator) A has a nontrivial null space. That is, there exists at least one nonzero vector x_0 such that $Ax_0 = 0$. This being true, we can always multiply x_0 by an arbitrary non-zero scalar, and the result will still be in the null space of A. We are then allowed to add to any solution of $Ax = y$ such a vector in N(A), and a different solution will result. That is,

$$A(x + \alpha x_0) = Ax + \alpha Ax_0 = Ax + 0 = Ax = y$$

In general, we can think of solutions of algebraic systems of equations in a way similar to the solutions of nonhomogeneous differential equations: Each complete solution x consists of two parts: the *particular part* x_p for which $Ax_p = y$ and the *homogeneous part* for which $Ax_h = 0$. The general solution is of the form $x = x_p + x_h$. Often, the nullity of A is even greater than one, so that there is an entire multidimensional subspace N(A) from which homogeneous terms can be found.

Finding the Null Space

In practice, when solving such a set of linear algebraic equations, the null space of the matrix A can be easily determined by computer[M] using an analysis package such as MATLAB. However, it is useful to learn the "manual" procedure for finding this null space and for solving the equation $Ax = y$ in general.[M]

`null(A)`

`\(backslash)`

To begin, we will determine the row-reduced echelon form of the matrix A. The *row-reduced echelon form*[M] is a simplification of the matrix through a series of so-called "elementary row operations," which simplify the matrix but leave the null space intact. The elementary row operations that are allowed are:

`rref(A)`

1. Multiply any row by a nonzero scalar.
2. Interchange any two rows.
3. Add a multiple of one row to another row.

The goal of performing these operations is first to reduce the matrix to an upper triangular matrix, forcing zeros to appear in the portion of the matrix below the main diagonal. (The diagonal of a nonsquare matrix is simply the sequence of elements, a_{11}, a_{22}, ...). After this is performed, a matrix of the following shape will result:

$$\begin{bmatrix} 0 & \boxed{\times} & \times & \times & \times & \times \\ 0 & 0 & 0 & \boxed{\times} & \times & \times \\ 0 & 0 & 0 & 0 & \boxed{\times} & \times \\ 0 & 0 & 0 & 0 & 0 & 0 \end{bmatrix}$$

In this matrix, the boxes indicate the first nonzero element in each row. This is called the *pivot* element, and the number of pivot elements (in this case, three) is the rank of the matrix. This matrix has the same null space as the original matrix, but this null space is easier to determine. (In fact, each subset of rows in the row-reduced echelon form spans the same space as the corresponding rows did originally, but if row interchange operations are used, the columns of the reduced matrix do not necessarily span the same "column space.") Hence, if this matrix were the A-matrix in a problem $Ax = y$, then $n = 6$, and the dimension of the null space would be $q(A) = n - r(A) = 6 - 3 = 3$. Because the pivot elements occur in columns two, four, and five, the variables x_2, x_4, and x_5 will be called *pivot* variables. The other variables, x_1, x_3, and x_6, are referred to as *free* variables.

null(A)

To generate a basis[M] for the N(A), we generate a linearly independent set of free variable vectors and solve the system of row-reduced equations for the pivot variables. The easiest way to generate this independent set is to choose one free variable as a '1,' the rest '0's, and solve for the pivot variables. This set of variables is a basis vector for N(A). We then set the next free variable equal to '1,' the others to '0,' and again solve, getting another basis vector for the null space. This process continues until we get $q(A)$ such vectors, after which we have found the basis for N(A). In common terminology, we will have found N(A) itself.

It is helpful to continue this discussion with the aid of an example. Suppose we are interested in the particular and homogeneous solutions of the equation

$$\begin{bmatrix} 1 & -1 & 2 \\ -1 & 2 & 0 \end{bmatrix} \begin{bmatrix} x_1 \\ x_2 \\ x_3 \end{bmatrix} = \begin{bmatrix} 8 \\ 2 \end{bmatrix}$$

We can find the ranks of matrices A and W by constructing W and row reducing it first:

$$W = [A \mid y] = \left[\begin{array}{ccc|c} 1 & -1 & 2 & 8 \\ -1 & 2 & 0 & 2 \end{array}\right]$$

The process of row reduction is fairly easy in this example. If we replace row 2 by the sum of row 1 and row 2, we arrive at the reduced form:

$$W' = \left[\begin{array}{ccc|c} \boxed{1} & -1 & 2 & 8 \\ 0 & \boxed{1} & 2 & 10 \end{array}\right] \tag{3.41}$$

We immediately see that there are two pivot elements (two nonzero rows), and hence, $r(W) = 2$. It is also apparent that the result would be the same if we disregard the right-most column, giving $r(A) = 2$ as well. Therefore, $r(A) = r(W) = 2$, and there must be at least one solution. Furthermore, $n = 3$, so $r(A) < n$, and there will be an infinite number of solutions, where

$\dim(\mathrm{N}(A)) = q(A) = n - r(A) = 3 - 2 = 1$. The free variable is recognized as x_3, so to generate a basis for the null space, consider the homogeneous situation

$$\left[\begin{array}{ccc|c} 1 & -1 & 2 & 0 \\ 0 & 1 & 2 & 0 \end{array}\right] \tag{3.42}$$

or think of this matrix as the set of equations

$$\begin{aligned} x_1 + (-1)x_2 + 2x_3 &= 0 \\ x_2 + 2x_3 &= 0 \end{aligned}$$

Set $x_3 = 1$, then solve for the pivot variables: $x_2 = -2$, and $x_1 = -4$. Thus, the vector $\boldsymbol{x}_0 = [-4 \quad -2 \quad 1]^{\mathrm{T}}$ is a basis vector for N(A), as is any scalar multiple of $\boldsymbol{x}_0$.

To complete the problem, we must now find the particular part of the general solution. Toward this end, we can pursue the row reduction of the W matrix further. Consider taking (3.41) and performing additional elementary row operations until all elements above any pivot elements become zero. This is the "complete" row reduction and can also be used to determine the existence of any solutions. Specifically for (3.41), we replace row 1 by the sum of row 1 and row 2. This gives

$$W'' = \left[\begin{array}{ccc|c} 1 & 0 & 4 & 18 \\ 0 & 1 & 2 & 10 \end{array}\right]$$

Because x_3 is a free variable, a solution to this equation, rewritten as

$$\begin{aligned} x_1 + 4x_3 &= 18 \\ x_2 + 2x_3 &= 10 \end{aligned}$$

is

$$\begin{aligned} x_1 &= 18 - 4x_3 \\ x_2 &= 10 - 2x_3 \end{aligned}$$

From this expression, we see that the homogeneous solution is "shifted" in space by a vector $\boldsymbol{x}_p = [18 \quad 10 \quad 0]^{\mathrm{T}}$, and this is the particular solution to the original problem, i.e.,

$$x = x_p + x_0 = \begin{bmatrix} 18 \\ 10 \\ 0 \end{bmatrix} + \alpha \begin{bmatrix} -4 \\ -2 \\ 1 \end{bmatrix}$$

Two further examples are provided below to further illustrate the possible situations:

Example 3.8: Unique and Nonexistent Solutions

Find every solution to the systems of equations, if they exist:

$$a) \begin{bmatrix} 1 & 2 & 3 \\ 2 & 1 & 2 \\ 3 & 2 & 1 \end{bmatrix} x = \begin{bmatrix} 4 \\ 7 \\ 1 \end{bmatrix} \quad \text{and} \quad b) \begin{bmatrix} 1 & 2 \\ 3 & 4 \\ 5 & 6 \end{bmatrix} x = \begin{bmatrix} 2 \\ 3 \\ -4 \end{bmatrix}$$

Solution:

a) We construct

$$W = \left[\begin{array}{ccc|c} 1 & 2 & 3 & 4 \\ 2 & 1 & 2 & 7 \\ 3 & 2 & 1 & 1 \end{array}\right]$$

for which complete row reduction (including elimination of nonzero values above pivot variables) produces

$$W'' = \left[\begin{array}{ccc|c} 1 & 0 & 0 & 2.125 \\ 0 & 1 & 0 & -4.5 \\ 0 & 0 & 1 & 3.625 \end{array}\right]$$

Because there are three pivot rows in both A'' and W'', then $r(A) = r(W) = 3 = n$, implying that there is a unique solution. This solution is apparent in the complete row reduction as $x = \begin{bmatrix} 2.125 & -4.5 & 3.625 \end{bmatrix}^T$.

b) We have a complete row reduction of

$$W'' = \left[\begin{array}{cc|c} 1 & 0 & 0 \\ 0 & 1 & 0 \\ 0 & 0 & 1 \end{array}\right]$$

We can see that the A'' portion (to the left of the partition) of this matrix has two pivot rows, and hence $r(A)=2$, while the whole matrix has three pivot rows, so $r(W)=3$. Thus $r(A)\neq r(W)$, and there is no solution to this problem.

In the special case that A is square $(n\times n)$ and $r(A)=r(W)=n$, then the solution is usually found with the much simpler technique $\boldsymbol{x}=A^{-1}\boldsymbol{y}$. This is possible, of course, because the full rank of the square A-matrix implies that its inverse exists. In the case that A is not square, we must consider two possible cases: either $m<n$ or $m>n$. These situations are addressed next.

More Unknowns Than Equations

In the case that $m<n$, we have more unknown elements of the vector $\boldsymbol{x}$ than we have equations to determine these unknowns. This is known as an *underdetermined* system and requires some additional analysis. When $m<n$, there is no unique solution, because we know that $r(A)\leq\min(m,n)$ and $(\min(m,n)=m)<n$. Thus, when solving for one solution to $A\boldsymbol{x}=\boldsymbol{y}$, we have the option of determining which of the infinitely many solutions to determine. We will seek the solution with the minimum norm from among all possibilities. Minimum-norm solutions are often sought when vector $\boldsymbol{x}$ represents a physical quantity with a cost associated with it, such as a control input, kinetic energy, or fuel consumption.

We will actually minimize one-half of the *squared* norm of $\boldsymbol{x}$, which will have a minimum at the same values as the norm itself. To do this, we introduce the concept of constrained minimization and cast the problem as follows: Find, among all $\boldsymbol{x}$, the value of $\boldsymbol{x}$ that minimizes $\frac{1}{2}\boldsymbol{x}^{\mathrm{T}}\boldsymbol{x}$ subject to the constraint that $A\boldsymbol{x}=\boldsymbol{y}$, or, equivalently, $A\boldsymbol{x}-\boldsymbol{y}=0$. To minimize a function subject to an equality constraint, a variable called a *LaGrange multiplier* is used. Using the LaGrange multiplier γ, a function called a hamiltonian is formed:

$$H=\frac{1}{2}\boldsymbol{x}^{\mathrm{T}}\boldsymbol{x}+\gamma^{\mathrm{T}}(A\boldsymbol{x}-\boldsymbol{y})$$

The $(m\times 1)$ vector LaGrange multiplier serves to create a scalar secondary optimization criterion from the vector expression $A\boldsymbol{x}-\boldsymbol{y}$. Hence, the hamiltonian is a scalar quantity that we wish to minimize over all possible $\boldsymbol{x}$ and γ (because γ is as yet undetermined). Therefore, because we are minimizing over two variables, we must use two derivatives and set each equal to zero:

$$\frac{\partial^{\mathrm{T}}H}{\partial\boldsymbol{x}}=\boldsymbol{x}+A^{\mathrm{T}}\gamma\equiv 0 \tag{3.43}$$

and

$$\frac{\partial^{\mathrm{T}} H}{\partial \gamma} = A\boldsymbol{x} - \boldsymbol{y} \equiv 0 \qquad (3.44)$$

Note that the second of these partial derivatives returns exactly the original constraint equation. This is expected, of course, since that constraint must always be maintained. Rearranging (3.43) and multiplying both sides by the quantity A, we get

$$A\boldsymbol{x} = -AA^{\mathrm{T}}\gamma$$

However, Equation (3.44) provides that $A\boldsymbol{x} = \boldsymbol{y}$, so we have

$$\boldsymbol{y} = -AA^{\mathrm{T}}\gamma$$

or

$$\gamma = -\left(AA^{\mathrm{T}}\right)^{-1}\boldsymbol{y} \qquad (3.45)$$

Substituting this expression back into (3.43), we have the solution

$$\boldsymbol{x} = A^{\mathrm{T}}\left(AA^{\mathrm{T}}\right)^{-1}\boldsymbol{y} \qquad (3.46)$$

This is the "shortest" vector $\boldsymbol{x}$ such that $A\boldsymbol{x} = \boldsymbol{y}$.

We should note that in (3.46), the inverse of the matrix product (AA^{T}) will only be guaranteed in the event that A is of full rank—in this case, if $r(A) = m$. In the event that $r(A) < m$, then some of the original equations to be solved are linearly dependent on the others and are therefore redundant. One can rectify that situation by simply discarding the linearly dependent equations, one at a time, until $r(A) = m$.

`pinv(A)`

The quantity $A^{\mathrm{T}}(AA^{\mathrm{T}})^{-1}$ in (3.44) is called a *pseudoinverse*^{M} of A (specifically, the unweighted Moore-Penrose pseudoinverse), denoted A^{+}. Pseudoinverses are special classes of generalized inverses and are discussed in more detail in Appendix A. Pseudoinverses are different from true inverses in that they are not unique, so we should keep in mind that they are providing only one of possibly many solutions. For example, for any nonsingular symmetric matrix B, the pseudoinverse solution

$$x = A_B^+ y \triangleq BA^T (ABA^T)^{-1} y$$

will also satisfy the equation $Ax = y$.

More Equations Than Unknowns

We will now consider the situation in which $m > n$. This occurs when there are more equations than unknowns, which is referred to as an *overdetermined* system. This is the more unusual of the two situations because in order for a solution to exist, we are asking that the vector y, which has relatively many components, be a linear combination of relatively few columns of the A-matrix. Nevertheless, if we find that $r(A) = r(W)$, then a solution must exist. We will approach the search for a solution differently from the underdetermined case, but we will find a solution with similar qualities.

In this case, we will define an "error" signal:

$$e = y - Ax \tag{3.47}$$

Now we will seek a solution to the problem of minimizing one-half the squared-norm of e over all possible vectors x, i.e., find x to minimize $\frac{1}{2}e^T e$. If indeed a solution exists, then this error should, of course, be zero.

To perform the minimization, we first substitute in the definition of the error, then simplify:

$$\begin{aligned} \frac{1}{2}e^T e &= \frac{1}{2}(y - Ax)^T (y - Ax) \\ &= \frac{1}{2}\left(y^T - x^T A^T\right)(y - Ax) \\ &= \frac{1}{2}\left(y^T y - x^T A^T y - y^T Ax + x^T A^T Ax\right) \end{aligned} \tag{3.48}$$

The two middle terms in the parentheses above are transposes of each other, but they are also scalars. They are, therefore, equal. Therefore

$$\frac{1}{2}e^T e = \frac{1}{2}\left(y^T y - 2x^T A^T y + x^T A^T Ax\right)$$

Taking a derivative of this and setting equal to zero,

$$\frac{\partial^T}{\partial x}\left(\frac{1}{2}e^T e\right) = \frac{1}{2}\left(-2A^T y + 2A^T Ax\right) = 0$$

so

$$A^{\mathrm{T}} y = A^{\mathrm{T}} A x$$

or

$$x = \left(A^{\mathrm{T}} A\right)^{-1} A^{\mathrm{T}} y \tag{3.49}$$

One can easily notice the similarity between this result and (3.46). In fact, the expression $\left(A^{\mathrm{T}} A\right)^{-1} A^{\mathrm{T}}$ is another type of pseudoinverse.

Example 3.9: Pseudoinverse Solution

Find the solution to

$$\begin{bmatrix} 2 & 2 \\ 1 & 2 \\ 1 & 0 \end{bmatrix} x = \begin{bmatrix} 4 \\ 3 \\ 1 \end{bmatrix}$$

Solution:

One can check that $r(A) = r(W)$, which is also obvious by inspection, since the vector y is easily seen to be the sum of the two columns of A. However, using the pseudoinverse as an exercise, we obtain

$$x = \left(A^{\mathrm{T}} A\right)^{-1} A^{\mathrm{T}} y = \begin{bmatrix} 1/3 & -1/3 & 2/3 \\ 0 & 1/2 & -1/2 \end{bmatrix} \begin{bmatrix} 4 \\ 3 \\ 1 \end{bmatrix} = \begin{bmatrix} 1 \\ 1 \end{bmatrix}$$

To further demonstrate that this is an exact solution, we can compute the error:

$$e = y - Ax = \begin{bmatrix} 4 \\ 3 \\ 1 \end{bmatrix} - \begin{bmatrix} 2 & 2 \\ 1 & 2 \\ 1 & 0 \end{bmatrix} \begin{bmatrix} 1 \\ 1 \end{bmatrix} = \begin{bmatrix} 0 \\ 0 \\ 0 \end{bmatrix}$$

which, of course, has zero norm.

Example 3.10: Controllability of Discrete-Time Systems

In this example we will consider a discrete-time linear system described by the state equations

$$\begin{aligned} \mathbf{x}(k+1) &= A\mathbf{x}(k) + Bu(k) \\ y(k) &= C\mathbf{x}(k) + Du(k) \end{aligned} \tag{3.50}$$

where $\mathbf{x}(k)$ denotes an n-dimensional state vector for a system at time step k and $u(k)$ is a scalar input term. Under what conditions will it be possible to force the state vector to be zero, $\mathbf{x}(k) = 0$, for some k, by applying a particular sequence of inputs $u(0), u(1), \ldots, u(k-1)$? The result will be important for the analysis of controllability in Chapter 8.

Solution:

We will first "execute" a few steps in the recursion of this equation:

At time $k = 0$:

$$\mathbf{x}(1) = A\mathbf{x}(0) + Bu(0) \tag{3.51}$$

At time $k = 1$:

$$\begin{aligned} \mathbf{x}(2) &= A\mathbf{x}(1) + Bu(1) \\ &= A\left[A\mathbf{x}(0) + Bu(0)\right] + Bu(1) \\ &= A^2\mathbf{x}(0) + ABu(0) + Bu(1) \end{aligned} \tag{3.52}$$

At time $k = 2$:

$$\begin{aligned} \mathbf{x}(3) &= A\mathbf{x}(2) + Bu(2) \\ &= A\left[A^2\mathbf{x}(0) + ABu(0) + Bu(1)\right] + Bu(2) \\ &= A^3\mathbf{x}(0) + A^2Bu(0) + ABu(1) + Bu(2) \end{aligned} \tag{3.53}$$

We can see that at any give time k, in order to get $\mathbf{x}(k) = 0$, we will have

$$\begin{aligned} 0 &= \mathbf{x}(k) \\ &= A^k\mathbf{x}(0) + A^{k-1}Bu(0) + A^{k-2}Bu(1) + \cdots + ABu(k-2) + Bu(k-1) \end{aligned} \tag{3.54}$$

This can be rearranged as

$$-A^k \mathbf{x}(0) = \left[B \;\vdots\; \cdots \;\vdots\; A^{k-2}B \;\vdots\; A^{k-1}B \right] \begin{bmatrix} u(k-1) \\ \vdots \\ u(1) \\ u(0) \end{bmatrix} \tag{3.55}$$

If we introduce the notation

$$P \stackrel{\Delta}{=} \left[B \;\vdots\; \cdots \;\vdots\; A^{k-2}B \;\vdots\; A^{k-1}B \right]$$

and

$$\mathbf{u} \stackrel{\Delta}{=} \begin{bmatrix} u(k-1) \\ \vdots \\ u(1) \\ u(0) \end{bmatrix}$$

then we have

$$-A^k \mathbf{x}(0) = P\mathbf{u}$$

In this expression, we note that $-A^k \mathbf{x}(0) \in \Re^n$, $P \in \Re^{n \times k}$, and $\mathbf{u} \in \Re^k$. This is now in the form of a simultaneous system of linear algebraic equations to solve. If $-A^k \mathbf{x}(0) \in \mathrm{R}(P)$, there exists a solution. Because we are allowing arbitrary values for both A and $\mathbf{x}(0)$, the only way to guarantee the existence condition is for $\mathrm{R}(P) = \Re^n$, or, equivalently, for $r(P) = n$. Because the P-matrix is $n \times k$, this will require that $k \geq n$, i.e., we must wait for at least n samples to be computed, but in no case will $r(P) > n$. Note that we do not necessarily require that the solution $\mathbf{u}$ be unique, only that at least one solution exists.

A system that satisfies this criterion is said to be *controllable*. (Though there is more to controllability than just this. See Chapter 8.)

Example 3.11: Observability of Discrete-Time Systems

In this example, we look at a "dual" problem. Instead of finding an input that results in a particular state, we wish to determine an initial state given a sequence of inputs and outputs. Suppose after k iterations of Equation (3.50), we have recorded the sequence of applied inputs $u(k), \ldots, u(0)$, and the measured sequence of responses $y(k), \ldots, y(0)$. Under what conditions will it be possible to determine the value of the unknown initial state vector $\mathbf{x}(0)$?

Solution:

As we did in the previous example, we begin by listing several steps (up to $k-1$) in the recursive execution of the state equations in (3.50).

$$\begin{aligned}
y(0) &= Cx(0) + Du(0) \\
y(1) &= Cx(1) + Du(1) \\
&= C[Ax(0) + Bu(0)] + Du(1) \\
&= CAx(0) + CBu(0) + Du(1)
\end{aligned}$$

$$\begin{aligned}
y(2) &= Cx(2) + Du(2) \\
&= C\left[A^2x(0) + ABu(0) + Bu(1)\right] + Du(2) \\
&= CA^2x(0) + CABu(0) + CBu(1) + Du(2) \\
&\vdots \\
y(k-1) &= CA^{k-1}x(0) + CA^{k-2}Bu(0) + \cdots + CBu(k-2) + Du(k-1)
\end{aligned}$$

Rearranging these equations into vector-matrix form,

$$\begin{bmatrix} y(0) \\ y(1) \\ y(2) \\ \vdots \\ y(k-1) \end{bmatrix} = \begin{bmatrix} C \\ CA \\ CA^2 \\ \vdots \\ CA^{k-1} \end{bmatrix} x(0) + \begin{bmatrix} D & 0 & 0 & \cdots & 0 \\ CB & D & 0 & \cdots & 0 \\ CAB & CB & D & \cdots & 0 \\ \vdots & & & \ddots & \\ CA^{k-2}B & & \cdots & & D \end{bmatrix} \begin{bmatrix} u(0) \\ u(1) \\ u(2) \\ \vdots \\ u(k-1) \end{bmatrix}$$

The term containing the large matrix and the input terms $u(\cdot)$ can be moved to the left side of the equal sign and combined with the vector containing the y's to give a new vector called Ψ_{k-1}:

$$\Psi_{k-1} \triangleq \begin{bmatrix} y(0) \\ y(1) \\ y(2) \\ \vdots \\ y(k-1) \end{bmatrix} - \begin{bmatrix} D & 0 & 0 & \cdots & 0 \\ CB & D & 0 & \cdots & 0 \\ CAB & CB & D & \cdots & 0 \\ \vdots & & & \ddots & \\ CA^{k-2}B & & \cdots & & D \end{bmatrix} \begin{bmatrix} u(0) \\ u(1) \\ u(2) \\ \vdots \\ u(k-1) \end{bmatrix} = \begin{bmatrix} C \\ CA \\ CA^2 \\ \vdots \\ CA^{k-1} \end{bmatrix} x(0) \tag{3.56}$$

Let

$$Q \triangleq \begin{bmatrix} C \\ CA \\ CA^2 \\ \vdots \\ CA^{k-1} \end{bmatrix} \tag{3.57}$$

Then in Equation (3.56), we note that $\Psi_{k-1} \in \Re^k$, $Q \in \Re^{k \times n}$, and $\boldsymbol{x}(0) \in \Re^n$. So to solve the equation

$$\Psi_{k-1} = Q\boldsymbol{x}(0)$$

for the initial condition $\boldsymbol{x}(0)$, we must have $\Psi_{k-1} \in \mathrm{R}(Q)$. Unlike in the previous example, we want a unique solution in this example, because the system can only have started in a single initial state. The condition for such a unique solution is

$$r([Q \mid \Psi_{k-1}]) = r(Q) = n \tag{3.58}$$

which, of course, will again require that $k \geq n$. If we take $k = n$ and if $r(Q) = n$, then because $\Psi_{n-1} \in \Re^n$, we will have $r([Q \mid \Psi_{k-1}]) = r(Q)$. In that situation, then, the critical question is whether $r(Q) = n$ for Q defined in (3.57). A system that satisfies this criterion is said to be *observable*. See Chapter 8.

3.3.2 Least-Squared Error Solutions and Experimental Data Analysis

A by-product of the solution to the overdetermined problem is the solution of a related problem: finding the vector $\boldsymbol{x}$ that is *closest to* being a solution to $\boldsymbol{y} = A\boldsymbol{x}$ where no exact solution exists, i.e., when $r(A) \neq r(W)$. We may wish to know the vector that minimizes the squared-error criterion (3.48). That vector will be the orthogonal projection of $\boldsymbol{y}$ onto the subspace of $\Re^m$ spanned by the columns of matrix A, because $\boldsymbol{y}$ itself does not lie in this subspace. Because the derivation resulting in the pseudoinverse solution of Equation (3.49) does not require that the error $\boldsymbol{e} = \boldsymbol{y} - A\boldsymbol{x}$ be zero, the same formula will serve to find this *least-squared error* solution.

A common situation in which we might seek a least-squared solution is when experimental data is gathered from a physical process. The process might be truly linear, such as the relationship between position and constant velocity, $x = vt + x_0$, or it may be an unknown process that we wish to model as a linear process, such as when we perform linear regression on a set of data points. In either of these situations, inaccuracies in measurements and unmodeled effects will often prevent the set of linear equations from having a solution, according to the criterion $r(A) = r(W)$. Nevertheless, we want a solution and a measure of the error introduced by that solution. Such a solution is given by Equation (3.49) above.

For example, suppose we have control of two experimental parameters a_1 and a_2. Suppose further that we suspect that a physically observed quantity y depends linearly on these two parameters as $a_1x_1 + a_2x_2 = y$. If we perform m experimental trials, changing a_1 and a_2 each time and recording the resulting output y, we will gather the following data:

$$\begin{aligned} a_{11}x_1 + a_{12}x_2 &= y_1 \\ a_{21}x_1 + a_{22}x_2 &= y_2 \\ &\vdots \\ a_{m1}x_1 + a_{m2}x_2 &= y_m \end{aligned}$$

which can be written in vector-matrix notation as

$$A\mathbf{x} = \begin{bmatrix} a_{11} & a_{12} \\ a_{21} & a_{22} \\ \vdots & \vdots \\ a_{m1} & a_{m2} \end{bmatrix} \begin{bmatrix} x_1 \\ x_2 \end{bmatrix} = \begin{bmatrix} y_1 \\ y_2 \\ \vdots \\ y_m \end{bmatrix} = \mathbf{y}$$

If m is large because many experimental trials were performed, then any noise in the measurements will make it extremely unlikely that $r(A) = r(W)$. This would require that a vector of many components be a linear combination of only two vectors, i.e., a member of a two-dimensional space. Even if there is little or no noise in the data, numerical inaccuracies inherent in the computation of matrix rank can lead to the conclusion that there is no exact solution. In such a case, we use the least-squared error solution in (3.49) and the error equation (3.47) to provide an accurate measurement.

Example 3.12: Analyzing Data All at Once

A student in the networks laboratory analyzes an amplifier circuit and determines that, theoretically, the observed output current i_{out} should be proportional to a linear combination of two input voltages, v_x and v_y. She then applies two variable voltage sources as v_x and v_y and records the current i_{out} for each set of inputs. The data she records is shown in the table below:

v_x	-5	-5	-5	-5	-5	-2	-2	-2	...
v_y	-5	-2	0	2	5	2	0	5	...
i_{out}	17.5	-3.5	-17.5	-32	-50	-21	-7	-42	...

v_x	0	0	2	2	2	5	5	5
v_y	2	5	-5	0	5	-5	0	5
i_{out}	-14	-35	42	7	-28	52.5	15	-17.5

Find the unknown gain constants in the following model for the relationships between the variables.

$$i_{out} = g_1 v_x + g_2 v_y \tag{3.59}$$

Solution:

We can model the descriptive equation in vector-matrix form as:

$$i_{out} = \begin{bmatrix} v_x & v_y \end{bmatrix} \begin{bmatrix} g_1 \\ g_2 \end{bmatrix}$$

There are a total of 16 input/output data sets, so we have

$$\boldsymbol{i}_{16\times 1} = V_{16\times 2}\ \boldsymbol{g}_{2\times 1}$$

Using a computer to perform rank tests, we find that $r(V) = 2$, and $r([V \mid \boldsymbol{i}]) = 3$, so there is no exact solution. The best solution that can be found is the least-squared error solution provided by

$$\boldsymbol{g} = \left(V^{\mathrm{T}} V\right)^{-1} V^{\mathrm{T}} \boldsymbol{i} = \begin{bmatrix} 3.4 \\ -7.0 \end{bmatrix} \tag{3.60}$$

(to two significant digits).

Analyzing the error in this result, we find that

$$\|\boldsymbol{e}\| = \|\boldsymbol{i} - V\boldsymbol{g}\| = 3.2$$

To see visually the accuracy of the plane given by Equation (3.59) with the estimated parameters g_i from Equation (3.60), a plot of the estimated plane (3.59) and the data points given in the problem statement are shown in Figure 3.11. Clearly, the parameters g_i give a reasonable fit to the data.

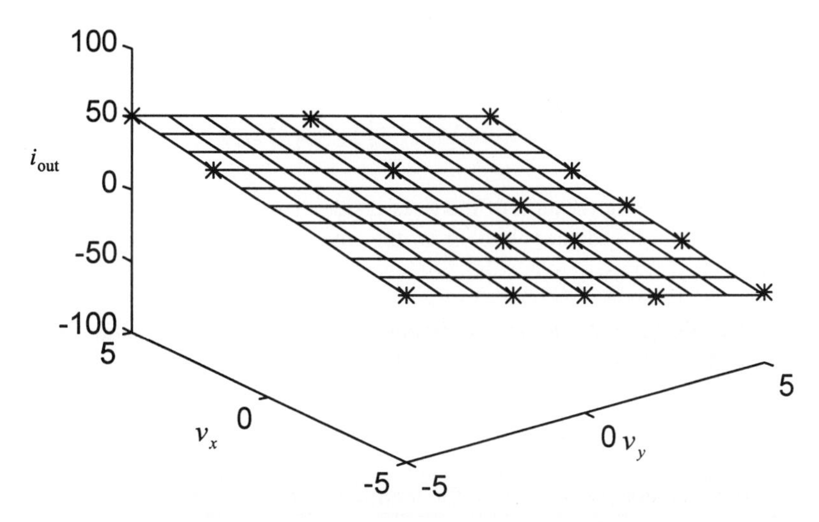

Figure 3.11 Measured data (*) and fitted plane for Example 3.12.

Sequentially Processed Data: Recursive Least-Squares

Note that in the example above, the matrix A will grow larger as m increases, i.e., as more data is taken. Because of this, the experimenter is required to store all the data in a single array and then compute the pseudoinverse on this large array (matrix). If it is then desired that more data points be taken, perhaps because the error $\boldsymbol{e}$ is unacceptably high, the old array must be retrieved, augmented with new rows representing new data, and the pseudoinverse recomputed, disregarding the information provided by the previously performed pseudoinverse. Rather than "waste" this previous calculation, a recursive formulation of the least-squared error solution can be derived.

After some minimal number k of data points has been gathered, then an initial computation will produce a solution $\boldsymbol{x}_k = (A_k^{\mathrm{T}} A_k)^{-1} A_k^{\mathrm{T}} \boldsymbol{y}_k$. If, in the example above, the first $k = n = 2$ trials were performed with linearly independent sets of a-parameters, then a simple inverse will provide an initial solution, $\boldsymbol{x}_k = A_k^{-1} \boldsymbol{y}_k$. When the $(k+1)^{\text{st}}$ data point is collected, we can augment the A-matrix with a new row, denoted a:

$$A_{k+1} = \begin{bmatrix} A_k \\ \hline a \end{bmatrix}$$

Then the pseudoinverse solution can be written as

$$\begin{aligned} x_{k+1} &= \left(A_{k+1}^{\mathrm{T}} A_{k+1}\right)^{-1} A_{k+1}^{\mathrm{T}} y_{k+1} \\ &= \left(\begin{bmatrix} A_k^{\mathrm{T}} & \vdots & a^{\mathrm{T}} \end{bmatrix} \begin{bmatrix} A_k \\ \hline a \end{bmatrix}\right)^{-1} \begin{bmatrix} A_k^{\mathrm{T}} & \vdots & a^{\mathrm{T}} \end{bmatrix} \begin{bmatrix} y_k \\ \hline y_{k+1} \end{bmatrix} \\ &= \left(A_k^{\mathrm{T}} A_k + a^{\mathrm{T}} a\right)^{-1} \left[A_k^{\mathrm{T}} y_k + a^{\mathrm{T}} y_{k+1}\right] \end{aligned} \tag{3.61}$$

For help inverting the sum of matrices inside the parentheses above, we use the following matrix identity known as the matrix inversion lemma (see also Appendix A):

> LEMMA: For compatibly dimensioned matrices B, C, D, and E, the following identity holds:
>
> $$\left(B^{-1} + CDE\right)^{-1} = B - BC\left(EBC + D^{-1}\right)^{-1} EB \tag{3.62}$$

For our situation, we can make the identifications $B^{-1} \Leftrightarrow A_k^{\mathrm{T}} A_k$, $C \Leftrightarrow a^{\mathrm{T}}$, $D \Leftrightarrow I$, and $E \Leftrightarrow a$. Therefore, by direct substitution,

$$x_{k+1} = \left\{\left(A_k^{\mathrm{T}} A_k\right)^{-1} - \left(A_k^{\mathrm{T}} A_k\right)^{-1} a^{\mathrm{T}} \left[a\left(A_k^{\mathrm{T}} A_k\right)^{-1} a^{\mathrm{T}} + 1\right]^{-1} a\left(A_k^{\mathrm{T}} A_k\right)^{-1}\right\} \cdot \left[A_k^{\mathrm{T}} y_k + a^{\mathrm{T}} y_{k+1}\right] \tag{3.63}$$

By comparison with (3.61), we see that the quantity in braces in (3.63) is equal to $(A_{k+1}^{\mathrm{T}} A_{k+1})^{-1}$, so at the $(k+2)^{\text{nd}}$ sample, it will be used as $(A_k^{\mathrm{T}} A_k)^{-1}$ is in (3.61). The quantity in braces has already been computed in the previous iteration, so the only inverse that is actually computed in (3.63) is the term in square brackets in the center. As long as only a single additional row of data is collected in each sample, this bracketed term will be a scalar, thus requiring no additional matrix inversions at all. Similarly, the bracketed term at the end, $[A_k^{\mathrm{T}} y_k + a^{\mathrm{T}} y_{k+1}]$, will become $A_{k+1}^{\mathrm{T}} y_{k+1}$ for the next computation. No matrices will grow in size. Equation (3.63) is thus the *recursive least-squared error* solution and is a convenient way to incorporate new information into previously computed results.

Example 3.13: Recursive Least-Squares

Suppose that we want to find the constants g_1 and g_2 in the equation

$$y = g_1 x_1 + g_2 x_2$$

given the set of measured data

input x_1	x_2	*output* y
x_{11}	x_{21}	y_1
x_{12}	x_{22}	y_2
x_{13}	x_{23}	y_3
x_{14}	x_{24}	y_4
⋮	⋮	⋮

We will demonstrate the algorithm by which the recursive least-squares formulation (3.63) generates estimates of these parameters. Because we have two measured variables, x_1 and x_2, and two unknown parameters, we can form an initial estimate after only two measurements, i.e., by using the first two lines of the table above. If only the first two measurements are used, then

$$\begin{bmatrix} y_1 \\ y_2 \end{bmatrix} = \begin{bmatrix} x_{11} & x_{21} \\ x_{12} & x_{22} \end{bmatrix} \begin{bmatrix} g_1 \\ g_2 \end{bmatrix}$$

has the solution

$$g_0 = \begin{bmatrix} x_{11} & x_{21} \\ x_{12} & x_{22} \end{bmatrix}^{-1} \begin{bmatrix} y_1 \\ y_2 \end{bmatrix} \stackrel{\Delta}{=} A_1 y_1$$

Now define the starting points for the recursive loop that results from (3.63). Assign

$$A := (A_1^{\mathrm{T}} A_1)^{-1} \quad \text{and} \quad B := A_1^{\mathrm{T}} y_1$$

Because the recursion starts with the *third* row of measured data from the table, set $i = 3$ and then execute the pseudocode algorithm shown here:

```
LOOP:
 set a := i^th row of input data, [ x_1i  x_2i ]   ;get new independent variables (input)
 set y := i^th row of output data, y_i             ;get new output data (measurement)
 set A := A − Aa^T(aAa^T+1)^-1 aA                  ;compute term in braces from (3.63)
 set B := B + a^T y                                ;compute bracketed term from (3.63)
 compute x_i = AB                                  ;compute an updated estimate
 i := i + 1                                        ;increment measurement count
 goto LOOP                                         ;continue until all the data is used
```

Note that in this algorithm, the terms referred to as A and B, which are, respectively, the term in braces and the term in brackets in (3.63), are recursively assigned. Note also that the inversion performed in the fourth line is scalar inversion.

3.4 Summary

Again, as in Chapter 2, this chapter has presented some of the mathematical basics necessary for understanding state spaces. Of primary importance here was the concept of a linear operator. Rather than thinking in terms of matrix-vector multiplication as used in matrix theory, we present the linear operator approach. This gives a much more geometrical understanding of some of the basic arithmetic computations we have been performing all along. Without this approach, certain linear system theory and control system concepts cannot be fully understood. Although we are perhaps premature in introducing them, the controllability and observability examples were presented in this chapter for just that purpose: to show how the operator interpretation of linear equations leads to a general test for these properties of state space systems. Solutions of linear equations through numerical methods or gaussian elimination cannot lend this insight.

Other important concepts introduced in this chapter were:

- All linear operators can be written as matrices, and all matrices can be thought of as operators. So although we stress the geometric understanding of linear operators, it should be clear that our old computational tools and habits are still useful.

- Linear operators themselves are written in specific bases, either explicitly defined or understood from their context. These bases may be changed with matrix multiplication operations, i.e., Equation (3.24).

- Operators themselves (equivalently, their matrix representations) form vector spaces and have norms of their own. Adjoint operators are the operator-theoretical manifestations of symmetric and hermitian matrices.

- The process of solving linear algebraic equations can be seen as the search for a preimage of a given vector in the range space of an operator. If the given vector is not in the range space of the operator, least-squared error solutions can be obtained with the projection operator, resulting in a pseudoinverse. In case there are many solutions, a similar pseudoinverse gives the smallest solution.

- Recursive solutions to experimental data analysis problems were presented for two reasons: they will be familiar to readers who continue to Chapter 11, which includes recursive estimators. Also, they are efficient tools for processing real-time data.

Up to this point, the mathematical preliminaries we have discussed have intentionally been kept rather general. Beginning in the next chapter, we will restrict our discussion of vector spaces and linear operators to state spaces and operators that operate on, and into, the same state space. Such operators will then be used to help solve and understand the state space equations developed in Chapter 1.

3.5 Problems

3.1 Determine which of the following operations on vectors are linear operators:

a) $$Ax = x + \begin{bmatrix} 2 \\ 1 \\ -1 \end{bmatrix}, \text{ where } x \in \Re^3$$

b) $$Ax = \begin{bmatrix} 1 \\ 0 \\ 0 \end{bmatrix}, \text{ where } x \in \Re^3$$

c) $$Ax = \int_{-\infty}^{\infty} f(\sigma)x(t-\sigma)d\sigma, \text{ where } f(\cdot) \text{ is a continuous function,}$$

and x is in the vector space of continuous functions.

d) $$Ax = \begin{bmatrix} x_1^2 \\ x_2^2 \\ x_3^2 \end{bmatrix}, \text{ where } x = \begin{bmatrix} x_1 \\ x_2 \\ x_3 \end{bmatrix}$$

e) $$Ax = \begin{bmatrix} \sin x_1 \\ \cos x_2 \\ 0 \end{bmatrix}, \text{ where } x = \begin{bmatrix} x_1 \\ x_2 \\ x_3 \end{bmatrix}$$

f) $$Ax = \begin{bmatrix} x_1 + 2x_2 + 3x_3 \\ x_1 + x_2 + x_3 \\ x_1 \end{bmatrix}, \text{ where } x = \begin{bmatrix} x_1 \\ x_2 \\ x_3 \end{bmatrix}$$

3.2 Show that $\dfrac{d\left(A^{-1}(t)\right)}{dt} = -A^{-1}(t)\dfrac{dA(t)}{dt}A^{-1}(t)$.

3.3 It is known that if $n \times n$ matrices A^{-1} and B^{-1} exist, then $(AB)^{-1} = B^{-1}A^{-1}$. Does $(A+B)^{-1}$ necessarily exist?

3.4 Let A be a linear operator on the space $\Re^3$ that transforms the vectors

$$\{x_1, x_2, x_3\} = \left\{ \begin{bmatrix} 0 \\ 0 \\ 1 \end{bmatrix}, \begin{bmatrix} 0 \\ -1 \\ 2 \end{bmatrix}, \begin{bmatrix} 1 \\ -1 \\ 2 \end{bmatrix} \right\}$$

into vectors

$$\{y_1, y_2, y_3\} = \left\{ \begin{bmatrix} -2 \\ 4 \\ 1 \end{bmatrix}, \begin{bmatrix} 1 \\ 1 \\ 0 \end{bmatrix}, \begin{bmatrix} 0 \\ 3 \\ -1 \end{bmatrix} \right\}$$

respectively, i.e., $Ax_i = y_i$ for $i = 1, 2, 3$.

a) Find the matrix representation of A in the standard basis $\{e_1, e_2, e_3\}$.

b) Find the matrix representation of A in the basis consisting of the vectors $\{x_1, x_2, x_3\}$.

3.5 Consider two bases for space $\Re^3$:

$$\{\boldsymbol{a}_1, \boldsymbol{a}_2, \boldsymbol{a}_3\} = \left\{ \begin{bmatrix} 2 \\ 1 \\ 4 \end{bmatrix}, \begin{bmatrix} 3 \\ -2 \\ -2 \end{bmatrix}, \begin{bmatrix} -4 \\ 2 \\ 1 \end{bmatrix} \right\} \qquad \{\boldsymbol{b}_1, \boldsymbol{b}_2, \boldsymbol{b}_3\} = \left\{ \begin{bmatrix} -2 \\ 3 \\ 1 \end{bmatrix}, \begin{bmatrix} -4 \\ -3 \\ -2 \end{bmatrix}, \begin{bmatrix} 5 \\ -2 \\ 0 \end{bmatrix} \right\}$$

If a linear operator is given in the $\{\boldsymbol{a}_1, \boldsymbol{a}_2, \boldsymbol{a}_3\}$ basis by

$$A = \begin{bmatrix} 8 & -2 & -1 \\ 4 & -2 & -3 \\ 2 & -3 & -3 \end{bmatrix}$$

find the representation for this operator in the basis $\{\boldsymbol{b}_1, \boldsymbol{b}_2, \boldsymbol{b}_3\}$. If we are given a vector $\boldsymbol{x} = \begin{bmatrix} 2 & -1 & -4 \end{bmatrix}^{\mathrm{T}}$ in the basis $\{\boldsymbol{a}_1, \boldsymbol{a}_2, \boldsymbol{a}_3\}$, determine the representation for $\boldsymbol{x}$ in the basis $\{\boldsymbol{b}_1, \boldsymbol{b}_2, \boldsymbol{b}_3\}$.

3.6 Consider the vectors in $\Re^3$:

$$\boldsymbol{x}_1 = \begin{bmatrix} -2 \\ -1 \\ 2 \end{bmatrix} \qquad \boldsymbol{x}_2 = \begin{bmatrix} 1 \\ 1 \\ -1 \end{bmatrix} \qquad \boldsymbol{x}_3 = \begin{bmatrix} 8 \\ 5 \\ -7 \end{bmatrix}$$

and the vectors in $\Re^4$:

$$\boldsymbol{y}_1 = \begin{bmatrix} 1 \\ 3 \\ -1 \\ 0 \end{bmatrix} \qquad \boldsymbol{y}_2 = \begin{bmatrix} 2 \\ 2 \\ 1 \\ 2 \end{bmatrix} \qquad \boldsymbol{y}_3 = \begin{bmatrix} 5 \\ -6 \\ 3 \\ 10 \end{bmatrix}$$

where $T\boldsymbol{x}_i = \boldsymbol{y}_i$, for $i = 1, 2, 3$.

a) Write the matrix of the operator $T: \Re^3 \to \Re^4$ in the basis $\{\boldsymbol{x}_1, \boldsymbol{x}_2, \boldsymbol{x}_3\}$ for $\Re^3$ and the standard basis for $\Re^4$.

b) Write the matrix of the operator T in the standard basis in $\Re^3$ and the standard basis in $\Re^4$.

c) Find the rank and degeneracy of T.

d) Find a basis for the null space of T.

e) Find a basis for the range space of T.

3.7 The matrix of a linear operator $T:\Re^3 \to \Re^3$ given in the standard basis $\{e_1, e_2, e_2\} = \left\{ \begin{bmatrix} 1 \\ 0 \\ 0 \end{bmatrix}, \begin{bmatrix} 0 \\ 1 \\ 0 \end{bmatrix}, \begin{bmatrix} 0 \\ 0 \\ 1 \end{bmatrix} \right\}$ (for both the domain and the range) is

$$T = \begin{bmatrix} 12 & -2 & 1 \\ -3 & -5 & 2 \\ 4 & -6 & -4 \end{bmatrix}$$

Find the matrix for this operator if both bases are

$$\{x_1, x_2, x_3\} = \left\{ \begin{bmatrix} -3 \\ 2 \\ -3 \end{bmatrix}, \begin{bmatrix} -1 \\ 1 \\ -1 \end{bmatrix}, \begin{bmatrix} 3 \\ -2 \\ 2 \end{bmatrix} \right\}$$

3.8 Consider linear operator $A:\mathbf{X} \to \mathbf{Y}$. Use the definition of a subspace to show that null set of A, i.e., the set of all vectors $\{x|\ Ax = 0\}$ is indeed a subspace of $\mathbf{X}$.

3.9 Let P be the plane in $\Re^3$ defined by $2x + 2y + 1z = 0$, and let ℓ be the vector $\ell = \begin{bmatrix} 8 & -2 & 2 \end{bmatrix}^T$. Denote by $A:\Re^3 \to \Re^3$ the projection operator that projects vectors in $\Re^3$ onto the plane P, not orthogonally, but along the vector ℓ (that is, the projections can be pictured as "shadows" of the original vectors onto P, where the light source is at an infinite distance in the direction of ℓ). See Figure P3.9.

a) Find the matrix of operator A in the basis $\{f_1, f_2, f_3\}$, where f_1 and f_2 are any noncollinear vectors in the plane P, and $f_3 = \ell$.

b) Find the matrix of operator A in the standard basis of $\Re^3$.

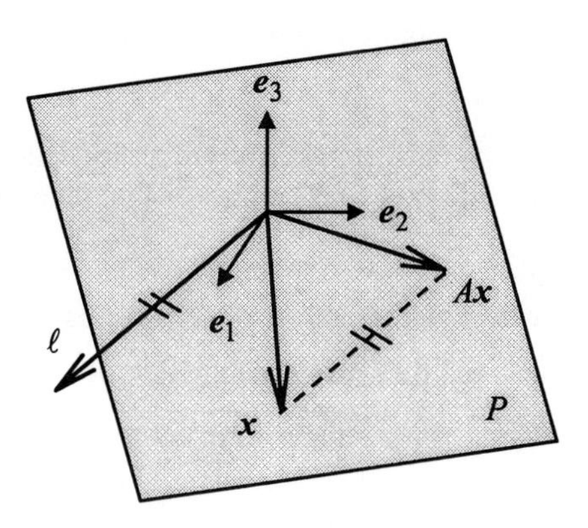

P3.9

3.10 Let $R: \Re^3 \to \Re^3$ denote the three-dimensional rotation operator that rotates vectors by an angle $\pi/3$ radians about the axis $x = y = z$ in $\Re^3$ (counterclockwise as viewed by an observer at $\begin{bmatrix}1 & 1 & 1\end{bmatrix}^T$ looking back at the origin). See Figure P3.10.

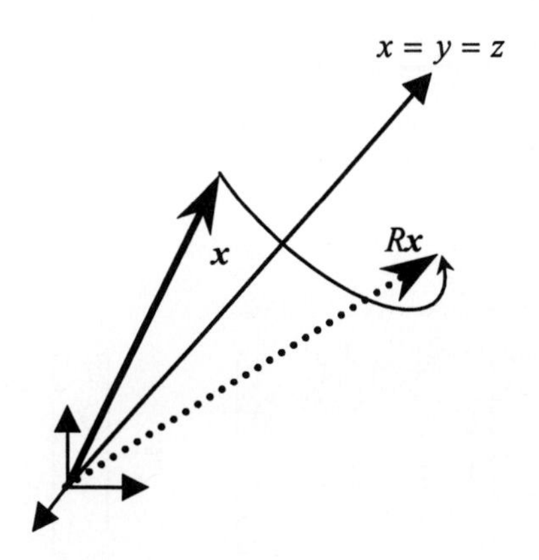

P3.10

a) Show that R is a linear operator.

b) Find the matrix representation for R in the basis $\{f_1, f_2, f_3\}$, which are mutually orthogonal unit vectors, and f_3 is in the direction of the axis of rotation.

c) Find the matrix representation for R in the standard basis.

3.11 Consider the space P of polynomials of order less than four, over the field of reals. Let $D: P \to P$ be a linear operator that takes a vector $\mathbf{p}(s)$ and returns $\mathbf{p}''(s) + 3\mathbf{p}'(s) + 2\mathbf{p}(s)$.

a) Find the matrix representation of D in the basis $\{1, s, s^2, s^3\}$.

b) Find the matrix representation of D in the basis $\{s+1, 2s+3, s^2+5s+4, s^3+s\}$.

3.12 The orthogonal complement $A^{\perp}$ of a linear space A is defined as being the set $\{\mathbf{x} | \langle \mathbf{x}, \mathbf{y}_i \rangle = 0 \text{ for all } \mathbf{y}_i \in A\}$, i.e., every vector in $A^{\perp}$ is orthogonal to every vector in A. Show that $\mathrm{N}^{\perp}(A) = \mathrm{R}(A^*)$ and $\mathrm{R}^{\perp}(A) = \mathrm{N}(A^*)$, where A^* is the adjoint operator of A.

3.13 Let $A = \begin{bmatrix} 2 & -2 \\ 1 & 4 \end{bmatrix}$. Find all 2×2 matrices X such that $AX = 0$.

3.14 For the operators represented by the matrices given in Problem 2.10, find bases for the null space and the range space.

3.15 For the pair of matrices

$$A = \begin{bmatrix} 2 & 3 & 1 & 4 & -9 \\ 1 & 1 & 1 & 1 & -3 \\ 1 & 1 & 1 & 2 & -5 \\ 2 & 2 & 2 & 3 & -8 \end{bmatrix} \qquad \mathbf{y} = \begin{bmatrix} 17 \\ 6 \\ 8 \\ 14 \end{bmatrix}$$

determine all possible solutions to the system $A\mathbf{x} = \mathbf{y}$.

3.16 Determine the values of α such that the system of linear equations

$$\begin{aligned} 3x_1 + 2x_2 + x_3 + 9x_4 &= 4 \\ -4x_1 + 2x_2 - 6x_3 - 12x_4 &= -3 + 7\alpha \\ x_1 + 4x_2 - 3x_3 + 3x_4 &= -2 \end{aligned}$$

has exact solutions, and find the general solution to the system.

3.17 Find the best solution, in the least-squared error sense, to the system of equations:

$$\begin{aligned} -2 &= x_1 - 2x_2 \\ 5 &= x_1 - 2x_2 \\ 1 &= -2x_1 + x_2 \\ -3 &= x_1 - 3x_2 \end{aligned}$$

3.18 Determine whether the three lines intersect at a common point

$$\begin{aligned} x - 2y &= -5 \\ 4x - 1y &= 8 \\ 2x &= 6 \end{aligned}$$

3.19 Find the particular and homogeneous solutions to the system of equations:

$$\begin{bmatrix} 2 & 2 & -3 \\ 4 & 8 & -2 \end{bmatrix} \begin{bmatrix} x_1 \\ x_2 \\ x_3 \end{bmatrix} = \begin{bmatrix} -2 \\ -12 \end{bmatrix}$$

3.20 Determine whether the following system has an exact solution, and if so, find it.

$$\begin{aligned} x_1 + 5x_2 + 7x_3 + 4x_4 &= 7 \\ 2x_1 + 4x_2 + 8x_3 + 5x_4 &= 2 \\ 3x_1 + 3x_2 + 9x_3 + 6x_4 &= -3 \end{aligned}$$

3.21 Use the recursive least-squares method (see page 133) to solve Example 3.12.

3.22 If the *unweighted* pseudoinverse solution

$$x = A^{+}y \triangleq A^{\mathrm{T}}(AA^{\mathrm{T}})^{-1}y$$

minimizes the value of $\frac{1}{2}x^{\mathrm{T}}x$ subject to the constraint $Ax = y$, what does the *weighted* pseudoinverse solution

$$x = A_B^{+}y \triangleq BA^{\mathrm{T}}(ABA^{\mathrm{T}})^{-1}y$$

minimize under the same constraint?

3.6 References and Further Reading

Our perspective on linear operators as given here is a reflection of references [12], [13], and [14]. These sources provide a background on operators in a general setting, rather than simply as matrices. While [12] and [13] are somewhat more mathematical (and [12] includes some material on topology and metric spaces), [14] offers a more applied approach. For computational references, see [6] and [7].

Our approach to projections and their applications is similar to [3] and is motivated by their applications in approximation theory and optimization theory. See [4] for further details on the use of projections in optimization.

The idea of an adjoint operator can be further pursued in [2] and [8]. This will lead to the concept of an *adjoint system*, which has implications in the duality of control systems concepts such as controllability and observability (Chapter 8), and regulators and observers (Chapter 10).

For further information in system identification with input/output data, [10] is a standard reference, but it does not emphasize the recursive least-squares approach that we introduce here. For such a treatment, see [9]. For further details on the various types and applications of generalized inverse matrices (including pseudoinverses), see [1].

The robotics applications can be pursued in [5] and [11]; the latter emphasizes the redundancy problem in robot kinematics, in which an underdetermined system must be solved.

[1] Boullion, Thomas L., and Patrick L. Odell, *Generalized Inverse Matrices*, Wiley Interscience, 1971.

[2] Brockett, Roger W., *Finite Dimensional Linear Systems*, John Wiley & Sons, 1970.

[3] Brogan, William L., *Modern Control Theory*, 3rd edition, Prentice-Hall, 1991.

[4] Dorny, C. Nelson, *A Vector Space Approach to Models and Optimization*, John Wiley & Sons, 1975.

[5] Fu, K. S., R. C. Gonzalez, and C. S. G. Lee, *Robotics: Control, Sensing, Vision, and Intelligence*, McGraw-Hill, 1987.

[6] Gantmacher, Feliks R., *The Theory of Matrices, Vols I and II*, Chelsea Publishing Co., 1959.

[7] Golub, Gene H., and Charles F. Van Loan, *Matrix Computations*, Johns Hopkins University Press, 1989.

[8] Kailath, Thomas, *Linear Systems*, Prentice-Hall, 1980.

[9] Lawson, Charles L., and Richard J. Hanson, *Solving Least Squares Problems*, Prentice-Hall, 1974.

[10] Ljung, Lennart, and Torsten Söderström, *Theory and Practice of Recursive Identification*, MIT Press, 1986.

[11] Nakamura, Yoshihiko, *Advanced Robotics: Redundancy and Optimization*, Addison-Wesley, 1991.

[12] Naylor, Arch W., and George R. Sell, *Linear Operator Theory in Engineering and Science*, Springer-Verlag, 1982.

[13] Shilov, Georgi E., *Linear Algebra*, Dover, 1977.

[14] Strang, Gilbert, *Linear Algebra and Its Applications*, Academic Press, 1980.

4

Eigenvalues and Eigenvectors

Because in this book we are interested in state spaces, many of the linear operations we consider will be performed on vectors within a single space, resulting in transformed vectors within that same space. Thus, we will consider operators of the form $A: X \rightarrow X$. A special property of these operators is that they lead to special vectors and scalars known as *eigenvectors* and *eigenvalues*. These quantities are of particular importance in the stability and control of linear systems. In this chapter, we will discuss eigenvalues and eigenvectors and related concepts such as singular values.

4.1 A-Invariant Subspaces

In any space and for any operator A, there are certain subspaces in which, if we take a vector and operate on it using A, the result remains in that subspace. Formally,

> ***A-Invariant Subspace:*** Let $\mathbf{X}_1$ be a subspace of linear vector space $\mathbf{X}$. This subspace is said to be *A-invariant* if for every vector $x \in \mathbf{X}_1$, $Ax \in \mathbf{X}_1$. When the operator A is understood from the context, then $\mathbf{X}_1$ is sometimes said to be simply "invariant." (4.1)

Finite-dimensional subspaces can always be thought of as lines, planes, or hyperplanes that pass through the origin of their parent space. In the next section, we will consider A-invariant subspaces consisting of lines through the origin, i.e., the *eigenvectors*.

4.2 Definitions of Eigenvectors and Eigenvalues

Recall that a linear operator A is simply a rule that assigns a new vector Ax to an old vector x. In general, operator A can take arbitrary actions on vector x, scaling it and "moving" it throughout the space. However, there are special situations in which the action of A is simply to scale the vector x for some particular vectors x as pictured in Figure 4.1.

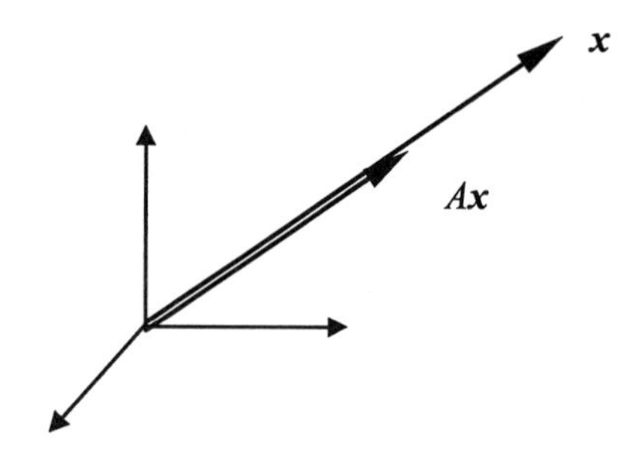

Figure 4.1 Scaling action of an operator acting on vector **x**.

If we denote the scaling factor in this situation by λ, then we have the relationship

$$Ax = \lambda x \tag{4.2}$$

Note that this relationship will not hold for all vectors x and all scalars λ, but only for special specific instances. These are the *eigenvalues*[M] and *eigenvectors*.[M] Notice that the eigenvectors clearly define one-dimensional A-invariant subspaces. In fact, the span of any collection of different eigenvectors will be an A-invariant subspace as well.

```
eig(A)
```

> ***Eigenvalues and Eigenvectors:*** In the relationship $Ax = \lambda x$, the nonzero values of x are *eigenvectors*, and the corresponding values for λ (which *may* be zero) are the *eigenvalues*. (4.3)

Note that Equation (4.2) above also implies that

$$\begin{aligned} 0 &= \lambda x - Ax \\ &= (\lambda I - A)x \end{aligned} \tag{4.4}$$

We can therefore interpret an eigenvector as being a vector from the null space of $\lambda I - A$ corresponding to an eigenvalue λ.

A term that we will use in future chapters when collectively referring to an operator's eigenvalues is *spectrum*. The spectrum of an operator is simply the set of all its eigenvalues.

Example 4.1: Electric Fields

In an *isotropic* dielectric medium, the relationship between the electric field vector E and the electric flux density (also known as the *displacement vector*) is known to be

$$D = \varepsilon E \tag{4.5}$$

where the quantity ε is the *dielectric constant*. However, there are dielectrics that are *anisotropic*, which means that their dielectric constants depend on direction. For these materials,

$$D_i = \sum_{j=1}^{3} \varepsilon_{ij} E_j \tag{4.6}$$

or, in matrix form,

$$\begin{bmatrix} D_1 \\ D_2 \\ D_3 \end{bmatrix} = \begin{bmatrix} \varepsilon_{11} & \varepsilon_{12} & \varepsilon_{13} \\ \varepsilon_{21} & \varepsilon_{22} & \varepsilon_{23} \\ \varepsilon_{31} & \varepsilon_{32} & \varepsilon_{33} \end{bmatrix} \begin{bmatrix} E_1 \\ E_2 \\ E_3 \end{bmatrix}$$

or

$$D = \begin{bmatrix} \varepsilon_{11} & \varepsilon_{12} & \varepsilon_{13} \\ \varepsilon_{21} & \varepsilon_{22} & \varepsilon_{23} \\ \varepsilon_{31} & \varepsilon_{32} & \varepsilon_{33} \end{bmatrix} E \tag{4.7}$$

For a particular medium with a known set of dielectric constants ε_{ij}, $i, j = 1, 2, 3$, find the directions, if any, in which the electric field and the flux density are collinear.

Solution:

If the electric field and the flux density are collinear, then they are proportional to one another, so we could say that $D = \lambda E$ giving

$$\lambda E = \begin{bmatrix} \varepsilon_{11} & \varepsilon_{12} & \varepsilon_{13} \\ \varepsilon_{21} & \varepsilon_{22} & \varepsilon_{23} \\ \varepsilon_{31} & \varepsilon_{32} & \varepsilon_{33} \end{bmatrix} E \tag{4.8}$$

Recognizing that this is an eigenvalue-eigenvector problem, we can say that the electric field and flux density vectors are collinear only in the directions given by the eigenvectors of the matrix of dielectric constants (and the constants of proportionality λ will be the corresponding eigenvalues).

4.3 *Finding Eigenvalues and Eigenvectors*

The first step in determining eigenvalues and eigenvectors is to force $\lambda I - A$ to have a nontrivial null space. The easiest way to do this is to determine the matrix representation for A and set $|\lambda I - A| = 0$. The result for an $n \times n$ A-matrix will be an nth order polynomial equation in the variable λ. By the fundamental theorem of algebra, this polynomial will have exactly *n* roots, and therefore each matrix of dimensions $n \times n$, or equivalently, each *n*-dimensional operator, will have exactly *n* eigenvalues (although we will find later that this does *not* imply that there will also be *n* eigenvectors). These, however, may be repeated and/or complex. Complex eigenvalues, of course, will appear in conjugate pairs.

Remark: The reader may have noticed that Equation (1.49) implies that, at least for a SISO system for which the division operation is defined, the denominator of the transfer function of a system is the nth-order polynomial $|sI - A|$. This is, of course, the same polynomial whose roots give us the eigenvalues. The temptation is therefore to guess that the poles and eigenvalues are the same, but this is *not* the case. Although all the roots of $|sI - A|$ are eigenvalues, not all eigenvalues are poles of the transfer function. We will see the reason for this in Chapter 8.

Example 4.2: Simple Eigenvalue Problem

Find the eigenvalues for the operator represented by matrix

$$A = \begin{bmatrix} 3 & -2 \\ -1 & 4 \end{bmatrix}$$

Solution:

Using the determinant rule,

$$\begin{aligned}|\lambda I - A| &= \begin{vmatrix}\lambda - 3 & 2 \\ 1 & \lambda - 4\end{vmatrix} \\ &= (\lambda - 3)(\lambda - 4) - 2 \\ &= \lambda^2 - 7\lambda + 10 \\ &= (\lambda - 2)(\lambda - 5)\end{aligned}$$

Therefore, $|\lambda I - A| = 0$ gives $\lambda_1 = 2$ and $\lambda_2 = 5$ as the two eigenvalues.

To determine the eigenvectors $\boldsymbol{x}_1$ and $\boldsymbol{x}_2$ corresponding, respectively, to these eigenvalues, we realize that $\boldsymbol{x}_i \in \mathrm{N}(\lambda_i I - A)$ by construction, so we simply find a vector in this null space[M] to act as eigenvector $\boldsymbol{x}_i$.

`null(A)`

Example 4.3: Simple Eigenvector Problem

Determine the eigenvectors for the matrix operator given in the previous example.

Solution:

Treating each eigenvalue separately, consider λ_1 and seek solutions to:

$$(\lambda_1 I - A)\boldsymbol{x} = \begin{bmatrix}-1 & 2 \\ 1 & -2\end{bmatrix}\boldsymbol{x} = 0$$

It is obvious that the matrix $\lambda_1 I - A$ has a single linearly independent column and therefore has rank one. The dimension of its null space is therefore one and there will be a single vector solution, which is $\boldsymbol{x}_1 = \alpha\begin{bmatrix}2 & 1\end{bmatrix}^{\mathrm{T}}$ for any scalar α. For the other eigenvalue,

$$(\lambda_2 I - A)\boldsymbol{x} = \begin{bmatrix}2 & 2 \\ 1 & 1\end{bmatrix}\boldsymbol{x} = 0$$

yields $\boldsymbol{x}_2 = \beta\begin{bmatrix}1 & -1\end{bmatrix}^{\mathrm{T}}$ for any scalar β.

By convention, we often eliminate the arbitrary constant from the eigenvectors by normalizing them. For example, the two eigenvectors for the example above would be given as

$$x_1 = \begin{bmatrix} 2/\sqrt{5} \\ 1/\sqrt{5} \end{bmatrix} \qquad x_2 = \begin{bmatrix} 1/\sqrt{2} \\ -1/\sqrt{2} \end{bmatrix}$$

It is understood that any multiple of such normalized eigenvectors will also be an eigenvector.

Example 4.4: Complex Eigenvalues and Eigenvectors

Suppose we have an operator that rotates vectors in $\Re^3$ clockwise about some other given vector $\boldsymbol{v}$ by an angle θ, as shown in Figure 4.2. This is a linear operator. If the axis of rotation is the direction of the vector $\boldsymbol{r} = \begin{bmatrix} 1 & 1 & 1 \end{bmatrix}^T$ and the angle of rotation is $\theta = 60°$ it can be shown that the matrix representation is ([4]):

$$A = \begin{bmatrix} 2/3 & -1/3 & 2/3 \\ 2/3 & 2/3 & -1/3 \\ -1/3 & 2/3 & 2/3 \end{bmatrix}$$

Find the eigenvalues and eigenvectors for this operator.

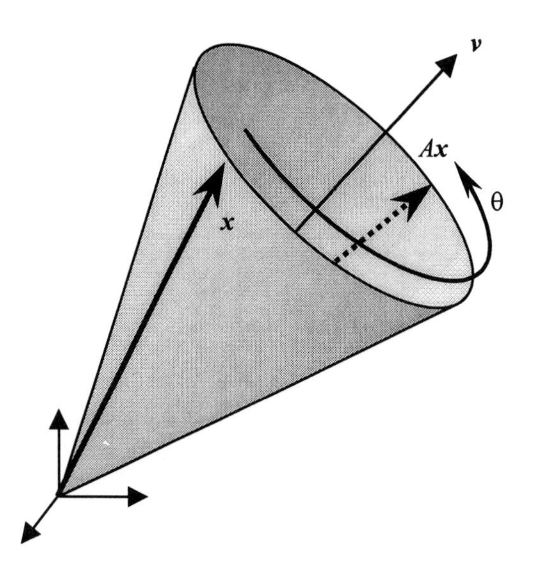

Figure 4.2 A vector rotating about a fixed, arbitrary axis. The rotating vector sweeps out the surface of a cone centered on the axis of rotation.

Solution:

Before proceeding with the computation of the eigenvalues and eigenvectors, consider the original definition. An eigenvector is a vector that is merely scaled by the action of the operator. In this situation, it is geometrically obvious that the only vector that will not "move" under this three-dimensional rotation operator will be a vector along the axis of rotation itself. That is, any vector of the form $\alpha \boldsymbol{v}$ will only rotate about itself, and in fact it will be scaled only by a factor of unity, i.e., its length will not change. We would expect any other vector in the space to rotate on the surface of the cone defined by the description of the operator. Thus, we can logically reason the existence of an eigenvalue of 1 and a corresponding eigenvector of $\alpha \boldsymbol{v}$.

By computing $|\lambda I - A|$ and finding the roots of the resulting polynomial, we find that the eigenvalues are $\lambda_1 = 1$, $\lambda_2 = \frac{1}{2} + j\frac{\sqrt{3}}{2}$, and $\lambda_3 = \frac{1}{2} - j\frac{\sqrt{3}}{2}$, where $j = \sqrt{-1}$. Furthermore, by finding the null spaces of the (complex) matrices $\lambda_i I - A$, we can compute the corresponding eigenvectors as

$$\boldsymbol{x}_1 = \begin{bmatrix} \frac{1}{\sqrt{3}} \\ \frac{1}{\sqrt{3}} \\ \frac{1}{\sqrt{3}} \end{bmatrix} \qquad \boldsymbol{x}_2 = \begin{bmatrix} \frac{1}{2} + j\frac{\sqrt{3}}{2} \\ \frac{1}{2} - j\frac{1}{2\sqrt{3}} \\ j\frac{1}{\sqrt{3}} \end{bmatrix} \qquad \boldsymbol{x}_3 = \begin{bmatrix} \frac{1}{2} - j\frac{\sqrt{3}}{2} \\ \frac{1}{2} + j\frac{1}{2\sqrt{3}} \\ -j\frac{1}{\sqrt{3}} \end{bmatrix}$$

The first point to notice here is that the complex eigenvectors corresponding to the complex pair of eigenvalues is itself a complex conjugate pair. Second, we see that the observation about the real eigenvalue and its corresponding eigenvector is indeed true. However we did not predict the existence of the complex pair of eigenvalues and eigenvectors. In general, it can be said that complex eigenvalues and eigenvectors do not conform to the same geometric interpretation as real-valued eigenvalues and eigenvectors. Nevertheless, they are just as important for most purposes, including stability theory and control systems that we study in later chapters.

Example 4.5: Eigenvalues and Eigenvectors of Operators on Function Spaces

Let **V** denote the linear vector space of polynomials in x of degree ≤ 2. Also consider a linear operator T that takes an arbitrary vector $p(x)$ and transforms it into the vector

$$(Tp)(x) = \frac{d}{dx}[(1-x)p(x)] + 2p(0)$$

Find the operator's eigenvalues and eigenvectors if it is expressed in the basis $\{e_i\} = \{1, x, x^2\}$.

Solution:

It is possible to find the eigenvalues and eigenvectors of such an operator directly, by manipulating the polynomials in the equation $(Tp)(x) = \lambda p(x)$. However, it is considerably easier to first express the operator as a matrix in the basis given, and then use existing numerical tools to compute the eigenvalues and eigenvectors.

To find the linear operator's matrix representation, we determine its effects on the basis vectors;

$$T(1) = \frac{d}{dx}[(1-x)\cdot 1] + 2\cdot(1) = 1 \qquad (= [1 \quad 0 \quad 0]^T)$$

$$T(x) = \frac{d}{dx}[(1-x)\cdot x] + 2\cdot(0) = 1 - 2x \qquad (= [1 \quad -2 \quad 0]^T)$$

$$T(x^2) = \frac{d}{dx}\left[(1-x)\cdot x^2\right] + 2\cdot(0) = 2x - 3x^2 \qquad (= [0 \quad 2 \quad -3]^T)$$

Therefore,

$$T = \begin{bmatrix} 1 & 1 & 0 \\ 0 & -2 & 2 \\ 0 & 0 & -3 \end{bmatrix}$$

In such a case, where the matrix is upper- or lower-triangular, it can be easily shown that the elements on the diagonal will be the eigenvalues. We therefore have by inspection, $\lambda_1 = 1, \lambda_2 = -2, \text{ and } \lambda_3 = -3$. For the unnormalized eigenvectors,

$$v_1 = \mathrm{N}(1\cdot I - A) = \begin{bmatrix} 1 \\ 0 \\ 0 \end{bmatrix} \quad v_2 = \mathrm{N}(-2\cdot I - A) = \begin{bmatrix} 1 \\ -3 \\ 0 \end{bmatrix} \quad v_3 = \mathrm{N}(-3\cdot I - A) = \begin{bmatrix} 1 \\ -4 \\ 2 \end{bmatrix} \tag{4.9}$$

Note that written as polynomials in their original space **V**, these vectors are $v_1 = 1$, $v_2 = 1 - 3x$, and $v_3 = 1 - 4x + 2x^2$. We can use these expressions to verify that the concept of eigenvalues and eigenvectors is not restricted to matrices by performing operator T on them directly as polynomials:

$$T(1)=1 \qquad (=\lambda_1 v_1)$$

$$T(1-3x)=6x-2=-2(1-3x) \qquad (=\lambda_2 v_2)$$

$$T(1-4x+2x^2)=-6x^2+12x-3=-3(1-4x+2x^2) \qquad (=\lambda_3 v_3)$$

4.4 The Basis of Eigenvectors

In this section, we will show the result of changing the basis of an operator to the basis formed by a set of n eigenvectors. Such bases will be shown to have simplifying effects on the matrix representations of operators. These simplified forms of operators are known as *canonical*[M] *forms*. However, to form such a basis of eigenvectors, we must first know whether or not a complete set of n linearly independent eigenvectors exists, and if they do not, what we do about it.

`canon(sys,type)`

4.4.1 Changing to the Basis of Eigenvectors

For the time being, we will assume that our matrix of interest has a complete set of n linearly independent eigenvectors. This will allow us to change the basis of the operator to the basis of eigenvectors without any trouble. To do so by following the procedure of Section 2.2.4, we will need the matrix that relates the old basis vectors, which we will assume are the standard basis vectors, to the new basis vectors, which are the eigenvectors. As we found in Example 2.7, it is considerably easier to write expressions for the new basis vectors in terms of the old basis vectors. That is, if $\{e_1,\ldots,e_n\}$ represents the old (standard) basis and $\{x_1,\ldots,x_n\}$ represents the set of n eigenvectors, then it is obvious that

$$x_i=\left[e_1 \mid e_2 \mid \cdots \mid e_n\right]\begin{bmatrix} x_{1i} \\ \hline x_{2i} \\ \hline \vdots \\ \hline x_{ni} \end{bmatrix}$$

or

$$\left[x_1 \mid x_2 \mid \cdots \mid x_n\right]=\left[e_1 \mid e_2 \mid \cdots \mid e_n\right]\left[\begin{array}{c|c|c|c} x_{11} & x_{12} & \cdots & x_{1n} \\ x_{21} & x_{22} & \cdots & x_{2n} \\ \vdots & \vdots & \ddots & \vdots \\ x_{n1} & x_{n2} & \cdots & x_{nn} \end{array}\right]$$

where x_{ji} is the jth component of vector $\mathbf{x}_i$. Thus, as we did in Example 2.7, we construct the matrix

$$M = B^{-1} = \begin{bmatrix} x_{11} & x_{12} & \cdots & x_{1n} \\ x_{21} & x_{22} & \cdots & x_{2n} \\ \vdots & \vdots & \ddots & \vdots \\ x_{n1} & x_{n2} & \cdots & x_{nn} \end{bmatrix} = \begin{bmatrix} \mathbf{x}_1 & \mathbf{x}_2 & \cdots & \mathbf{x}_n \end{bmatrix} \tag{4.10}$$

The matrix M is known as the *modal matrix*, for reasons that will become clear in Chapter 6.

Then, from Equation (3.25), we find that

$$\hat{A} = M^{-1}AM \tag{4.11}$$

Before presenting an example, it is useful to predict what form this "new" operator matrix $\hat{A}$ will take. It is known that the ith column of an operator matrix consists of the representation of the effect of that operator acting on the ith basis vector. It is also known that when an operator operates on an eigenvector, the result is a scaled version of that eigenvector, i.e., $A\mathbf{x} = \lambda\mathbf{x}$. Therefore, when the basis is the set of eigenvectors, then the operator acting on the basis vectors will give vectors with components only along those same basis vectors. The ith column of A should have nonzero entries in only the ith position, i.e., the matrix $\hat{A}$ will be *diagonal.*

Example 4.6: Diagonalization of Operators

Change the matrix operator

$$A = \begin{bmatrix} 3 & 0 & 2 \\ 0 & 3 & -2 \\ 2 & -2 & 1 \end{bmatrix}$$

which is expressed in the standard basis, into the basis of its own eigenvectors.

Solution:

We first find the eigenvalues:

$$0 = |\lambda I - A| = \begin{vmatrix} \lambda - 3 & 0 & -2 \\ 0 & \lambda - 3 & 2 \\ -2 & 2 & \lambda - 1 \end{vmatrix}$$

$$= \lambda^3 - 7\lambda^2 + 7\lambda + 15$$

$$= (\lambda - 5)(\lambda - 3)(\lambda + 1)$$

so $\lambda_1 = 5$, $\lambda_2 = 3$, and $\lambda_3 = -1$. Finding the (normalized) eigenvectors,

$$\boldsymbol{x}_1 = \mathrm{N}(\lambda_1 I - A) = \begin{bmatrix} 1/\sqrt{3} \\ -1/\sqrt{3} \\ 1/\sqrt{3} \end{bmatrix} \qquad \boldsymbol{x}_2 = \mathrm{N}(\lambda_2 I - A) = \begin{bmatrix} 1/\sqrt{2} \\ 1/\sqrt{2} \\ 0 \end{bmatrix}$$

$$\boldsymbol{x}_3 = \mathrm{N}(\lambda_3 I - A) = \begin{bmatrix} 1/\sqrt{6} \\ -1/\sqrt{6} \\ 2/\sqrt{6} \end{bmatrix}$$

We therefore have the modal matrix

$$M = \begin{bmatrix} 1/\sqrt{3} & 1/\sqrt{2} & 1/\sqrt{6} \\ -1/\sqrt{3} & 1/\sqrt{2} & -1/\sqrt{6} \\ 1/\sqrt{3} & 0 & 2/\sqrt{6} \end{bmatrix}$$

and the transformed operator matrix

$$\hat{A} = M^{-1}AM = \begin{bmatrix} 5 & 0 & 0 \\ 0 & 3 & 0 \\ 0 & 0 & -1 \end{bmatrix}$$

which is, of course, diagonal.

4.4.2 Repeated Eigenvalues

In the examples considered so far, each eigenvalue has been distinct. The first case in which we will encounter matrices that do *not* have a complete set of n independent eigenvectors is when one or more of the eigenvalues are repeated. When an eigenvalue is repeated, i.e., the same λ is a multiple root of $|\lambda I - A|$, some new situations can arise. If, for example, eigenvalue λ_j is repeated m_j times, we say that m_j is the *algebraic multiplicity* of λ_j. Two cases can arise:

1. If we compute the nullity of $\lambda_j I - A$ and find it to be $q(\lambda_j I - A) = n - r(\lambda_j I - A) = m_j$, then by definition, the dimension of $N(\lambda_j I - A)$ is m_j. It would therefore be possible to find m_j linearly independent eigenvectors associated with the same eigenvalue λ_j. This would be done in the usual way, generating eigenvectors $x_i, \ldots, x_{m_j} \in N(\lambda_j I - A)$ as in the previous chapter, probably with the help of a computer.[M]

null(A)

2. If $q(\lambda_j I - A) < m_j$, then we cannot find m_j eigenvectors because $N(\lambda_j I - A)$ is not big enough. We therefore conclude that although every $n \times n$ matrix (or n-dimensional operator) has n eigenvalues, there are *not* always a full set of n eigenvectors.

Case 1 represents an eigenvector problem that is no more difficult than if all the eigenvalues were distinct. Such a situation defines an A-invariant subspace of dimension greater than one associated with a single eigenvalue as follows:

> ***Eigenspace:*** The set of all eigenvectors corresponding to an eigenvalue λ_i forms a basis for a subspace of **X**, called the *eigenspace* of λ_i. This eigenspace also happens to be the null space of a transformation defined as $\lambda_i I - A$. (4.12)

It is clear that the eigenspace of λ_i is A-invariant and has dimension equal to $q(\lambda_i I - A)$. The above definition is sometimes restricted to the real eigenvectors of A when **X** itself is real. This allows us to avoid defining complex subspaces of real vector spaces.

However, in case 2, we are faced with certain problems related to the use of eigenvectors. Most importantly, we clearly cannot diagonalize the A-matrix by changing to the basis of eigenvectors because we will not have a sufficient set of n eigenvectors that we need to construct an $n \times n$ transformation matrix. To resolve this difficulty, we introduce the "next best thing" to an eigenvector, called *a generalized eigenvector.**

---

* Our use of the term "generalized eigenvector" is the classical one. In other texts (including MATLAB's manual entry under EIG), a generalized eigenvector is a vector $\boldsymbol{x}$ that solves the so-called "generalized eigenvalue problem," i.e., $(A-\lambda B)\boldsymbol{x} = 0$. The two usages are unrelated.

4.4.3 Generalized Eigenvectors

When an eigenvalue λ_i has an algebraic multiplicity $m_i > 1$, it may have any number of eigenvectors g_i, where $g_i \le m_i$, equal to the nullity of $\lambda_i I - A$:

$$g_i = q(\lambda_i I - A) \tag{4.13}$$

This number will be referred to as the geometric multiplicity of λ_i because it represents the dimension of the eigenspace of λ_i. Considering all possible eigenvalues, we will have a total of $\sum_i g_i \le n$ eigenvectors for an n-dimensional operator A.

When $\sum_i g_i < n$, we will have an insufficient number of eigenvectors to construct a modal matrix as in (4.10) and thereby diagonalize the operator A. However, we can define so-called "generalized eigenvectors" that serve a similar purpose. The modal matrix thus constructed will not diagonalize the operator, but it will produce a similar form known as the *Jordan canonical form*. Canonical forms, such as the diagonal form, which is a special case of a Jordan form, are simply standardized matrix structures that have particularly convenient forms for different purposes. We will see the usefulness of canonical forms in future chapters.

We begin discussing generalized eigenvectors by recalling that if $\mathbf{x}_1$ is a *regular* eigenvector corresponding to eigenvalue λ_1, then $A\mathbf{x}_1 = \lambda_1 \mathbf{x}_1$. If for this eigenvalue $g_1 < m_1$ and if we can find a nontrivial solution $\mathbf{x}_2$ to the equation

$$A\mathbf{x}_2 = \lambda_1 \mathbf{x}_2 + \mathbf{x}_1 \tag{4.14}$$

that is linearly independent of $\mathbf{x}_1$, then $\mathbf{x}_2$ will be a generalized eigenvector. If $m_1 - g_1 = 1$, then $\mathbf{x}_2$ is the only generalized eigenvector necessary. If $m_1 - g_1 = p_1$, where $p_1 > 1$, then further generalized eigenvectors can be found from the "chain"

$$\begin{aligned} A\mathbf{x}_3 &= \lambda_1 \mathbf{x}_3 + \mathbf{x}_2, \\ A\mathbf{x}_4 &= \lambda_1 \mathbf{x}_4 + \mathbf{x}_3, \\ &\vdots \end{aligned}$$

until p_1 such vectors are found.

This definition for generalized eigenvectors therefore suggests one method for computing them, which is commonly called the *bottom-up* method:

1. For repeated eigenvalue λ_i with algebraic multiplicity m_i and geometric multiplicity g_i given by (4.13), first find g_i regular eigenvectors $\boldsymbol{x}_{i1},\ldots,\boldsymbol{x}_{ig_i}$.
2. For each of these regular eigenvectors, compute nontrivial solutions to

$$(A-\lambda_i I)\boldsymbol{x}_{j+g_i}=\boldsymbol{x}_j \tag{4.15}$$

for $j=i1,\ldots,ig_i$, which can be recognized as a rearranged version of Equation (4.14). If the computed $\boldsymbol{x}_{j+g_i}$ is linearly independent of all previously computed regular and generalized eigenvectors, then it is a generalized eigenvector.
3. To complete the set of $p_i=m_i-g_i$ generalized eigenvectors, continue to compute linearly independent solutions of

$$(A-\lambda_i I)\boldsymbol{x}_{j+(k+1)g_i}=\boldsymbol{x}_{j+kg_i},$$
$$\vdots$$

for all $j=i1,\ldots,ig_1$ and $k=1,2,\ldots$ until all generalized eigenvectors are found.

This technique is illustrated in the following example.

Example 4.7: A Small Number of Generalized Eigenvectors

Determine all eigenvalues, regular eigenvectors, and generalized eigenvectors for the operator represented by the matrix

$$A=\begin{bmatrix}1 & 2 & 0\\ 0 & 1 & 0\\ -3 & 3 & 5\end{bmatrix}$$

Solution:

First computing the eigenvalues:

$$0 = |\lambda I - A| = \begin{vmatrix} \lambda - 1 & -2 & 0 \\ 0 & \lambda - 1 & 0 \\ 3 & -3 & \lambda - 5 \end{vmatrix}$$
$$= (\lambda - 5)(\lambda - 1)^2$$

so $\lambda_1 = 5$ and $\lambda_2 = 1\ (m_2 = 2)$. For eigenvalue $\lambda_1 = 5$, we will have only a single regular eigenvector, computed as $x_1 = \begin{bmatrix} 0 & 0 & 1 \end{bmatrix}^T$ from the solution of the equation

$$(5 \cdot I - A)x_1 = \begin{bmatrix} 4 & -2 & 0 \\ 0 & 4 & 0 \\ 3 & -3 & 0 \end{bmatrix} x_1 = 0$$

For $\lambda_2 = 1\ (m_2 = 2)$, we find that

$$g_2 = q(1 \cdot I - A) = q\left(\begin{bmatrix} 0 & -2 & 0 \\ 0 & 0 & 0 \\ 3 & -3 & -4 \end{bmatrix} \right) = 1$$

so that there will be a single regular eigenvector, calculated as $x_2 = \begin{bmatrix} 4/5 & 0 & 3/5 \end{bmatrix}^T$ from the equation

$$(1 \cdot I - A)x_2 = 0$$

We therefore require $p_2 = m_2 - g_2 = 1$ generalized eigenvectors for λ_2. To compute it, we construct the linear algebraic equation

$$(A - 1 \cdot I)x_3 = x_2$$

or

$$\begin{bmatrix} 0 & 2 & 0 \\ 0 & 0 & 0 \\ -3 & 3 & 4 \end{bmatrix} x_3 = \begin{bmatrix} 4/5 \\ 0 \\ 3/5 \end{bmatrix} \tag{4.16}$$

Equation (4.16) will of course have a homogeneous solution, which we may discard, because it will be exactly the regular eigenvector x_2. The particular solution, as obtained using the methods of Section 3.3.1, is $x_3 = [-\frac{3}{5} \;\; \frac{2}{5} \;\; -\frac{3}{5}]^T$. This is linearly independent of x_2 and is therefore the generalized eigenvector (the *only* one). Note that unlike regular eigenvectors, generalized eigenvectors are particular solutions to linear equations and therefore cannot be multiplied by an arbitrary constant. Nor are they unique.

Chains of Generalized Eigenvectors

It may be noticed that the procedure listed above has no stopping condition. That is, the search simply continues until all generalized eigenvectors are found, but no guidance is provided as to which indices j and k will provide these linearly independent vectors. A more procedural method can be devised if one considers the "chaining" of eigenvectors in more detail. To generate this algorithmic procedure, we first introduce the concept of the *index* of an eigenvalue.

> ***Index of Eigenvalue:*** The *index* of a repeated eigenvalue λ_i, denoted η_i, is the smallest integer η such that
>
> $$r(A - \lambda_i I)^{\eta} = n - m_i \tag{4.17}$$
>
> where n is the dimension of the space (size of the matrix), and m_i is the algebraic multiplicity of λ_i.

The rest of the algorithm is then calculated as follows:

1. For eigenvalue λ_i with index η_i, find all linearly independent solutions to the simultaneous set of matrix equations

$$\begin{aligned} (A - \lambda_i I)^{\eta_i} x &= 0 \\ (A - \lambda_i I)^{\eta_i - 1} x &\neq 0 \end{aligned} \tag{4.18}$$

Each such solution will start a different "chain" of generalized eigenvectors. Because $r(A - \lambda_i I)^{\eta_i} = n - m_i$, we know there will be no more than $n - (n - m_i) = m_i$ such solutions, each of which is a generalized eigenvector. Denote these solutions $v_1^1, \ldots, v_{m_i}^1$. [Note that there might be fewer than m_i such vectors, because although there are exactly m_i solutions to $(A-\lambda_i I)^{\eta_i}\boldsymbol{x}=0$, some of them might not satisfy $(A-\lambda_i I)^{\eta_i-1}\boldsymbol{x}\neq 0$. For generality, we will assume there are m_i solutions to (4.18)].

2. Begin generating further generalized eigenvectors by computing the chain for each $j=1,\dots,m_i$:

$$\begin{aligned}(A-\lambda_i I)\boldsymbol{v}_j^1 &= \boldsymbol{v}_j^2\\ (A-\lambda_i I)\boldsymbol{v}_j^2 &= \boldsymbol{v}_j^3\\ (A-\lambda_i I)\boldsymbol{v}_j^3 &= \boldsymbol{v}_j^4\\ &\vdots\end{aligned}$$

until we get to the point that

$$(A-\lambda_i I)\boldsymbol{v}_j^{\eta_i}=0$$

which of course indicates that $\boldsymbol{v}_j^{\eta_i}$ is a *regular* eigenvector. The chains thus end when regular eigenvectors are reached.

3. The *length* of the chains thus found will be η_i. There also may be chains of shorter length. If the chains of length η_i do not produce the full set of p_i generalized eigenvectors, begin the procedure again by finding all solutions to the set of equations

$$\begin{aligned}(A-\lambda_i I)^{\eta_i-1}\boldsymbol{x}&=0\\ (A-\lambda_i I)^{\eta_i-2}\boldsymbol{x}&\neq 0\end{aligned} \tag{4.19}$$

and repeating the procedure. This will produce chains of length η_i-1. Continue until all generalized eigenvectors have been found.

4. Repeat for any other repeated eigenvalues.

The method outlined in these steps is usually referred to as the *top-down* method for finding eigenvectors. Although it is more computationally intensive and difficult to perform "by hand," it is easily programmed on a computer and has the added advantage that it can simultaneously find the regular eigenvectors as well. The previous method is known as the bottom-up method because it starts with known regular eigenvectors. The following examples illustrate the top-down method for simple matrices and more complex matrices.

Example 4.8: Top-Down Method for a Small Matrix

Use the top-down method to find the generalized eigenvector for the same 3×3 matrix as in Example 4.7.

Solution:

Recall that for this example, the eigenvalues were $\lambda_1 = 5$ and $\lambda_2 = 1\ (m_2 = 2)$, and the (regular) eigenvector corresponding to $\lambda_1 = 5$ was $\boldsymbol{x}_1 = \begin{bmatrix} 0 & 0 & 1 \end{bmatrix}^{\mathrm{T}}$. Now consider the repeated eigenvalue $\lambda_2 = 1\ (m_2 = 2)$. To find the index η_2, first note that $n - m_2 = 3 - 2 = 1$:

$$r\left((A - 1 \cdot I)^1\right) = r\left(\begin{bmatrix} 0 & 2 & 0 \\ 0 & 0 & 0 \\ -3 & 3 & 4 \end{bmatrix}\right) = 2 \qquad (\neq n - m_2)$$

$$r\left((A - 1 \cdot I)^2\right) = r\left(\begin{bmatrix} 0 & 0 & 0 \\ 0 & 0 & 0 \\ -12 & 6 & 16 \end{bmatrix}\right) = 1 \qquad (= n - m_2)$$

Therefore $\eta_2 = 2$.

So now we seek all solutions to the equations

$$(A - 1 \cdot I)^2 \boldsymbol{x} = \begin{bmatrix} 0 & 0 & 0 \\ 0 & 0 & 0 \\ -12 & 6 & 16 \end{bmatrix} \boldsymbol{x} = 0$$

$$(A - 1 \cdot I)^1 \boldsymbol{x} = \begin{bmatrix} 0 & 2 & 0 \\ 0 & 0 & 0 \\ -3 & 3 & 4 \end{bmatrix} \boldsymbol{x} \neq 0$$

To solve such a problem, we simply seek a vector $\boldsymbol{x} \in \mathrm{N}\left((A - 1 \cdot I)^2\right)$ such that $\boldsymbol{x} \notin \mathrm{N}\left((A - 1 \cdot I)^1\right)$. This is most easily done by computer. In this case, it is easily observed that the vector $\boldsymbol{v}_1^1 = \begin{bmatrix} 1 & 2 & 0 \end{bmatrix}^{\mathrm{T}}$ is one such solution. Now to generate the chain:

$$(A-1\cdot I)v_1^1 = v_1^2 = \begin{bmatrix} 0 & 2 & 0 \\ 0 & 0 & 0 \\ -3 & 3 & 4 \end{bmatrix}\begin{bmatrix} 1 \\ 2 \\ 0 \end{bmatrix} = \begin{bmatrix} 4 \\ 0 \\ 3 \end{bmatrix}$$

$$(A-1\cdot I)v_1^2 = v_1^3 = \begin{bmatrix} 0 & 2 & 0 \\ 0 & 0 & 0 \\ -3 & 3 & 4 \end{bmatrix}\begin{bmatrix} 4 \\ 0 \\ 3 \end{bmatrix} = \begin{bmatrix} 0 \\ 0 \\ 0 \end{bmatrix}$$

So clearly the chain ends with the vector $v_1^2 = \begin{bmatrix} 4 & 0 & 3 \end{bmatrix}^T$, which will be the *regular* eigenvector corresponding to $\lambda_2 = 1$. This eigenvector, which we prefer to denote as x_2 after normalization, will appear exactly the same as eigenvector x_2 in Equation (4.16) of Example 4.7. The generalized eigenvector chained to this regular eigenvector is $v_1^1 = \begin{bmatrix} 1 & 2 & 0 \end{bmatrix}^T$. Note that this generalized eigenvector is *not* the same as the generalized eigenvector computed in the bottom-up example. Generalized eigenvectors are not unique.

Example 4.9: Multiple Chains of Generalized Eigenvectors

Find all eigenvalues, eigenvectors, and generalized eigenvectors for the operator represented by the matrix:

$$A = \begin{bmatrix} 3 & -1 & 1 & 1 & 0 & 0 \\ 1 & 1 & -1 & -1 & 0 & 0 \\ 0 & 0 & 2 & 0 & 1 & 1 \\ 0 & 0 & 0 & 2 & -1 & -1 \\ 0 & 0 & 0 & 0 & 1 & 1 \\ 0 & 0 & 0 & 0 & 1 & 1 \end{bmatrix}$$

Solution:

The first thing one should notice about this example is that the last two rows are identical. The matrix, therefore, will be rank-deficient and must have at least one zero eigenvalue, because $|A| = |0\cdot I - A| = 0$. (In general, the A matrix itself will have rank deficiency equal to the geometric multiplicity of any zero eigenvalues.) However, to determine all the eigenvalues, we find all the zeros to the polynomial

$$|\lambda I - A| = (\lambda-2)^5\lambda$$

which provide eigenvalues of $\lambda_1 = 0$ and $\lambda_2 = 2$ $(m_2 = 5)$. For $\lambda_1 = 0$, we can compute the corresponding eigenvector by finding a vector in the (one-dimensional) null space of $A - 0 \cdot I$ $(= A)$. This will give $x_1 = \begin{bmatrix} 0 & 0 & 0 & 0 & 1 & -1 \end{bmatrix}^T$ (unnormalized).

For $\lambda_2 = 2$ $(m_2 = 5)$, we begin by computing the value $n - m_2 = 6 - 5 = 1$. The index η_2 must be found from the sequence of calculations

$$\begin{aligned} r(A - 2 \cdot I) &= 4 \qquad (\neq n - m_2) \\ r\left((A - 2 \cdot I)^2\right) &= 2 \qquad (\neq n - m_2) \\ r\left((A - 2 \cdot I)^3\right) &= 1 \qquad (= n - m_2) \end{aligned}$$

so $\eta_2 = 3$. This implies that there will be a chain of generalized eigenvectors, of length three, ending with a regular eigenvector. Because $m_2 = 5$, we also know there must also be either a second chain of length two, or two more chains each of length one. To resolve this question, we note that because $r(A - 2 \cdot I) = 4$, the eigenspace of λ_2 will be of dimension $n - 4 = 2$, so there can only be two regular eigenvectors and, hence, two chains. We will therefore expect to find a second chain of length two.

Returning to the computation of the first chain, we must find a solution to the pair of equations

$$\begin{aligned} (A - 2 \cdot I)^3 \boldsymbol{x} &= 0 \\ (A - 2 \cdot I)^2 \boldsymbol{x} &\neq 0 \end{aligned} \tag{4.20}$$

where

$$(A - 2 \cdot I)^3 = \begin{bmatrix} 0 & 0 & 0 & 0 & 0 & 0 \\ 0 & 0 & 0 & 0 & 0 & 0 \\ 0 & 0 & 0 & 0 & 0 & 0 \\ 0 & 0 & 0 & 0 & 0 & 0 \\ 0 & 0 & 0 & 0 & -4 & 4 \\ 0 & 0 & 0 & 0 & 4 & -4 \end{bmatrix} \qquad (A - 2 \cdot I)^2 = \begin{bmatrix} 0 & 0 & 2 & 2 & 0 & 0 \\ 0 & 0 & 2 & 2 & 0 & 0 \\ 0 & 0 & 0 & 0 & 0 & 0 \\ 0 & 0 & 0 & 0 & 0 & 0 \\ 0 & 0 & 0 & 0 & 2 & -2 \\ 0 & 0 & 0 & 0 & -2 & 2 \end{bmatrix}$$

One such solution is the vector $v_1^1 = [0 \quad 0 \quad 0 \quad 1 \quad 1 \quad 1]^T$. Beginning the chain, we find

$$
\begin{aligned}
v_1^2 &= (A - 2 \cdot I)v_1^1 = [1 \quad -1 \quad 2 \quad -2 \quad 0 \quad 0]^T \\
v_1^3 &= (A - 2 \cdot I)v_1^2 = [2 \quad 2 \quad 0 \quad 0 \quad 0 \quad 0]^T \\
v_1^4 &= (A - 2 \cdot I)v_1^3 = [0 \quad 0 \quad 0 \quad 0 \quad 0 \quad 0]^T
\end{aligned}
$$

Therefore, v_1^3 is a regular eigenvector chained, in order, to generalized eigenvectors v_1^2 and v_1^1. The reader can verify that any other solutions to the system of equations (4.20) produce only vectors that are linearly dependent on the vectors that form this chain. However, one need not perform this check, because it is known that there can only be one chain of length three.

To find the chain of length two, we compute the solution to the equations

$$
\begin{aligned}
(A - 2 \cdot I)^2 x &= 0 \\
(A - 2 \cdot I)^1 x &\neq 0
\end{aligned}
\tag{4.21}
$$

one of which is $v_2^1 = [0 \quad 0 \quad 0 \quad 0 \quad 1 \quad 1]^T$. This produces the chain

$$
\begin{aligned}
v_2^2 &= (A - 2 \cdot I)v_2^1 = [0 \quad 0 \quad 2 \quad -2 \quad 0 \quad 0]^T \\
v_2^3 &= (A - 2 \cdot I)v_2^2 = [0 \quad 0 \quad 0 \quad 0 \quad 0 \quad 0]^T
\end{aligned}
$$

so the second chain consists of regular eigenvector v_2^2 and generalized eigenvector v_2^1. It is very important to note that there are other solutions to Equation (4.21), but that all of them result in chains that consist of vectors that are linearly dependent on those previously found. It is necessary that the modal matrix we construct be nonsingular, so all regular and generalized eigenvectors must by linearly independent. We discard all the dependent solutions.

4.4.4 When n Independent Eigenvectors Exist

It is clear by now that the case of n independent eigenvectors makes construction of the modal matrix much simpler. There are certain matrices for which this will always be true. As we have already noted, the first is when all the eigenvalues of the matrix are distinct. By the construction of the null space of $\lambda I - A$, as in Equation (4.4), we know that each distinct eigenvalue will have associated with it at least one eigenvector. When all n eigenvalues of an operator are distinct, we can also rely on the following result:

THEOREM: The eigenvectors of an operator A, all of whose eigenvalues are distinct, are linearly independent. (4.22)

PROOF: We will construct this proof by contradiction. Let matrix A have distinct eigenvalues, i.e., $\lambda_i \neq \lambda_j$ for all $i \neq j$. Suppose further that there are only k linearly independent eigenvectors among the complete set of n. Without loss of generality, let these k linearly independent eigenvectors be the first k. Then for all $k < j \leq n$, eigenvector $\boldsymbol{x}_j$ can be written as a linear combination of vectors in $\{\boldsymbol{x}_1, \ldots, \boldsymbol{x}_k\}$. Write this linear combination as $\boldsymbol{x}_j = \sum_{i=1}^{k} a_{ij}\boldsymbol{x}_i$, where not all the coefficients a_{ij} are zero. Therefore,

$$A\boldsymbol{x}_j = \lambda_j \boldsymbol{x}_j = \sum_{i=1}^{k} a_{ij}\lambda_j \boldsymbol{x}_i \tag{4.23}$$

In addition,

$$A\boldsymbol{x}_j = A\left(\sum_{i=1}^{k} a_{ij}\boldsymbol{x}_i\right) = \sum_{i=1}^{k} a_{ij} A\boldsymbol{x}_i = \sum_{i=1}^{k} a_{ij}\lambda_i \boldsymbol{x}_i \tag{4.24}$$

Comparing (4.23) to (4.24), we have

$$\sum_{i=1}^{k} a_{ij}\lambda_j \boldsymbol{x}_i = \sum_{i=1}^{k} a_{ij}\lambda_i \boldsymbol{x}_i$$

giving,

$$\sum_{i=1}^{k} a_{ij}\left(\lambda_i - \lambda_j\right)\boldsymbol{x}_i = 0$$

Because this summation only runs up to k, i.e., through the independent $\boldsymbol{x}_i$'s, equality to zero implies that all coefficients $a_{ij}\left(\lambda_i - \lambda_j\right)$ are zero. However, because it is known that not all a_{ij}'s are zero, then for some $i \neq j$, $\lambda_i = \lambda_j$. Because this contradicts our original assumption, we conclude that all the vectors x_i, $i = 1, \ldots, n$ are linearly independent.

There is a second situation in which we can guarantee the existence of n independent eigenvectors, and that is when the matrix is hermitian (refer to Section 3.2.2). In this case, we first show, using the following theorem, that the eigenvectors corresponding to different eigenvalues are orthogonal, and that all eigenvalues are real.

THEOREM: If matrix A is hermitian, all its eigenvalues will be real, and it has a full set of n orthogonal eigenvectors, regardless of the existence of any repeated eigenvalues. (4.25)

PROOF: We first demonstrate that the eigenvalues for A are all real. We already know that each eigenvalue will have at least one eigenvector, so we can write $Ae_i = \lambda_i e_i$, or $Ae_i - \lambda_i e_i = 0$. Then taking the inner product of each side of this equation with e_i, we get the equality

$$\langle e_i, Ae_i - \lambda_i e_i \rangle = 0 \tag{4.26}$$

Then with the following manipulations:

$$\begin{aligned} 0 &= \langle e_i, Ae_i \rangle - \langle e_i, \lambda_i e_i \rangle \\ &= \langle Ae_i, e_i \rangle - \lambda_i \langle e_i, e_i \rangle \\ &= \langle \lambda_i e_i, e_i \rangle - \lambda_i \langle e_i, e_i \rangle \\ &= \bar{\lambda}_i \langle e_i, e_i \rangle - \lambda_i \langle e_i, e_i \rangle \\ &= (\bar{\lambda}_i - \lambda_i) \langle e_i, e_i \rangle \end{aligned} \tag{4.27}$$

where the transition from the first to the second line in this sequence is possible because of the self-adjointness of the operator A. Now because $\langle e_i, e_i \rangle > 0$, we must have $\bar{\lambda}_i - \lambda_i = 0$, which is only possible if λ_i is real.

Now take two different eigenvalues λ_i and λ_j. Consider the inner product $\langle e_i, Ae_j \rangle$:

$$\begin{aligned}
\langle e_i, Ae_j \rangle &= \langle e_i, \lambda_j e_j \rangle \\
\langle Ae_i, e_j \rangle &= \lambda_j \langle e_i, e_j \rangle \\
\langle \lambda_i e_i, e_j \rangle &= \lambda_j \langle e_i, e_j \rangle \\
\bar{\lambda}_i \langle e_i, e_j \rangle &= \lambda_j \langle e_i, e_j \rangle \\
\lambda_i \langle e_i, e_j \rangle &= \lambda_j \langle e_i, e_j \rangle
\end{aligned}$$

implying $(\lambda_i - \lambda_j)\langle e_i, e_j \rangle = 0$. But we have stipulated that $\lambda_i \neq \lambda_j$, so $\langle e_i, e_j \rangle = 0$ and the eigenvectors are thus orthogonal.

Now we are prepared to show that hermitian matrices have n linearly independent eigenvectors.

THEOREM: Hermitian matrices have a complete set of n regular (i.e., not generalized) eigenvectors. (4.28)

PROOF: Suppose a hermitian (self-adjoint) matrix A has a (real) eigenvalue λ_i of arbitrary algebraic multiplicity. Suppose further that it has a generalized eigenvector, that is, a vector x_{i+1} chained to a regular eigenvector x_i such that $Ax_{i+1} - \lambda_i x_{i+1} = x_i$. Taking the inner product of both sides of this equality with x_i,

$$\begin{aligned}
\langle Ax_{i+1} - \lambda_i x_{i+1}, x_i \rangle &= \langle x_i, x_i \rangle \\
&= \langle Ax_{i+1}, x_i \rangle - \langle \lambda_i x_{i+1}, x_i \rangle \\
&= \langle x_{i+1}, Ax_i \rangle - \lambda_i \langle x_{i+1}, x_i \rangle \\
&= \langle x_{i+1}, \lambda_i x_i \rangle - \lambda_i \langle x_{i+1}, x_i \rangle \\
&= \lambda_i \langle x_{i+1}, x_i \rangle - \lambda_i \langle x_{i+1}, x_i \rangle \\
&= 0
\end{aligned}$$

But if x_i is an eigenvector, we cannot have $\langle x_i, x_i \rangle = 0$. Therefore, we have shown by contradiction that there are no generalized eigenvectors. If there are no generalized eigenvectors, then there must be n regular eigenvectors.

Example 4.10: Moments of Inertia

Consider a system of point masses m_i, $i = 1, \cdots, n$, each located by a vector $r_i = \begin{bmatrix} x_i & y_i & z_i \end{bmatrix}^{\mathrm{T}}$, that are fixed relative to one another. Such a system has an *inertia matrix* defined as

$$J = \begin{bmatrix} J_{xx} & J_{xy} & J_{xy} \\ J_{yx} & J_{yy} & J_{yz} \\ J_{zx} & J_{zy} & J_{zz} \end{bmatrix}$$

$$\triangleq \begin{bmatrix} \sum_{i=1}^{n} m_i\left(y_i^2 + z_i^2\right) & -\sum_{i=1}^{n} m_i x_i y_i & -\sum_{i=1}^{n} m_i x_i z_i \\ -\sum_{i=1}^{n} m_i y_i x_i & \sum_{i=1}^{n} m_i\left(x_i^2 + z_i^2\right) & -\sum_{i=1}^{n} m_i y_i z_i \\ -\sum_{i=1}^{n} m_i z_i x_i & -\sum_{i=1}^{n} m_i z_i y_i & \sum_{i=1}^{n} m_i\left(x_i^2 + y_i^2\right) \end{bmatrix} \tag{4.29}$$

(A similar set of definitions applies to rigid bodies, with incremental masses and integrals used instead of point masses and summations.) With this definition and the angular velocity vector ω, the angular momentum of the system Ω can be written as $\Omega = J\omega$ and the kinetic energy can be written as $K = \frac{1}{2}\omega^{\mathrm{T}} J\omega$. Find a set of three coordinate axes for the system such that if the angular velocity is along one of those coordinate axes, the angular momentum will be along one of them as well.

Solution:

The problem statement implies that the angular velocity is to be proportional to the angular momentum, i.e., $\Omega = \lambda\omega$, where λ is a constant of proportionality. This in turn implies that $J\omega = \lambda\omega$, or again, the eigenvalue-eigenvector problem $(J - \lambda I)\omega = 0$. If we solve this problem to find the eigenvectors υ_i, $i = 1, 2, 3$, and use them as new basis vectors to diagonalize the system, we will obtain a new inertia matrix

$$\bar{J} = \begin{bmatrix} J_1 & 0 & 0 \\ 0 & J_2 & 0 \\ 0 & 0 & J_3 \end{bmatrix} \tag{4.30}$$

where the quantities J_i, $i = 1, 2, 3$, are known as the *principal moments of inertia*. The eigenvectors are called *principal axes of inertia*. (How can we be sure three such axes exist?) If we denote by ω_i, $i = 1, 2, 3$, the components of the angular velocity about the three principal axes, then we will have

$$\Omega = \bar{J}\omega = \begin{bmatrix} J_1\omega_1 \\ J_2\omega_2 \\ J_3\omega_3 \end{bmatrix} \tag{4.31}$$

and

$$K = \tfrac{1}{2}\omega^{\mathrm{T}}\bar{J}\omega = \tfrac{1}{2}\left(J_1\omega_1^2 + J_1\omega_2^2 + J_1\omega_3^2\right) \tag{4.32}$$

The principal axes of an object will always be equal to any axes of symmetry of the body.

4.4.5 Jordan Canonical Forms

We have mentioned that the motivation for computing generalized eigenvectors is that we can use them in the modal matrix when we have an insufficient number of regular eigenvectors. We have seen that when the modal matrix, whose columns constitute a new basis of eigenvectors, contains only regular eigenvectors, then the matrix operator becomes a diagonal matrix of eigenvalues. This is because $A\boldsymbol{x} = \lambda\boldsymbol{x}$ for regular eigenvectors $\boldsymbol{x}$.

When generalized eigenvectors are included in the modal matrix, the resulting matrix operator in the new basis will not be diagonal because for generalized eigenvectors $\boldsymbol{x}_i$ chained to regular eigenvector $\boldsymbol{x}_\eta$, we have

$$\begin{aligned} A\boldsymbol{x}_\eta &= \lambda\boldsymbol{x}_\eta \\ A\boldsymbol{x}_{\eta-1} &= \lambda\boldsymbol{x}_{\eta-1} + \boldsymbol{x}_\eta \\ &\vdots \\ A\boldsymbol{x}_2 &= \lambda\boldsymbol{x}_2 + \boldsymbol{x}_3 \\ A\boldsymbol{x}_1 &= \lambda\boldsymbol{x}_1 + \boldsymbol{x}_2 \end{aligned}$$

From these relationships we define a *Jordan block* as a submatrix with the following structure:

$$\begin{matrix} \lambda & 1 & 0 & 0 & \cdots & 0 \\ 0 & \lambda & 1 & 0 & & 0 \\ 0 & 0 & \lambda & \ddots & \ddots & \vdots \\ 0 & 0 & 0 & \ddots & 1 & 0 \\ 0 & 0 & 0 & \ddots & \lambda & 1 \\ 0 & 0 & 0 & \cdots & 0 & \lambda \end{matrix}$$

The size of this block will be the length of the chain that defines it. This implies that the size of the largest Jordan block will be the size of the longest chain of eigenvectors, which has been defined as the index η of the eigenvalue. In a transformed A-matrix, there will be as many Jordan blocks corresponding to a particular eigenvalue as there are chains of eigenvectors corresponding to that eigenvalue, and their size will vary according to the lengths of the respective chains. Thus, there will be as many Jordan blocks for an eigenvalue as there are regular eigenvectors associated with that eigenvalue, which we know is the geometric multiplicity g_i. There also may be one or more Jordan blocks corresponding to different eigenvalues, and they will, of course, be calculated independently of the Jordan blocks for other eigenvalues. In the most trivial case, a matrix may have a complete set of n regular eigenvectors, so all its Jordan blocks, whether they belong to distinct eigenvalues or not, will be of size one, which gives the diagonal matrix in the new basis. This is why we say that the diagonal form is a special case of a Jordan form.

When transforming a matrix into its Jordan form^M, it is important to group regular eigenvectors and their chained generalized eigenvectors into the modal matrix in the reverse order in which the chain was produced using the top-down method. The regular eigenvector is first, followed sequentially by chained generalized eigenvectors. This will produce a coherent Jordan block in the transformed matrix. The blocks themselves can be arranged in the transformed matrix by rearranging the order in which the chains are placed as columns in the modal matrix. We will see in Chapters 6 and 8 the usefulness of Jordan forms. The following example illustrates these properties on the previously presented examples.

`jordan(A)`

Example 4.11: Jordan Forms for Some Previous Examples

Transform the A-matrices used in the previous two examples into their Jordan forms by constructing modal matrices with the regular and generalized eigenvectors.

$$A_1 = \begin{bmatrix} 1 & 2 & 0 \\ 0 & 1 & 0 \\ -3 & 3 & 5 \end{bmatrix} \qquad A_2 = \begin{bmatrix} 3 & -1 & 1 & 1 & 0 & 0 \\ 1 & 1 & -1 & -1 & 0 & 0 \\ 0 & 0 & 2 & 0 & 1 & 1 \\ 0 & 0 & 0 & 2 & -1 & -1 \\ 0 & 0 & 0 & 0 & 1 & 1 \\ 0 & 0 & 0 & 0 & 1 & 1 \end{bmatrix}$$

Solution:

All of the eigenvalues, eigenvectors, and generalized eigenvectors have already been computed for these two examples. For matrix A_1, we construct the modal matrix with these vectors:

$$M_1 = \left[x_1 \mid v_1^2 \mid v_1^1 \right] = \begin{bmatrix} 0 & 4 & 1 \\ 0 & 0 & 2 \\ 1 & 3 & 0 \end{bmatrix}$$

$$= \left[\begin{matrix} \text{regular} \\ \text{eigenvector} \end{matrix} \;\middle|\; \begin{matrix} \text{regular} \\ \text{eigenvector} \end{matrix} \;\middle|\; \begin{matrix} \text{generalized} \\ \text{eigenvector} \end{matrix} \right]$$

then

$$\hat{A}_1 = M_1^{-1} A_1 M_1 = \left[\begin{array}{c|cc} 5 & 0 & 0 \\ \hline 0 & 1 & 1 \\ 0 & 0 & 1 \end{array} \right]$$

For matrix A_2, the following (sequentially ordered) chains of eigenvectors and generalized eigenvectors were found:

$$\underset{(\lambda_1=0)}{x_1} = \begin{bmatrix} 0 \\ 0 \\ 0 \\ 0 \\ 1 \\ -1 \end{bmatrix} \qquad \{v_1^3, v_1^2, v_1^1\} = \left[\begin{array}{c|c|c} 2 & 1 & 0 \\ 2 & -1 & 0 \\ 0 & 2 & 0 \\ 0 & -2 & 1 \\ 0 & 0 & 1 \\ 0 & 0 & 1 \end{array} \right] \qquad \{v_2^2, v_2^1\} = \left[\begin{array}{c|c} 0 & 0 \\ 0 & 0 \\ 2 & 0 \\ -2 & 0 \\ 0 & 1 \\ 0 & 1 \end{array} \right]$$

therefore if we form the modal matrix

$$M_2 = \begin{bmatrix} 0 & 2 & 1 & 0 & 0 & 0 \\ 0 & 2 & -1 & 0 & 0 & 0 \\ 0 & 0 & 2 & 0 & 2 & 0 \\ 0 & 0 & -2 & 1 & -2 & 0 \\ 1 & 0 & 0 & 1 & 0 & 1 \\ -1 & 0 & 0 & 1 & 0 & 1 \end{bmatrix}$$

we get

$$\hat{A}_2 = M_2^{-1} A_2 M_2 = \left[\begin{array}{c|ccc|cc} 0 & 0 & 0 & 0 & 0 & 0 \\ \hline 0 & 2 & 1 & 0 & 0 & 0 \\ 0 & 0 & 2 & 1 & 0 & 0 \\ 0 & 0 & 0 & 2 & 0 & 0 \\ \hline 0 & 0 & 0 & 0 & 2 & 1 \\ 0 & 0 & 0 & 0 & 0 & 2 \end{array}\right]$$

which is seen to be a Jordan canonical form with a single 1×1 Jordan block corresponding to $\lambda_1 = 0$, and two Jordan blocks, of sizes 3×3 and 2×2, which is the order in which the eigenvector chains were arranged in modal matrix M_2.

4.5 Singular Values

Eigenvalues and eigenvectors of matrices have important application in the stability and control of systems of differential equations, as we will see in Chapters 7 and 10. For many of our purposes, the eigenvalues and eigenvectors sufficiently characterize the numerical properties of our operators. However, they are defined only for square matrices that represent operators $\Re^n \to \Re^n$. They cannot be computed for nonsquare matrices.

In this section we will introduce the concept of *singular values* and *singular value decompositions* (SVD). These tools will apply with equal validity for square and nonsquare matrices alike. They have many purposes, both directly, as in modern methods for robust controller design, and indirectly, as the underlying numerical algorithm in such everyday computations as rank determination.

> ***Singular Value Decomposition:*** Let A be an arbitrary $m \times n$ matrix. There exist two orthonormal matrices, U and V, such that

$$A = U\Sigma V^{\mathrm{T}} \tag{4.33}$$

where

$$\Sigma = \left[\begin{array}{c|c} S & 0 \\ \hline 0 & 0 \end{array}\right] \tag{4.34}$$

and S is the diagonal matrix, $S = diag\{\sigma_1, \sigma_2, \ldots, \sigma_r\}$. The r values $\sigma_1, \sigma_2, \ldots, \sigma_r$ are arranged such that $\sigma_1 \geq \sigma_2 \geq \cdots \geq \sigma_r > 0$. Together with the values $\sigma_{r+1} = \cdots = \sigma_n = 0$ they are known as *singular values* for matrix A. The factorization (4.33) is the *singular value decomposition*[M] of A.

svd(A)

4.5.1 Properties of Singular Value Decompositions

The first thing to note about an SVD is that the value r, i.e., the number of nonzero singular values, is equal to the rank of matrix A. A matrix therefore has as many zero singular values as its rank deficiency. These singular values are also the positive square roots of the eigenvalues of $A^{\mathrm{T}}A$. The columns of matrix U are the *left singular vectors*, are orthonormal, and are the eigenvectors of AA^{T}. The columns of V are the *right singular vectors*, are also orthonormal, and are the eigenvectors of $A^{\mathrm{T}}A$. However, while all these properties might suggest an easy way to compute an SVD, this is *not* a numerically accurate way to do it. In practice, there are several numerically robust methods for iteratively computing the singular value decomposition (see [6], [9], and [12]). We will not be further concerned with the actual computation of SVDs, but rather with their application in linear systems problems.

Example 4.12: Singular Value Decomposition

For the rank-2 matrix, give an SVD.

$$A = \begin{bmatrix} 4 & 8 & -2 & 0 \\ 3 & 3 & -3 & 1 \\ 1 & 5 & 1 & -1 \end{bmatrix}$$

Solution:

As mentioned, SVDs are never actually computed by hand, but rather with some numerically accurate computer programs. Therefore, we will give here, for purposes of illustrating such a decomposition, only the result:

$$A = U\Sigma V^{\mathrm{T}}$$
$$= \begin{bmatrix} 0.8165 & 0 & -0.5774 \\ 0.4082 & -0.7071 & 0.5774 \\ 0.4082 & 0.7071 & 0.5774 \end{bmatrix} \begin{bmatrix} 11.225 & 0 & 0 & 0 \\ 0 & 3.7417 & 0 & 0 \\ 0 & 0 & 0 & 0 \end{bmatrix} \cdot \begin{bmatrix} 0.4362 & -0.378 & 0.8138 & -0.066 \\ 0.8729 & 0.378 & -0.2825 & 0.1243 \\ -0.2182 & 0.7559 & 0.4977 & 0.365 \\ 0 & -0.378 & -0.1009 & 0.9203 \end{bmatrix}^{\mathrm{T}}$$

Notice that there are indeed two nonzero singular values, according to the rank of A.

4.5.2 Some Applications of Singular Values

Listed in this section are some common uses for singular values and SVDs [6]. It is assumed that a good computer program is available for computing the SVD.

Norm Computations

In Chapter 3 it was noted that although the euclidean, or 2-norm, of a matrix operator was relatively easy to define, it was not at that time clear how such a norm was actually computed. Generally, 2-norms are computed with the aid of the singular values:

$$\|A\|_2 = \sigma_1 \tag{4.35}$$

which is the largest singular value. This can actually be readily shown by considering Equation (3.36):

$$\begin{aligned} \|A\|_2 &= \sup_{x \neq 0} \frac{\|Ax\|}{\|x\|} \\ &= \sup_{x \neq 0} \frac{\langle Ax, Ax \rangle^{1/2}}{\langle x, x \rangle^{1/2}} = \sup_{x \neq 0} \frac{\langle x, A^T Ax \rangle^{1/2}}{\langle x, x \rangle^{1/2}} \end{aligned} \tag{4.36}$$

for real matrix A. Now because $A^{\mathrm{T}}A$ is obviously symmetric, we can consider using an orthonormal set $\{e_i\}$ of its eigenvectors as a basis in which to expand vector x, $x = \sum_{i=1}^{n} x_i e_i$, where each eigenvector e_i corresponds to an eigenvalue σ_i^2 of $A^{\mathrm{T}}A$ (notice the notation for these eigenvalues). Doing this gives

$$
\begin{aligned}
\|A\|_2 &= \sup_{x \neq 0} \frac{\left\langle \sum_{j=1}^{n} x_j e_j, A^{\mathrm{T}}A \sum_{i=1}^{n} x_i e_i \right\rangle^{1/2}}{\langle x, x \rangle^{1/2}} \\
&= \sup_{x \neq 0} \frac{\left\langle \sum_{j=1}^{n} x_j e_j, \sum_{i=1}^{n} x_i A^{\mathrm{T}}A e_i \right\rangle^{1/2}}{\left[\sum_{i=1}^{n} x_i^2 \right]^{1/2}} \\
&= \sup_{x \neq 0} \frac{\left\langle \sum_{j=1}^{n} x_j e_j, \sum_{i=1}^{n} x_i \sigma_i^2 e_i \right\rangle^{1/2}}{\left[\sum_{i=1}^{n} x_i^2 \right]^{1/2}} \\
&= \sup_{x \neq 0} \frac{\left[\sum_{i=1}^{n} x_i \sigma_i^2 \left\langle \sum_{j=1}^{n} x_j e_j, e_i \right\rangle \right]^{1/2}}{\left[\sum_{i=1}^{n} x_i^2 \right]^{1/2}} \\
&= \sup_{x \neq 0} \frac{\left[\sum_{i=1}^{n} x_i^2 \sigma_i^2 \right]^{1/2}}{\left[\sum_{i=1}^{n} x_i^2 \right]^{1/2}} \\
&= \sigma_1
\end{aligned}
$$

where σ_1 is the largest singular value of A. One can verify that the supremum (i.e., the least upper bound) achieved in the final line of this development occurs when $x = e_1$, i.e., the right singular vector of A (the eigenvector of $A^T A$) corresponding to singular value σ_1^2.

If A is a square invertible matrix (full rank), then we also have

$$\left\| A^{-1} \right\|_2 = \frac{1}{\sigma_n} \tag{4.37}$$

The Frobenius norm can also be computed using singular values:

$$\begin{aligned} \|A\|_F &= \left[\sigma_1^2 + \cdots + \sigma_n^2\right]^{1/2} \\ &= \left[\sigma_1^2 + \cdots + \sigma_r^2\right]^{1/2} \end{aligned} \tag{4.38}$$

Rank Determination and Condition Number

The SVD is by far the most common method for determining the rank of a matrix. As previously stated, the rank of a matrix is the number of nonzero singular values for that matrix.

Determinating the condition[M] number of a matrix is a related function. The *condition number* is a relative measure of how close a matrix is to rank deficiency and can be interpreted as a measure of how much numerical error is likely to be introduced by computations involving that matrix. For a square matrix, the condition number is a measure of how close the matrix is to being singular. The condition number is defined as $cond(A) = \sigma_1 / \sigma_n$, i.e., the largest singular value over the smallest. By this definition, it is clear that a singular matrix, which has at least one zero singular value, will have an infinite or undefined condition number. A matrix with a large condition number is said to be *ill-conditioned*, and a matrix with a small condition number (which is desirable) is said to be *well-conditioned*.

`cond(A)`

When matrix A is used in a system of linear simultaneous equations, $Ax = y$, the condition number is an approximation for the amount by which errors in A or y might be amplified in the solution for x.

Matrix Inverses

If A is a square matrix, then of course U and V will also be square matrices. In such a case, SVDs are useful in computing the inverse of nonsingular matrix A. Because the inverse of an orthonormal matrix is equal to its transpose, we have

$$A^{-1} = \left(U\Sigma V^{\mathrm{T}}\right)^{-1} = V\Sigma^{-1}U^{\mathrm{T}}$$
$$= V \cdot diag(\tfrac{1}{\sigma_1},\ldots,\tfrac{1}{\sigma_n}) \cdot U^{\mathrm{T}} \tag{4.39}$$

From this expression it can be seen how the "size" of the inverse depends on the singular values and why (4.37) is true.

Computing Subspaces

By partitioning the SVD into blocks corresponding to the location of the zero and nonzero singular values, we can exploit some special properties of the matrices U and V:

$$A = U\Sigma V^{\mathrm{T}} = \begin{bmatrix} U_1 & U_2 \end{bmatrix} \begin{bmatrix} S & 0 \\ 0 & 0 \end{bmatrix} \begin{bmatrix} V_1^{\mathrm{T}} \\ V_2^{\mathrm{T}} \end{bmatrix} \tag{4.40}$$

where the matrices are partitioned compatibly. It can be shown that the columns of matrix U_1 form an orthonormal basis of the range space of A, and the columns of V_2 form an orthonormal basis for the null space of A.

Projection Operators

Using the same partitions for the U and V matrices as above, the SVD is an easy way to compute projection operators. For the projection operator that projects vectors onto the range space of A, we may use:

$$P_{\mathrm{R}(A)} = U_1 U_1^{\mathrm{T}} = AA^{+} \tag{4.41}$$

and for the projection onto the null space of A, we may use

$$P_{\mathrm{N}(A)} = V_2 V_2^{\mathrm{T}} = I - A^{+}A \tag{4.42}$$

Solving Simultaneous Equations

Consider the equation $A\boldsymbol{x} = \boldsymbol{y}$, where $\boldsymbol{x} \in \Re^n$, $\boldsymbol{y} \in \Re^m$, and $A: \Re^n \to \Re^m$. As we have noted above, if A is nonsingular, then a solution to this system in terms of the SVD can be given as

$$\boldsymbol{x} = V \cdot \Sigma^{+} \cdot U^{\mathrm{T}} \cdot \boldsymbol{y} \tag{4.43}$$

where

$$\Sigma^{+} = \left[\begin{array}{c|c} S^{-1} & 0 \\ \hline 0 & 0 \end{array}\right]$$

and the zero blocks are sized appropriately according to the dimensions of U and V.

This formula also applies in the two cases considered in the previous chapter, i.e., the overdetermined and underdetermined cases, regardless of whether a unique solution exists, an infinite number of solutions exist, or no exact solutions exist. If there are zero singular values, we replace the corresponding

$$\frac{1}{\sigma_i} = \frac{1}{0}$$

terms by *zero*, and use exactly the same formula, i.e., (4.43). Equation (4.43) therefore gives the pseudoinverse solution to such systems, automatically implementing the left- or right-pseudoinverse as appropriate. In the event that no exact solution exists, (4.43) will give the solution of least-squared error. In the event that multiple solutions exist, (4.43) returns the solution with minimum norm. Adding to $\mathbf{x}$ any linear combination of the columns of V_2 will then be adding homogeneous terms from the null space of A.

Example 4.13: Singular Value Decompositions in Robotics

In this extended example, we will illustrate the use of SVDs in two aspects of a simple robotics problem, both involving the inverse kinematics of a simple robot. Another example will be given in Section 5.3.3. The problem concerns the determination of joint angles for a three-link robot. The reader need not have any previous exposure to robotics.

The robotic *forward* kinematics problem is simply the problem of determining the position and orientation of the tip of the robot, given the joint angles. The *inverse* kinematics problem is the problem of determining the joint angles for a given tip position. In our problem, we wish to determine the vector of angular joint velocities

$$\dot{\Theta} = \begin{bmatrix} \dot{\theta}_1 \\ \dot{\theta}_2 \\ \dot{\theta}_3 \end{bmatrix}$$

such that the tip of the third link travels with a prescribed velocity in a two-dimensional coordinate system XY (see Figure 4.3).

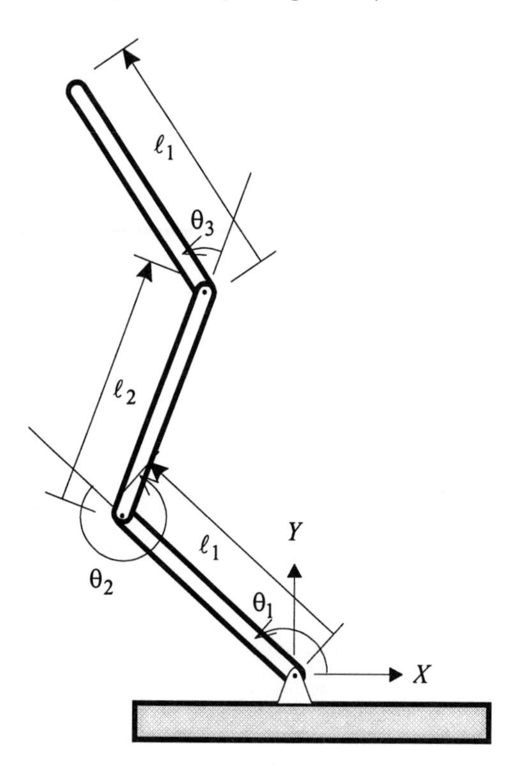

Figure 4.3 A three-link robot arm in a plane. Coordinates of the tip of the third link are measured in the Cartesian system XY, while individual joint angles are measured positively counterclockwise from the axis of the preceding link, except for the first link, which is measured from the horizontal.

The forward kinematic equations for this robot arm can be written from basic trigonometry as

$$\begin{aligned}\begin{bmatrix} x \\ y \end{bmatrix} &= \begin{bmatrix} l_1 \cos(\theta_1) + l_2 \cos(\theta_1 + \theta_2) + l_3 \cos(\theta_1 + \theta_2 + \theta_3) \\ l_1 \sin(\theta_1) + l_2 \sin(\theta_1 + \theta_2) + l_3 \sin(\theta_1 + \theta_2 + \theta_3) \end{bmatrix} \\ &\triangleq f(\Theta) = \begin{bmatrix} f_1(\Theta) \\ f_2(\Theta) \end{bmatrix}\end{aligned} \tag{4.44}$$

This vector function $f \in \Re^2$ of the vector of joint angles $\Theta \in \Re^3$ specifies the two coordinates of the tip of link three in terms of the three joint angles.

Often it is the roboticist's job to specify the three motor velocities

$$\dot{\Theta} = \begin{bmatrix} \dot{\theta}_1 \\ \dot{\theta}_2 \\ \dot{\theta}_3 \end{bmatrix}$$

necessary to produce a particular desired velocity

$$\dot{X}_{desired} = \begin{bmatrix} \dot{x}_{desired} \\ \dot{y}_{des.ired} \end{bmatrix}$$

of the tip of the robot. This represents the inverse velocity kinematics problem, which can be simplified with some new notation:

$$\begin{aligned} \dot{X}_{desired} &= \frac{d}{dt} f(\Theta) = \frac{\partial f(\Theta)}{\partial \Theta} \frac{d\Theta}{dt} \\ &\triangleq J(\Theta)\dot{\Theta} \end{aligned} \tag{4.45}$$

where $J(\Theta) \triangleq \partial f(\Theta)/\partial\Theta$ is known as the *jacobian* of the nonlinear transformation f. It is, in this case, a 2×3 matrix that depends nonlinearly on the joint angles Θ. Using the formula given in Appendix A for the derivative of a vector with respect to another vector, we can obtain for this example:

$$J(\Theta) = \begin{bmatrix} -l_1 S_1 - l_2 S_{12} - l_3 S_{123} & -l_2 S_{12} - l_3 S_{123} & -l_3 S_{123} \\ l_1 C_1 + l_2 C_{12} + l_3 C_{123} & l_2 C_{12} + l_3 C_{123} & l_3 C_{123} \end{bmatrix} \tag{4.46}$$

where $C_1 \triangleq \cos(\theta_1)$, $S_{12} \triangleq \sin(\theta_1 + \theta_2)$, etc. If at any given time we can accurately measure the angles Θ and thus compute $J(\Theta)$, then (4.45) can be considered a linear system of two equations and three unknowns.

Of course, if the jacobian $J(\Theta)$ were square and nonsingular, it could be inverted, and hence, (4.45) could be solved to determine the necessary joint velocities $\dot{\Theta}$. However, in this case the jacobian is nonsquare. Hence, a pseudoinverse solution is possible.

$$\dot{\Theta} = J^{+}(\Theta)\dot{X}_{desired} \tag{4.47}$$

or, as computed using the singular value decomposition,

$$\dot{\Theta} = V \cdot \Sigma^{+} \cdot U^{\mathrm{T}} \cdot \dot{\mathrm{X}}_{desired} \tag{4.48}$$

where $J = U\Sigma V^{\mathrm{T}}$.

Note that, according to the discussion of this pseudoinverse in Chapter 3, this solution, in the case of multiple possible solutions to underdetermined systems, minimizes a secondary criterion, namely the squared norm of the vector $\dot{\Theta}$, or, equivalently, $\frac{1}{2}\dot{\Theta}^{\mathrm{T}}\dot{\Theta}$. This has the effect of minimizing the vector of joint velocities for a given endpoint velocity.

When the jacobian of such a system is rank deficient, a particular physical phenomenon occurs. This is known as kinematic *singularity** and results in a robot that, in that joint angle configuration Θ, is unable to produce endpoint velocities in one or more directions of its workspace. For example, as seen in Figure 4.4, such a robot whose joints are all aligned is unable to instantaneously produce velocities along the axis of its aligned links.

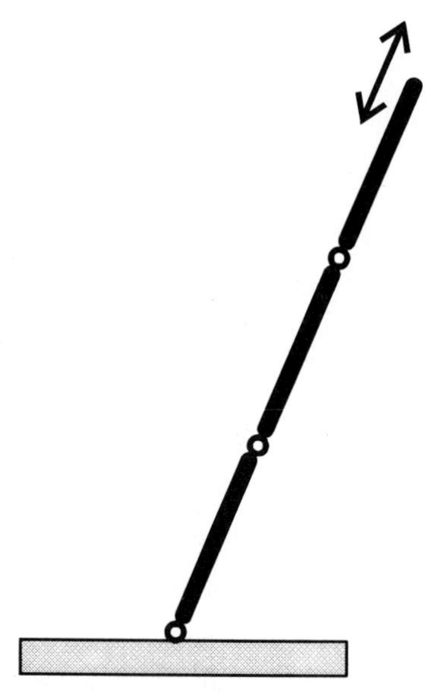

Figure 4.4 A three-link planar robot in a singular configuration. It is unable to instantaneously produce velocity in the direction of the arrow.

However, the concept of singularity can also be considered a matter of degree, i.e., in some configurations a robot may be closer to singularity than in other configurations. This can be displayed graphically with what are known as

* Note that the usage of this term is consistent with our notion of matrix singularity. When the nonsquare jacobian is rank-deficient, the pseudoinverse will not exist because the matrix JJ^{T} will be square and rank-deficient and, thus, singular.

manipulability ellipses in the robotics field. A manipulability ellipse is an ellipse, usually drawn with its center at the tip of the robot, whose axes depict the ability of the robot to produce velocity in the directions of those axes, with a unit joint angle velocity. For example, the three configurations in Figure 4.5 are shown with approximate manipulability ellipses attached.

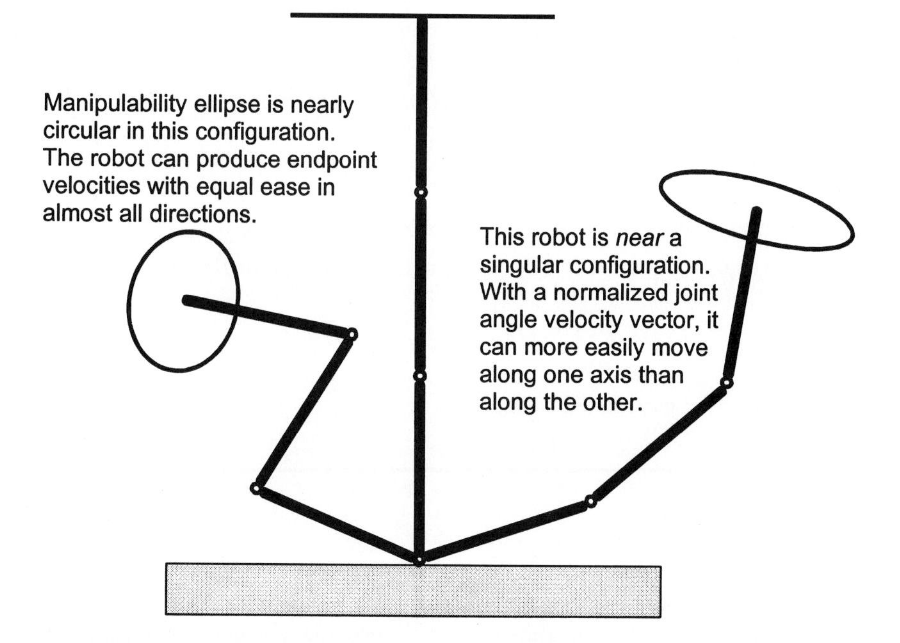

Figure 4.5 Three configurations of a robot with different degrees of manipulability.

The manipulability ellipses may be drawn by computing the SVD of the jacobian. The axis lengths of the ellipses are equal to the singular values of J, and the directions of those axes are equal to the left singular vectors (columns of U). These dimensions will be further illustrated in Example 5.1. In practice, the pseudoinverse solution itself may be augmented by a term that is proportional to the direction of the gradient of a manipulability measure, $M(\Theta)$, e.g.,

$$M(\Theta) = \sqrt{\left|JJ^{\mathrm{T}}\right|} = \prod_{i=1}^{n} \sigma_i \tag{4.49}$$

where the σ_i's are the singular values of J. This has the effect of maximizing the manipulator's distance from the nearest singularity. However, such a term must be first projected onto the null space of the jacobian, lest it prevent the robot from reaching the desired velocity $\dot{X}_{desired}$. That is, one can use

$$\begin{aligned}\dot{\Theta} &= J^{+}(\Theta)\dot{X}_{desired} + P_{N(J)}\nabla M(\Theta) \\ &= J^{+}(\Theta)\dot{X}_{desired} + \left[I - J^{+}(\Theta)J(\Theta)\right]\nabla M(\Theta)\end{aligned} \tag{4.50}$$

which, when multiplied by the jacobian, returns

$$\begin{aligned}J(\Theta)\dot{\Theta} &= J(\Theta)J^{+}(\Theta)\dot{X}_{desired} + \left[J(\Theta) - J(\Theta)J^{+}(\Theta)J(\Theta)\right]\nabla M(\Theta) \\ &= \dot{X}_{desired}\end{aligned}$$

demonstrating that the homogeneous term does not affect $\dot{X}_{desired}$. Referring to (4.43) and (4.42), Equation (4.50) can be computed using the SVD:

$$\dot{\Theta} = V \cdot \Sigma^{+} \cdot U^{T} \cdot \dot{X}_{desired} + V_2 V_2^{T} \nabla M(\Theta) \tag{4.51}$$

with the SVD of J partitioned as in (4.40).

4.6 Summary

In this chapter, we have presented new material concerning eigenvalues, eigenvectors, and singular values. Again we stress the geometric interpretation of these concepts: that eigenvectors are A-invariant subspaces. The convenience of knowing these invariant subspaces will become apparent in Chapter 6. Furthermore, the notion of a singular value is introduced, partly as a counterpart to eigenvalues and eigenvectors for nonsquare matrices, but mostly because they are finding increasing use in modern control theory, robust and adaptive control, and signal processing. The fact that the largest singular value provides the norm of a matrix is equally valid in time- and frequency-domain, and this has recently led to significant developments in control theory. In particular, in the field of optimal control, wherein a cost criterion is to be minimized by a stabilizing control input, the use of the singular value has provided a means to minimize frequency-domain criteria, whereas previously, optimal control had been primarily performed in time-domain.

Other important developments in this chapter are:

- Eigenvectors and eigenvalues, like other concepts introduced in previous chapters, are not to be considered properties of *matrices* but rather of *operators*. This brings us back to the notion of invariance, i.e., a subspace on which an operator has only a scaling effect.

- We investigated the basis of eigenvectors as a change-of-basis matrix. This results in Jordan forms in general, and diagonal forms in special cases. The importance of these forms is partly illustrated in Exercise 4.14, where it is shown how diagonalization aids the solution of differential systems. This will be further emphasized in Chapter 6.

- The concept of generalized eigenvectors was discussed in some detail. The use of generalized eigenvectors is limited except for computing Jordan forms, because they are not invariant subspaces, as are regular eigenvectors.

- As we have mentioned, the singular value, briefly introduced here, is indispensible for many matrix computations and numerical methods. Algorithms for computing them are numerically stable and are the basis for many of the other computations performed by computer-aided engineering software packages. Quantities such as rank, condition, and the solutions to simultaneous equations are often computed via an underlying SVD.

Chapter 5 is the final chapter in Part I. It presents a mathematical treatment of other kinds of functions on matrices, i.e., not necessarily linear operators. These functions will be required when we solve state space systems, analyze their stability, and develop control and estimation methods for them.

4.7 Problems

4.1 Let T be an operator that rotates vectors in $\Re^3$ by an angle 45° counterclockwise about the vector $x = [1 \quad 0 \quad 1]^T$ and magnifies them by a factor of four. Determine all real-valued eigenvectors and their corresponding eigenvalues.

4.2 Find all eigenvalues and eigenvectors for the matrix

$$A = \begin{bmatrix} 7 & 2 & 2 & 0 \\ -2 & 2 & -1 & 0 \\ -7 & -4 & 0 & 1 \\ 2 & 1 & 1 & 3 \end{bmatrix}$$

4.3 Find the eigenvalues and corresponding eigenvectors of the following matrices:

$$\text{a)} \begin{bmatrix} 1 & -2 & 0 \\ -1 & 2 & -1 \\ 0 & -1 & 1 \end{bmatrix} \quad \text{b)} \begin{bmatrix} 1 & 3 & 3 \\ 3 & 1 & 3 \\ -3 & -3 & -5 \end{bmatrix} \quad \text{c)} \begin{bmatrix} 0 & 1 \\ -\omega_n^2 & 0 \end{bmatrix}$$

4.4 Prove that for matrix A with eigenvalues λ_i, $|A| = \prod_{i=1}^{n} \lambda_i$.

4.5 Find the eigenvalues, eigenvectors (real and/or generalized), and Jordan form for the matrices:

$$\text{a)} \begin{bmatrix} 8 & -8 & -2 \\ 4 & -3 & -2 \\ 3 & -4 & 1 \end{bmatrix} \quad \text{b)} \begin{bmatrix} 1 & 0 & -4 \\ 0 & 3 & 0 \\ -2 & 0 & -1 \end{bmatrix} \quad \text{c)} \begin{bmatrix} 2 & 1 & 1 \\ 0 & 3 & 1 \\ 0 & -1 & 1 \end{bmatrix}$$

4.6 Let $\mathbf{V}$ be the vector space of polynomials of degree less than or equal to two, with inner product

$$\langle f, g \rangle = \bar{f}(0)g(0) + \bar{f}(1)g(1) + \bar{f}(2)g(2)$$

Also let T be a linear operator defined by

$$(Tp)(x) = \frac{d}{dx}(xp(x))$$

a) Find the matrix for the operator T in the basis $\{1, x, x^2\}$.

b) Find the eigenvalues and eigenvectors of T.

c) Determine whether or not T is self-adjoint, i.e., its adjoint is the same as itself. Why or why not?

4.7 Find the eigenvalues and eigenvectors of the operator

$$A = \begin{bmatrix} 24 & 18 & 9 \\ -28 & 26 & 4 \\ 56 & -28 & 4 \end{bmatrix}$$

Then determine coefficients a_i such that

$$\begin{bmatrix} 4 \\ 4 \\ -2 \end{bmatrix} = \sum_{i=1}^{3} a_i \mathbf{e}_i$$

where the $\mathbf{e}_i$'s are eigenvectors of A.

4.8 Find a similarity transformation that reduces the following matrices to their Jordan forms. Identify the Jordan blocks.

$$A = \begin{bmatrix} 7 & -11 & 30 \\ -14 & 13 & -43 \\ -8 & 10 & 29 \end{bmatrix} \qquad B = \begin{bmatrix} 0 & 1 \\ -1 & -2 \end{bmatrix}$$

4.9 Find the eigenvalues of the matrix

$$\begin{bmatrix} \alpha & \omega \\ -\omega & \alpha \end{bmatrix}$$

4.10 Let **V** be the linear space of polynomials of degree less than or equal to two, with the inner product defined as

$$\langle f, g \rangle = \int_{-1}^{1} \bar{f}(x) g(x)\, dx$$

Consider the linear operator

$$(Tp)(x) = \tfrac{1}{2} x(1-x) p''(x) + x p'(x) + p(x) + x^2 p(0)$$

a) Find the matrix of T with respect to the basis $\{1, x, x^2\}$.

b) Find the eigenvalues and eigenvectors of T.

c) Determine whether T is hermitian.

4.11 Let $\mathbf{W}$ be the linear space of real-valued functions that are integrable on the interval $x \in [0,1]$. Let operator T be defined by

$$(Tf)(x) = \int_0^1 (x - 2t) f(t)\, dt$$

a) Show that T is a linear operator.

b) Find the range space $R(T)$.

c) Find all nonzero eigenvalues of T and their corresponding eigenvectors.

d) Find a vector $p(t) \in \mathbf{W}$ of degree two such that $Tp = 0$.

4.12 Let $\mathbf{V}$ be the space of functions spanned by $\cos t$ and $\sin t$. An operator T is defined as

$$(Tf)(t) = \int_0^{2\pi} \sin(t - x) f(x)\, dx$$

a) Find the matrix of the operator T in the basis $\{\cos t, \sin t\}$.

b) Find the eigenvalues and eigenvectors of T.

4.13 Suppose two square matrices A and $\hat{A}$ are similar to one another; i.e., $\hat{A} = M^{-1}AM$ for some orthonormal matrix M. Show explicitly that the eigenvalues of $\hat{A}$ are the same as those of A. Do the two matrices have the same eigenvectors?

4.14 For the electrical circuit in Example 1.2, find a set of coordinates that transforms the describing equations into a pair of *decoupled* first-order differential equations, i.e., two first-order equations that can be solved independently of one another. Are these coordinates physically meaningful? (If necessary, use some nominal component values to simplify the algebra.)

4.15 Write MATLAB (or code in another language) to detect the presence of and find the generalized eigenvectors for an arbitrary matrix. Use it to find the Jordan form for the matrices given in Example 4.11.

4.8 References and Further Reading

Much of the material in this chapter is standard and can be found in any text on matrix computations, such as [5], including the control systems texts [1] and [2]. Explicit discussion of the computation of generalized eigenvectors can be difficult to find; ours is closely related to [1]. For further discussion of eigenvalues and eigenvectors for function spaces, see [3] and [8]. For more examples systems with *eigenfunctions*, see [8].

Our treatment of SVDs should be regarded as merely an overview. Good introductions to the theory of SVDs can be found in [6] and [12]. The use of SVDs for the solution of least-squares problems is found in [7]. Computation of the SVDs is discussed in some detail in [6], [12], and [13]. Their application in robotics, as we introduced by example, can be found in [9], [10], and [11] (for more information on robot kinematics itself, see [4] and [10]).

[1] Brogan, William L., *Modern Control Theory*, 3rd edition, Prentice-Hall, 1991.

[2] Chen, Tsi-Chong, *Linear System Theory and Design*, Holt, Rinehart, and Winston, 1984.

[3] Coddington, Earl A., and Norman Levinson, *Theory of Ordinary Differential Equations*, McGraw-Hill, 1955.

[4] Fu, K. S., R. C. Gonzalez, and C. S. G. Lee, *Robotics: Control, Sensing, Vision, and Intelligence*, McGraw-Hill, 1987.

[5] Golub, Gene H., and Charles F. Van Loan, *Matrix Computations*, Johns Hopkins University Press, 1989.

[6] Klema, Virginia, and Alan J. Laub, "The Singular Value Decomposition: Its Computation and Some Applications", *IEEE Transactions on Automatic Control*, vol. AC-25, no. 2, April 1980, pp. 63-79.

[7] Lawson, Charles L., and Richard J. Hanson, *Solving Least Squares Problems*, Prentice-Hall, 1974.

[8] Lebedev, N. N., I. P. Skalskaya, and Y. S. Uflyand, *Worked Problems in Applied Mathematics*, Dover, 1965.

[9] Maciejewski, Anthony A., and Charles A. Klein, "The Singular Value Decomposition: Computation and Applications to Robotics," *International Journal of Robotics Research*, vol. 8, no. 6, December 1989, pp. 164-176.

[10] Nakamura, Yoshihiko, *Advanced Robotics: Redundancy and Optimization*, Addison-Wesley, 1991.

[11] Nakamura, Yoshihiko, and Yingti Xu, "Geometrical Fusion Method for Multi-Sensor Robotic Systems," *IEEE International Conference on Robotics and Automation*, 1989, pp. 668-673.

[12] Press, William H. Brian P. Flannery, Saul A. Teukolsky, and William T. Vetterling, *Numerical Recipes in C: The Art of Scientific Computing*, Cambridge University Press, 1988.

[13] Ulrey, Renard R., Anthony A. Maciejewski, and Howard J. Siegel, "Parallel Algorithms for Singular Value Decomposition," *Proc. 8th International Parallel Processing Symposium*, April 1994, pp. 524-533.

5

Functions of Vectors and Matrices

We have already seen that the linear operator is a function on a vector space. The linear operator transforms a vector from one space into another space, $A: \Re^n \to \Re^m$, or possibly from a space into itself. Among this group of operators is a special class of operators known as *linear functionals*, which are simply linear operators that map vectors into the real numbers, $\Re$. Linear functionals serve some special purposes in vector spaces, and it is worth considering them in more detail here. In addition, three other classes of functions are discussed: the multilinear functional, the quadratic form, and matrix functions. Multilinear functionals, such as bilinear forms, also map vectors into the reals, while quadratic forms map vectors into reals, but are not linear. The more general matrix functions are functions, not necessarily linear, that map matrices (or matrix forms of operators) into other matrices. Matrix functions will be especially important in the solution of state equations as discussed in Chapter 6.

5.1 Linear Functionals

As mentioned above, the linear functional is simply a linear function of a vector:

> ***Linear Functional:*** A *linear functional* $f(x) \in \Re$ maps a vector $x \in \mathbf{X}$ into the set of real numbers such that:

$$\begin{aligned} f(cx) &= cf(x) \\ f(x+y) &= f(x)+f(y) \end{aligned} \tag{5.1}$$

where c is a scalar in the field of $\mathbf{X}$ and $\boldsymbol{y} \in \mathbf{X}$.

Consider that the n-dimensional vector space $\mathbf{X}$ is provided with a basis $\{\boldsymbol{e}_i\}$. Then

$$f(\boldsymbol{x}) = f(\sum_{i=1}^{n} x_i \boldsymbol{e}_i) = \sum_{i=1}^{n} x_i f(\boldsymbol{e}_i) \tag{5.2}$$

We can determine the effect of a linear functional on a vector by first determining its effect on the basis vectors. If we know the functional's effect on the basis vectors, we can use this result to operate on arbitrary vectors with little computation, which we have done in Section 3.1.2 with linear operators. We can think of the coefficients x_i as being the "components" of the functional in basis $\{\boldsymbol{e}_i\}$.

5.1.1 Changing the Basis of a Functional

In the event that the basis of a vector space changes and we wish to determine the effect of a functional f in the new basis $\{\hat{\boldsymbol{e}}_i\}$, we use the relationship between basis vectors as expressed in Section 2.2.4:

$$\boldsymbol{e}_j = \sum_{i=1}^{n} b_{ij} \hat{\boldsymbol{e}}_i \tag{5.3}$$

or in matrix form:

$$\left[\boldsymbol{e}_1 \mid \boldsymbol{e}_2 \mid \cdots \mid \boldsymbol{e}_n\right] = \left[\hat{\boldsymbol{e}}_1 \mid \hat{\boldsymbol{e}}_2 \mid \cdots \mid \hat{\boldsymbol{e}}_n\right] B \tag{5.4}$$

Then

$$\begin{aligned} f(\boldsymbol{x}) &= \sum_{j=1}^{n} x_j f(\sum_{i=1}^{n} b_{ij} \hat{\boldsymbol{e}}_i) \\ &= \sum_{j=1}^{n} x_j \sum_{i=1}^{n} b_{ij} f(\hat{\boldsymbol{e}}_i) \\ &= \sum_{i=1}^{n} \left(\sum_{j=1}^{n} b_{ij} x_j \right) f(\hat{\boldsymbol{e}}_i) \\ &\triangleq \sum_{i=1}^{n} \hat{x}_i f(\hat{\boldsymbol{e}}_i) \end{aligned} \tag{5.5}$$

where the $\hat{x}_i$'s are considered the component of the functional in the new basis. By equating the third and fourth lines of (5.5), it can be seen that

$$\hat{x}_i = \sum_{j=1}^{n} b_{ij} x_j$$

or in vector-matrix form:

$$\hat{x} = Bx \tag{5.6}$$

which is the same result as Equation 2.33. This gives us the formula for changing the basis of the components.

5.2 Multilinear Functionals

As might be guessed from the name, a multilinear functional is a functional that is linear in a number of different vectors, which would be expressed as $A(x_1, x_2, \ldots, x_k)$. The most common form of the multilinear functional is the so-called "bilinear form." (The term *form* is generally used in finite-dimensional spaces, whereas the term *functional* is sometimes restricted to infinite-dimensional spaces such as function spaces.) We will discuss only these bilinear forms here. Other types of multilinear functionals can be found in books on tensors [1]. The bilinear form is a functional that is linear in two different vectors, according to the following definition:

> ***Bilinear Form:*** A *bilinear form B* is a functional that acts on *two* vectors from space **X** in such a way that if x, y, and z are vectors in **X**, and α is a scalar in the field of **X**, then

$$\begin{aligned} B(x+y,z) &= B(x,z)+B(y,z) \\ B(x,z+y) &= B(x,z)+B(x,y) \\ B(x,\alpha y) &= \alpha B(x,y) \\ B(\alpha x,y) &= \alpha B(x,y) \end{aligned} \tag{5.7}$$

True to its name, the bilinear form $B(x,y)$ is linear in x and in y. If the last condition is replaced by the condition

$$B(\alpha x, y) = \overline{\alpha} B(x,y) \tag{5.8}$$

then the form B is known strictly as a *sesquilinear form*, but this distinction is often neglected.

In an n-dimensional linear vector space, bilinear forms can be specified with a set of n^2 components, because each vector $\boldsymbol{x}$ and $\boldsymbol{y}$ can be expanded into, at most, n independent components. These n^2 numbers may be written as an $n \times n$ matrix, where the $(i,j)^{th}$ element of this matrix is:

$$b_{ij} = B(\boldsymbol{e}_i, \boldsymbol{e}_j) \tag{5.9}$$

so the bilinear form itself may be written as:

$$B(\boldsymbol{x}, \boldsymbol{y}) = \bar{\boldsymbol{x}}^{\mathrm{T}} \left[b_{ij} \right] \boldsymbol{y} \stackrel{\Delta}{=} \bar{\boldsymbol{x}}^{\mathrm{T}} B \boldsymbol{y} \tag{5.10}$$

Such a bilinear form is said to be *symmetric* if $B(\boldsymbol{x}, \boldsymbol{y}) = B(\boldsymbol{y}, \boldsymbol{x})$, and hermitian if $B(\boldsymbol{x}, \boldsymbol{y}) = \overline{B(\boldsymbol{y}, \boldsymbol{x})}$, in which case it can be shown that the matrix representing the form in (5.10) is symmetric (hermitian) in the usual matrix sense.

5.2.1 Changing the Basis of a Bilinear Form

If the basis of the vector space is changed, we would expect that the matrix representing the bilinear form will change as well. To determine how it changes denote by B the matrix for the bilinear form in the original basis $\{\boldsymbol{e}_i\}$, and let $\hat{B}$ denote the form in a new basis $\{\hat{\boldsymbol{e}}_j\}$. Assume that the relationship between the old basis and the new basis is known, as it was in Chapter 2 [see Equation (2.28)]:

$$\boldsymbol{e}_i = \sum_{k=1}^{n} \beta_{ki} \hat{\boldsymbol{e}}_k \tag{5.11}$$

(Here we are using the symbol β_{ki} to avoid confusion with the elements of the bilinear form b.) Substituting this into the expression for the $(i,j)^{th}$ element of the matrix representing B, i.e., Equation (5.9), we obtain:

$$\begin{aligned} b_{ij} &= B(\boldsymbol{e}_i, \boldsymbol{e}_j) \\ &= B\left(\sum_{l=1}^{n} \beta_{li} \hat{\boldsymbol{e}}_l, \sum_{m=1}^{n} \beta_{mj} \hat{\boldsymbol{e}}_m \right) \end{aligned}$$

$$= \sum_{l=1}^{n} \beta_{li} \sum_{m=1}^{n} \beta_{mj} B(\hat{e}_l, \hat{e}_m)$$

$$\triangleq \sum_{l=1}^{n} \beta_{li} \sum_{m=1}^{n} \beta_{mj} \hat{b}_{lm}$$

One can see that if the coefficients β_{ij} are arranged in a matrix β, then the summations above result in the relationship

$$B = \beta^{\mathrm{T}} \hat{B} \beta \tag{5.12}$$

which is somewhat different, yet similar to the similarity transformation that performs a change of basis on the matrix of a linear operator, as in Equation (3.24). Note, though, that if the change of basis matrix β is orthonormal, so that its inverse is equal to its transpose, the transformations are performed with the same operations.

5.2.2 Bilinear Forms as Inner Products

The bilinear form written as (5.10) for any n-dimensional space can also be written as $B(x, y) = \langle x, By \rangle$. This suggests that the bilinear forms and inner products are equivalent. However, the bilinear form cannot be expressed as the matrix product $B(x, y) = \bar{x}^{\mathrm{T}} By$ for function spaces, which are infinite dimensional. For example, consider the space of real-valued functions defined on the interval $t \in [0,1]$. A valid bilinear form for this space is

$$B(f(t), g(t)) = \int_0^1 f(t) g(t) w(t)\, dt \tag{5.13}$$

One can note that this bilinear form also appears to be a valid inner product on the space. This is generally true with one condition: inner products are bilinear forms defined on real vector spaces provided that $B(x, x) > 0$ holds for $x \neq 0$ and $B(x, x) = 0$ holds for $x = 0$. This condition is discussed in more detail in the next section.

If this were a space of complex-valued functions, then the inner product would be a sesquilinear form:

$$B(f(t), g(t)) = \int_0^1 \bar{f}(t) g(t) w(t)\, dt \tag{5.14}$$

In (5.13) and (5.10), the function $w(t)$ is known as a *weighting function*. We have used $w(t)=1$ in past examples of inner products, but it may be introduced into the form (inner product) for special purposes, such as to make an operator self-adjoint by changing the way the inner product is computed.

5.3 Quadratic Forms

We can define quadratic forms in the same way that we defined bilinear forms except that we shall use the same vector twice. For example, given a matrix A, a *quadratic* form for a finite-dimensional (complex-valued) vector may be expressed as

$$A(\boldsymbol{x},\boldsymbol{x}) = \bar{\boldsymbol{x}}^{\mathrm{T}} A\boldsymbol{x} \tag{5.15}$$

or

$$A(\boldsymbol{x},\boldsymbol{x}) = \langle \boldsymbol{x}, A\boldsymbol{x} \rangle \tag{5.16}$$

In most cases, it is convenient to speak only of symmetric or hermitian quadratic forms. This is because if $\bar{\boldsymbol{x}}^{\mathrm{T}} A\boldsymbol{x}$ is a scalar, then

$$\bar{\boldsymbol{x}}^{\mathrm{T}} A\boldsymbol{x} = \overline{\left(\bar{\boldsymbol{x}}^{\mathrm{T}} A\boldsymbol{x}\right)}^{\mathrm{T}} = \bar{\boldsymbol{x}}^{\mathrm{T}} \bar{A}^{\mathrm{T}} \boldsymbol{x}$$

Therefore,

$$\bar{\boldsymbol{x}}^{\mathrm{T}} A\boldsymbol{x} = \frac{1}{2}\left(\bar{\boldsymbol{x}}^{\mathrm{T}} A\boldsymbol{x} + \bar{\boldsymbol{x}}^{\mathrm{T}} \bar{A}^{\mathrm{T}} \boldsymbol{x}\right) = \bar{\boldsymbol{x}}^{\mathrm{T}} \left(\frac{A+\bar{A}^{\mathrm{T}}}{2}\right) \boldsymbol{x} \tag{5.17}$$

which is, by inspection, hermitian. Thus, when we speak of quadratic forms, we will always assume that we are working with a symmetric matrix.

We will be using quadratic forms as "weighted" squared norms; this approach is similar to the way we treated bilinear forms as inner products. However, there is a restriction on the use of $\langle \boldsymbol{x}, A\boldsymbol{x} \rangle^{1/2}$ as a norm because any norm we define must satisfy the conditions set forth in Chapter 2, including the requirement that they be nonnegative, and zero only when $\boldsymbol{x}=0$. For a quadratic form in general, this may not be the case. In fact, we may have one of five possibilities:

1. If $\bar{\boldsymbol{x}}^{\mathrm{T}} A\boldsymbol{x} > 0$ for all $\boldsymbol{x} \neq 0$, the (symmetric) quadratic form (5.15) is said to be *positive definite*.

2. If $\bar{x}^{\mathrm{T}} A x \geq 0$ for all $x \neq 0$, the (symmetric) quadratic form (5.15) is said to be *positive semidefinite.*
3. If $\bar{x}^{\mathrm{T}} A x < 0$ for all $x \neq 0$, the (symmetric) quadratic form (5.15) is said to be *negative definite.*
4. If $\bar{x}^{\mathrm{T}} A x \leq 0$ for all $x \neq 0$, the (symmetric) quadratic form (5.15) is said to be *negative semidefinite.*
5. If $\bar{x}^{\mathrm{T}} A x > 0$ for some $x \neq 0$ and $\bar{x}^{\mathrm{T}} A x < 0$ for other $x \neq 0$, the (symmetric) quadratic form (5.15) is said to be *indefinite.*

These same terms are often applied to the matrix A itself, not just the quadratic form it defines.

5.3.1 Testing for Definiteness

There are a number of tests to determine whether a matrix it is positive definite, negative definite, etc. However, we will consider only the most commonly used test, which is given by the following theorem:

> THEOREM: A quadratic form (or equivalently, the symmetric matrix A defining it) is *positive (negative) definite* if *all* of the eigenvalues of A are positive (negative) or have positive (negative) real parts. The quadratic form is *positive (negative) semidefinite* if all of the eigenvalues of A are nonnegative (nonpositive) or have nonnegative (nonpositive) real parts. (5.18)

> PROOF: Given that we are working with a symmetric matrix, we can use the result of the previous chapter to realize that we can compute an orthonormal basis of the eigenvectors of A. This gives us the orthonormal modal matrix M with which we can change variables, $x = M\hat{x}$. Substituting this into (5.15),

$$
\begin{aligned}
\bar{x}^{\mathrm{T}} A x &= \bar{\hat{x}}^{\mathrm{T}} M^{\mathrm{T}} A M \hat{x} \\
&= \bar{\hat{x}}^{\mathrm{T}} diag\{\lambda_1, \ldots, \lambda_n\} \hat{x} \\
&= \sum_{i=1}^{n} |x_i|^2 \lambda_i
\end{aligned}
$$

> where $diag\{\,\}$ is a matrix with the elements of the set argument on the diagonal and zeros elsewhere. It is apparent from this expression that for $\bar{x}^{\mathrm{T}} A x$ to be positive (nonnegative) for *any* $x \neq 0$, then $\lambda_i > 0$ $(\lambda_i \geq 0)$ for each i.

Similarly, for this expression to be negative (nonpositive) for *any* $\mathbf{x} \neq 0$, then $\lambda_i < 0$ $(\lambda_i \leq 0)$ for each i.

While this theorem does not cover the indefinite case, it should be clear that an indefinite quadratic form is one whose matrix A has some positive eigenvalues and some negative eigenvalues.

There is another common test for definiteness that is often used for small matrices and is based on the *principal minors*. *Minors* of a matrix are determinants formed by striking out an equal number of rows and columns from a square matrix. *Principal minors* of a matrix are the minors whose diagonal elements are also diagonals of the original matrix. *Leading principal minors* are principal minors that contain *all* the diagonal elements of the original matrix, up to n, the size of the minor. For example, in Figure 5.1 below we indicate all the leading principal minors of an arbitrary matrix.

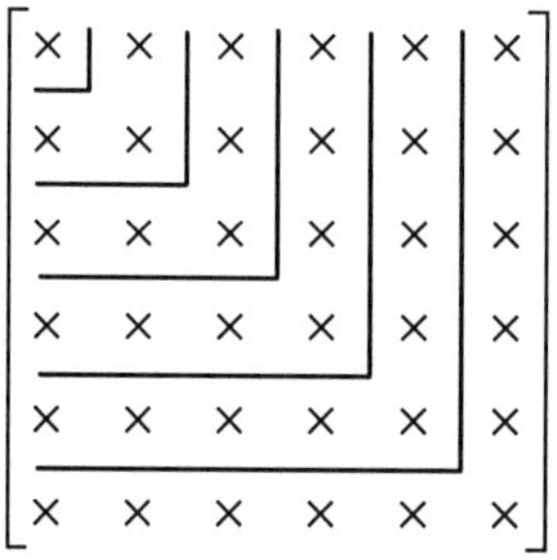

$$\begin{bmatrix} \times & \times & \times & \times & \times & \times \\ \times & \times & \times & \times & \times & \times \\ \times & \times & \times & \times & \times & \times \\ \times & \times & \times & \times & \times & \times \\ \times & \times & \times & \times & \times & \times \\ \times & \times & \times & \times & \times & \times \end{bmatrix}$$

Figure 5.1 Leading principal minors are formed by making determinants of the leading square sub-matrices of the original matrix.

With this established, we can test a matrix for positive definiteness by establishing that all the leading principal minors are positive. The matrix will be positive semidefinite if all the leading principal minors are nonnegative. The matrix will be negative definite if the leading principal minors alternate in sign, beginning with a negative term as we start with the 1×1 case. (How would you state the rule for negative semidefiniteness?)

5.3.2 Changing the Basis of a Quadratic Form

Analogous to the basis change for bilinear forms and entirely consistent with the basis change performed in the example above, we can change the matrix for a quadratic form when the basis of the underlying space is changed. The derivation is the same as in Section 5.2.1 and provides the same result. If A is the matrix for a quadratic form on a space using basis $\{e_i\}$ and $\hat{A}$ is the matrix for the same quadratic form on the same space, but using $\{\hat{e}_i\}$ for the basis, then the two forms are related by

$$A = \beta^{\mathrm{T}} \hat{A} \beta$$

where β is the change of basis matrix denoted by B in Equation (2.33).

5.3.3 Geometry of Quadratic Forms

An interesting property of quadratic forms is that they may be efficiently used to represent the equations of geometric objects known as *quadrics*. An example already seen in Chapter 4 is the ellipsoid. If $\boldsymbol{x} \in \Re^n$ and the $n \times n$ matrix A is positive definite and symmetric, then

$$\boldsymbol{x}^{\mathrm{T}} A \boldsymbol{x} = 1 \tag{5.19}$$

is the equation of an ellipsoid in $\Re^n$. If A is not positive definite, then other geometric possibilities arise, such as conics and hyperboloids, which are also quadrics.

To get a graphical image of the ellipsoid, we must draw the surface of the equation $\boldsymbol{x}^{\mathrm{T}} A \boldsymbol{x} = 1$. In order to see the relative dimensions of the ellipsoid and not scale the entire ellipsoid, we normalize its size by considering only the values of $\boldsymbol{x}^{\mathrm{T}} A \boldsymbol{x}$ such that $\boldsymbol{x}^{\mathrm{T}} \boldsymbol{x} = 1$. Now to get a picture of what the ellipsoid looks like, i.e., the directions of its principal axes and how far the surface of the ellipsoid extends in each of those directions, we will find the minimum and maximum distance of the surface of the ellipsoid from the origin. We will thus extremize $\boldsymbol{x}^{\mathrm{T}} A \boldsymbol{x}$ subject to $\boldsymbol{x}^{\mathrm{T}} \boldsymbol{x} = 1$.

We solve this optimization problem in a manner similar to the underdetermined system problem of Chapter 3. Introducing a scalar LaGrange multiplier γ, we construct the hamiltonian

$$H = \boldsymbol{x}^{\mathrm{T}} A \boldsymbol{x} + \gamma \left(1 - \boldsymbol{x}^{\mathrm{T}} \boldsymbol{x}\right) \tag{5.20}$$

Now taking the appropriate derivatives,

$$\frac{\partial^{\mathrm{T}} H}{\partial \boldsymbol{x}} = 2A\boldsymbol{x} - 2\gamma \boldsymbol{x} = 0 \tag{5.21}$$

and

$$\frac{\partial H}{\partial \gamma} = 1 - \boldsymbol{x}^{\mathrm{T}} \boldsymbol{x} = 0 \tag{5.22}$$

Simplifying (5.21), we arrive at

$$(A - \gamma I)\boldsymbol{x} = 0 \tag{5.23}$$

which implies that the LaGrange multiplier is equal to an eigenvalue and $\boldsymbol{x}$ is the corresponding eigenvector (recall that if A is positive definite, there will be n of these). Thus the extrema of the ellipsoid occur at the eigenvectors. [Equation (5.22) merely reinforces the normalization of the size of the ellipsoid.] If indeed $\boldsymbol{x}$ is a normalized eigenvector corresponding to eigenvalue γ, then at this point,

$$\begin{aligned} \boldsymbol{x}^{\mathrm{T}} A\boldsymbol{x} &= \boldsymbol{x}^{\mathrm{T}} \gamma \boldsymbol{x} \\ &= \gamma \boldsymbol{x}^{\mathrm{T}} \boldsymbol{x} \\ &= \gamma \|\boldsymbol{x}\|^2 \\ &= \gamma \end{aligned}$$

This means that the extremal value taken by the ellipsoid at this point is the eigenvalue γ. This will be true for the distance from the origin to the surface of the ellipsoid along any of the principal axes. Note that this result is true for a symmetric, positive definite matrix only; for a nonsymmetric matrix the principal axes are the left singular vectors and the axis lengths are the singular values.

The example below gives an application in which the computation of an ellipsoid is useful for analysis of physical systems. The subsequent diagram (Figure 5.2) then illustrates the geometry of the ellipse and the related eigenvalues and eigenvectors of matrix A.

Example 5.1: Statistical Error Analysis

Ellipsoids such as those depicted in Figure 4.5 in Example 4.13 are useful graphical representations for quantities that have different magnitudes in different directions. In that example, the ellipses represented the extent to which a robot is able to produce motion in specific directions of its workspace. A similar representation is commonly used for statistical error analysis wherein the error expected in a computed quantity may vary depending on the source of that error.

For example, consider that in a system with several sensors (such as the three-link robot example) wherein the joint angle sensors are noisy devices, each producing a measurement written as

$$\hat{\theta} = \theta + \delta\theta \tag{5.24}$$

where θ is the true value of the angle and $\delta\theta$ is an error term. If we assume that the error terms for the angle sensors are uncorrelated to each other, and if these errors are zero-mean, then we can define the covariance matrix of the vector $\Theta = [\theta_1 \quad \theta_2 \quad \cdots \quad \theta_n]^{\mathrm{T}}$ of angle measurements as:

$$\begin{aligned} Q &\triangleq \mathrm{E}\left[(\delta\Theta - \mathrm{E}[\delta\Theta])(\delta\Theta - \mathrm{E}[\delta\Theta])^{\mathrm{T}}\right] \\ &= \mathrm{E}\left[\delta\Theta \cdot \delta\Theta^{\mathrm{T}}\right] \\ &= diag\left(\nu_1^2, \ldots, \nu_n^2\right) \end{aligned} \tag{5.25}$$

where the ν_i^2 terms are the variances of the n individual sensor noises, $\delta\Theta$ is a vector of angle errors as in Equation (5.24), and E is the expected value operator.

Using these noisy angle measurements to determine a different physical quantity, such as the tip of the third link of the robot, Equation (4.44) would give

$$\begin{aligned} \mathrm{X} &= \boldsymbol{f}(\hat{\Theta}) \\ &= \boldsymbol{f}(\Theta + \delta\Theta) \end{aligned} \tag{5.26}$$

Assuming the quantities $\delta\theta_i$ are small, we can expand (5.26) into its Taylor series equivalent to give

$$\mathrm{X} \approx \boldsymbol{f}(\Theta) + J(\Theta)\delta\Theta \tag{5.27}$$

where the jacobian $J(\Theta)$ is defined as $J(\Theta) \triangleq \partial \boldsymbol{f}(\Theta)/\partial\Theta$. If we take the expected value of X to be $\mathrm{E}[\mathrm{X}] = \mathrm{E}[\boldsymbol{f}(\hat{\Theta})] = \boldsymbol{f}(\Theta)$, then the covariance of the computed endpoint location is (see [6], [8], and [10]):

$$\begin{aligned} R \triangleq \mathrm{cov}[\mathrm{X}] &\approx \mathrm{E}\left[(\mathrm{X} - \mathrm{E}[\mathrm{X}])(\mathrm{X} - \mathrm{E}[\mathrm{X}])^{\mathrm{T}}\right] \\ &= \mathrm{E}\left[\left(\boldsymbol{f}(\hat{\Theta}) - \boldsymbol{f}(\Theta)\right)\left(\boldsymbol{f}(\hat{\Theta}) - \boldsymbol{f}(\Theta)\right)^{\mathrm{T}}\right] \\ &= \mathrm{E}\left[(J(\Theta)\delta\Theta)(J(\Theta)\delta\Theta)^{\mathrm{T}}\right] \\ &= \mathrm{E}\left[J(\Theta)\delta\Theta\delta\Theta^{\mathrm{T}} J(\Theta)^{\mathrm{T}}\right] \\ &= J(\Theta)\mathrm{E}\left[\delta\Theta\delta\Theta^{\mathrm{T}}\right]J^{\mathrm{T}}(\Theta) \\ &= J(\Theta)QJ^{\mathrm{T}}(\Theta) \end{aligned} \tag{5.28}$$

By definition, this covariance matrix R will be positive definite because it is assumed that there is at least some noise in each sensor.

REMARK: Similarly, an analogous analysis may be performed for the robot that indicates to what extent an error in the motor positions affects the endpoint position. This sort of sensitivity analysis is often critical in engineering design wherein it is desired that the inevitably noisy physical devices must produce minimal effect on the desired outcome.

Suppose

$$R = \begin{bmatrix} 3.25 & 1.3 \\ 1.3 & 1.75 \end{bmatrix}$$

The eigenvalues for this matrix are $\lambda_1 = 4$ and $\lambda_1 = 1$, so it is clearly positive definite. Corresponding to these eigenvalues are the normalized eigenvectors, arranged into the modal matrix:

$$M = [e_1 \mid e_2] = \begin{bmatrix} \sqrt{3}/2 & -1/2 \\ 1/2 & \sqrt{3}/2 \end{bmatrix}$$

Therefore, the ellipsoid (in two dimensions) given by the equation $x^{\mathrm{T}} R x = 1$ appears in Figure 5.2. We can interpret this diagram by saying that we expect that the error variance is about four times higher in the e_1 direction than in the e_2 direction, which is to say that the sensor noise affects measurements in one direction more so than in the other direction. If in fact the ellipsoid for the robot example is this eccentric (or more so), we could conclude that the robot is more precise in one direction than in another, a phenomenon that occurs frequently in realistic systems.

Example 5.2: Conic Sections

It can be easily shown that the equation for the general quadric shape:

$$ax^2 + dy^2 + fz^2 + 2bxy + 2cxz + 2eyz + 2jx + 2ky + 2lz + q = 0 \qquad (5.29)$$

can be expressed in matrix-vector form assuming physical coordinates are arranged into the vector $X \triangleq [x \quad y \quad z]^{\mathrm{T}}$ as:

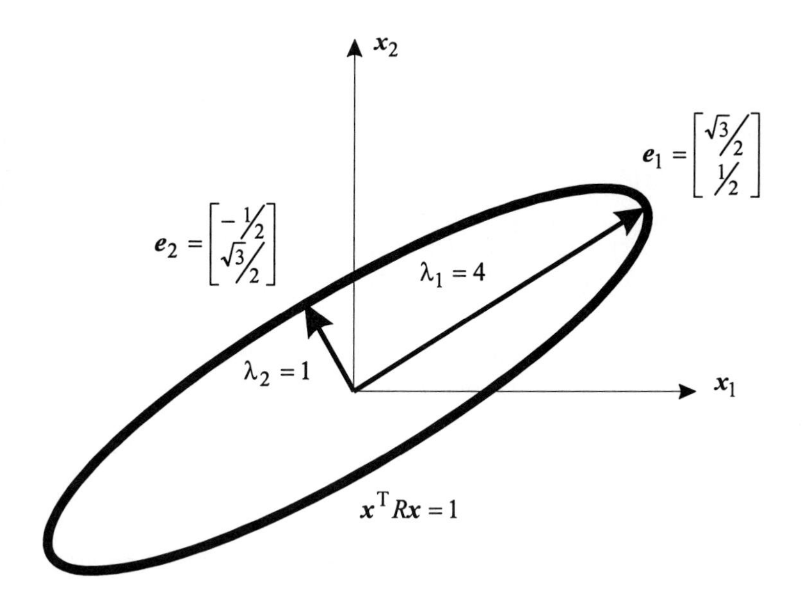

Figure 5.2 Graphical representation of the significance of the positive definite covariance matrix R in the robot example.

$$
\begin{aligned}
0 &= X^{\mathrm{T}} \begin{bmatrix} a & b & c \\ b & d & e \\ c & e & f \end{bmatrix} X + 2\begin{bmatrix} j & k & l \end{bmatrix} X + q \\
&\triangleq X^{\mathrm{T}} A X + 2BX + q
\end{aligned}
\tag{5.30}
$$

or with $Y \triangleq \begin{bmatrix} x & y & z & 1 \end{bmatrix}^{\mathrm{T}}$ as:

$$
\begin{aligned}
0 &= Y^{\mathrm{T}} \begin{bmatrix} a & b & c & j \\ b & d & e & k \\ c & e & f & l \\ j & k & l & q \end{bmatrix} Y \\
&\triangleq Y^{\mathrm{T}} \Lambda Y
\end{aligned}
\tag{5.31}
$$

Give conditions on the matrices A, B, and Λ, and the value of the scalar q such that Equations (5.30) and (5.31) represent various conic sections.

Solution:

It is readily apparent that if matrix A is diagonal, $B = \mathbf{0}$, and $q < 0$, then Equation (5.30) represents an ellipsoid whose axes are aligned with the coordinate axes x, y, and z. If in addition one of the values of a, d, or f is zero, the result is an elliptic cylinder. If $A \equiv 0$, Equation (5.30) represents either a plane or a line, depending on whether one or two of j, k, and l are zero.

However, many other conic sections can be represented with the matrix expressions in (5.30) and (5.31). Among these are:

1. If $rank(A) = 3$, $rank(\Lambda) = 4$, the eigenvalues of A all have the same sign, and $\det(\Lambda) > 0$, then the resulting shape is an ellipsoid that has been rotated so that its major axes are the eigenvectors of A and the values of B represent an offset from the origin. (Consider how to make the equations represent a sphere.)
2. If $rank(A) = 3$, $rank(\Lambda) = 4$, the eigenvalues of A have different signs, and $\det(\Lambda) > 0$, then the resulting shape is a hyperboloid of one sheet.
3. If $rank(A) = 3$, $rank(\Lambda) = 4$, the eigenvalues of A have different signs, and $\det(\Lambda) < 0$, then the resulting shape is a hyperboloid of two sheets.
4. If $rank(A) = 3$, $rank(\Lambda) = 3$, and the eigenvalues of A have different signs, the resulting shape is a quadric cone.
5. If $rank(A) = 2$, $rank(\Lambda) = 3$, and the eigenvalues of A have the same sign, the resulting shape is an elliptic cylinder (that is generally tilted in space).

There are numerous other possibilities, including "imaginary" shapes such as $x^2 + y^2 + z^2 = -1$.

5.4 *Functions of Matrices*

In this section we turn our attention to matrix functions that return matrix results. We have already seen simple examples, such as the addition of two operators (matrices), which is performed element-by-element, and the multiplication of two operators, which is *not* performed element-by-element. Multiplication of a matrix by itself leads to the further definitions:

$$\begin{aligned}
AA &= A^2 \\
A\cdots A \ (n \text{ times}) &= A^n \\
\left(A^n\right)^m &= A^{nm} \\
A^n A^m &= A^{(n+m)} \\
\left(A^{-1}\right)^n &= A^{-n}
\end{aligned} \tag{5.32}$$

(The last line in this list is not always accepted as proper notation.) By adding and subtracting scalar multiples of powers of A, we can generate polynomials in A. As a matter of notation, if we can define an nth-order (monic) polynomial

$$\begin{aligned}
p(\lambda) &= \lambda^n + \alpha_{n-1}\lambda^{n-1} + \cdots + \alpha_1\lambda^1 + \alpha_0 \\
&= (\lambda - p_1)(\lambda - p_2)\cdots(\lambda - p_n)
\end{aligned} \tag{5.33}$$

then that same polynomial evaluated on the matrix A is^{M}:

`polyvalm(p,A)`

$$\begin{aligned}
p(A) &= A^n + \alpha_{n-1}A^{n-1} + \cdots + \alpha_1 A^1 + \alpha_0 I \\
&= (A - p_1 I)(A - p_2 I)\cdots(A - p_n I)
\end{aligned} \tag{5.34}$$

5.4.1 Cayley-Hamilton Theorem

Polynomials evaluated on matrices prove to have the following interesting property:

> THEOREM (Cayley-Hamilton theorem): If the characteristic polynomial of an arbitrary $n \times n$ matrix A is denoted by $\phi(\lambda)$, computed as $\phi(\lambda) = |A - \lambda I|$, then A satisfies its own characteristic equation, denoted by $\phi(A) = 0$. (5.35)

As a result of this theorem, we will never have reason to compute matrix powers higher than n because if

$$\phi(A) = A^n + \phi_{n-1}A^{n-1} + \cdots + \phi_1 A + \phi_0 I = 0$$

then

$$A^n = -\phi_{n-1}A^{n-1} - \cdots - \phi_1 A - \phi_0 I \tag{5.36}$$

Example 5.3: Application of the Cayley-Hamilton Theorem to Matrix Inverses

Let A be an $n \times n$ nonsingular matrix. Use the Cayley-Hamilton theorem to express the inverse A^{-1} without using powers of A higher than $n-1$.

Solution:

Let the characteristic equation for A be written as

$$\begin{aligned}\phi(\lambda) &= \lambda^n + \phi_{n-1}\lambda^{n-1} + \cdots + \phi_1\lambda + \phi_0 = 0 \\ &= \prod_{i=1}^{n}(\lambda - \lambda_i)\end{aligned} \tag{5.37}$$

Then because A is nonsingular, we know it cannot have any zero eigenvalues. Consideration of the polynomial expansion of (5.37) reveals that the coefficient ϕ_0 in the characteristic equation in (5.37) will be the product of all eigenvalues, which is therefore nonzero as well. Because matrix A satisfies its own characteristic equation,

$$A^n + \phi_{n-1}A^{n-1} + \cdots + \phi_1 A + \phi_0 I = 0 \tag{5.38}$$

can be solved by division by the nonzero ϕ_0:

$$I = -\frac{1}{\phi_0}\left(\phi_1 A + \phi_2 A^2 + \cdots + \phi_{n-1}A^{n-1} + A^n\right)$$

Multiplying both sides through by A^{-1} we get

$$A^{-1} = -\frac{1}{\phi_0}\left(\phi_1 I + \phi_2 A^1 + \cdots + \phi_{n-1}A^{n-2} + A^{n-1}\right)$$

We will want to produce functions of A of more generality, not limited to polynomials or inverses. Yet, we will want to retain the simplicity of matrix powers for computational purposes. For this we must establish the power series defined on a matrix. Recall that scalar functions are said to be *analytic* if and only if they can (at least locally) be represented by a convergent power series. The Taylor series is such a power series. The Taylor series for an analytic function $f(\lambda)$ is

$$f(\lambda) = \sum_{i=0}^{\infty} \frac{1}{i!} \frac{d^i f}{d\lambda^i}\bigg|_{\lambda=0} \lambda^i \tag{5.39}$$

We can thus define the same function acting on a matrix A as

$$f(A) = \sum_{i=0}^{\infty} \frac{1}{i!} \frac{d^i f}{d\lambda^i}\bigg|_{\lambda=0} A^i \tag{5.40}$$

Some functions whose Taylor series expansions are familiar can be computed with matrix arguments quite easily, e.g.,

$$\begin{aligned} \sin(A) &= A - \frac{A^3}{3!} + \frac{A^5}{5!} - \frac{A^7}{7!} + \cdots \\ \cos(A) &= I - \frac{A^2}{2!} + \frac{A^4}{4!} - \frac{A^6}{6!} + \cdots \\ e^{At} &= I + At + \frac{A^2 t^2}{2!} + \frac{A^3 t^3}{3!} + \cdots \end{aligned} \tag{5.41}$$

5.4.2 Using the Cayley-Hamilton Theorem to Compute Matrix Functions

The power series expansions in Equation (5.41), while convergent, are unsuitable for use in applications where closed-form functions are necessary. The best one could do with these expansions is to use an approximation obtained by truncating the series after some minimal number of terms. However, the Cayley-Hamilton theorem suggests a useful simplification. Because, as the theorem states, the matrix satisfies its own characteristic polynomial, $\phi(A) = 0$, we can always express the nth power of a matrix by a sum of its lower powers, 0 through $n-1$, as in Equation (5.38). This implies that if we express any analytic function of matrix A as its Taylor series, all powers of A higher than $n-1$ will be reducible to a sum of powers *up to* $n-1$. Then because the Taylor series converges, any analytic function $f(A)$ may be written

$$f(A) = \alpha_{n-1} A^{n-1} + \cdots + \alpha_1 A + \alpha_0 I \tag{5.42}$$

provided we can find the appropriate values of the n constant coefficients α_i, $i = 1, \ldots, n$. To assist in the computation of these constants, the following theorem is helpful:

THEOREM: Suppose $f(\lambda)$ is an arbitrary function of scalar variable λ and $g(\lambda)$ is an $(n-1)^{st}$ order polynomial in λ. If $f(\lambda_i) = g(\lambda_i)$ for every eigenvalue λ_i of the $n \times n$ matrix A, then $f(A) = g(A)$. (5.43)

This theorem implies the useful property that if two functions, one polynomial and one perhaps not, agree on all the eigenvalues of a matrix, then they agree on the matrix itself. This can be used in conjunction with the Cayley-Hamilton theorem, as in the following example, to compute closed-form matrix functions.[M]

`funm(A,'fun')`

Example 5.4: Closed-Form Matrix Functions

Compute the values of $\sin A$, A^5, and e^{At} for the matrix

$$A = \begin{bmatrix} -3 & 1 \\ 0 & -2 \end{bmatrix}$$

Solution:

Because matrix A is triangular, its eigenvalues are obvious: $\lambda_1 = -3$ and $\lambda_2 = -2$. Defining an arbitrary $(n-1)^{st}$ order polynomial as

$$g(\lambda) = \alpha_1 \lambda + \alpha_0$$

we will match values of this function with values of the three unknown functions to determine constants α_0 and α_1.

First, we solve the set of equations $f(\lambda_i) = g(\lambda_i)$ for $i = 1, \ldots, n$.

$$\begin{aligned} \sin(-3) &= \alpha_1(-3) + \alpha_0 \\ \sin(-2) &= \alpha_1(-2) + \alpha_0 \end{aligned} \tag{5.44}$$

to get $\alpha_1 = -0.768$ and $\alpha_0 = -2.45$. Thus,

$$g(\lambda) = -0.768\lambda - 2.45$$

Therefore,

$$f(A) = \sin A = g(A) = -0.768\,A - 2.45\,I = \begin{bmatrix} -0.141 & -0.768 \\ 0 & -0.909 \end{bmatrix}$$

Performing the same procedure on the function A^5,

$$\begin{aligned} (-3)^5 &= \alpha_1(-3) + \alpha_0 \\ (-2)^5 &= \alpha_1(-2) + \alpha_0 \end{aligned} \tag{5.45}$$

gives $\alpha_1 = 211$ and $\alpha_0 = 390$. This means

$$f(A) = A^5 = 211\,A + 390\,I = \begin{bmatrix} -243 & 211 \\ 0 & 32 \end{bmatrix}$$

For the function e^{At}, the equations

$$\begin{aligned} e^{-3t} &= \alpha_1(-3) + \alpha_0 \\ e^{-2t} &= \alpha_1(-2) + \alpha_0 \end{aligned} \tag{5.46}$$

are more difficult to solve in the sense that the constants α_0 and α_1 must now be functions of t as well. This can be done with symbolic math packages[M] such as Mathematica® or Maple®, or in simple cases such as this, they can be solved by hand:

```
expm(A)
```

$$\begin{bmatrix} e^{-3t} \\ e^{-2t} \end{bmatrix} = \begin{bmatrix} -3 & 1 \\ -2 & 1 \end{bmatrix} \begin{bmatrix} \alpha_1 \\ \alpha_0 \end{bmatrix}$$

so

$$\begin{aligned} \begin{bmatrix} \alpha_1 \\ \alpha_0 \end{bmatrix} &= \begin{bmatrix} -3 & 1 \\ -2 & 1 \end{bmatrix}^{-1} \begin{bmatrix} e^{-3t} \\ e^{-2t} \end{bmatrix} \\ &= \begin{bmatrix} -e^{-3t} + e^{-2t} \\ -2e^{-3t} + 3e^{-2t} \end{bmatrix} \end{aligned}$$

Then

$$e^{At} = \left(-e^{-3t} + e^{-2t}\right)\begin{bmatrix} -3 & 1 \\ 0 & -2 \end{bmatrix} + \left(-2e^{-3t} + 3e^{-2t}\right)\begin{bmatrix} 1 & 0 \\ 0 & 1 \end{bmatrix}$$

$$= \begin{bmatrix} e^{-3t} & -e^{-3t} + e^{-2t} \\ 0 & e^{-2t} \end{bmatrix}$$

REMARK: In the above examples, the matrix A conveniently had distinct eigenvalues, so the set of equations $f(\lambda_i) = g(\lambda_i)$ as in (5.44), (5.45), and (5.46), constituted a sufficient set of n independent equations to solve. In the event that the eigenvalues are repeated, one can use the original equation $f(\lambda_i) = g(\lambda_i)$ and its derivative equation $df/d\lambda_i = dg/d\lambda_i$ as an independent set. Higher multiplicities will require higher derivatives.

Example 5.5: Matrix Functions with Repeated Eigenvalues

To illustrate the procedure for finding independent equations when an eigenvalue is repeated, we will find the value of e^{At} when

$$A = \begin{bmatrix} -1 & 1 \\ 0 & -1 \end{bmatrix}$$

Solution:

Clearly, $\lambda_1 = \lambda_2 = -1$, so we will get only a single independent equation from

$$f(\lambda) = g(\lambda) = \alpha_1 \lambda + \alpha_0$$
$$e^{-t} = \alpha_1(-1) + \alpha_0$$

Using the derivative (with respect to λ, not t) of this equation, we get

$$f'(\lambda) = g'(\lambda) = \alpha_1$$
$$te^{-t} = \alpha_1$$

Obviously, this differentiation simplifies the solution process, giving us $\alpha_1 = te^{-t}$ and $\alpha_0 = te^{-t} + e^{-t}$. Therefore,

$$\begin{aligned} e^{At} &= te^{-t}\begin{bmatrix} -1 & 1 \\ 0 & -1 \end{bmatrix} + (te^{-t} + e^{-t})\begin{bmatrix} 1 & 0 \\ 0 & 1 \end{bmatrix} \\ &= \begin{bmatrix} -te^{-t} & te^{-t} \\ 0 & -te^{-t} \end{bmatrix} + \begin{bmatrix} te^{-t} + e^{-t} & 0 \\ 0 & te^{-t} + e^{-t} \end{bmatrix} \\ &= \begin{bmatrix} e^{-t} & te^{-t} \\ 0 & e^{-t} \end{bmatrix} \end{aligned}$$

Using Jordan Forms for Matrix Functions

The important point to note in the above examples is that the function of the matrix is not simply performed element-by-element. However, in the event that the matrix A is diagonal, then the function *will be* performed element-by-element. This property can be useful if we have a modal matrix with which to diagonalize an operator, compute a function of that diagonal matrix, and then transform the function back to the original basis. If the modal matrix M is used to diagonalize matrix A, then we have

$$A = M\hat{A}M^{-1} \tag{5.47}$$

Exploiting the Taylor series expansion of the function $f(A)$ of this matrix,

$$\begin{aligned} f(A) &= \sum_{i=0}^{\infty} \frac{1}{i!} \left. \frac{d^i f}{d\lambda^i} \right|_{\lambda=0} A^i \\ &= \sum_{i=0}^{\infty} \frac{1}{i!} \left. \frac{d^i f}{d\lambda^i} \right|_{\lambda=0} \left(M\hat{A}M^{-1} \right)^i \end{aligned}$$

This series contains powers of $M\hat{A}M^{-1}$ that can be reduced by realizing that

$$\begin{aligned} \left(M\hat{A}M^{-1} \right)^2 = \left(M\hat{A}M^{-1} \right)\left(M\hat{A}M^{-1} \right) = M\hat{A}M^{-1}M\hat{A}M^{-1} = M\hat{A}^2 M^{-1} \\ \vdots \\ \left(M\hat{A}M^{-1} \right)^k = M\hat{A}^k M^{-1} \end{aligned} \tag{5.48}$$

Therefore,

$$
\begin{aligned}
f(A) &= \sum_{i=0}^{\infty} \frac{1}{i!} \frac{d^i f}{d\lambda^i}\bigg|_{\lambda=0} \left(M\hat{A}M^{-1}\right)^i \\
&= M\left(\sum_{i=0}^{\infty} \frac{1}{i!} \frac{d^i f}{d\lambda^i}\bigg|_{\lambda=0} \hat{A}^i \right) M^{-1} \\
&= Mf(\hat{A})M^{-1}
\end{aligned}
\tag{5.49}
$$

Thus, if $\hat{A}$ is diagonal and therefore $f(\hat{A})$ is easily computed element-by-element, then the similarity transformation $f(A) = Mf(\hat{A})M^{-1}$ can be performed to find $f(A)$.

Note that this simplification holds not just for diagonalizable matrices, but for any similarity transformation on matrix A. If A is not diagonalizable, i.e., if its representation is a Jordan form, then $f(A)$ can still be found from $f(A) = Mf(\hat{A})M^{-1}$, and $f(\hat{A})$ can be computed by considering one Jordan block at a time. However, the function f *cannot* be computed element-by-element within a Jordan block. Nevertheless, we can find an easy way to compute the formulae for such a function.

Consider that $\hat{A}$ is a single $n \times n$ Jordan block. Then $\hat{A}$ corresponds to an eigenvalue λ_i that is repeated at least n times. Therefore, the part of the characteristic polynomial considering $\hat{A}$ alone is:

$$
\phi(\lambda) = \left|\lambda - I\hat{A}\right| = (\lambda - \lambda_i)^n = 0 \tag{5.50}
$$

This would imply that $(\lambda - \lambda_i)^k = 0$ for $k = 1, \ldots, n$. To find function $f(\hat{A})$, we would use an $(n-1)^{\text{st}}$-order polynomial $g(\lambda)$, which can be written without loss of generality as

$$
g(\lambda) = \alpha_{n-1}(\lambda - \lambda_i)^{n-1} + \cdots + \alpha_2(\lambda - \lambda_i)^2 + \alpha_1(\lambda - \lambda_i)^1 + \alpha_0 \tag{5.51}
$$

Because all the eigenvalues are the same within this Jordan block, we will need the equation $f(\lambda) = g(\lambda)$ and $n-1$ of its derivatives in order to find the constants α_j, $j = 0, \ldots, n-1$. Because of (5.51), though, these equations are simple to solve:

$$
\begin{array}{lcl}
g(\lambda_i) = f(\lambda_i) = \alpha_0 & \to & \alpha_0 = f(\lambda_i) \\
g'(\lambda_i) = f'(\lambda_i) = \alpha_1 & \to & \alpha_1 = f'(\lambda_i) \\
g''(\lambda_i) = f''(\lambda_i) = 2\alpha_2 & \to & \alpha_2 = \frac{1}{2} f''(\lambda_i) \\
g'''(\lambda_i) = f'''(\lambda_i) = (3)(2)\alpha_3 & \to & \alpha_3 = \frac{1}{3!} f'''(\lambda_i) \\
\vdots & \vdots & \vdots \\
g^{(n-1)}(\lambda_i) = f^{(n-1)}(\lambda_i) = (n-1)!\alpha_{n-1} & \to & \alpha_{n-1} = \frac{1}{(n-1)!} f^{(n-1)}(\lambda_i)
\end{array}
$$

where the notation $f^{(n-1)}(\lambda_i)$ stands for $\left. \dfrac{d^{n-1} f(\lambda)}{d\lambda^{n-1}} \right|_{\lambda=\lambda_i}$.

Therefore,

$$
f(\hat{A}) = \frac{1}{(n-1)!} f^{(n-1)}(\lambda_i)\left(\hat{A} - \lambda_i I\right)^{n-1} + \cdots + f'(\lambda_i)\left(\hat{A} - \lambda_i I\right) + f(\lambda_i) I \tag{5.52}
$$

To simplify this, we can realize that

$$
\hat{A} - \lambda_i I = \begin{bmatrix} 0 & 1 & 0 & \cdots & 0 \\ & & 1 & & \vdots \\ \vdots & & \ddots & \ddots & 0 \\ & & & & 1 \\ 0 & & \cdots & & 0 \end{bmatrix}
$$

$$
\left(\hat{A} - \lambda_i I\right)^2 = \begin{bmatrix} 0 & 0 & 1 & \cdots & 0 \\ & & 0 & \ddots & \vdots \\ \vdots & & \ddots & \ddots & 1 \\ & & & & 0 \\ 0 & & \cdots & & 0 \end{bmatrix}
$$

$$
\vdots
$$

$$
\left(\hat{A} - \lambda_i I\right)^{n-1} = \begin{bmatrix} 0 & 0 & \cdots & 0 & 1 \\ & & 0 & \ddots & 0 \\ \vdots & & \ddots & \ddots & \vdots \\ & & & & 0 \\ 0 & & \cdots & & 0 \end{bmatrix}
$$

and finally

$$\left(\hat{A}-\lambda_i I\right)^n = \begin{bmatrix} 0 & \cdots & 0 \\ \vdots & \ddots & \vdots \\ 0 & \cdots & 0 \end{bmatrix} \tag{5.53}$$

Therefore, (5.52) becomes

$$f(\hat{A}) = \begin{bmatrix} f(\lambda_i) & f'(\lambda_i) & \frac{1}{2}f''(\lambda_i) & \cdots & \frac{1}{(n-1)!}f^{(n-1)}(\lambda_i) \\ 0 & f(\lambda_i) & f'(\lambda_i) & \ddots & \vdots \\ & 0 & f(\lambda_i) & \ddots & \frac{1}{2}f''(\lambda_i) \\ \vdots & & \ddots & \ddots & f'(\lambda_i) \\ 0 & & \cdots & 0 & f(\lambda_i) \end{bmatrix} \tag{5.54}$$

When this is computed, the similarity transformation $f(A) = Mf(\hat{A})M^{-1}$ may be applied as before.

Example 5.6: Matrix Exponential of Jordan Form

Find the value of e^{At} where $A = \begin{bmatrix} -3 & -4 \\ 1 & 1 \end{bmatrix}$.

Solution:

We will do this example using the Jordan form of A. We first find that the eigenvalues of A are $\lambda_1 = \lambda_2 = -1$ and that the geometric multiplicity of this eigenvalue is one, implying the existence of one regular eigenvector and one generalized eigenvector. Carrying out a top-down procedure for finding these vectors, we compute a modal matrix of

$$M = \begin{bmatrix} -4 & 0 \\ 2 & 1 \end{bmatrix}$$

so that

$$\hat{A} = M^{-1}AM = \begin{bmatrix} -1 & 1 \\ 0 & -1 \end{bmatrix}$$

Now because $f(\lambda) = e^{-t}$, we may use (5.54) to compute

$$f(\hat{A}) = \begin{bmatrix} f(-1) & f'(-1) \\ 0 & f(-1) \end{bmatrix} = \begin{bmatrix} e^{-t} & te^{-t} \\ 0 & e^{-t} \end{bmatrix}$$

which agrees with the result in the previous example. Finally,

$$f(A) = Mf(\hat{A})M^{-1} = \begin{bmatrix} -4 & 0 \\ 2 & 1 \end{bmatrix} \begin{bmatrix} e^{-t} & te^{-t} \\ 0 & e^{-t} \end{bmatrix} \begin{bmatrix} -1/4 & 0 \\ 1/2 & 1 \end{bmatrix}$$

$$= \begin{bmatrix} e^{-t} - 2te^{-t} & -4te^{-t} \\ te^{-t} & e^{-t} + 2te^{-t} \end{bmatrix}$$

5.4.3 Minimal Polynomials

Although the Cayley-Hamilton theorem guarantees that there exists an equation, namely, the characteristic equation, that is satisfied by any matrix A, it is sometimes possible for other polynomial equations to be satisfied by A. If such a polynomial equation is of lower order, it may economize on some matrix calculations, as we have seen in the examples.

> ***Minimal Polynomial:*** The minimal polynomial of matrix A is the polynomial $\phi_m(\lambda)$ of lowest order such that $\phi_m(A) = 0$.
> (5.55)

By using the Jordan form, minimal polynomials are relatively easy to find. Recall that the characteristic polynomial of A can be written as

$$\phi(\lambda) = \prod_{i=1}^{n} (\lambda - \lambda_i) \tag{5.56}$$

for all n eigenvalues λ_i, $i = 1, \ldots, n$, some of which are possibly repeated. If all the eigenvalues are distinct, then the characteristic polynomial will be exactly the same as the minimal polynomial. If, however, there are some Jordan blocks in the Jordan form $\hat{A}$ of A, then the minimal polynomial will be of lower order

$$\phi_m(\lambda) = \prod_{i=1}^{k} (\lambda - \lambda_i)^{\eta_i}$$

or

$$\phi_m(\hat{A}) = \prod_{i=1}^{k} (\hat{A} - \lambda_i I)^{\eta_i} \tag{5.57}$$

where, as in Chapter 4, η_i is the index of each eigenvalue λ_i. Here, the summation is performed over the k *distinct* values for the eigenvalues. The discussion of the index revealed that η_i is the size of the largest Jordan block associated with λ_i. The sequence of computations ending with Equation (5.53) would indicate that if J_i is an $\eta_i \times \eta_i$ Jordan block belonging to eigenvalue λ_i, then

$$
\begin{aligned}
(\lambda_i I - J_i)^n &\neq 0 \quad \text{if } n < \eta_i \\
(\lambda_i I - J_i)^n &= 0 \quad \text{if } n \geq \eta_i
\end{aligned}
$$

So we must include η_i factors in (5.56) for each repeated eigenvalue λ_i, but additional factors would be redundant. Because $f(A) = Mf(\hat{A})M^{-1}$, it is a valid approach to determine the minimal polynomial based on the Jordan form; similar matrices will have the same characteristic polynomial and the same minimal polynomial.

Example 5.7: Minimal Polynomials for Jordan Forms

Determine the minimal polynomials for the following matrices, which are already in Jordan form:

$$
A_1 = \begin{bmatrix} -5 & 1 & 0 \\ 0 & -5 & 1 \\ 0 & 0 & -5 \end{bmatrix} \quad A_2 = \begin{bmatrix} -5 & 1 & 0 \\ 0 & -5 & 0 \\ 0 & 0 & -5 \end{bmatrix} \quad A_3 = \begin{bmatrix} -5 & 1 & 0 & 0 & 0 \\ 0 & -5 & 1 & 0 & 0 \\ 0 & 0 & -5 & 0 & 0 \\ 0 & 0 & 0 & -5 & 0 \\ 0 & 0 & 0 & 0 & -2 \end{bmatrix}
$$

Solution:

$$
\begin{aligned}
\phi_{m1}(\lambda) &= (\lambda + 5)^3 \\
\phi_{m2}(\lambda) &= (\lambda + 5)^2 \\
\phi_{m3}(\lambda) &= (\lambda + 5)^3(\lambda + 2)
\end{aligned}
$$

5.5 *Summary*

Entire books are written and entire courses are taught on functional analysis. The introduction here is intended only to give some support to the concepts of bilinear and quadratic forms, which are important in the stability studies we will discuss in Chapter 7. The other matrix functions studied here, particularly the matrix exponential e^{At} and its computation through the Cayley-Hamilton theorem will be used in the next chapter on state space system solutions. Summarizing the main points in this chapter:

- A broad class of matrix (or vector) functions known as the functional produces scalar results from matrix (or vector) arguments. These may be linear, bilinear, sesquilinear, or nonlinear, of which the most important example is the quadratic form.

- Arbitrary analytic functions may be defined for matrices, which can be expressed with the help of power series expansions and the Cayley-Hamilton theorem. This theorem spares us the need to ever work with matrix powers higher than $n-1$, where n is the size of the matrix.

- The matrix exponential may be computed given the knowledge of a matrix's eigenvalues because we have a theorem that indicates that two functions that agree on the (scalar) eigenvalues of a matrix also agree on the matrix itself.

- Through an example, we have seen another application for singular values that also indicates the use of quadratic forms and the geometry they imply for their arguments.

The next part of the book begins with Chapter 6. Beginning in this chapter, we concentrate again on state space descriptions of physical systems governed by linear differential equations. All of the tools necessary for the analysis, design, and control of such systems are found in Chapters 1 through 5. Chapter 6 itself concerns the analytical solution to state variable equations, emphasizing SISO linear, time-invariant systems. We will also revisit the basis of eigenvectors in a discussion of the decomposition of system solutions along a natural basis set.

5.6 *Problems*

5.1 Let $\{f_1, f_2, f_3\}$ and $\{g_1, g_2, g_3\}$ be two bases for $\Re^3$, such that

$$\begin{aligned} g_1 &= 2f_1 + f_2 + f_3 \\ g_2 &= -1f_1 + 3f_2 + 2f_3 \\ g_3 &= 2f_1 - f_2 + f_3 \end{aligned}$$

a) If the matrix of a linear operator in this space is given in basis $\{f_1, f_2, f_3\}$ by

$$T = \begin{bmatrix} 8 & -20 & 24 \\ -4 & 20 & -2 \\ -6 & -10 & 32 \end{bmatrix}$$

find the matrix in the basis $\{g_1, g_2, g_3\}$.

b) If we are given a vector $x = -4f_1 + 4f_2 + 4f_3$, find the components of x in the basis $\{g_1, g_2, g_3\}$.

c) Given a linear functional ϕ on the space that is specified by the values $\phi(f_1) = -4, \phi(f_2) = 4$, and $\phi(f_3) = 4$, find the three numbers that specify the functional in the basis $\{g_1, g_2, g_3\}$.

5.2 Consider two bases for space $\Re^3$:

$$\{a_1, a_2, a_3\} = \left\{ \begin{bmatrix} 2 \\ 1 \\ 4 \end{bmatrix}, \begin{bmatrix} 3 \\ -2 \\ -2 \end{bmatrix}, \begin{bmatrix} -4 \\ 2 \\ 1 \end{bmatrix} \right\} \quad \{b_1, b_2, b_3\} = \left\{ \begin{bmatrix} -2 \\ 3 \\ 1 \end{bmatrix}, \begin{bmatrix} -4 \\ -3 \\ -2 \end{bmatrix}, \begin{bmatrix} 5 \\ -2 \\ 0 \end{bmatrix} \right\}$$

If a linear functional on this space is given by $\gamma(a_1) = 2$, $\gamma(a_2) = -1,$, and $\gamma(a_3) = -4$, find the representation for the functional in the basis $\{b_1, b_2, b_3\}$.

5.3 Consider linear vector space V with standard basis $\{e_1, e_2, e_3\}$. Let bilinear form $B(x, y)$ be given in this space by the coefficients

$$B(e_i, e_j) = \begin{bmatrix} 2 & 1 & 0 \\ 1 & 4 & 1 \\ 0 & 1 & 4 \end{bmatrix}$$

Let a linear operator T be given in the same basis as

$$T = \begin{bmatrix} 1 & 2 & 0 \\ 0 & 1 & 2 \\ 2 & 0 & 1 \end{bmatrix}$$

a) Show that $\langle x, y \rangle = B(x, y)$ is a valid inner product for this space.

b) Find the adjoint of T with respect to the inner product $\langle x, y \rangle = B(x, y)$.

5.4 Let $p(t)$ be a polynomial, A a linear operator on vector space **V**, λ an eigenvalue of A, and x its corresponding eigenvector. Show that the same eigenvector x will also be an eigenvector of the operator $p(A)$ and that the eigenvalue it corresponds to will be $p(\lambda)$.

5.5 Let M be an $n \times n$ hermitian matrix. Let e_i, $i = 1, \ldots, n$, be a set of linearly independent vectors. If it is known that $e_i^* M e_i > 0$ for all $i = 1, \ldots, n$, is it necessarily true that M is a positive-definite matrix?

5.6 Determine a closed-form expression for e^A, where $A = \begin{bmatrix} -1 & -2 \\ -3 & -6 \end{bmatrix}$.

5.7 If $A = \begin{bmatrix} -4 & 2 & 5 \\ 1 & -1 & -1 \\ -1 & 2 & 2 \end{bmatrix}$, find A^{10}.

5.8 Compute e^{At} for the following two matrices:

a) $\begin{bmatrix} -14 & -2 & -14 \\ -7 & -3 & -8 \\ 11 & 2 & 11 \end{bmatrix}$ b) $\begin{bmatrix} -60 & 30 & 210 \\ 40 & -20 & -140 \\ -20 & 10 & 70 \end{bmatrix}$

5.9 Find e^{At} for the given matrices:

a) $\begin{bmatrix} -4 & 0 & 0 \\ 0 & -7 & 1 \\ 0 & -9 & -1 \end{bmatrix}$ b) $\begin{bmatrix} \alpha & \omega \\ -\omega & \alpha \end{bmatrix}$ c) $\begin{bmatrix} -8 & 7 \\ -4 & 3 \end{bmatrix}$

5.10 Find a closed-form expression for A^k, $k \ge 1$,

where $A = \begin{bmatrix} 0 & 0 & 0 \\ 9 & 23 & 30 \\ -7 & -18 & -23.5 \end{bmatrix}$.

5.7 References and Further Reading

As in the previous chapter, the material given here on functions of matrices is available from any standard text on matrix computations, such as [4] and [5]. Multilinear functionals can be treated in increased generality as *tensors*, and this material is discussed in [1].

Because of its critical importance, the matrix exponential e^{At} is the subject of particular attention, being discussed in [7] and [9]. Its computation from the Jordan form is given in [2] and [3]. Note, though, that like eigenvalues, eigenvectors, and singular values, these "pencil-and-paper" computation methods are presented to reinforce an understanding of the nature and behavior of the matrix exponential. They are *not* the best ways to compute the exponential numerically. For details on numerical issues, see [7].

The robot joint-error analysis example can be pursued further through [6], [8], and [10].

[1] Bishop, Richard L., and Samuel I. Goldberg, *Tensor Analysis on Manifolds*, Dover, 1980.

[2] Brogan, William L., *Modern Control Theory*, 3rd edition, Prentice-Hall, 1991.

[3] Chen, Tsi-Chong, *Linear System Theory and Design*, Holt, Rinehart, and Winston, 1984.

[4] Gantmacher, Feliks R., *The Theory of Matrices, Vols I and II*, Chelsea Publishing Co., 1959.

[5] Golub, Gene H., and Charles F. Van Loan, *Matrix Computations*, Johns Hopkins University Press, 1989.

[6] Menq, Chia-Hsiang, and Jin-Hwan Borm, "Statistical Measure and Characterization of Robot Errors," *IEEE International Conference on Robotics and Automation*, 1988, pp. 926-931.

[7] Moler, Cleve B., and C. F. Van Loan, "Nineteen Dubious Ways to Compute the Exponential of a Matrix," *SIAM Review*, vol. 20, 1978, pp. 810-836.

[8] Nakamura, Yoshihiko, and Yingti Xu, "Geometrical Fusion Method for Multi-Sensor Robotic Systems," *IEEE International Conference on Robotics and Automation*, 1989, pp. 668-673.

[9] Vidyasagar, M., "A Characterization of e^{At} and a Constructive Proof of the Controllability Condition," *IEEE Transactions on Automatic Control*, vol. AC-16, no. 4, 1971, pp. 370-371.

[10] Yufeng, Long, and Lu Jianqin, "On the Optimum Tolerances of Structural Parameters and Kinematic Parameters of Robot Manipulators," *Mechanism and Machine Theory*, vol. 28, no. 6, 1993, pp. 819-824.

Part II

Analysis and Control of State Space Systems

6

Solutions to State Equations

It was stated in Chapter 1 that the state of a system at a given time is sufficient information to determine the state at all future times, assuming that the system dynamics (the system matrices *A*, *B*, *C*, and *D*) and input are all known. Often, the reason for studying a linear system is that one or more of these quantities is not known. Sometimes the system matrices are unknown or only approximated, as with adaptive and robust control. Random signals are present in the input and output equation, requiring stochastic control. In the most common case, only a desired state or output is specified, and it is the task of the engineer to "design" the input by introducing compensators.

In this chapter, we consider only the simplest situations, wherein all terms of the state equations are known, and we seek an analytical solution to these state equations. This process is useful for our understanding of the behavior and properties of state equations, but often we cannot write such solutions in practice, because we usually have insufficient information to do so.

6.1 Linear, Time-Invariant (LTI) Systems

Recall the state equations for a linear, time-invariant (LTI) state space system[M]:

```
ss(A,B,C,D)
```

$$\begin{aligned} \dot{x} &= Ax + Bu, \qquad x(t_0) = x_0 \\ y &= Cx + Du \end{aligned} \tag{6.1}$$

The difficulty in solving this system is the first equation, $\dot{x} = Ax + Bu$, because it is a differential equation. When we determine $x(t)$, it becomes a straightforward matter to substitute it into the second equation to determine $y(t)$. Note that by denoting our input, output, and feed-forward matrices as capital letters (*B*, *C*, and *D*), we are implying that these manipulations hold for MIMO as well as SISO situations.

We will use the method of introducing an integrating factor in the solution of first order differential equations of the form:

$$\dot{x}(t) - Ax(t) = Bu(t) \tag{6.2}$$

Multiplying (6.2) by the factor e^{-At} will result in a "perfect" differential on the left side:

$$\begin{gathered} e^{-At}\left[\dot{x}(t) - Ax(t) = Bu(t)\right] \\ e^{-At}\dot{x}(t) - e^{-At}Ax(t) = e^{-At}Bu(t) \\ \frac{d}{dt}\left[e^{-At}x(t)\right] = e^{-At}Bu(t) \end{gathered} \tag{6.3}$$

(Note that A commutes with e^{At}.)

Integrating both sides of this equation over dummy variable τ from t_0 to t,

$$e^{-At}x(t) - e^{-At_0}x(t_0) = \int_{t_0}^{t} e^{-A\tau}Bu(\tau)\,d\tau \tag{6.4}$$

Finally, moving the initial condition term to the right-hand side and multiplying both sides of the result by e^{At} gives:

$$\begin{aligned} x(t) &= e^{At}e^{-At_0}x(t_0) + e^{At}\int_{t_0}^{t} e^{-A\tau}Bu(\tau)\,d\tau \\ &= e^{A(t-t_0)}x(t_0) + \int_{t_0}^{t} e^{A(t-\tau)}Bu(\tau)\,d\tau \end{aligned} \tag{6.5}$$

Actually, although we have noted that this result is based upon an assumption of time-invariance, we require only that the matrix A be time-invariant, because the integrating factor would not have worked properly otherwise. If B were time-varying, we would simply write:

$$x(t) = e^{A(t-t_0)}x(t_0) + \int_{t_0}^{t} e^{A(t-\tau)}B(\tau)u(\tau)\,d\tau \tag{6.6}$$

This is the familiar *convolution integral* solution from basic linear systems. Its derivation here for vectors emphasizes one of the primary motivations for state space analysis: many of the procedures and results for vectors are direct extensions of first-order (scalar) cases.

Completing the problem by computing $y(t)$, we obtain M:

```
lsim(sys,u,
  T,X0)
```

$$y(t) = C(t)e^{A(t-t_0)}x(t_0) + C(t)\int_{t_0}^{t} e^{A(t-\tau)}B(\tau)u(\tau)\,d\tau + Du(t) \tag{6.7}$$

Here again, we have allowed C and D to be functions of time because they do not interfere with the actual solution of the differential equation.

The solution (6.7) makes apparent the importance of the matrix exponential e^{At} studied in the last chapter. For LTI systems, this will become known as the state transition-matrix for reasons that will become apparent in Section 6.4.2.

Example 6.1: Simple LTI System

A simple free-floating object, moving in a straight line, might be described by Newton's law with the equation $F = m\ddot{x}$. Express the system as a pair of state equations and solve using $F(t) = e^{-at}$, $x(0) = x_0$, and $\dot{x}(0) = 0$.

Solution:

Using phase variables, we can define $x_1(t) = x(t)$ and $x_2(t) = \dot{x}(t)$, giving the state equations

$$\begin{bmatrix} \dot{x}_1 \\ \dot{x}_2 \end{bmatrix} = \begin{bmatrix} 0 & 1 \\ 0 & 0 \end{bmatrix} \begin{bmatrix} x_1 \\ x_2 \end{bmatrix} + \begin{bmatrix} 0 \\ 1/m \end{bmatrix} F(t)$$

$$y = \begin{bmatrix} 1 & 0 \end{bmatrix} \begin{bmatrix} x_1 \\ x_2 \end{bmatrix}$$

This system is known as a double-integrator. The A-matrix is observed to be in Jordan form, where the two eigenvalues are obviously both zero. We can use the methods of the previous chapter to compute e^{At}, but for this case, a Taylor series expansion has only two nonzero terms since $A^k = 0$ for $k \geq 2$. Therefore,

$$e^{At} = I + At = \begin{bmatrix} 1 & t \\ 0 & 1 \end{bmatrix}$$

Using (6.7), we compute

$$
\begin{aligned}
y(t) &= \begin{bmatrix}1 & 0\end{bmatrix} e^{A(t-0)} x(0) + \begin{bmatrix}1 & 0\end{bmatrix} \int_0^t e^{A(t-\tau)} \begin{bmatrix} 0 \\ 1/m \end{bmatrix} e^{-a\tau}\, d\tau \\
&= \begin{bmatrix}1 & 0\end{bmatrix} \begin{bmatrix} 1 & t \\ 0 & 1 \end{bmatrix} \begin{bmatrix} x_0 \\ 0 \end{bmatrix} + \begin{bmatrix}1 & 0\end{bmatrix} \int_0^t \begin{bmatrix} 1 & t-\tau \\ 0 & 1 \end{bmatrix} \begin{bmatrix} 0 \\ 1/m \end{bmatrix} e^{-a\tau}\, d\tau \\
&= x_0 + \begin{bmatrix}1 & 0\end{bmatrix} \int_0^t \begin{bmatrix} (t-\tau)1/m \\ 1/m \end{bmatrix} e^{-a\tau}\, d\tau \\
&= x_0 + \frac{1}{m} \int_0^t (t-\tau) e^{-a\tau}\, d\tau \\
&= x_0 + \frac{1}{m} \left[-\frac{1}{a} t e^{-a\tau} + \frac{1}{a} \tau e^{-a\tau} + \frac{1}{a^2} e^{-a\tau} \right] \Bigg|_{\tau=0}^{\tau=t} \\
&= \left(x_0 - \frac{1}{a^2 m} \right) + t \frac{1}{am} + \frac{1}{a^2 m} e^{-at}
\end{aligned}
$$

One may verify that this is the correct solution via the usual procedures for solving simple differential equations.

6.2 Homogeneous Systems

In a homogeneous system, of course, the only component of a system's response is the zero-input, or initial condition response. This part of the total response is not easily seen in frequency-domain analysis. Without an input, of course, the state equations are

$$
\begin{aligned}
\dot{x} &= Ax \\
y &= Cx
\end{aligned}
$$

and the state vector solution for an LTI system is simply

$$
x(t) = e^{A(t-t_0)} x(t_0) \tag{6.8}
$$

or

$$
y(t) = C e^{A(t-t_0)} x(t_0)
$$

Obviously, the matrix exponential e^{At} plays a role in both the zero-state and the zero-input parts of the total response. To see the effect of this matrix exponential, it is useful to draw sketches of the solutions in the space, beginning with various initial conditions. This is most easily done in two dimensions, for which such plots are common and are called *phase portraits*.

6.2.1 Phase Portraits

A phase portrait is strictly defined as a graph of several zero-input responses on a plot of the *phase-plane*, $\dot{x}(t)$ versus $x(t)$, these being known as phase variables. However the term has become commonly used to denote any sketch of zero-input solutions on the plane of the state variables, regardless of whether they are phase variables or not.

To create a phase portrait, one simply chooses initial conditions to represent wide areas of interest in the $x_1 - x_2$ plane, solves the system^M according to (6.8), and sketches the result as a function of time, starting from t_0. For example, consider the system

`initial(sys,X0)`

$$\dot{x} = A_1 x = \begin{bmatrix} -1 & 0 \\ 0 & -4 \end{bmatrix} x \tag{6.9}$$

A phase portrait for this system is given in Figure 6.1. In the figure, sufficient initial conditions are chosen around the edge of the graph to accurately interpolate the solution for any initial condition in between. The solution drawn from each initial condition is sketched as a directed curve, as indicated by the arrow showing the progression of positive time. These curves are known as *phase trajectories*.

With experience, the general shape of the trajectories can be imagined without a detailed plot. Generally, it is the qualitative shape of the trajectories that is important in a phase portrait. (This is particularly true with nonlinear systems, whose phase portraits can sometimes be constructed entirely qualitatively or with piecewise analysis of their dynamics.) For example, certain qualitative features of Figure 6.1 might be predicted from our knowledge of the state space.

Consider the A-matrix given in (6.9) as a linear operator, taking vectors x into vectors $\dot{x}$. Thus, given any position $x(t)$ on the plot, the equation $\dot{x} = Ax$ gives us the *tangent vector* to the phase trajectory at that point. This is the direction in which the trajectory is evolving. In the figure, we have illustrated this point by indicating a vector, $x(t) \approx [-3 \quad -1]^T$, and the tangent to the curve at that point, $\dot{x}(t) = A_1 x = [3 \quad 4]^T$. Of course, this *velocity* will change as the chosen point x changes.

We can compute that this system has eigenvalues $\sigma_1 = \{-1, -4\}$ (a set known as the *spectrum* of A), and corresponding eigenvectors of

$$\{e_1, e_2\} = \left\{ \begin{bmatrix} 1 \\ 0 \end{bmatrix}, \begin{bmatrix} 0 \\ 1 \end{bmatrix} \right\} \tag{6.10}$$

Knowing that these two vectors are invariant subspaces, we conclude that if an initial condition $x(t_0)$ lies on either of these lines, then so will the vector $\dot{x}(t_0) = Ax(t_0)$ at any t_0. Therefore, if a chosen point lies on one of the eigenvectors, identified in the graph, the trajectory emanating from that point will lie on that same line forever. Thus, we can always get a start at constructing a phase portrait by drawing in the invariant subspaces, i.e., the straight lines, on the plot.

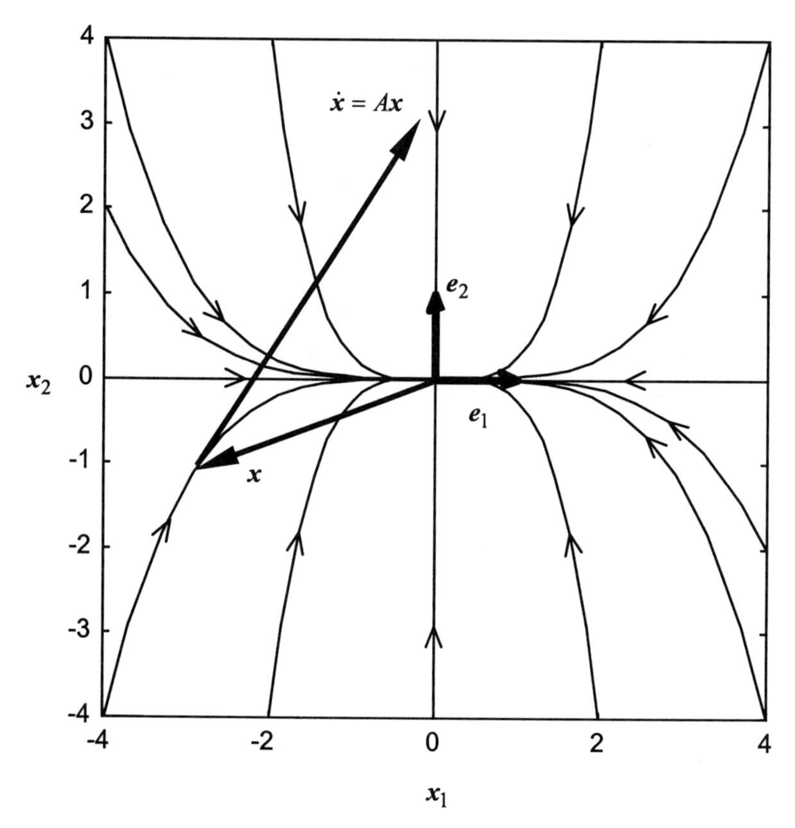

Figure 6.1 Phase portrait for the homogeneous system given in Equation (6.9). For this type of portrait, wherein all trajectories asymptotically approach the origin without encircling it, the origin is known as a *stable node*.

More will be said about the phase trajectories on these portraits in the next section. Until then we will present further examples of phase portraits for the homogeneous equation $\dot{x} = Ax$.

In order to relate such phase trajectories to the more familiar step responses, Figure 6.2 shows the step response[M] for the system

`step(sys)`

$$\dot{x} = A_1 x + b_1 u = \begin{bmatrix} -1 & 0 \\ 0 & -4 \end{bmatrix} x + \begin{bmatrix} 1 \\ 4 \end{bmatrix} u$$
$$y = \begin{bmatrix} 1 & 1 \end{bmatrix} x \tag{6.11}$$

where the output equation was selected arbitrarily and the matrix b_1 in (6.11) was chosen to give unity DC gain to each state variable, i.e., so that they both asymptotically approach 1. This figure further illustrates the relative speed of the two state variables. Note how the "faster" of the two variables converges to its final value sooner than the "slower" one, and of course the output is simply the sum of the two inputs.

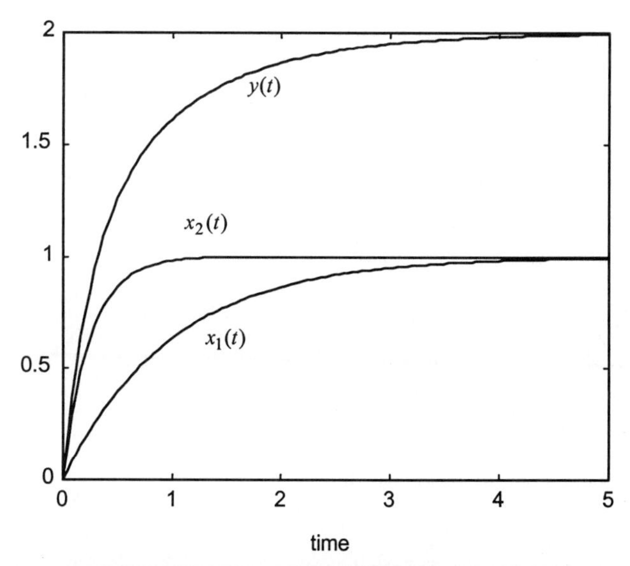

Figure 6.2 Step response for the system described by Equation (6.11). The state variables and output are plotted as functions of time.

Figure 6.3 shows a portrait that is similar to Figure 6.1 except that it appears rotated and distorted. This is the portrait for the homogeneous (no input) system with A-matrix

$$A_2 = \begin{bmatrix} -2 & -1 \\ -2 & -3 \end{bmatrix} \tag{6.12}$$

with the following spectrum and eigenvectors (expressed as columns in a modal matrix):

$$\sigma_2 = \{-1 \quad -4\} \qquad M_2 = \begin{bmatrix} 1 & 1 \\ -1 & 2 \end{bmatrix}$$

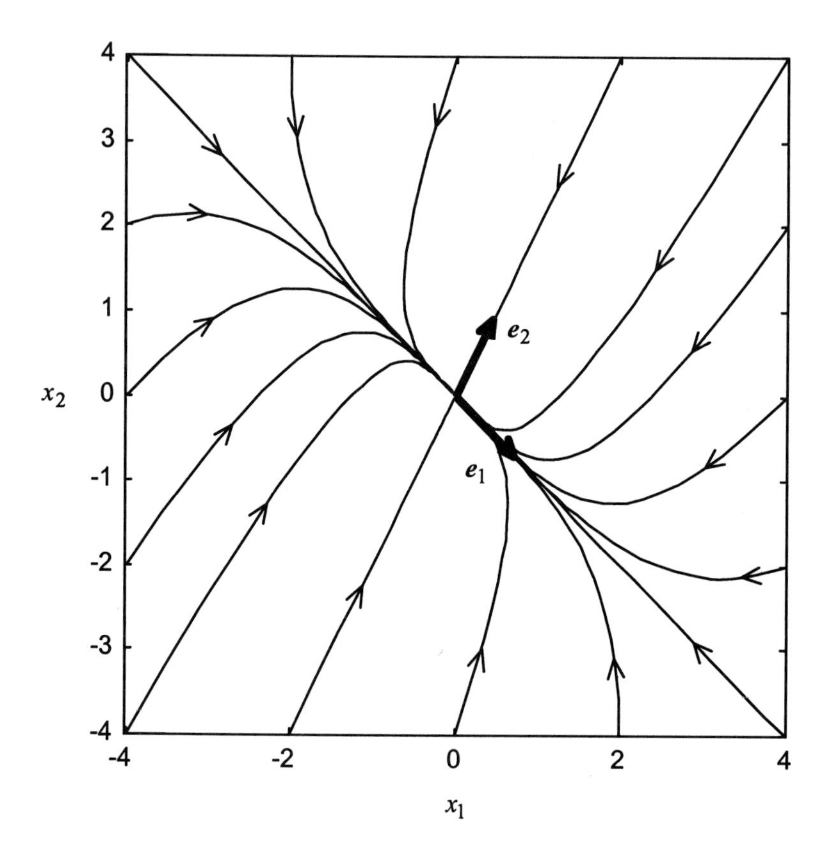

Figure 6.3 Phase portrait for the homogeneous system given by the matrix in (6.12). Note that this origin is also a *stable node.*

As before, we see that the eigenvectors represent invariant subspaces on which trajectories stay forever. This portrait was produced by performing the similarity transformation

$$A_2 = M_2 A_1 M_2^{-1} \tag{6.13}$$

From our knowledge of similarity transformations, we recognize this as nothing but a change of basis of the state space. We can therefore regard Figure 6.3 as containing the same information as Figure 6.1, but in different coordinates. When this same similarity matrix is applied to Equation (6.11), the result is

$$\dot{x} = M_2 A_1 M_2^{-1} x + M_2 b_1 u = \begin{bmatrix} -2 & -1 \\ -2 & -3 \end{bmatrix} x + \begin{bmatrix} 5 \\ 7 \end{bmatrix} u$$
$$y = CM_2^{-1} x = \begin{bmatrix} 1 & 0 \end{bmatrix} x \qquad (6.14)$$

which has the step response shown below in Figure 6.4. Note that the state variables no longer reach unity asymptotically. This is because the similarity transformation affects the DC gain of the state variables. However, as we would expect, the output signal $y(t)$ remains exactly the same as in Figure 6.2, because similarity transformations do not affect the input/output performance, only its internal representation.

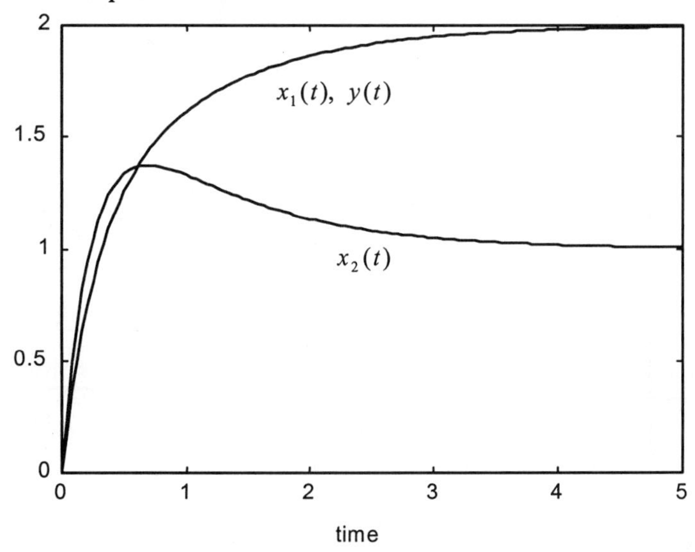

Figure 6.4 Step response for the system described by Equation (6.14). The state variables and output are plotted as a function of time.

Figure 6.5 is somewhat different in the sense that the trajectories do not all tend toward the origin. This portrait is based upon the following system:

$$A_3 = \begin{bmatrix} -2 & 0 \\ 0 & 1 \end{bmatrix} \qquad \sigma_3 = \{-2 \quad 1\} \qquad M_3 = \begin{bmatrix} 1 & 0 \\ 0 & 1 \end{bmatrix} \qquad (6.15)$$

It is again a diagonal system with the eigenvectors constituting the standard basis. However, in this system, one of the corresponding eigenvalues is positive. Being diagonal, it is easy to decompose the system into decoupled parts:

$$\begin{bmatrix} \dot{x}_1 \\ \dot{x}_2 \end{bmatrix} = \begin{bmatrix} -2 & 0 \\ 0 & 1 \end{bmatrix} \begin{bmatrix} x_1 \\ x_2 \end{bmatrix} = \begin{bmatrix} -2x_1 \\ x_2 \end{bmatrix} \tag{6.16}$$

which of course has solutions that can be found independently of one another:

$$\begin{bmatrix} x_1(t) \\ x_2(t) \end{bmatrix} = \begin{bmatrix} e^{-2(t-t_0)} & 0 \\ 0 & e^{(t-t_0)} \end{bmatrix} \begin{bmatrix} x_1(t_0) \\ x_2(t_0) \end{bmatrix} = \begin{bmatrix} e^{-2(t-t_0)} x_1(t_0) \\ e^{(t-t_0)} x_2(t_0) \end{bmatrix}$$

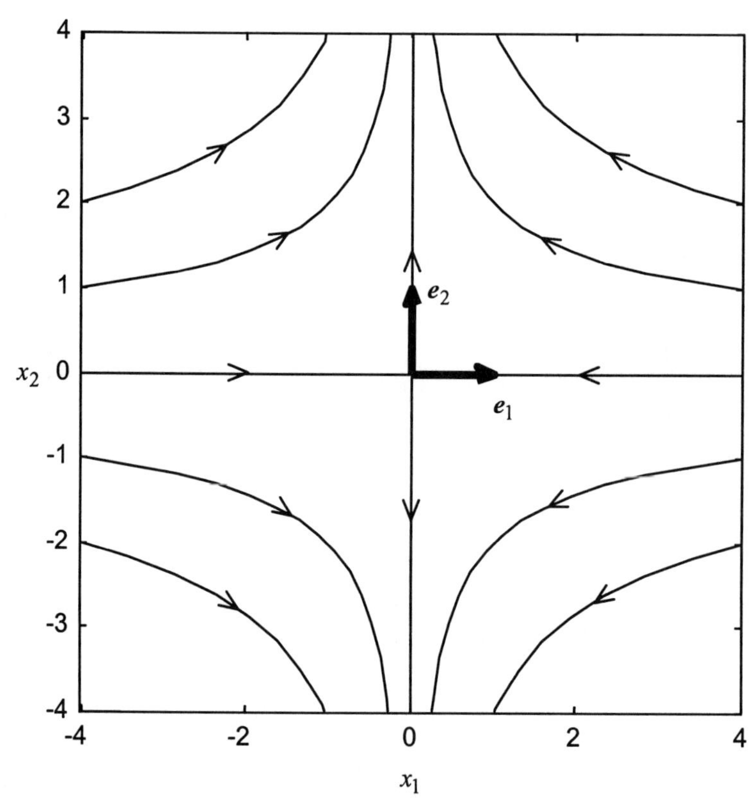

Figure 6.5 Phase portrait for the homogeneous system given by (6.15). In cases such as this wherein the trajectories approach the origin from one direction but diverge from it in the other, the origin is known as a *saddle point*.

The positive exponent in $x_2(t)$ indicates that a solution starting on the invariant space $\boldsymbol{e}_2$, while it will stay there, will nevertheless diverge from the origin. On the other hand, the negative exponent makes $x_1(t)$ tend toward zero.

The trajectories in between, while not lying on invariant subspaces, must interpolate the invariant trajectories, giving the shapes seen in the figure.

Performing a similarity transformation on the system described by (6.15), which is similar to (6.13) but with a different modal matrix, we can arrive at a form of (6.15) with a new basis:

$$A_4 = \begin{bmatrix} -3 & -2 \\ 2 & 2 \end{bmatrix} \qquad \sigma_4 = \{-2 \quad 1\} \qquad M_4 = \begin{bmatrix} -2 & 1 \\ 1 & -2 \end{bmatrix} \tag{6.17}$$

The phase portrait of this transformed system is shown in Figure 6.6. Again, motion along the two invariant subspaces tends toward different directions, and the remaining trajectories interpolate smoothly.

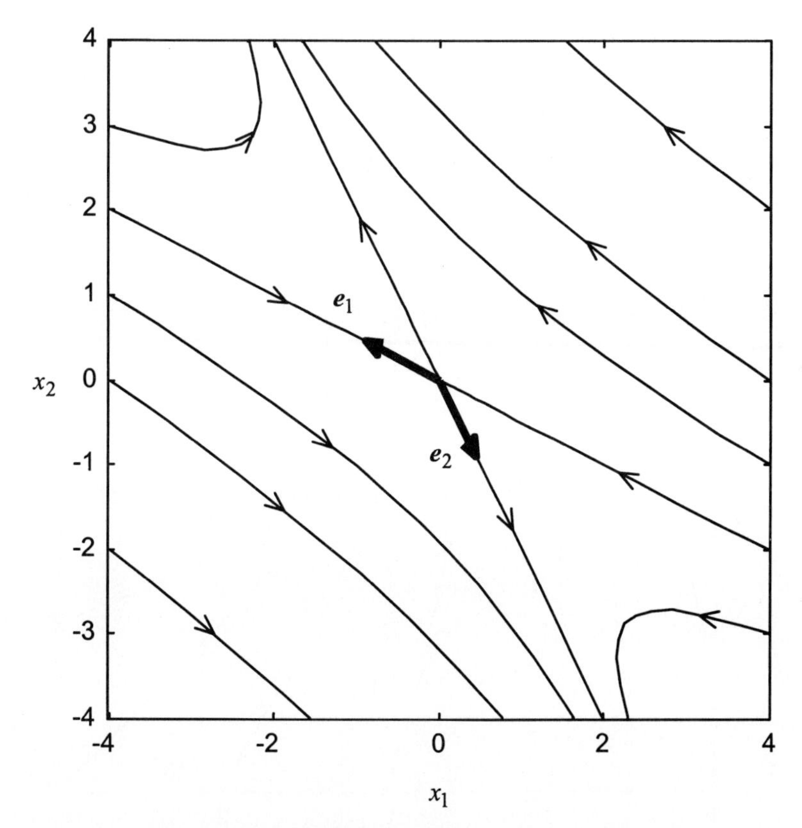

Figure 6.6 Phase portrait for the homogeneous system given by (6.17). The origin in this figure is a *saddle point*.

It should be pointed out that except at the singular points (for LTI systems, the origin only), two trajectories cannot cross or meet at a point. This would imply that at that point, there are two independent solutions to the differential equation, which is prohibited by the uniqueness theorem. In our linear case, we also expect the trajectories to be smooth, which is seen in all the plots. Smoothness is guaranteed by the existence of a unique tangent vector given by the equation $\dot{x} = Ax$.

The nature of the phase portrait when there is only a single eigenvector, as we might expect in a system having a generalized eigenvector, is investigated next. An example of such a system is:

$$A_5 = \begin{bmatrix} -1 & 1 \\ -1 & -3 \end{bmatrix} \qquad \sigma_5 = \{-2 \quad -2\} \qquad e_1 = \begin{bmatrix} -1 \\ 1 \end{bmatrix} \tag{6.18}$$

This system, represented by the phase portrait in Figure 6.7, indeed shows only a single invariant subspace. This eigenvector, which corresponds to a negative eigenvalue of -2, tends toward the origin as expected.

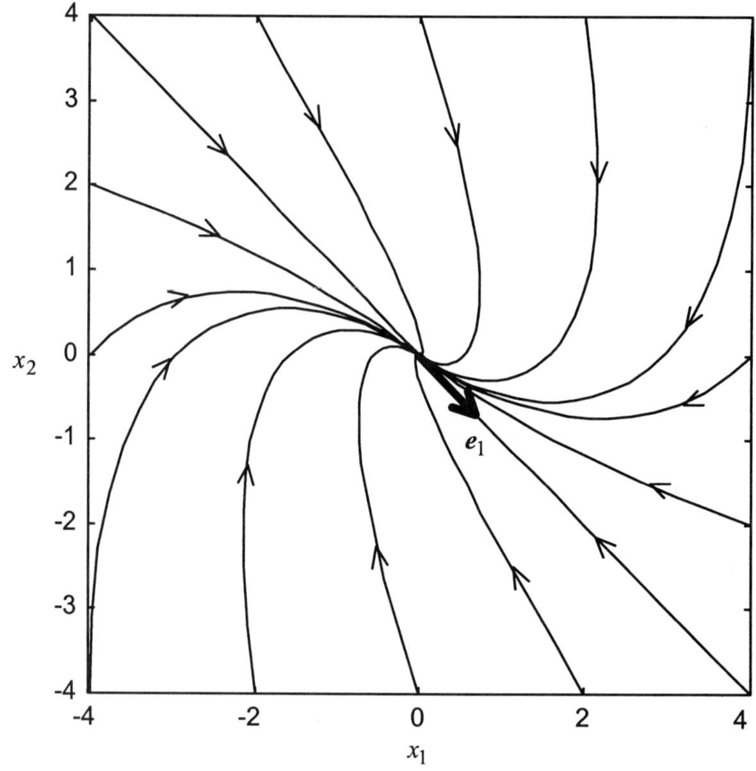

Figure 6.7 Phase portrait for the homogeneous system given by (6.18). The origin in this system is called a *stable node* because again the trajectories approach the origin without encircling it.

The other trajectories also tend toward the origin, but do so by partially spiraling inward. Because trajectories cannot cross, none of the spiraling trajectories may rotate more than 180° before asymptotically reaching the origin.

Figure 6.8 shows the initial condition response (zero inputs) for the same system, i.e., from (6.18). Note that qualitatively, the plot appears to be similar to previous time responses, e.g., Figure 6.4, except that now *both* curves show one inflection point.

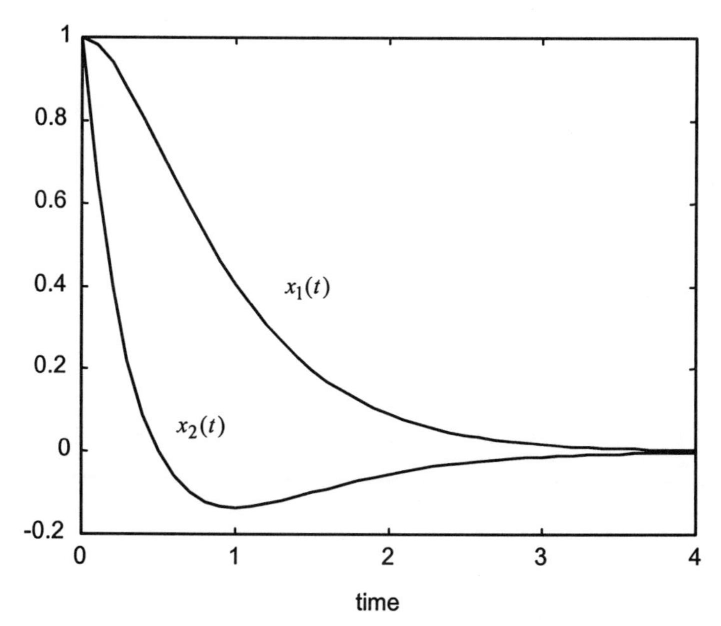

Figure 6.8 Initial condition response for the system described by Equation (6.14). The state variables are plotted as a function of time.

Transforming the system described by (6.18) into its Jordan form, we get

$$A = \begin{bmatrix} -2 & 1 \\ 0 & -2 \end{bmatrix} \qquad \sigma_6 = \{-2 \quad -2\} \qquad e_1 = \begin{bmatrix} 1 \\ 0 \end{bmatrix} \tag{6.19}$$

The phase portrait is given in Figure 6.9. As we might guess, the single regular eigenvector in this portrait lies along a coordinate axis because the Jordan form produces a system of differential equations, one of which is decoupled from the other.

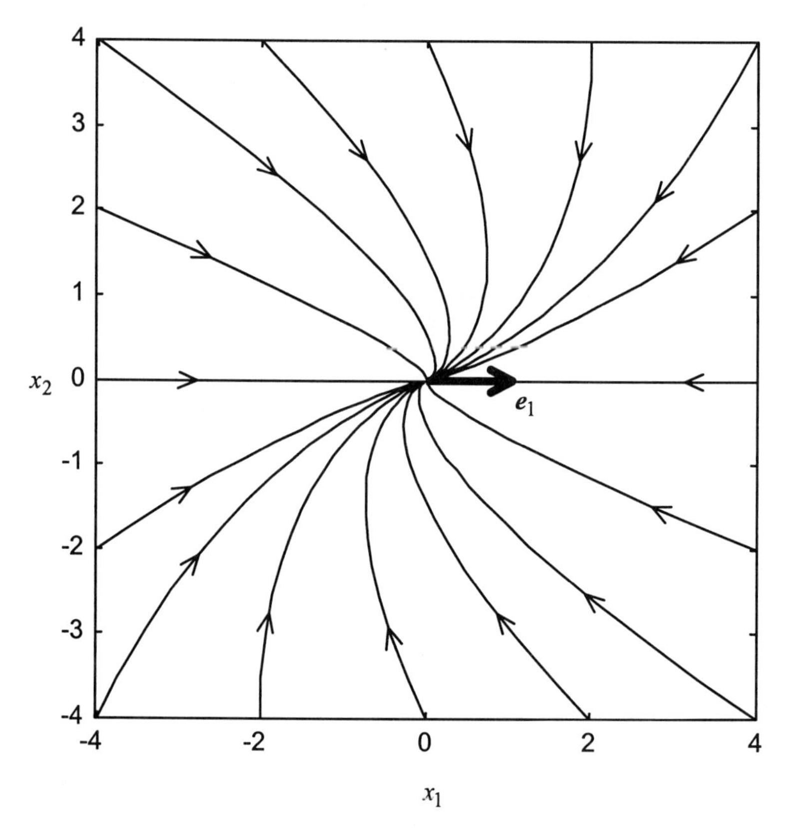

Figure 6.9 Phase portrait for the homogeneous system given by (6.19). The origin is a *stable node*.

In the case of complex eigenvectors, we mentioned in Section 4.3 that the geometric interpretation of invariant subspaces does not directly apply because we are interested in only the real solutions to the system. Therefore, for a system such as

$$A_7 = \begin{bmatrix} 2 & -3 \\ 6 & -4 \end{bmatrix} \quad \sigma_7 = \{-1+j3 \quad -1-j3\} \quad M_7 = \begin{bmatrix} -1+j1 & -1-j1 \\ j2 & -j2 \end{bmatrix} \tag{6.20}$$

we see in the portrait in Figure 6.10 that no invariant subspaces appear. This being the case, the spiraling trajectories rotate around the origin forever, asymptotically approaching it. (The trajectories are spiraling *inward* because of the negative real part of the eigenvalues; a system with eigenvalues containing positive real parts would spiral outward.)

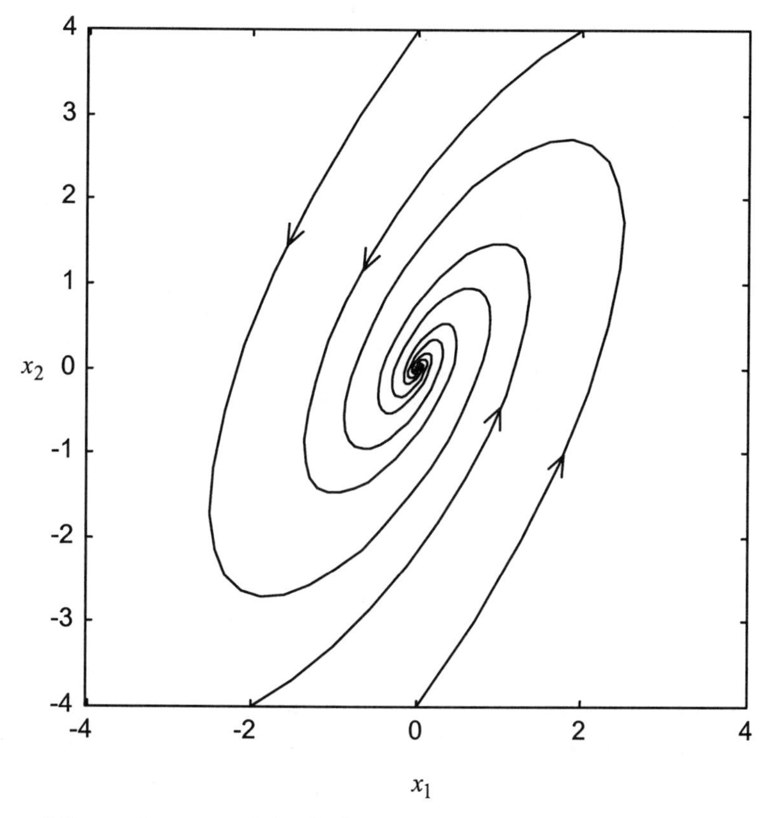

Figure 6.10 Phase portrait for the homogeneous system given by (6.20). This type of portrait shows an origin that is known as a *stable focus*.

The time-domain response of the system described by (6.20) is shown in the initial condition response of Figure 6.11. The oscillations shown in the phase portrait of Figure 6.10 are clearly seen.

As a final example, consider the system given by

$$A_8 = \begin{bmatrix} -1 & 2 \\ -5 & 1 \end{bmatrix} \qquad \sigma_8 = \{j3 \quad -j3\} \qquad M_8 = \begin{bmatrix} 3+j & 3-j \\ j5 & -j5 \end{bmatrix} \tag{6.21}$$

Relative to the previous example, we predict that this system will have no real invariant subspaces, and the trajectories will therefore be free to encircle the origin (without crossing each other). However, because there are *no* real parts of the eigenvalues, we cannot have solutions that decay to zero or tend to infinity. The resulting solutions are depicted in Figure 6.12. We know from our conventional solutions to differential equations or from our explicit computation of e^{At} in Chapter 5 that the solutions given by (6.8) for (6.21) will be nondecaying sinusoids, just as those of (6.20) were decaying sinusoids. These periodic solutions appear as ellipses in phase portraits, as in Figure 6.12. A time response of such a system would show non-decaying oscillations, as opposed to the decaying oscillations of Figure 6.11. It is perhaps interesting to note that the principal axes of these ellipses, the directions of extremal displacement, occur along the *singular vectors* of matrix A. Refer to Examples 4.13 and 5.1 for a discussion of singular vectors and ellipse geometry.

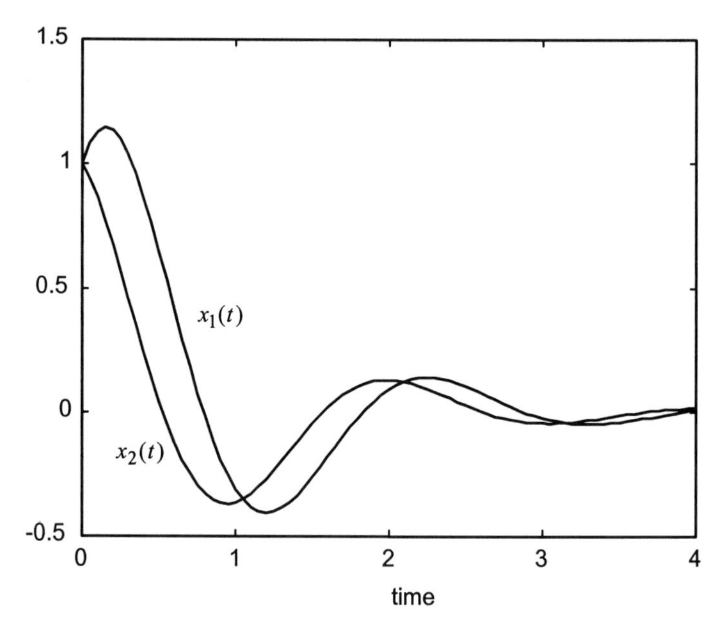

Figure 6.11 Initial condition response for the system described by Equation (6.20). The state variables are plotted as a function of time.

Without a decay or exponential growth in the solution, one might ask how it is possible to determine the directions of the arrows in Figure 6.12. The simplest way is to choose a single sample point, such as $x(t_0) = [1 \quad 0]^T$, shown in the figure, and determine the direction of the tangent

$$\dot{x}(t_0) = A_8 x(t_0) = \begin{bmatrix} -1 \\ -5 \end{bmatrix}$$

also shown in the figure. Given the direction of rotation of this single trajectory, the direction of all the others must be the same.

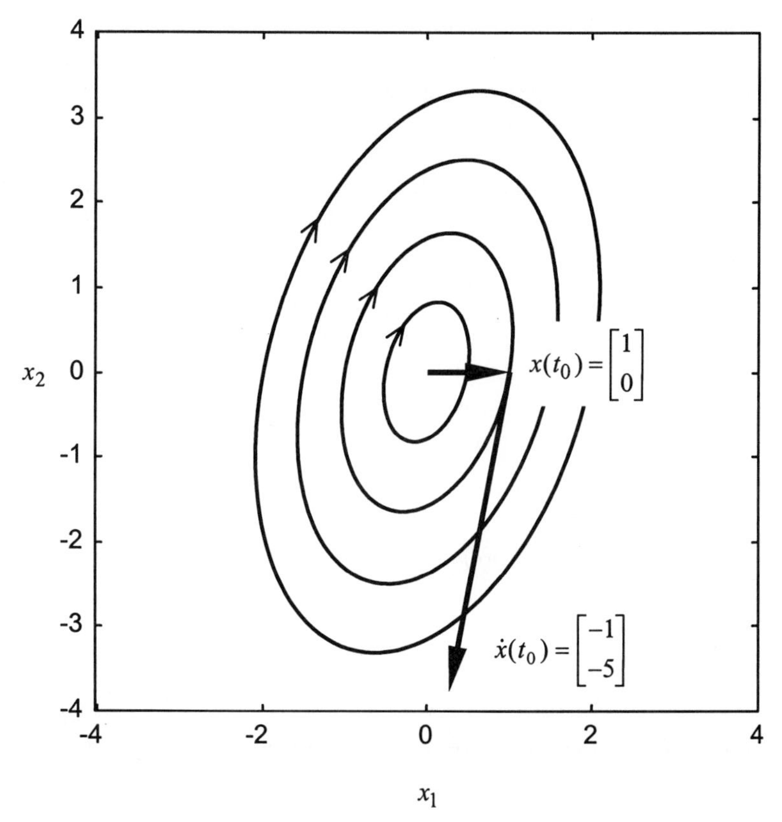

Figure 6.12 Phase portrait for the homogeneous system given by (6.21). The vector $\dot{x}(t_0)$ is shortened to fit on the plot. The origin in this portrait is called a *center*.

Although we have been presenting two-dimensional phase portraits here, the phase trajectory can be drawn in three dimensions, and the concept extends to arbitrary dimensions. Although our ability to draw in these higher dimensions diminishes, the intuitive geometric picture they provide does not. Our discussion of solutions using phase trajectories will continue in the next section and we will refer to them again in the next chapter on stability.

6.3 System Modes and Decompositions

In this section we will again consider LTI systems that are solved using Equation (6.6). Of course, for any t, vector $x(t)$ is an element in the (linear) state space, a vector space with all the attendant rules and properties. What we demonstrate in this section is the utility of considering the basis of eigenvectors for the state space when computing $x(t)$. The decomposition of $x(t)$ in this way is widely used in engineering analysis, especially for large-scale problems such as the deflection of flexible structures, vibrations, and the reduction of large state space to smaller approximations.

Let $\{e_i\}$ be the set of n linearly independent eigenvectors for a system, including, if necessary, generalized eigenvectors. Because this set may be used as a basis for the state space, we can uniquely decompose $x(t)$ as

$$x(t) = \sum_{i-1}^{n} \xi_i(t) e_i \tag{6.22}$$

where the coefficients in this expansion $\xi_i(t),\ i = 1, \ldots, n$ are functions of time. We can perform the same decomposition on the input terms in (6.1), allowing B to vary with time:

$$B(t)u(t) = \sum_{i=1}^{n} \beta_i(t) e_i \tag{6.23}$$

Substituting these expansions into the original state equation:

$$\sum_{i=1}^{n} \dot{\xi}_i(t) e_i = \sum_{i=1}^{n} \xi_i(t) A e_i + \sum_{i=1}^{n} \beta_i(t) e_i$$

or, by rearranging and supposing for now that $\{e_i\}$ consists of a complete set of n regular eigenvectors, we can obtain

$$\begin{aligned} \sum_{i=1}^{n} \left(\dot{\xi}_i(t) - \xi_i(t) A - \beta_i(t) \right) e_i &= \sum_{i=1}^{n} \dot{\xi}_i(t) e_i - \xi_i(t) \lambda_i e_i - \beta_i(t) e_i \\ &= 0 \end{aligned} \tag{6.24}$$

where λ_i is the eigenvalue corresponding to regular eigenvector e_i. Because the set $\{e_i\}$ is linearly independent, the terms in parentheses in (6.24) must also be identically equal to zero, or

$$\dot{\xi}_i(t) = \xi_i(t) \lambda_i + \beta_i(t) \qquad i = 1, \ldots, n \tag{6.25}$$

Equation (6.25) constitutes a set of n independent, first-order LTI differential equations. In the case that $\{e_i\}$ contains some generalized eigenvectors, (6.25) would take the same form for some i (those i to which no generalized eigenvectors are chained). However, when the generalized eigenvector $\boldsymbol{e}_{i+1}$ is chained to (regular or generalized) eigenvector $\boldsymbol{e}_i$, (6.25) would change to

$$\dot{\xi}_i(t) = \xi_i(t)\lambda_i + \xi_{i+1}(t) + \beta_i(t) \qquad i = 1, \ldots, n \tag{6.26}$$

There will be a total of $n - \sum_j g_j$ such coupled equations, where the summation of geometric multiplicities g_j is taken over all distinct eigenvalues.

The terms $\xi_i(t)\boldsymbol{e}_i$ in (6.22) and (6.24) are known as system *modes*, and (6.22) is the *modal decomposition* of the solution $x(t)$. The modes are equivalent to "new" state variables. They are, in fact, the same state variables we obtained when we make the change of basis $x = M\xi$, using the modal matrix *M* that results in our familiar similarity transformation, i.e.,

$$\begin{aligned} \dot{\xi} &= M^{-1}AM\xi + M^{-1}Bu \\ y &= CM\xi + Du \end{aligned} \tag{6.27}$$

This fact can be seen by expressing Equation (6.22) as

$$\begin{aligned} x(t) &= \sum_{i=1}^{n} \xi_i(t)\boldsymbol{e}_i \\ &= M \begin{bmatrix} \xi_1(t) \\ \vdots \\ \xi_n(t) \end{bmatrix} \end{aligned} \tag{6.28}$$

This is the source of the name "modal" in the term *modal matrix*. It is the matrix that decomposes a system so that decoupled equations are solved to produce system modes. Matrix $M^{-1}AM = J$ is simply the Jordan form discussed in Chapter 4.

As we did in Chapter 4, if we first convert the system to its basis of eigenvectors, then we can exploit the simpler form of (6.27) to generate the state-vector solution:

$$\begin{aligned} \xi(t) &= e^{J(t-t_0)}\xi(t_0) + \int_{t_0}^{t} e^{J(t-\tau)} M^{-1}B(\tau)u(\tau)\, d\tau, \\ \xi(t_0) &= M^{-1}x(t_0) \end{aligned} \tag{6.29}$$

where $\xi \triangleq [\xi_1 \quad \cdots \quad \xi_n]^{\mathrm{T}}$. Equation (6.29) will be easier to solve because of the simple structure of the Jordan form (see Section 5.4.2). After obtaining such a solution, the solution in the original basis may be computed via

$$x(t) = M\xi(t)$$

Modal Decompositions in Infinite-Dimensional Spaces

Modal decompositions are often useful in infinite-dimensional spaces, such as those produced by models of distributed-parameter systems. For example, the time- and space-varying displacement of flexible beams, strings, and plates are often expressed as an infinite summation of shape-functions multiplied by time-functions. These are sometimes called *eigenfunctions*, which are simply the modes as described above. For example, the displacement u of a point on a beam might then be expressed as a summation of products of time- and space-dependent functions:

$$u(x,t) = \sum_{i=1}^{\infty} X_i(x)T_i(t) \tag{6.30}$$

The infinite series arises because the describing equation for beam displacement is a partial differential equation, rather than an ordinary differential equation such as the ones with which we have been working. When ordered in decreasing magnitude of $X_i(x)$, it is common practice to truncate the series after the few most significant terms, thereby approximating an infinite dimensional system with a finite dimensional one that can be analyzed via matrix arithmetic.

For example, the equations that describe the displacement $u(x,t)$ of a stretched string are

$$\frac{\partial^2 u}{\partial x^2} - \frac{\rho}{T}\frac{\partial^2 u}{\partial t^2} = 0 \tag{6.31}$$

where ρ is the density of the string and T is its tension. If the string is of length ℓ, is initially displaced at a location $x = c$ to a height $u(c,0) = h$, and is subsequently released, it can be shown [6] that

$$u(x,t) = k\sum_{n=1}^{\infty} \frac{1}{n^2} \sin\left(\frac{n\pi c}{\ell}\right) \sin\left(\frac{n\pi x}{\ell}\right) \cos\left(\frac{n\pi\sqrt{T}t}{\ell\sqrt{\rho}}\right) \tag{6.32}$$

where k is a constant that depends on the system parameters ℓ, c, and h. Although there are infinitely many modes [or "shapes", i.e., $\sin(n\pi x/\ell)$] that contribute to this solution, their magnitudes decrease as $1/n^2$, so after some finite number, a suitable approximation can be obtained by truncating the series.

Because we are not able to delve into the solution techniques for (6.31), we cannot further discuss infinite-dimensional systems. However even in finite-dimensional linear systems, this kind of modal expansion helps us understand solutions by considering them one component at a time.

6.3.1 A Phase Portrait Revisited

In light of our knowledge of system modes, we now reconsider the information provided by the phase portraits and provide further clues to their qualitative construction. Consider again the first example system, given by

$$\dot{x} = A_1 x = \begin{bmatrix} -1 & 0 \\ 0 & -4 \end{bmatrix} x \tag{6.33}$$

with eigenvalues and eigenvectors computed in (6.10). An enlarged view of the upper-right quadrant of the phase portrait for that system is given again in Figure 6.13, where the trajectories on the eigenvectors are drawn with arrows pointing in the direction of positive time. The natural question arises when using phase portraits as a qualitative solution tool, "How do we know which direction the trajectories will take, i.e., how do we know they appear concave up as in Figure 6.13, rather than facing toward the sides? Could not trajectories approach the origin from the vertical direction instead of seeming to flatten out and approach horizontally, as the graphs show?"

To resolve these questions and provide a tool for guessing the trajectory's shapes, we show an initial condition, $x(t)$, and the tangent to the trajectory at that point, $\dot{x}(t) = Ax(t)$. This tangent vector is decomposed along the two eigenvectors, e_1 and e_2. Analytically, we know from the preceding section that

$$\begin{aligned} \dot{x}(t) = Ax(t) &= A\sum_{i=1}^{2} \xi_i(t) e_i \\ &= \sum_{i=1}^{2} \xi_i(t) A e_i \\ &= \sum_{i=1}^{2} \xi_i(t) \lambda_i e_i \\ &= -e^{-t} e_1 - 4e^{-4t} e_2 \end{aligned} \tag{6.34}$$

where the last line easily results from our A-matrix being already diagonal. From this expansion, the lengths of the components along each of the eigenvector directions are clear. For the time selected, the component along (negative) e_2 is larger that the component along (negative) e_1. The trajectory at this time, then, is evolving more along e_2 than e_1.

Taking a macroscopic view of the plot, we can see that the eigenvalue $\lambda_2 = -4$ is larger in magnitude ("faster" in time constant terms) than the eigenvalue $\lambda_1 = -1$. The result is that, until it decays to negligible proportions relative to the first mode (the component along e_1), the second mode (the component of motion along e_2) is larger. When sketching the plot, then, we naturally expect the vertical component of motion to dominate for small time, and the reverse for large time. Hence, we expect the curvature of the graph to be the concave up (and down) shape as observed.

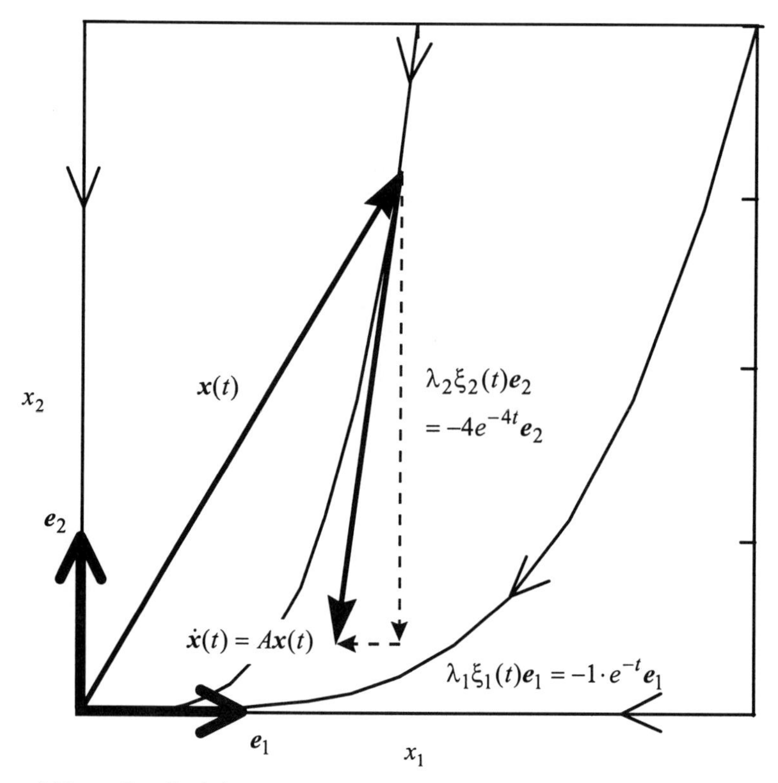

Figure 6.13 Detail of phase portrait for the homogeneous system given by (6.9).

Example 6.2: Sketching a Phase Portrait Using Qualitative Analysis

A two-dimensional LTI homogeneous system is known to have eigenvalues $\sigma = \{-10, -2\}$ and the corresponding eigenvectors

$$\{e_1, e_2\} = \left\{ \begin{bmatrix} 3 \\ 1 \end{bmatrix}, \begin{bmatrix} 1 \\ 3 \end{bmatrix} \right\} \tag{6.35}$$

Sketch the phase portrait for the system.

Solution:

The first step in determining the nature of the trajectories is to sketch the invariant subspaces, i.e., the straight lines that lie along the given eigenvectors. Noticing that both eigenvalues are negative real numbers, we expect all trajectories to approach the origin from any location. Because the eigenvalue λ_1 is faster than λ_2, we expect the trajectories for small time to experience more change in the direction of e_1 than in the direction of e_2. After a long time, the first mode will have decayed, and the trajectories' direction will be dominated by the second mode, i.e., along e_2. We then have sufficient information to sketch the phase portrait shown in Figure 6.14.

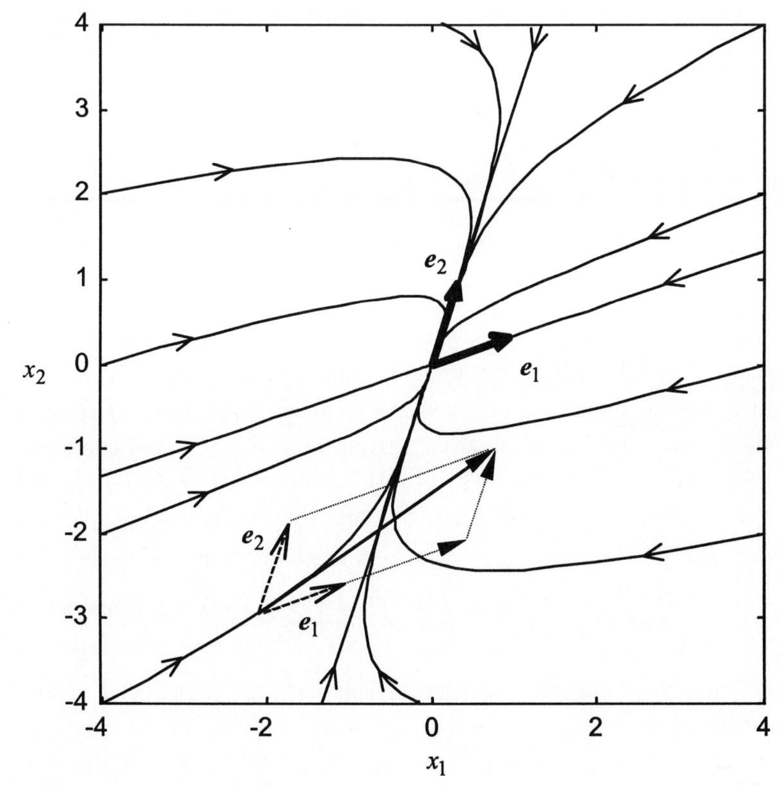

Figure 6.14 Phase portrait for the example system given by the eigenvectors in (6.35).

The figure also shows the decomposition of a trajectory into its modes. For the sample time selected, it can be seen that there is more motion along the first mode than along the second. At a later time (farther along that trajectory), there will be more motion along the second mode.

6.4 The Time-Varying Case

The matrix exponential integrating factor method, used in Equation (6.3), will not work when the A-matrix is a function of time. When we consider the linear time-varying system

$$\dot{x}(t) = A(t)x(t) + B(t)u(t) \tag{6.36}$$

we must either solve the system using a different technique or else find a different integrating factor. In the next two sections, special matrices are derived, called the fundamental solution matrix and the state transition matrix. The fundamental solution matrix performs the function of the integrating factor, while the overall solution is in terms of the state transition matrix. As we will see, though, the state transition matrix is difficult to compute.

6.4.1 State Fundamental Solution Matrix

Consider a homogeneous version of (6.36):

$$\dot{x}(t) = A(t)x(t) \tag{6.37}$$

We know from Chapter 2 that the set of solutions to (6.37) constitutes a linear vector space. It can be observed that this space is n-dimensional by considering a basis set of n linearly independent initial condition vectors $\{x_{i0}\}$, $i = 1,\ldots,n$. If we restrict our attention to matrices $A(t)$ that are smooth, then it is possible to guarantee that the solutions to (6.37) given an arbitrary initial condition, $x(t_0)$, are unique. Then the set $\{x_i(t)\}$, where $\dot{x}_i(t) = A(t)x_i(t)$ and $x_i(t_0) = x_{i0}$, defines n solutions for (6.37) that are linearly independent on $[t_0, t]$. Any additional solution $\xi(t)$ must be a linear combination of the $x_i(t)$ because $\xi(t_0) = \sum_{i=1}^{n} \alpha_i x_{i0}$ implies that $\xi(t) = \sum_{i=1}^{n} \alpha_i x_i(t)$ for some set of scalars α_i, $i = 1,\ldots,n$.

Organizing these linearly independent solutions, we construct a matrix $X(t)$ as follows:

$$X(t) = \begin{bmatrix} x_1(t) & x_2(t) & \cdots & x_n(t) \end{bmatrix}$$

Clearly, $\dot{X}(t) = A(t)X(t)$. Such a matrix $X(t)$ is known as a *fundamental solution matrix.*

Using the matrix identity revealed in Problem 3.2:

$$\frac{dX^{-1}(t)}{dt} = -X^{-1}(t)\frac{dX(t)}{dt}X^{-1}(t)$$

$$= -X^{-1}(t)\overbrace{A(t)X(t)}\underbrace{X(t)X^{-1}(t)}_{=I}$$

$$= -X^{-1}(t)A(t)$$

From this result, we can now show that the matrix $X^{-1}(t)$ qualifies as a valid integrating factor for the state equations in (6.37).

We may use this integrating factor in the nonhomogeneous equations in (6.36):

$$X^{-1}(t)[\dot{x}(t) - A(t)x(t) = B(t)u(t)]$$

$$X^{-1}(t)\dot{x}(t)\underbrace{-X^{-1}(t)A(t)}x(t) = X^{-1}(t)B(t)u(t)$$

$$X^{-1}(t)\dot{x}(t) + \frac{dX^{-1}(t)}{dt}x(t) = X^{-1}(t)B(t)u(t)$$

$$\frac{d}{dt}\left[X^{-1}(t)x(t)\right] = X^{-1}(t)B(t)u(t)$$

Now integrating both sides of the bottom line above yields

$$X^{-1}(t)x(t) - X^{-1}(t_0)x(t_0) = \int_{t_0}^{t} X^{-1}(\tau)B(\tau)u(\tau)\,d\tau$$

or

$$x(t) = X(t)X^{-1}(t_0)x(t_0) + \int_{t_0}^{t} X(t)X^{-1}(\tau)B(\tau)u(\tau)\,d\tau \tag{6.38}$$

This appears to be a general solution to the time-varying state equations, but, unfortunately, computing this solution is not so easy. After all, if we could easily find the n linearly independent solutions necessary for the construction of $X(t)$, we would not have needed to derive (6.38). In most cases, we have no general method for computing the fundamental solution matrix $X(t)$. The ease of computing components $x_i(t)$ depends a great deal on the exact form of the time functions within $A(t)$. Like nonlinear systems, such time-varying systems tend to be solved individually, as circumstances allow. One can see from any ordinary differential equations text that certain classes of time-varying equations are solved, e.g., Bessel, LeGendre, or Hermite equations, but certainly not all of them can be solved.

6.4.2 The State-Transition Matrix

If $X(t)$ is not easily calculated, then of what useful purpose is Equation (6.38)? First, we will use its existence to prove the existence of a different matrix. Note that in (6.38), the fundamental solution matrix only appears as a product of the form $X(t)X^{-1}(\tau)$. This product

$$\Phi(t,\tau) \stackrel{\Delta}{=} X(t)X^{-1}(\tau) \tag{6.39}$$

is known as the *state-transition matrix*, and it lends insight into the solutions of time-varying systems and time-invariant systems. With this notation, (6.38) becomes

$$x(t) = \Phi(t,t_0)x(t_0) + \int_{t_0}^{t} \Phi(t,\tau)B(\tau)u(\tau)\,d\tau \tag{6.40}$$

Computing the State-Transition Matrix

While $X(t)$ can indeed be difficult to find, there does exist an iterative method for the computation of $\Phi(t,\tau)$, known as the Peano-Baker integral series, which we provide here without proof [7]:

$$\Phi(t,\tau) = I + \int_{\tau}^{t} A(\sigma_1)d\sigma_1 + \int_{\tau}^{t} A(\sigma_1)\int_{\tau}^{\sigma_1} A(\sigma_2)d\sigma_2 d\sigma_1 + \cdots \tag{6.41}$$

A difficulty with this technique is the necessary repeated integrations. In addition, it is unlikely that the series will converge to a closed form. Often, this method is convenient when the matrix $A(t)$ is inherently *nilpotent*. A nilpotent matrix is one such that $A^p \equiv 0$ for all $p > q$. An example of a nilpotent matrix is a triangular matrix with zeros on the diagonal. As we saw in Section 5.4.2, repeated multiplications of such matrices by themselves eventually results in the zero matrix, not because of the numbers in the matrix, but because of its inherent physical structure.

A second method applies only in the special circumstance that $A(t)A(\tau) = A(\tau)A(t)$. If this is the case, then

$$\Phi(t,\tau) = \exp\left[\int_{\tau}^{t} A(\sigma)\,d\sigma\right] \tag{6.42}$$

Although it is rare for this condition to be satisfied for general $A(t)$, it does hold when A is constant or when $A(t)$ is diagonal, in which cases there are more direct methods for finding the solution of the system. One should be warned that the expression in (6.42) is *not* equivalent to e^{At}, and must often be computed with a Taylor series expansion of the matrix exponential.

Example 6.3: State-Transition Matrix Using Series Expansion

For a system with the A-matrix given in Example 6.1, find the state-transition matrix $\Phi(t,\tau)$ using the Peano-Baker series.

Solution.

It should first be noted that the state transition matrix for Example 6.1 is apparent from the solution. Computing it explicitly,

$$\begin{aligned}
\Phi(t,\tau) &= I + \int_{\tau}^{t} \begin{bmatrix} 0 & 1 \\ 0 & 0 \end{bmatrix} d\sigma_1 + \int_{\tau}^{t} \begin{bmatrix} 0 & 1 \\ 0 & 0 \end{bmatrix} \int_{\tau}^{\sigma_1} \begin{bmatrix} 0 & 1 \\ 0 & 0 \end{bmatrix} d\sigma_2\, d\sigma_1 + \cdots \\
&= \begin{bmatrix} 1 & 0 \\ 0 & 1 \end{bmatrix} + \left.\begin{bmatrix} 0 & \sigma_1 \\ 0 & 0 \end{bmatrix}\right|_{\tau}^{t} + \int_{\tau}^{t} \begin{bmatrix} 0 & 1 \\ 0 & 0 \end{bmatrix} \left.\begin{bmatrix} 0 & \sigma_2 \\ 0 & 0 \end{bmatrix}\right|_{\tau}^{\sigma_1} d\sigma_1 + \cdots \\
&= \begin{bmatrix} 1 & 0 \\ 0 & 1 \end{bmatrix} + \begin{bmatrix} 0 & t-\tau \\ 0 & 0 \end{bmatrix} + 0 + \cdots \\
&= \begin{bmatrix} 1 & t-\tau \\ 0 & 1 \end{bmatrix}
\end{aligned}$$

Here, it is apparent that the second- and higher-order integrations will yield zero. Note the agreement between this result and the matrix seen in the integrand of the solution of Example 6.1.

Properties of the State-Transition Matrix

By expressing an arbitrary solution as a linear combination of fundamental solutions, $x(t) = X(t)x(t_0)$, we can perform some algebraic manipulations to find that

$$\begin{aligned} x(t) &= X(t)X^{-1}(\tau)X(\tau)x(t_0) \\ &= X(t)X^{-1}(\tau)x(\tau) \\ &= \Phi(t,\tau)x(\tau) \end{aligned} \tag{6.43}$$

Therefore, the state-transition matrix $\Phi(t,\tau)$ is a linear operator that takes a vector $x(\tau)$ and produces a vector $x(t)$ (i.e., a solution at time t). Hence the name for the matrix; it performs the *transition* from a state at one time to a state at another. This property can be extended to successive time instants by realizing that

$$\begin{aligned} \Phi(t_2,t_1)\Phi(t_1,t_0) &= X(t_2)X^{-1}(t_1)X(t_1)X^{-1}(t_0) \\ &= X(t_2)X^{-1}(t_0) \\ &= \Phi(t_2,t_0) \end{aligned} \tag{6.44}$$

Similarly, it is easily shown that $\Phi^{-1}(t,\tau) = \Phi(\tau,t)$.

Furthermore, by differentiating,

$$\begin{aligned} \frac{d\Phi(t,\tau)}{dt} &= \frac{d\left[X(t)X^{-1}(\tau)\right]}{dt} = \frac{dX(t)}{dt}X^{-1}(\tau) \\ &= A(t)X(t)X^{-1}(\tau) \\ &= A(t)\Phi(t,\tau) \end{aligned} \tag{6.45}$$

The state transition matrix and the fundamental solution matrix $X(t)$ are solutions for the same homogeneous differential equation, (6.37). This suggests a numerical method of solving (6.45), numerically integrating (over time t), subject to the initial condition that $\Phi(\tau,\tau) = I$, in order to estimate $\Phi(t,\tau)$. This is not done in practice very often, mostly because the system input $u(t)$ is itself not known a priori. Therefore, such an *open-loop* solution of (6.40) is not useful.

When the system is time-invariant, we know from a comparison of (6.6) and (6.38) that $X(t)=e^{At}$, yielding

$$\Phi(t,t_0)=e^{At}e^{-At_0}=e^{A(t-t_0)}$$

Remember from the definition that for a time-invariant system, the solution depends only on the difference $t-t_0$. Fortunately, in this situation we have already demonstrated analytical methods for generating the state-transition matrix e^{At} (see Chapter 5). Note that this situation is entirely consistent with the properties of state-transition matrices derived for the more general time-varying case above. In particular, we have already shown that in a homogeneous system,

$$x(t)=e^{A(t-t_0)}x(t_0)$$

which is the time-invariant counterpart to (6.43). If we returned to the phase plane depictions, we could verify that for an initial condition $x(t_0)$, multiplication by the state transition matrix $e^{A(t-t_0)}$ would produce a different vector $x(t)$ for whatever time t we choose.

We should remark at this point that phase portraits for time-varying systems are of little use. Recall that the phase portrait for a system depends on the eigenvalues and eigenvectors of the A-matrix. When the A-matrix is time-varying, the eigenvalues and eigenvectors will therefore also depend on time. In phase portraits, time is a parameter on the trajectories. The plots, therefore, cannot be constructed for a time-varying A-matrix. In fact, we will observe in the next chapter that the eigenvalues of time-varying systems are themselves of limited use. One cannot always even predict the stability of a system by calculating the eigenvalues at a particular instant in time.

6.5 *Solving Discrete-Time Systems*

As we have mentioned, there are natural systems that are modeled in discrete-time. In a linear systems context, this means that the inputs are applied, and the states change, at discrete intervals of period T. Such a system may be modeled with the equations

$$\begin{aligned} x(k+1) &= A_d(k)x(k)+B_d(k)u(k) \\ y(k) &= C_d(k)x(k)+D_d(k)u(k) \end{aligned} \tag{6.46}$$

Here we have implicitly assumed by our notation that the discrete-time system matrices A_d, B_d, C_d, and D_d may be functions of time k. In a time-invariant system, they will not.

However, often, because the universe is modeled (at least by most engineers) as evolving in continuous-time, the equations in (6.46) more often result from the discretization of a continuous-time system such as

$$\begin{aligned} \dot{x}(t) &= A(t)x(t) + B(t)u(t) \\ y(t) &= C(t)x(t) + D(t)u(t) \end{aligned} \tag{6.47}$$

We can arrive at the equivalent (6.46) from (6.47) by considering a discrete-time approximation to the state-equation solution (6.40).

6.5.1 Discretization

Consider the kth time instant, wherein $t = kT$. At T units of time later, $t = (k+1)T$. In order to get an accurate discrete-time equivalent, we will assume that period T is much smaller than the Shannon period for the input signal $u(t)$ (see [4], pp. 79-82). If this is the case, we can approximate the input as $u(t) \approx u(kT)$, or simply $u(k)$, over the entire interval $kT \le t < (k+1)T$. Making the input constant over the sampling interval allows us to remove the input term from the integrand in the solution for (6.40):

$$x(k+1) = \Phi(k+1,k)x(k) + \int_k^{k+1} \Phi(k+1,\tau)B(\tau)\,d\tau\; u(k) \tag{6.48}$$

In this form, if $\Phi(t,\tau)$ and $B(\tau)$ are known, the solution for the state vector could be computed each time an input is applied. The matrices

$$A_d(k) \triangleq \Phi(k+1,k) \qquad B_d(k) \triangleq \int_k^{k+1} \Phi(k+1,\tau)B(\tau)\,d\tau$$

can be computed from knowledge of the system, and (6.46) is the result. The output equation, being an algebraic equation, follows from a knowledge of $C(t)$ and $D(t)$. Also note that (6.48) provides the values of the state vector at the chosen sample instants. At intermediate times, the state variables may take on other values, resulting in a "ripple" effect if the variables are plotted as functions of continuous time. For discrete-time analysis, though, it is only the values at the sample instants that we are interested in.

```
c2d(sysc,Ts,
method)
```

Note that (6.48) is not the solution to the state equations in (6.46). Rather, (6.48) is a discretization^{M} of the state equations in (6.47). As we saw in Examples 3.10 and 3.11, the solutions of discrete-time systems are often inductively obtained by iterating over a few times intervals on (6.46).

6.5.2 Discrete-Time State-Transition Matrix

Consider the first equation of (6.46). If we write a few terms in the computation of $x(k)$, given an initial point $x(j)$, we get:

$$\begin{aligned}
x(j+1) &= A_d(j)x(j) + B_d(j)u(j) \\
x(j+2) &= A_d(j+1)x(j+1) + B_d(j+1)u(j+1) \\
&= A_d(j+1)\left[A_d(j)x(j) + B_d(j)u(j)\right] + B_d(j+1)u(j+1) \\
&= A_d(j+1)A_d(j)x(j) + A_d(j+1)B_d(j)u(j) + B_d(j+1)u(j+1) \\
x(j+3) &= A_d(j+2)A_d(j+1)A_d(j)x(j) + A_d(j+2)A_d(j+1)B_d(j)u(j) \\
&\qquad + A_d(j+2)B_d(j+1)u(j+1) + B_d(j+2)u(j+2) \\
&\vdots
\end{aligned}$$

until, by induction,

$$x(k) = \left(\prod_{i=j}^{k-1} A_d(i)\right) x(j) + \sum_{i=j+1}^{k} \left(\prod_{q=i}^{k-1} A_d(q)\right) B(i-1)u(i-1) \tag{6.49}$$

where it is necessary to define

$$\prod_{q=k}^{k-1} A_d(q) \stackrel{\Delta}{=} I$$

Consider the situation in which the system is homogeneous, i.e., $u(k) \equiv 0$ for all k. Then we would have

$$x(k) = \left(\prod_{i=j}^{k-1} A_d(i)\right) x(j) \tag{6.50}$$

This formula implicitly defines the state-transition matrix for discrete-time systems:

$$\Psi(k,j) = \prod_{i=j}^{k-1} A_d(i) \tag{6.51}$$

As in the continuous-time case, Equation (6.50) makes it apparent that the state-transition matrix $\Psi(k,j)$ may be interpreted as the linear operator that takes a state vector at time j and returns the state vector at time k. This matrix is defined only for $k \geq j$ and shares most of the properties of the continuous-time state transition matrix, except for invertibility. From the structure of (6.51), it is clear that if any $A_d(i)$ is not invertible, which is entirely possible, then $\Psi(k,j)$ itself will not be invertible.

Using the notation of the discrete-time state-transition matrix for time-varying systems, i.e., Equation (6.51) above, then the expression (6.49) for the general solution of discrete-time systems is

$$x(k) = \Psi(k,j)x(j) + \sum_{i=j+1}^{k} \Psi(k,i)B(i-1)u(i-1) \tag{6.52}$$

where the initial condition is taken at $t = j$.

6.5.3 Time-Invariant Discrete-Time Systems

Suppose now that in the discrete-time system of (6.46), the matrices A_d and B_d were independent of time *k*. Then (6.48) would become

$$x((k+1)T) = e^{A[(k+1)T-kT]}x(kT) + \int_{kT}^{(k+1)T} e^{A[(k+1)T-\tau]}B(\tau)\,d\tau\; u(kT)$$

or

$$x(k+1) = e^{AT}x(k) + \int_{kT}^{(k+1)T} e^{A[(k+1)T-\tau]}B(\tau)\,d\tau\; u(k) \tag{6.53}$$

From this, we have the definitions

$$A_d \triangleq e^{AT} \qquad B_d \triangleq \int_{kT}^{(k+1)T} e^{A[(k+1)T-\tau]}B(\tau)\,d\tau$$

again giving a formula for obtaining (6.46). Once again, matrices A_d and B_d can be computed "off-line," i.e., without knowledge of the input. Note that matrix A_d is independent of *k*.

`lsim(sys,u,t)`

Furthermore, for a time-invariant system, (6.49) will become[M]

$$x(k) = (A_d)^{k-j}x(j) + \sum_{i=j+1}^{k} (A_d)^{k-i}B(i-1)u(i-1) \tag{6.54}$$

yielding the time-invariant, discrete-time state transition matrix

$$\Psi(k,j) = \Psi(k-j) = (A_d)^{k-j}$$

We should point out here that the discussion of modes and modal decompositions as presented in Section 6.3 applies here as well. Because the eigenvalues and eigenvectors of a matrix A or A_d are calculated in the same manner regardless of whether the system is discrete-time or continuous-time, the modal matrix M functions in the same way for both systems. If we use it to define a new state vector $x(k) = M\xi(k)$, then

$$\begin{aligned}\xi(k+1) &= M^{-1}A_d M\xi(k) + M^{-1}B_d(k)u(k) \\ &\triangleq \hat{A}_d\xi(k) + \hat{B}_d(k)u(k) \\ y(k) &= C_d\xi(k) + D_d u(k) \\ &\triangleq \hat{C}_d\xi(k) + \hat{D}_d u(k)\end{aligned} \tag{6.55}$$

The transformed matrix $\hat{A}_d$ will again be in its Jordan form.

Example 6.4: Discretization of a System

For the A and B matrices given below, discretize the system to get a discrete-time state variable description using the formula (6.53). Assume the system is sampled at $T = 0.1$ s.

$$A = \begin{bmatrix} -3 & 1 \\ 0 & -2 \end{bmatrix} \qquad B = \begin{bmatrix} 1 \\ 1 \end{bmatrix}$$

Solution

The A-matrix was also used in Example 5.4, so we know from that example that

$$e^{At} = \begin{bmatrix} e^{-3t} & -e^{-3t} + e^{-2t} \\ 0 & e^{-2t} \end{bmatrix}$$

Therefore,

$$A_d = e^{AT} = \begin{bmatrix} e^{-3T} & -e^{-3T} + e^{-2T} \\ 0 & e^{-2T} \end{bmatrix} = \begin{bmatrix} 0.741 & 0.0779 \\ 0 & 0.819 \end{bmatrix}$$

As for B_d, we have

$$
\begin{aligned}
B_d &= \int_{kT}^{(k+1)T} e^{A[(k+1)T-\tau]} B \, d\tau \\
&= \int_{kT}^{(k+1)T} \begin{bmatrix} e^{-3[(k+1)T-\tau]} & -e^{-3[(k+1)T-\tau]} + e^{-2[(k+1)T-\tau]} \\ 0 & e^{-2[(k+1)T-\tau]} \end{bmatrix} \begin{bmatrix} 1 \\ 1 \end{bmatrix} d\tau \\
&= \int_{kT}^{(k+1)T} \begin{bmatrix} e^{-2[(k+1)T-\tau]} \\ e^{-2[(k+1)T-\tau]} \end{bmatrix} d\tau = \frac{1}{2} \begin{bmatrix} e^{-2[(k+1)T-\tau]} \\ e^{-2[(k+1)T-\tau]} \end{bmatrix} \Bigg|_{kT}^{(k+1)T} \\
&= \frac{1}{2} \begin{bmatrix} e^{-2[(k+1)T-(k+1)T]} - e^{-2[(k+1)T-kT]} \\ e^{-2[(k+1)T-(k+1)T]} - e^{-2[(k+1)T-kT]} \end{bmatrix} \\
&= \frac{1}{2} \begin{bmatrix} 1-e^{-2T} \\ 1-e^{-2T} \end{bmatrix} = \begin{bmatrix} 0.0906 \\ 0.0906 \end{bmatrix}
\end{aligned}
$$

Therefore, the discrete-time approximation of the system, sampled at 10Hz, is

$$
\boldsymbol{x}(k+1) = \begin{bmatrix} 0.741 & 0.0779 \\ 0 & 0.819 \end{bmatrix} \boldsymbol{x}(k) + \begin{bmatrix} 0.0906 \\ 0.0906 \end{bmatrix} u(k)
$$

6.6 *Summary*

In this chapter, we have investigated the analytical solution of state space equations. In doing so, we have demonstrated one of the important features of state variable analysis, namely, that the first-order analysis methods of scalar differential equations can help us solve (first-order) vector differential equations. This is first apparent in the integrating factor technique used to solve the LTI systems. Because any linear system can be represented in state space, these solution methods are presented to give the reader a set of tools for solving state space equations that are similar to those used for solving scalar differential equations.

However, as we have stated, the solution methods developed here are not necessarily the most useful tools for control system design and analysis. This is because, as we have shown, we must know the applied input to find the solution of a system. In control systems design, the input signal is the goal and is not generally available a priori. Nevertheless, the solution technique has provided insight into the evolution of state variables, and provided a technique by which discrete-time systems can be generated as approximations of sampled continuous-time systems.

Other highlights of the chapter are:

- The solution of an LTI system has the same form, i.e., a convolution integral, as the solution of scalar systems. Only the computation of the matrix exponential e^{At} complicates the matter.
- For homogeneous systems, phase portraits can be useful visualization tools. It is recommended that the reader practice generating phase portraits for systems with different dynamic characteristics. As we have mentioned, the evolution of a system's solution trajectories is often understood using this visual imagery, even in higher dimensions. In nonlinear problems as well, phase portraits are indispensable tools and can be used to predict the existence of limit cycles (nonlinear oscillations), switching times, final values, and stability characteristics that are very often difficult to determine analytically.
- We have introduced the notion of a system mode. This is another state space tool that is often generalized into higher dimensions. Heat conduction problems, bending plates and beams, and many electromagnetic phenomena are described by partial differential equations that result in infinite-dimensional state spaces. By describing the solution of such systems as sums of modes, we can retain only the most dominant and significant components of the overall solutions. Others, because they are less significant, might be simply ignored. The modal decomposition of a system is also a convenient method for generating a qualitative picture of a phase portrait.
- For time-varying systems, state variable solutions are quite difficult to obtain. The state-transition matrix, while symbolically providing a simple integral solution, can be as elusive as the solution of a nonlinear equation. Only limited techniques are available for its construction, such as the Peano-Baker series, which itself can be difficult to compute, especially in closed-form
- In discrete-time, we have shown that the state variable solution and the state transition matrix are similar in appearance to their continuous-time counterparts.

Part of the discussion of the solutions generated in this chapter has hinted at the stability properties of systems. For example, in the generation of the phase portraits, we mentioned the tendency of a system mode to approach or diverge from the origin. We sometimes find that such modes are unstable, for obvious reasons. However, such a notion of stability ignores the fact that phase portraits, by their definition, have not accounted for the presence of the input signal. Indeed, there are a number of different perspectives on the stability of a system, and these will be discussed in the next chapter.

6.7 Problems

6.1 Solve the state variable system given in Problem 1.9 for $x(t)$, $t \geq 0$, given that $u(t) = e^{-3t}$.

6.2 Draw phase portraits for the systems with the following A-matrices:

a) $\begin{bmatrix} -8 & -6 \\ 0 & -2 \end{bmatrix}$ b) $\begin{bmatrix} -8 & -6 \\ 0 & 2 \end{bmatrix}$ c) $\begin{bmatrix} 0 & 4 \\ -4 & 0 \end{bmatrix}$

d) $\begin{bmatrix} 4 & -4 \\ 4 & 4 \end{bmatrix}$ e) $\begin{bmatrix} 1 & 0 \\ 0 & 0 \end{bmatrix}$ f) $\begin{bmatrix} 1 & 0 \\ 1 & 0 \end{bmatrix}$

g) $\begin{bmatrix} 0 & 0 \\ 0 & 0 \end{bmatrix}$ h) $\begin{bmatrix} -1 & 1 & 0 \\ 0 & -1 & 0 \\ 0 & 0 & -1 \end{bmatrix}$

6.3 Determine state-transition matrix $\Phi(t, \tau)$ for the following A-matrices:

a) $\begin{bmatrix} 0 & 1 & t^2 \\ 0 & 0 & t \\ 0 & 0 & 0 \end{bmatrix}$ b) $\begin{bmatrix} 0 & 0 \\ 2t & 2t \end{bmatrix}$ c) $\begin{bmatrix} e^{-2t} & 0 \\ 0 & 2t \end{bmatrix}$

6.4 Show how systems of the form

$$\dot{x}(t) = \begin{bmatrix} 1 & 0 \\ 1 & f(t) \end{bmatrix} x(t)$$

can be considered as decoupled scalar equations, the second of which has the solution of the first as an input term. Use this fact to determine an expression for $\Phi(t, \tau)$ for the second-order system.

6.5 Show that if $\Phi(t, t_0)$ is the state-transition matrix for A, then $A(t) = \dot{\Phi}(t, t_0)\big|_{t=t_0}$. Derive an analogous result for the discrete-time state transition matrix, i.e., given $\Psi(k, j)$, determine $A(k)$.

6.6 Show for a time-invariant system, that $e^{At} = L^{-1}\{(sI - A)^{-1}\}$, where L is the LaPlace transform operator.

6.7 Suppose $\Phi_A(t,\tau)$ is the state-transition matrix for $A(t)$. Define a (nonsingular) change of variables by $x(t) = M\xi(t)$ such that $\dot{\xi}(t) = M^{-1}AM\xi(t) \stackrel{\Delta}{=} \hat{A}\xi(t)$. Determine an expression for a new state-transition matrix $\Phi_{\hat{A}}(t,\tau)$ in terms of $\Phi_A(t,\tau)$. Does the result depend on whether M is time varying?

6.8 For every system $\dot{x}(t) = Ax(t)$ there is a defined system $\dot{p}(t) = -A^{\mathrm{T}} p(t)$, which is called the adjoint system. Show that if $\Phi_A(t,\tau)$ is the state-transition matrix for the original system, then the state-transition matrix for the adjoint system is $\Phi_A^{\mathrm{T}}(\tau,t)$.

6.9 Use Equation (6.7) to find an expression for the impulse response matrix, i.e., the solution to a LTI system whose input is the Dirac delta function $\delta(t)$.

6.10 Find the state-transition matrix $\Psi(k,0)$ for the following system:

$$x(k+1) = \begin{bmatrix} 3.5 & -2 \\ 6 & -3.5 \end{bmatrix} x(k)$$

6.11 For the following system, find an expression for $y(n)$ if the input $u(k) = 1$, $k = 0,1,\ldots,\infty$

$$x(k+1) = \begin{bmatrix} 0 & 1 \\ -0.5 & -1 \end{bmatrix} x(k) + \begin{bmatrix} 1 \\ 1 \end{bmatrix} u(k)$$

$$y(k) = \begin{bmatrix} 1 & 0 \end{bmatrix} x(k)$$

6.12 Discretize the system below using a sample time of $T = 1$ s.

$$\dot{x}(t) = \begin{bmatrix} -2 & 0 \\ 1 & 0 \end{bmatrix} x(t) + \begin{bmatrix} 1 \\ 0 \end{bmatrix} u(t)$$

$$y(t) = \begin{bmatrix} 0 & 1 \end{bmatrix} x(t) + 2u(t)$$

6.8 References and Further Reading

Finding solutions to the time-invariant linear state space system is fairly easy and can be found in any linear systems or control systems text. For time-varying systems, which require the state-transition matrix, finding solutions is more difficult. Good discussions of such systems, including the computation of the state-transition matrix, may be found in [2], [5], and [7]. References [2] and [7] offer some particularly interesting problems. For the discrete-time case, see [4] and [7].

The stretched-string problem and many other modal system problems, such as deformable beams and plates, electromagnetic, and thermal gradient systems can be found in [6], which contains a large number of worked problems. Other infinite dimensional systems are discussed in [3].

More details on the construction and interpretation of phase portraits, particularly for nonlinear systems, can be found in [1].

[1] Atherton, Derek P., *Nonlinear Control Engineering*, Van Nostrand Reinhold, 1982.

[2] Brockett, Roger W., *Finite Dimensional Linear Systems*, John Wiley & Sons, 1970.

[3] Coddington, Earl A., and Norman Levinson, *Theory of Ordinary Differential Equations*, McGraw-Hill, 1955.

[4] Franklin, Gene, and J. David Powell, *Digital Control of Dynamic Systems*, Addison-Wesley, 1981.

[5] Kailath, Thomas, *Linear Systems*, Prentice-Hall, 1980.

[6] Lebedev, N. N., I. P. Skalskaya, and Y. S. Uflyand, *Worked Problems in Applied Mathematics*, Dover, 1965.

[7] Rugh, Wilson, *Linear System Theory*, 2nd edition, Prentice-Hall, 1996.

7

System Stability

During the description of phase portrait construction in the preceding chapter, we made note of a trajectory's tendency to either approach the origin or diverge from it. As one might expect, this behavior is related to the *stability* properties of the system. However, phase portraits do not account for the effects of applied inputs, only for initial conditions. Furthermore, we have not clarified the definition of stability of systems such as the one in Equation (6.21) (Figure 6.12), which neither approaches nor departs from the origin. All of these distinctions in system behavior must be categorized with a study of stability theory. In this chapter, we will present definitions for different kinds of stability that are useful in different situations and for different purposes. We will then develop stability testing methods so that one can determine the characteristics of systems without having to actually generate explicit solutions for them.

7.1 Lyapunov Stability

The first "type" of stability we consider is the one most directly illustrated by the phase portraits we have been discussing. It can be thought of as a stability classification that depends solely on a system's initial conditions and not on its input. It is named for Russian scientist A. M. Lyapunov, who contributed much of the pioneering work in stability theory at the end of the nineteenth century. So-called "Lyapunov stability" actually describes the properties of a particular point in the state space, known as the *equilibrium point*, rather than referring to the properties of the system as a whole. It is necessary first to establish the terminology of such points.

7.1.1 Equilibrium Points

Many of the concepts introduced in this chapter will apply equally well to nonlinear systems as to linear systems. Even though much of the control system design we will perform will be applied to linear models of perhaps nonlinear systems, it is a useful skill to analyze the stability properties of nonlinear systems also. So consider the general form of an arbitrary system, perhaps nonlinear, and perhaps time-varying:

$$\dot{x}(t) = f(x(t), u(t), t) \tag{7.1}$$

Whereas we were interested in the behavior of the phase trajectories relative to only the origin in the previous chapter, for general nonlinear systems the point of interest is not necessarily the zero vector, but rather a general equilibrium point:

> ***Equilibrium point:*** An equilibrium point x_e is a constant vector such that if $x(t_0) = x_e$ and $u(t) \equiv 0$ in (7.1), then $x(t) = x_e$ for all $t \geq t_0$. (7.2)

Because an equilibrium point is a constant vector, we have

$$\dot{x}(t) = 0 = f(x_e, 0, t) \tag{7.3}$$

Equilibrium points are sometimes also referred to as *critical points* or *stationary points*. They represent a steady-state constant solution to the dynamic equation, (7.1). In discrete-time systems, an equilibrium point is a vector x_e such that $x(k+1) = x(k)\ (= x_e)$, rather than $\dot{x}(t) = 0$.

Considering now the special case of a linear system, an equilibrium point(s) can be computed from

$$\dot{x}(t) = 0 = Ax_e \tag{7.4}$$

From this expression, it is apparent that the origin $x_e = 0$ will always be an equilibrium point of a linear system. Other equilibrium points will be any point that lies in the null space^{M} of the system matrix A. Sets of equilibrium points will therefore constitute subspaces of the state space, and we can have no isolated equilibrium points outside the origin.

`null(A)`

Example 7.1: Equilibrium Points for Linear and Nonlinear Systems

Find the equilibrium points for the following systems:

$$\text{a)}\ \dot{x} = \begin{bmatrix} 1 & 3 \\ 3 & 9 \end{bmatrix} x + \begin{bmatrix} 0 \\ 1 \end{bmatrix} u \qquad \text{b)}\ x(k+1) = \begin{bmatrix} 2 & 0 \\ 0 & .5 \end{bmatrix} x(k) \qquad \text{c)}\ \dot{x} = -\sin x$$

Solution:

a) The b-matrix is irrelevant because we will assume a zero input. The A-matrix has a zero eigenvalue so it has a nontrivial null space, which can be found to lie along the line $x_e = [-3 \quad 1]^T$. This is the equilibrium set.

b) The discrete-time equilibrium implies that $x(k+1) = x(k) = Ax(k)$, so we seek the solutions to the system $(A - I)x_e = 0$. For the A-matrix given, $A - I$ has no zero eigenvalues, so it has no nontrivial null space. The only equilibrium solution is therefore $x_e = [0 \quad 0]^T$.

c) The third system is nonlinear. Setting $\dot{x} = 0 = -\sin x_e$, we see that $x_e = \pm n\pi$, where $n = 0, 1, \ldots$. This constitutes an infinite sequence of isolated equilibria, which we cannot encounter in a linear system.

We are now prepared for the stability definitions pertinent to equilibrium points. For the linear systems we treat here, we will speak of only isolated equilibria at the origin, but sets of equilibria such as lines through the origin may be interpreted similarly.

> ***Stability in the Sense of Lyapunov***: An equilibrium point x_e of the system $\dot{x} = A(t)x$ is stable in the sense of Lyapunov, or simply *stable*, if for any $\varepsilon > 0$, there exists a value $\delta(t_0, \varepsilon) > 0$ such that if $\|x(t_0) - x_e\| < \delta$, then $\|x(t) - x_e\| < \varepsilon$, regardless of t ($t > t_0$). The equilibrium point is *uniformly stable* in the sense of Lyapunov if $\delta = \delta(\varepsilon)$, i.e., if the constant δ does not depend on initial time t_0. (7.5)

As one might expect, time-invariant systems that are stable are uniformly stable because the initial time t_0 should not affect the qualitative behavior of the system. In words, the above definition means that for a stable in the sense of Lyapunov system, anytime we desire that the state vector remain within distance ε of the origin, it is possible to accomplish this by simply giving it an initial condition away from x_e by an amount no larger than δ. This scenario is depicted in Figure 7.1. Clearly, we must have $\delta < \varepsilon$ or else the system will have an initial condition that has already violated the desired bounds of the state vector.

Lyapunov stability is sometimes referred to as *internal stability* because its definition describes the behavior of the state variables rather than the output variables, which are by contrast considered *external*. We will see stability definitions related to system inputs and outputs in Section 7.3.

Note that the notion of Lyapunov stability does not require that the system in question approach the equilibrium (usually, the origin), only that it remain within the bounding radius ε. For a system that is Lyapunov stable, we further qualify the equilibrium point as *asymptotically stable* if the state vector tends toward zero, i.e., if $\|x(t) - x_e\| \to 0$ as $t \to \infty$. In addition, if we can find constants $\gamma > 0$ and $\lambda > 0$ such that for all $t > t_0$,

$$\|x(t) - x_e\| \le \gamma e^{-\lambda(t-t_0)} \|x(t_0) - x_e\|$$

then the equilibrium is said to be *exponentially stable*.

As one might suspect, an LTI system that is asymptotically stable is also exponentially stable because all solutions of LTI systems that approach the origin do so by exponential decay. This is not necessarily the case for time-varying systems and is certainly not always the case for nonlinear systems.

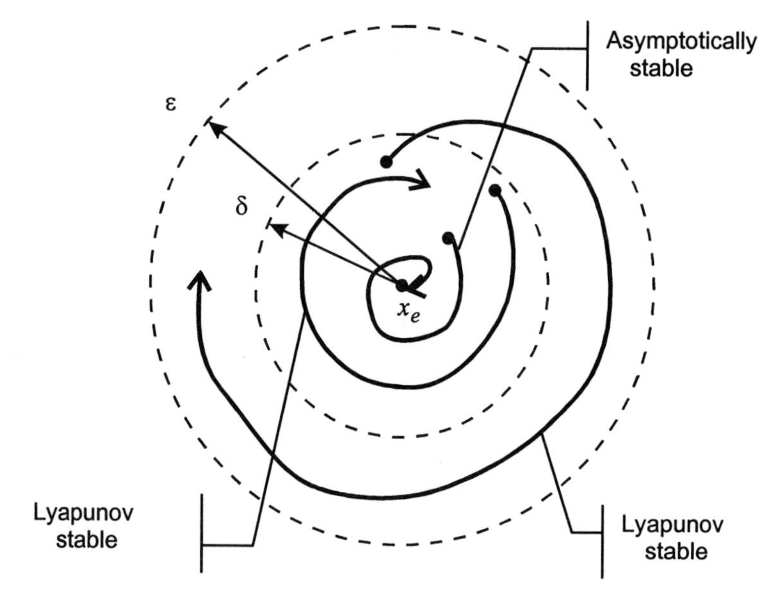

Figure 7.1 Representative trajectories illustrating Lyapunov stability. The picture shows a two-dimensional space such that the dotted circles indicate bounds on the 2-norm of the state vector.

We can also speak of systems that are stable *locally* or stable *globally* (also, stable *in the large*). An equilibrium that is stable globally is one such that the constant δ can be chosen arbitrarily large. This implies that if an equilibrium is asymptotically stable, then the trajectory from *any* initial condition will asymptotically approach it. This in turn implies that there can be only one equilibrium that is asymptotically stable in the large.

For linear systems, wherein equilibria are either isolated points or subspaces, all stability is global. There is no other equilibrium that a large initial condition might approach (or diverge from). However, for nonlinear systems, wherein multiple isolated equilibria might exist, one sometimes has to define regions of attraction around attractors, i.e., stable equilibria. These are sometimes called *basins of attraction* or *insets*, denoting the region in the state space from which an initial condition will approach the equilibrium point. Outside that region of attraction, the trajectory might go elsewhere. Again, that will not happen in linear systems.

As an illustration, the phase portrait of the nonlinear system

$$\ddot{\theta} + 0.4\dot{\theta} + \sin\theta = 0 \tag{7.6}$$

is shown in Figure 7.2.

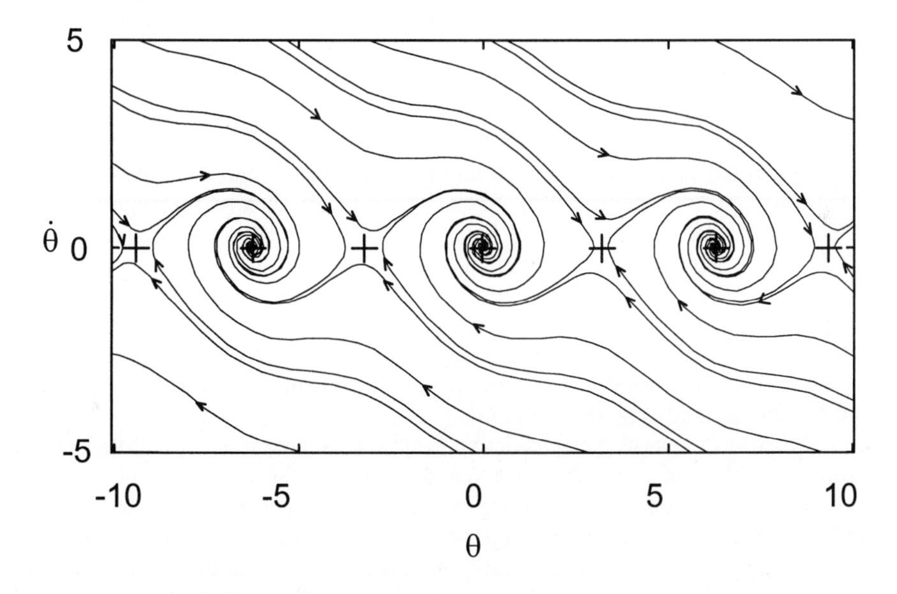

Figure 7.2 Phase portrait of the damped pendulum of Equation (7.6). The symbols (+) indicate the equilibria. (What would the phase portrait of an undamped pendulum look like?)

The graph plots trajectories on the $\dot{\theta}$ versus θ axis. The system (7.6) represents the equation of a damped pendulum, i.e., with viscous friction. If we compute the equilibria of this system by solving $\sin\theta_e = 0$, which is similar to

Example 7.1c), we find equilibrium points at $x_e = \pm n\pi$, where $n = 0,1,\ldots$ Of these, it is apparent from the figure that zero and the even multiples of π are stable, whereas the odd multiples of π are unstable. Note, in particular, that each stable equilibrium has a well-defined region of attraction, which defines the set from which an initial condition will produce a trajectory that approaches that equilibrium. For the equilibria that are stable, they are not *globally* stable. This being the equation for a pendulum, we can identify the stable equilibria as points at which the pendulum hangs directly down. The unstable equilibria are points at which the pendulum is balancing straight up, which are intuitively clearly unstable.

7.1.2 Classification of Equilibria

Given our definition of Lyapunov stability, we can reconsider the phase portraits of Chapter 6 in terms of these definitions. Because it is the eigenvalues of the A-matrix that determine the directions of the phase trajectories, it will be the eigenvalues that determine the type of stability exhibited.

In the captions to the phase portraits in Chapter 6, the origins of each plot were given labels, e.g., stable node, or focus. Each such label applies to any equilibrium of a phase portrait, and is a characteristic of the stability type of that equilibrium. Recalling the system examples from Chapter 6 that had two eigenvalues with negative real parts, i.e., the systems given by Equations (6.9), (6.12), (6.18), (6.19), and (6.20), we can immediately see from the corresponding figures that all trajectories asymptotically approach the origin (in fact, they do so exponentially). Such equilibria are clearly Lyapunov stable. Those that have at least one real invariant subspace (i.e., eigenvector), namely (6.9), (6.12), (6.18), and (6.19), have equilibria known as *stable nodes*. The equilibrium in the asymptotically stable system with complex conjugate eigenvalues (6.20) is known as a *stable focus*. The origin of the system described by (6.21) is stable but not asymptotically. It is known as a *center*. For systems with one negative and one positive eigenvalue, such as in (6.15) and (6.17), the origin is known as a *saddle point*.

Most of the phase portraits drawn (with the exception of the saddle points) show stable systems. Unstable systems, for example, those with *unstable nodes*, can easily be imagined by reversing the arrowheads on the portraits with stable nodes. Likewise, an unstable focus spirals out of the origin, the opposite of Figure 6.10. In the case of the center, there is no real part to the eigenvalues, so no such distinction needs to be made. However, with an appropriate choice of the A-matrix, the trajectories can be made to progress in the opposite direction from those shown (i.e., counterclockwise rather than clockwise).

7.1.3 Testing For Lyapunov Stability

While the definitions above give us the *meaning* of stability, they do little to help us *determine* the stability of a given system (except that it is now easy to anticipate the stability conditions for a time-invariant system based on the locations of its eigenvalues). In this section, we will give some results that will constitute stability tests, useful for examining a system at hand.

Considering a general time-varying zero-input system,

$$\dot{x}(t) = A(t)x(t) \tag{7.7}$$

we present the first result as a theorem:

> THEOREM: The zero-input system in (7.7) is stable in the sense of Lyapunov if and only if there exists a constant $\kappa(t_0) < \infty$ such that for all $t \geq t_0$, the state-transition matrix $\Phi(t, t_0)$ satisfies the relation
>
> $$\|\Phi(t, t_0)\| \leq \kappa(t_0) \tag{7.8}$$
>
> If the constant κ does *not* depend on t_0, then the origin of the system described by (7.7) will be uniformly stable in the sense of Lyapunov.
>
> PROOF: Recall that $x(t) = \Phi(t, t_0)x(t_0)$. Then by the Cauchy-Schwarz inequality (see Section 2.2.7), we have
>
> $$\|x(t)\| \leq \|\Phi(t, t_0)\| \|x(t_0)\| \tag{7.9}$$
>
> (Sufficiency) Suppose $\|\Phi(t, t_0)\| \leq \kappa(t_0)$. Given a value ε, we are free to choose the value $\delta(t_0, \varepsilon) = \varepsilon / \kappa(t_0)$. Then from (7.9),
>
> $$\|x(t)\| \leq \|\Phi(t, t_0)\| \|x(t_0)\| \leq \kappa(t_0) \|x(t_0)\| < \kappa(t_0) \frac{\varepsilon}{\kappa(t_0)}$$
>
> or
>
> $$\|x(t)\| < \varepsilon \text{ for all } t \geq t_0$$
>
> (Necessity) Assume that the system is stable and show that $\|\Phi(t, t_0)\|$ is bounded. Suppose that $\|\Phi(t, t_0)\|$ were *not* bounded. Then by definition, given $x(t_0)$ such that

$\|x(t_0)\| < \delta(t_0, \varepsilon)$ and an arbitrary value $\varepsilon < M < \infty$, there exists a time t such that

$$\|\Phi(t, t_0)x(t_0)\| > M$$

But since $x(t) = \Phi(t, t_0)x(t_0)$, this would imply

$$\|x(t)\| > M > \varepsilon$$

which violates the stability condition, a contradiction. Thus, necessity is proven.

For asymptotic stability, we give the following result:

THEOREM: The origin of the system (7.7) is uniformly asymptotically stable if and only if there exist two constants $\kappa_1 > 0$ and $\kappa_2 > 0$ such that

$$\|\Phi(t, t_0)\| \le \kappa_1 e^{-\kappa_2 (t - t_0)} \tag{7.10}$$

PROOF: (Sufficiency) If (7.10) holds and if $\|x(t_0)\| < \delta$, then

$$\|x(t)\| \le \|\Phi(t, t_0)\| \|x(t_0)\| < \kappa_1 e^{-\kappa_2 (t - t_0)} \delta$$

so clearly $\|x(t)\| \to 0$ as $t \to \infty$.

(Necessity) We will divide the time axis into equal steps of size τ. We will then look at the state of the system at a particular time t, considered to satisfy

$$t_0 + k\tau \le t < t_0 + (k+1)\tau \tag{7.11}$$

Knowing that the equilibrium is uniformly asymptotically stable (globally, because it is linear), then we know that for any starting point $\|x(t_0)\| \le \delta_1$, given an arbitrary value ε_1, there must exist a time τ such that $\|x(t_0 + \tau)\| < \varepsilon_1$. This being the case, we are free to choose $\|x(t_0)\| = 1$, and select $\varepsilon_1 = e^{-1}$. Then there will exist τ such that

$$\|x(t_0 + \tau)\| = \|\Phi(t_0 + \tau, t_0)x(t_0)\| < e^{-1}$$

By the definition of an operator norm,

$$\|\Phi(t_0 + \tau, t_0)\| = \sup_{\|x(t_0)\|=1} \|\Phi(t_0 + \tau, t_0)x(t_0)\| < e^{-1}$$

So now if we consider time $t_0 + k\tau$,

$$\begin{aligned} \|\Phi(t_0 + k\tau, t_0)\| &= \|\Phi(t_0 + k\tau, t_0 + (k-1)t_0) \\ &\quad \cdot \Phi(t_0 + (k-1)\tau, t_0 + (k-2)t_0) \cdots \\ &\quad \cdots \Phi(t_0 + 2\tau, t_0 + \tau)\Phi(t_0 + \tau, t_0)\| \\ &\le \|\Phi(t_0 + k\tau, t_0 + (k-1)t_0)\| \\ &\quad \cdot \|\Phi(t_0 + (k-1)\tau, t_0 + (k-2)t_0)\| \cdot \\ &\quad \cdots \|\Phi(t_0 + 2\tau, t_0 + \tau)\| \|\Phi(t_0 + \tau, t_0)\| \\ &< e^{-k} \end{aligned} \tag{7.12}$$

Now consider the quantity $\|\Phi(t, t_0 + k\tau)\|$. Because $\|\Phi(\cdot)\|$ is bounded, we know that $\|\Phi(t, t_0 + k\tau)\|$ is a finite number. Call it

$$\|\Phi(t, t_0 + k\tau)\| \stackrel{\Delta}{=} \kappa_1 e^{-1} \tag{7.13}$$

for some constant $\kappa_1 < \infty$. Now we have that

$$\|\Phi(t, t_0)\| = \|\Phi(t, t_0 + k\tau)\Phi(t_0 + k\tau, t_0)\|$$

so

$$\begin{aligned} \|\Phi(t, t_0)\| &\le \|\Phi(t, t_0 + k\tau)\| \cdot \|\Phi(t_0 + k\tau, t_0)\| \\ &< \kappa_1 e^{-1} e^{-k} = \kappa_1 e^{-(k+1)} = \kappa_1 e^{-(1/\tau)[t_0 + (k+1)\tau - t_0]} \end{aligned} \tag{7.14}$$

by (7.13) and (7.12). However, from (7.11) we have asserted that $t < t_0 + (k+1)\tau$, so

$$e^{-(t-t_0)} > e^{-[t_0+(k+1)\tau-t_0]}$$

and therefore,

$$\kappa_1 e^{-(1/\tau)[t_0+(k+1)\tau-t_0]} < \kappa_1 e^{-(1/\tau)(t-t_0)}$$

Now defining $\kappa_2 \stackrel{\Delta}{=} (1/\tau)$, we have proven necessity by showing that

$$\|\Phi(t,t_0)\| < \kappa_1 e^{-\kappa_2(t-t_0)}$$

As usual, we can do a little better than this for time-invariant systems. Given the proof above, the proofs of the following two theorems are relatively easy and are left as an exercise.

> THEOREM: The equilibrium of the LTI system $\dot{x} = Ax$ is Lyapunov stable if and only if all the eigenvalues of A have nonpositive real parts, and those that have zero real parts are nonrepeated. (7.15)

> THEOREM: The equilibrium of the LTI system $\dot{x} = Ax$ is asymptotically Lyapunov stable if and only if all the eigenvalues of A have negative real parts. (7.16)

It is these last two results that are the most commonly used to test for Lyapunov stability. However, as we will see in the next section, one must be careful not to apply the above results for time-invariant systems to time-varying systems.

Discrete-Time Systems

These results are directly extensible to discrete-time systems. In particular, all the boundedness conditions for the state-transition matrix $\Phi(t,t_0)$ extend as well to $\Psi(k,j)$. In the case of time-invariant systems, we can still derive eigenvalue conditions, but we replace all reference to the left half of the complex plane with the inside of the unit circle. Thus, the origin of a discrete-time LTI system is asymptotically stable if the eigenvalues of A_d are strictly inside the unit circle, and it is Lyapunov stable if the eigenvalues are inside or on the unit circle. Those eigenvalues that are on the unit circle, however, must not be repeated.

7.1.4 Eigenvalues of Time-Varying Systems

In general, one should avoid making conclusions about the stability of linear, time-varying systems based on the eigenvalues at any point in time. The following statements are true of time-varying systems:

- Time-varying systems whose "frozen-time" eigenvalues, i.e., eigenvalues computed at any single instant, all have negative real parts are not necessarily stable.

- If the eigenvalues of the matrix $A(t)+A^{\mathrm{T}}(t)$ have real parts that are always negative, the system is asymptotically stable (this is a sufficient but not necessary condition).

- If *all* the eigenvalues of $A(t)+A^{\mathrm{T}}(t)$ are *always* positive, the system is unstable [2].

- A time-varying system with one or more eigenvalues whose real part is positive is not necessarily unstable.

- If a time-varying system's eigenvalues have real parts such that $\mathrm{Re}(\lambda)<\gamma<0$ for all λ and all time *t*, and the system is sufficiently "slowly varying," then the system will be asymptotically stable. By slowly varying, we mean that there exists a value $\nu<\infty$ such that $\left\|\dot{A}(t)\right\|\leq\nu$ that ensures this stability. Likewise, a slowly varying system with an eigenvalue in the right half plane that *never* crosses the imaginary axis can be shown to be unstable.

It is not easy to prove all these statements without the theory presented in the next section. We will demonstrate one of these results using an example.

Example 7.2: An Unstable System with Negative Eigenvalues

Determine the eigenvalues and closed-form solution to the system given by

$$\begin{bmatrix}\dot{x}_1\\ \dot{x}_2\end{bmatrix}=\begin{bmatrix}-\alpha & e^{2\alpha t}\\ 0 & -\alpha\end{bmatrix}\begin{bmatrix}x_1\\ x_2\end{bmatrix} \tag{7.17}$$

where $\alpha>0$.

Solution:

From its triangular form, we know that the eigenvalues are $\lambda_1=\lambda_2=-\alpha$. In order to solve the system, we can recognize that the variable $x_2(t)$ is decoupled from $x_1(t)$ and has the solution $x_2(t) = e^{-\alpha t} x_2(0)$. Then the first equation can be written as

$$\dot{x}_1(t) = -\alpha x_1(t) + e^{\alpha t} x_2(0)$$

This equation can be shown (how?) to have the solution

$$x_1(t) = \left[x_1(0) - \frac{1}{1+\alpha} x_2(0) \right] e^{-\alpha t} + \frac{1}{1+\alpha} x_2(0) e^{\alpha t}$$

which is clearly unbounded because of the second term. This system is therefore an example of an unstable time-varying system with negative eigenvalues.

7.2 Lyapunov's Direct Method

A different method for testing the stability of linear, zero-input systems is called Lyapunov's direct method. It is based on the concept of energy and dissipative systems. The physically intuitive explanation is that an isolated system will have a certain amount of a nonnegative abstract quantity that is similar to energy. If the system has no input and its energy is always decreasing, we think of it as being stable. If its energy increases, we think of it as being unstable. However, because not all state space systems are descriptions of physical phenomena normally endowed with real energy, we generalize the concept. This is the basis for Lyapunov's direct method.

First, we introduce the above physical model with a mechanical example. Consider the mass-spring-damper system pictured in Figure 7.3. The dynamic equation that describes this system, in the absence of an input force, is

$$m\ddot{x}(t) + b\dot{x}(t) + kx(t) = 0 \tag{7.18}$$

The characteristic equation of such a system is therefore $ms^2 + bs + k = 0$. Knowledge of basic systems would indicate that if the three parameters m, b, and k are all of the same algebraic sign, then the roots of the characteristic polynomial will have negative real parts and the system will be stable in the sense of Lyapunov. Because the three parameters have physical interpretations of mass, damping, and restoring force constants, we know that all three will be positive. A state space version of (7.18) can be written by defining $x(t)$ and $\dot{x}(t)$ as the two state variables, giving

$$\begin{bmatrix} \dot{x}(t) \\ \ddot{x}(t) \end{bmatrix} = \begin{bmatrix} 0 & 1 \\ -k/m & -b/m \end{bmatrix} \begin{bmatrix} x(t) \\ \dot{x}(t) \end{bmatrix} \tag{7.19}$$

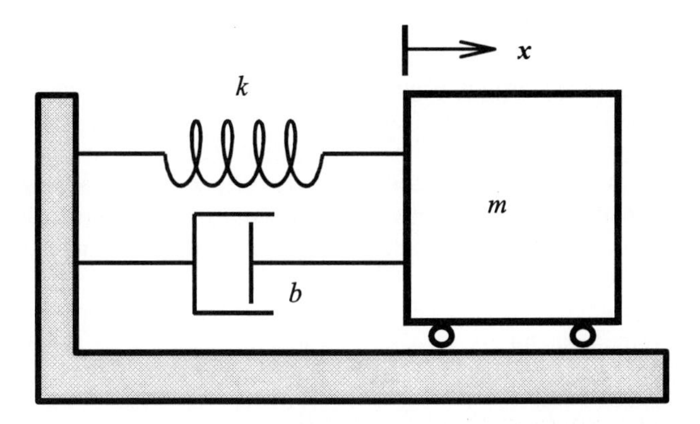

Figure 7.3 An unforced sliding system with a mass, spring, and damper. The position $x = 0$ is defined as the position of the mass when the string is unstretched and uncompressed.

Now consider the energy in the system, which we will call $V(x, \dot{x})$:

$$\begin{aligned} V(x,\dot{x}) &= V_{\text{kinetic}}(x,\dot{x}) + V_{\text{potential}}(x) \\ &= \frac{1}{2}\left(m\dot{x}^2 + kx^2\right) \end{aligned} \tag{7.20}$$

To determine how this energy changes over time, we compute

$$\dot{V}(x,\dot{x}) = m\dot{x}\ddot{x} + kx\dot{x}$$

Evaluating this energy change "along the trajectory," i.e., rearranging the original differential equation and substituting in for $\ddot{x}$,

$$\begin{aligned} \dot{V}(x,\dot{x}) &= \dot{x}\left(-b\dot{x} - kx\right) + kx\dot{x} \\ &= -b\dot{x}^2 \end{aligned} \tag{7.21}$$

If indeed $b > 0$, then the value of $\dot{V}(x,\dot{x})$ will be nonpositive, and will only be zero when $\dot{x} = 0$, i.e., when the system is stationary. If we should encounter a situation such as an initial condition where $\dot{x}_0 = 0$ and $x_0 \neq 0$, then we can check this condition against the invariant subspaces (eigenvectors) of the state space description (7.19). Because $\left[x(0) \quad \dot{x}(0)\right]^{\mathrm{T}} = \left[x_0 \quad 0\right]^{\mathrm{T}}$ is *not* an invariant subspace, the system will soon generate a nonzero velocity, and the energy will continue to decrease until $x(t) \to 0$ and $\dot{x}(t) \to 0$, i.e., move toward the equilibrium position.

Hence, the energy argument provides a valid means of testing this simple physical system for asymptotic stability. We can also argue that if, in this system, there were no damping, i.e., $b = 0$, then the energy is not decreasing but is instead constant. This situation would describe an oscillator whose energy is continually being exchanged between kinetic and potential forms (in the mass and spring, respectively). This would be a system that is Lyapunov stable. The electrical counterpart would be, of course, an RLC circuit in which the resistance is removed, leaving energy to bounce back and forth between the capacitor and the inductor.

7.2.1 Lyapunov Functions and Testing

In order to generalize this stability concept to more abstract, perhaps time-varying forms, we provide the following theorems. They are stated in terms of generic, possibly nonlinear systems before examining the linear and LTI cases later:

> THEOREM: The origin of the linear time-varying system $\dot{x}(t) = f(x(t),t)$ is Lyapunov stable if there exists a time-varying function $V(x,t)$ such that the following conditions are satisfied:
>
> 1. $V(x,t) \geq \gamma_1(x) > 0$ for all $x \neq 0$ and for all t. $V(x,t) = 0$ only when $x = 0$. The function $\gamma_1(x)$ is any continuous, nondecreasing function of the state x, but not time t explicitly, such that $\gamma_1(0) = 0$.
>
> 2. $\dot{V}(x,t) \leq -\gamma_2(x) < 0$ for all $x \neq 0$ and for all t, where
>
> $$\dot{V}(x,t) = \frac{\partial V(x,t)}{\partial x} f(x,t) + \frac{\partial V(x,t)}{\partial t} \tag{7.22}$$
>
> The function $\gamma_2(x)$ is a continuous, nondecreasing function of the state x, but not time t explicitly, such that $\gamma_2(0) = 0$. (7.23)

> THEOREM: The origin of the system is uniformly Lyapunov stable if, in addition to the conditions above, $\gamma_3(x) \geq V(x,t)$ for all x and all t. The function $\gamma_3(x)$ is a continuous,

nondecreasing function of the state *x*, but not time *t* explicitly, such that $\gamma_3(0) = 0$. (7.24)

THEOREM: The origin of the system is globally Lyapunov stable if, in addition to conditions 1 and 2 above, $V(x,t)$ is unbounded as $\|x\| \to \infty$. (7.25)

If all three theorems hold, the system is globally, uniformly, Lyapunov stable.

THEOREM: If condition 2 is altered as follows, then stability in the above theorems is strengthened to *asymptotic* stability:

2a. $\dot{V}(x,t) < -\gamma_2(x) < 0$ for all $x \neq 0$ and for all *t*. (7.26)

In these theorems, condition 1 is equivalent to the function $V(x,t)$ being *positive definite*, condition 2 is equivalent to $\dot{V}(x,t)$ being *negative semidefinite*, and condition 2a is equivalent to $\dot{V}(x,t)$ being *negative definite*. For linear systems, these terms are used in the same sense as in Chapter 5, where we considered the positive- or negative-definiteness of quadratic forms only.

Although we present no proofs of the above theorems, we assert that they simplify for time-invariant systems. For time-invariant systems (where all stability is uniform), we have the considerably simpler statements given in the following theorem. For the purposes of this theorem, we say that any function $V(\xi)$ is positive (semi)definite if $V(\xi) > 0$ $[V(\xi) \geq 0]$ for all $\xi \neq 0$ and $V(0) = 0$. A function $V(\xi)$ is said to be negative (semi)definite if $-V(\xi)$ is positive (semi-)definite.

THEOREM: The origin of a time-invariant system $\dot{x}(t) = f(x(t))$ is Lyapunov stable if there exists a positive-definite function $V(x)$ such that its derivative

$$\dot{V}(x) = \frac{dV(x)}{dx}\frac{dx}{dt} = \frac{dV(x)}{dx} f(x)$$

is negative semidefinite. If $\dot{V}(x)$ is negative definite, the origin is asymptotically stable. (7.27)

In all these definitions and theorems, discrete-time systems can be easily accommodated. The only difference from continuous-time systems is that instead of speaking of the derivative $\dot{V}(x)$ or $\dot{V}(x,t)$, we speak instead of the first difference: $\Delta V(x,k) = V(x(k+1),k+1) - V(x(k),k)$.

If the functions $V(x,t)$, or $V(x)$, exist as defined above, they are called *Lyapunov functions*. As we shall see, although it can be shown that stable or asymptotically stable systems do indeed have Lyapunov functions, the theorems do nothing to help us find them. For nonlinear systems, in particular, finding such a function in order to prove stability can be a chore for which experience is the best assistance. The usual procedure is to guess at a Lyapunov function (called a *candidate* Lyapunov function) and test it to see if it satisfies the above conditions. However, for linear systems, Lyapunov functions generally take quadratic forms.

Example 7.3: Lyapunov Function for a Nonlinear System

Use the candidate Lyapunov function

$$V(x_1,x_2) = \frac{1}{2}x_1^2 + x_2^2$$

to test the stability of the system

$$\begin{aligned} \dot{x}_1 &= -x_1 - 2x_2^2 \\ \dot{x}_2 &= x_1 x_2 - x_2^3 \end{aligned} \tag{7.28}$$

Solution:

First we note that the system is time-invariant, so we need not go to the trouble of finding the functions γ_1, etc. Instead, we proceed to the simplified tests, the first of which is positive definiteness of $V(x)$. Clearly, $V(x) > 0$ and $V(0) = 0$, so V is positive definite. Now we compute

$$\begin{aligned} \dot{V}(x) &= x_1\dot{x}_1 + 2x_2\dot{x}_2 \\ &= x_1\left(-x_1 - 2x_2^2\right) + 2x_2\left(x_1x_2 - x_2^3\right) \\ &= -x_1^2 - 2x_1x_2^2 + 2x_1x_2^2 - 2x_2^4 \\ &= -x_1^2 - 2x_2^4 \end{aligned}$$

The derivative $\dot{V}(x)$ is therefore negative definite, and we can claim that the system in (7.28) is asymptotically stable. Furthermore, because $V(x) \to \infty$ as $\|x\| \to \infty$, and the system is time-invariant, we can also say that the system is globally, uniformly asymptotically stable.

As the first step toward generating Lyapunov functions for linear systems, reconsider the linear system given in the mass-spring-damper example offered above. Recall that we chose as an "energy" function the true energy of the mechanical system

$$V(x,\dot{x})=\frac{1}{2}\left(m\dot{x}^2+kx^2\right)$$

This is a positive-definite function of the state of the system. Furthermore, we computed $\dot{V}(x,\dot{x})=-b\dot{x}^2$, which, if $b>0$, is strictly negative for all $\dot{x}\neq 0$, but may be zero for $x\neq 0$. We therefore do not make the claim that $\dot{V}$ is negative definite. Rather, we implicitly resorted in that section to LaSalle's theorem (see [12] and [13]), which is useful in many similar situations and which justifies our claim of asymptotic stability in this example.

> THEOREM (LaSalle's theorem): Within the region of the state space for which the derivative of the candidate Lyapunov function is such that $\dot{V}(x)\leq 0$, let Z denote the subset of the state space in which $\dot{V}(x)=0$. Within Z, let M denote the largest *invariant* subset. Then every initial state approaches M, even if V is not positive-definite. (7.29)

As we apply this theorem to the mass-spring-damper problem, we note that $Z=\{(x,\dot{x})\mid \dot{x}=0\}$. The only invariant subspaces are the eigenvectors e_1 and e_2. These invariant sets intersect with Z only at the origin, so M is just the origin. Therefore, as we concluded, the origin is asymptotically stable.

7.2.2 Lyapunov Functions for LTI Systems

In the case of linear systems, it is sufficient to consider quadratic forms as Lyapunov functions. This is largely because for linear systems, stability implies global stability, and the parabolic shape of a quadratic function satisfies all the criteria of the above theorems. We will begin considering the special case of linear systems with the time-invariant case. LTI systems, when examined with Lyapunov's direct method as described here, result in some special tests and equations.

Consider the LTI system

$$\dot{x}=Ax \tag{7.30}$$

and the candidate Lyapunov function $V(x)=x^{\mathrm{T}}Px$, where matrix P is positive definite. Then testing the system (7.30), we compute

$$\begin{aligned}\dot{V}(x) &= \dot{x}^{\mathrm{T}} Px + x^{\mathrm{T}} P\dot{x} \\ &= (Ax)^{\mathrm{T}} Px + x^{\mathrm{T}} P(Ax) \\ &= x^{\mathrm{T}} A^{\mathrm{T}} Px + x^{\mathrm{T}} PAx \\ &= x^{\mathrm{T}}\left(A^{\mathrm{T}} P + PA\right)x\end{aligned} \tag{7.31}$$

To satisfy the theorem, then, it is necessary that the matrix $A^{\mathrm{T}}P + PA$ be negative definite for asymptotic stability, and negative semidefinite for Lyapunov stability. That is, for some positive (semi)definite matrix Q, it is sufficient to show that

$$A^{\mathrm{T}}P + PA = -Q \tag{7.32}$$

to demonstrate stability of the homogeneous system in (7.30). Equation (7.32) is known as the Lyapunov equation.[M]

`lyap(A,C)`

However, one must be cautious in the application of this test because if the quantity $A^{\mathrm{T}}P + PA$ is *not* negative (semi)definite, nothing should be inferred about the stability. Recall that Lyapunov's direct method asserts that if a Lyapunov function is found, then the system is stable. It does *not* say that if a candidate Lyapunov function such as $x^{\mathrm{T}}Px$ fails, the system is not stable. So if we choose a positive-definite matrix P and we compute a Q that is *in*definite, we have shown nothing. (If Q is negative definite, however, instability can be shown.)

Instead, we resort to the reverse proccss: Wc select a positive (semi) definite Q and compute the solution P to the Lyapunov equation (7.32). This result is stated as a theorem:

> THEOREM: The origin of the system in (7.30) described by matrix A is asymptotically stable if and only if, given a positive definite matrix Q, the matrix Lyapunov equation (7.32) has a solution P that is positive definite. (7.33)
>
> PROOF: (Sufficiency) If P and Q are both positive definite, then it has already been shown by Lyapunov's direct method that the system will be asymptotically stable, because $x^{\mathrm{T}}Px$ will have been demonstrated to be a Lyapunov function.
>
> (Necessity) If the origin is known to be a stable equilibrium and Q is positive definite, we will prove necessity by first proving that P is the unique, positive definite result of

$$P = \int_0^\infty e^{A^\mathrm{T} t} Q e^{At} \, dt \tag{7.34}$$

To demonstrate this, we substitute (7.34) into the left-hand side of (7.32):

$$\begin{aligned}
A^\mathrm{T} P + PA &= \int_0^\infty A^\mathrm{T} e^{A^\mathrm{T} t} Q e^{At} \, dt + \int_0^\infty e^{A^\mathrm{T} t} Q e^{At} A \, dt \\
&= \int_0^\infty \left[A^\mathrm{T} e^{A^\mathrm{T} t} Q e^{At} + e^{A^\mathrm{T} t} Q e^{At} A \right] dt \\
&= \int_0^\infty \frac{d\left[e^{A^\mathrm{T} t} Q e^{At} \right]}{dt} dt \\
&= e^{A^\mathrm{T} t} Q e^{At} \Big|_0^\infty \\
&= 0 - Q \\
&= -Q
\end{aligned} \tag{7.35}$$

where convergence of the integral is guaranteed by the fact that the eigenvalues of A are in the left half of the complex plane.

To show that this solution is unique, suppose that there were a second solution $\hat{P}$ such that $A^\mathrm{T} \hat{P} + \hat{P} A = -Q$. Then we would have

$$(A^\mathrm{T} P + PA) - (A^\mathrm{T} \hat{P} + \hat{P} A) = 0$$

implying

$$A^\mathrm{T}\left(P - \hat{P}\right) + \left(P - \hat{P}\right) A = 0$$

Multiplying this equation by $e^{A^\mathrm{T} t}$ from the left and by e^{At} from the right,

$$e^{A^{\mathrm{T}}t}A^{\mathrm{T}}\left(P-\hat{P}\right)e^{At}+e^{A^{\mathrm{T}}t}\left(P-\hat{P}\right)Ae^{At}=\frac{d}{dt}\left[e^{A^{\mathrm{T}}t}\left(P-\hat{P}\right)e^{At}\right]$$
$$=0$$

Now integrating this equation over $[0, \infty)$,

$$0=\int_0^\infty d\left[e^{A^{\mathrm{T}}t}\left(P-\hat{P}\right)e^{At}\right]dt=e^{A^{\mathrm{T}}t}\left(P-\hat{P}\right)e^{At}\Big|_0^\infty$$
$$=0-\left(P-\hat{P}\right)$$
$$=\hat{P}-P$$

so we must have $P=\hat{P}$.

Finally, to demonstrate that P is positive definite, consider the quadratic form in terms of an arbitrary vector ξ:

$$\begin{aligned}\xi^{\mathrm{T}}P\xi&=\xi^{\mathrm{T}}\left[\int_0^\infty e^{A^{\mathrm{T}}t}Qe^{At}\,dt\right]\xi\\&=\int_0^\infty \xi^{\mathrm{T}}e^{A^{\mathrm{T}}t}Qe^{At}\xi\,dt\end{aligned}\tag{7.36}$$

Because matrix Q is positive definite, the integrand of (7.36) is positive definite, so

$$\int_0^\infty \xi^{\mathrm{T}}e^{A^{\mathrm{T}}t}Qe^{At}\xi\,dt>0$$

for all $\xi\neq 0$. Therefore, P will be positive definite and hence, will define a valid Lyapunov function. This therefore constitutes a constructive proof of the theorem.

We note that this theorem does not directly extend to Lyapunov stability; that is, choice of an arbitrary positive semidefinite Q will not necessarily generate a positive definite P for a system that is Lyapunov stable. However, by an extension of LaSalle's theorem above, we can provide the following generalization:

THEOREM: The origin of the system (7.30) described by matrix A is asymptotically stable if and only if, given a positive semidefinite matrix Q, such that $x^{T}Qx$ is not identically zero when evaluated on a nonzero trajectory of (7.30), the matrix Lyapunov equation (7.32) has a solution P that is positive definite. (7.37)

The proof of this theorem, while not given explicitly here, results from the application of LaSalle's theorem.

Example 7.4: Stability Test for a Parameterized System

An LTI system is given by

$$\dot{x} = \begin{bmatrix} -3k & 3k \\ 2k & -5k \end{bmatrix} x \tag{7.38}$$

Use the Lyapunov stability theorem to find a bound on k such that the origin of this system is asymptotically stable.

Solution:

While the Lyapunov stability theorem is not commonly used for stability testing (because it is much easier to simply test the eigenvalues), it can be used to determine stability bounds on matrix parameters such as this one. Because the theorem says that *any* positive definite matrix Q will suffice, it is sufficient to choose $Q = I$. Then because we know that the required solution P will have to be positive definite and, hence, symmetric, we construct the equation

$$\begin{bmatrix} -3k & 2k \\ 3k & -5k \end{bmatrix} \begin{bmatrix} p_{11} & p_{12} \\ p_{12} & p_{22} \end{bmatrix} + \begin{bmatrix} p_{11} & p_{12} \\ p_{12} & p_{22} \end{bmatrix} \begin{bmatrix} -3k & 3k \\ 2k & -5k \end{bmatrix} = -\begin{bmatrix} 1 & 0 \\ 0 & 1 \end{bmatrix}$$

Multiplying out the left-hand side, we arrive at three independent scalar equations (the two off-diagonal scalar equations will be identical):

$$\begin{aligned} -3kp_{11} + 2kp_{12} - 3kp_{11} + 2kp_{12} &= -6kp_{11} + 4kp_{12} = -1 \\ -3kp_{12} + 2kp_{22} + 3kp_{11} - 5kp_{12} &= 3kp_{11} - 8kp_{12} + 2kp_{22} = 0 \\ 3kp_{12} - 5kp_{22} + 3kp_{12} - 5kp_{22} &= 6kp_{12} - 10kp_{22} = -1 \end{aligned}$$

As these constitute three linear equations in three unknowns, they can be readily solved as

$$p_{11} = \frac{19}{72k} \qquad p_{12} = \frac{7}{48k} \qquad p_{22} = \frac{15}{80k}$$

To verify the positive definiteness of this solution, we form the determinant

$$|P| = \begin{vmatrix} \frac{19}{72k} & \frac{7}{48k} \\ \frac{7}{48k} & \frac{15}{80k} \end{vmatrix} = \frac{1}{(4k)^2} \begin{bmatrix} \frac{19}{18} & \frac{7}{12} \\ \frac{7}{12} & \frac{15}{20} \end{bmatrix}$$

To ensure that both leading principal minors are positive, we can choose $k > 0$. Therefore, we may conclude that the origin of the system is asymptotically stable for all $k > 0$.

Example 7.5: Discrete-Time LTI Lyapunov Equation

Derive the Lyapunov equation for a discrete-time homogeneous LTI system:

$$x(k+1) = Ax(k) \tag{7.39}$$

Solution:

Assuming that we can again use a quadratic form as the candidate Lyapunov function:

$$V(k) = x^{\mathrm{T}}(k)Px(k) \tag{7.40}$$

we can find the first difference as

$$\begin{aligned} \Delta V(x,k) &= V(k+1) - V(k) \\ &= x^{\mathrm{T}}(k+1)Px(k+1) - x^{\mathrm{T}}(k)Px(k) \\ &= x^{\mathrm{T}}(k)A^{\mathrm{T}}PAx(k) - x^{\mathrm{T}}(k)Px(k) \\ &= x^{\mathrm{T}}(k)\left(A^{\mathrm{T}}PA - P\right)x(k) \end{aligned}$$

So the term in parentheses must be a negative definite matrix if the quadratic form in (7.40) is to be a valid Lyapunov function. The discrete-time Lyapunov equation[M] is therefore

`dlyap(A,Q)`

$$A^{\mathrm{T}}PA - P = -Q \tag{7.41}$$

7.2.3 Unstable Systems

It is a common misconception that the theory of internal stability, and in particular Lyapunov stability, is limited in its ability to conclude that a system is unstable, i.e., not stable in the sense of Lyapunov. However, there are some results that one may use to label a system (actually, its equilibium) as unstable:

> THEOREM: The origin of the linear system in (7.7) is *unstable* if there exists a function $V(x,t)$ and a continuous nondecreasing function of the state only, $\gamma(x)$, such that $\gamma(x) > V(x,t)$ for all x and t, $\gamma(0) = 0$, and the following three conditions are satisfied:
>
> 1. $V(0,t) = 0$ for all $t \geq t_0$.
>
> 2. $V(x,t_0) > 0$ at any point arbitrarily close to the origin.
>
> 3. $\dot{V}(x,t)$ is positive definite in an arbitrary region around the origin. (7.42)

Note that the function $\gamma(x)$ is the same one that appears in the original theorems on Lyapunov stability, e.g., conditions 1 or 2. Its existence is unnecessary for time-invariant systems. We can illustrate the process on a trivial example.

Example 7.6: An Unstable System

Determine a function $V(x,t)$ that, according to the theorem above, demonstrates the instability of the following system:

$$\dot{x} = \begin{bmatrix} 2 & 0 \\ 0 & -1 \end{bmatrix} x$$

Solution:

It is clear from the self-evident eigenvalues that the system will not be Lyapunov stable. To illustrate the existence of $V(x,t)$ consider

$$V(x,t) = x_1^2 - x_2^2 = x^{\mathrm{T}} \begin{bmatrix} 1 & 0 \\ 0 & -1 \end{bmatrix} x \stackrel{\Delta}{=} x^{\mathrm{T}} P x \tag{7.43}$$

One should be careful not to refer to such a function as a Lyapunov function, because it is not; it is clearly not positive definite. However, it is zero at the origin, and is positive arbitrarily near the origin [consider approaching the origin from the $(x_1,0)$ direction]. Now compute

$$\begin{aligned}\dot{V}(x,t) &= 2x_1\dot{x}_1 - 2x_2\dot{x}_2\\ &= 2x_1(2x_1) - 2x_2(-x_2)\\ &= 4x_1^2 + 2x_2^2\end{aligned}$$

which is clearly positive definite. The system has therefore been shown to be unstable.

7.3 External Stability

In each of the stability definitions, theorems, and examples above, linear systems were considered in the absence of any input. However, any student of systems and controls knows that the input generally has an effect on the qualitative behavior of the output. From this perspective on the system, i.e., the input/output or *external* behavior, we will need new definitions and tests for system stability. Such tests may disregard the behavior of the internal dynamics (i.e., the state vector) altogether.

7.3.1 Bounded Input, Bounded Output Stability

Our working definition for external stability will use the concept of bounded input, bounded output (BIBO) stability. We will consider the boundedness of the input function $u(t)$ and the output function $y(t)$. For a given function of time $f(t)$, we consider the function to be bounded if for all time *t*, there exists a finite constant *M* such that $\|f(t)\| \le M$. We only consider the stability of systems of the general form

$$\begin{aligned}\dot{x}(t) &= A(t)x(t) + B(t)u(t)\\ y(t) &= C(t)x(t) + D(t)u(t)\end{aligned} \tag{7.44}$$

We further restrict our attention to systems wherein $A(t)$ and $B(t)$ are themselves bounded operators (otherwise instability might be obvious).

> ***BIBO Stability:*** The system in (7.44) is BIBO stable if for *any* bounded input $u(t)$, $\|u(t)\| \le M$, and for any initial condition $x_0 = x(t_0)$, there exists a finite constant $N(M, x_0, t_0)$ such that $\|y(t)\| \le N$ for all $t \ge t_0$. (7.45)

One must be careful when applying this definition. If *any* bounded input results in an unbounded output, then the system is considered not BIBO stable. Some texts will provide the definition of so-called "bounded input, bounded state" (BIBS) stability, which we give here for completeness, but which we will not discuss in much depth:

> ***BIBS Stability:*** The system in (7.44) is BIBS stable if for *any* bounded input $u(t)$, $\|u(t)\| \le M$, and for any initial condition $x_0 = x(t_0)$, there exists a finite constant $N_s(M, x_0, t_0)$ such that $\|x(t)\| \le N_s$ for all $t \ge t_0$. (7.46)

Testing for BIBO Stability

To investigate the conditions under which a system is BIBO stable, we first reconsider the expression for system output:

$$y(t) = C(t)\Phi(t,t_0)x(t_0) + \int_{t_0}^{t} C(t)\Phi(t,\tau)B(\tau)u(\tau)d\tau + D(t)u(t) \quad (7.47)$$

where, of course, $\Phi(t,\tau)$ is the state-transition matrix discussed in Chapter 6. In order to move the feedthrough term into the integrand, we can exploit the Dirac delta function to achieve

$$y(t) = C(t)\Phi(t,t_0)x(t_0) + \int_{t_0}^{t} \left[C(t)\Phi(t,\tau)B(\tau) + \delta(t-\tau)D(\tau) \right] u(\tau)\, d\tau \quad (7.48)$$

This allows us to refer to the bracketed term

$$H(t,\tau) \overset{\Delta}{=} C(t)\Phi(t,\tau)B(\tau) + \delta(t-\tau)D(\tau) \quad (7.49)$$

as the *impulse-response matrix*. Now considering the initial condition $x(t_0)$ to be the result of an infinite history of past inputs, we use the expression

$$y(t) = \int_{-\infty}^{t} H(t,\tau)u(\tau)\, d\tau$$

in the following theorem:

THEOREM: The linear system in (7.44) is BIBO stable if and only if there exists a finite constant N_H such that

$$\int_{-\infty}^{t} \|H(t,\tau)\| \, d\tau \le N_H \tag{7.50}$$

PROOF: We will prove this theorem by initially considering the scalar case wherein we have a scalar impulse-response matrix denoted $h(t,\tau)$ and $\|h(t,\tau)\| = |h(t,\tau)|$.

(Sufficiency) Suppose input $u(t)$ is bounded by finite scalar M. Then we have

$$|y(t)| \le \int_{-\infty}^{t} |h(t,\tau)| |u(\tau)| \, d\tau \le M \int_{-\infty}^{t} |h(t,\tau)| \, d\tau$$

Then if scalar N_H exists such that

$$\int_{-\infty}^{t} |h(t,\tau)| \, d\tau \le N_H \tag{7.51}$$

then clearly

$$|y(t)| \le M N_H < \infty$$

and BIBO stability is demonstrated.

(Necessity) If the system is BIBO stable, then for any bounded input $u(t)$, there will exist a finite constant N such that

$$|y(t)| \le N \tag{7.52}$$

To construct an argument by contradiction, suppose that (7.50) were *not* true, i.e., for *any* finite scalar N_H

$$\int_{-\infty}^{t} |h(t,\tau)| d\tau > N_H \tag{7.53}$$

Because we have asserted BIBO stability, we can choose our input function, so long as it is bounded. We select

$$u(t) = \text{sgn}[h(t,\tau)]$$

Then we have

$$|y(t)| = \left| \int_{-\infty}^{t} h(t,\tau) u(\tau)\, d\tau \right| = \int_{-\infty}^{t} |h(t,\tau)| d\tau \leq N \tag{7.54}$$

from (7.52). Yet this result contradicts (7.53), so necessity is demonstrated.

The extension of this proof for sufficiency to the vector and multivariable case is straightforward. The necessity part is based on the argument that *if a matrix operator is unbounded, then at least one of its individual entries must also be unbounded.* This is used to justify a term-by-term application of the proof above to the impulse-response matrix [15].

When using a result such as this, we should remember that the matrix norm may be computed in a number of ways, as discussed in Section 3.2.1. One of the most common techniques is to use the singular value decomposition, recalling that $\|M\| = \sigma_1$, i.e., the largest eigenvalue of $M^{\mathrm{T}} M$.

BIBO Stability for Discrete-Time Systems

As one might expect from the development of the BIBO stability definition, test, and proof above, the test for discrete-time systems is analogous to that for continuous-time systems. In order to state this result, we must determine an expression for the impulse-response matrix that is similar to (7.49) for continuous-time. Recalling Equation (6.49), we can construct the output of the generic (i.e., possibly time-varying) discrete-time system as

$$y(k) = C_d(k)\left(\prod_{i=j}^{k-1} A_d(i)\right)x(j)$$
$$+ \sum_{i=j+1}^{k} C_d(k)\left[\prod_{q=i}^{k-1} A_d(q)\right]B(i-1)u(i-1) + D_d(k)u(k)$$

or, considering the initial condition $x(j)$ to have arisen from an infinite past history of inputs,

$$y(k) = \sum_{i=-\infty}^{k} C_d(k)\left[\prod_{q=i}^{k-1} A_d(q)\right]B(i-1)u(i-1) + D_d(k)u(k)$$

Performing the same trick of moving the feedthrough term into the summation by introducing the discrete-time Dirac delta function,

$$y(k) = \sum_{i=-\infty}^{k} \left\{C_d(k)\left[\prod_{q=i}^{k-1} A_d(q)\right]B(i-1) + \delta(i-k-1)D_d(i-1)\right\}u(i-1) \tag{7.55}$$

Now the term in braces in (7.55) can be identified as the discrete-time impulse-response matrix, which we will denote $H_d(k,i)$.

$$H_d(k,i) = C_d(k)\left[\prod_{q=i}^{k-1} A_d(q)\right]B(i-1) + \delta(i-k-1)D_d(i-1) \tag{7.56}$$

With this notation, we can state the result as follows:

THEOREM: The discrete-time linear state space system

$$x(k+1) = A_d(k)x(k) + B_d(k)u(k)$$
$$y(k) = C_d(k)x(k) + D_d(k)u(k)$$

with impulse-response matrix $H_d(k,i)$ as given in (7.56) is BIBO stable if and only if there exists a finite constant N_d such that

$$\sum_{i=-\infty}^{k} \|H_d(k,i)\| \le N_d \tag{7.57}$$

for all k.

7.3.2 BIBO Stability for Time-Invariant Systems

Consider a state space system described by (7.44) that is time-invariant, i.e., one for which the system matrices are not explicit functions of time. Then, recalling the definition of the impulse-response matrix (7.49), we have

$$H(t,\tau) \stackrel{\Delta}{=} Ce^{A(t-\tau)}B + \delta(t-\tau)D \tag{7.58}$$

Then, by the theorem giving the condition for BIBO stability, we see that the LTI system will be BIBO stable if and only if there exists a finite constant N_H such that

$$\int_{-\infty}^{t} \|H(t,\tau)\| \, d\tau = \int_{-\infty}^{t} \left\| Ce^{A(t-\tau)}B + \delta(t-\tau)D \right\| d\tau \le N_H$$

The first remark we must make about such a system is that for testing purposes, the D-matrix is often not included in the integrand. Its contribution to the integral is clearly additive, and its integral

$$\int_{-\infty}^{t} \delta(t-\tau)D \, d\tau = D$$

will clearly be unbounded if and only if D itself is infinite. Such infinite feedthrough–gain systems rarely occur, and if they do, the reason for the overall system instability would be obvious (recall that we are considering only bounded matrices anyway).

It remains therefore to test

$$\int_{-\infty}^{t} \left\| Ce^{A(t-\tau)}B \right\| d\tau = C\left(\int_{-\infty}^{t} \left\| e^{A(t-\tau)} \right\| d\tau \right) B \tag{7.59}$$

for boundedness. Realizing that $e^{A(t-\tau)} = \Phi(t,\tau)$ for an LTI system, the middle factor in (7.59) is reminiscent of the test for Lyapunov stability given in (7.8). However, Lyapunov stability and BIBO stability of LTI systems are *not* equivalent. The reason for this is that one cannot neglect the effects of premultiplication by C and postmultiplication by B in (7.59). Suppose, for example, that

$$\left(\int_{-\infty}^{t}\left\|e^{A(t-\tau)}\right\|d\tau\right)B \in \mathrm{N}(C) \tag{7.60}$$

where $\mathrm{N}(C)$ is the null space of C. Then no matter what the size of the integral, Equation (7.60) would imply the boundedness of (7.59). Therefore, it would be possible for a system that is not asymptotically (or even Lyapunov) stable to indeed be BIBO stable. This applies to time-varying systems as well. However, the reverse is true, as it is relatively easy to show that asymptotic stability implies BIBO stability, i.e., boundedness of the state does imply boundedness of the output (assuming bounded system matrices).

This leads to a relatively simple conclusion about the concept of BIBS stability. One can readily demonstrate that a linear system is BIBS stable if, for all bounded inputs, there exists a finite constant N_s such that

$$\int_{t_0}^{t}\left\|\Phi(t,\tau)B(\tau)\right\|d\tau \le N_s \tag{7.61}$$

for all t.

A better way to approach BIBO stability is in the frequency-domain. One can show that for an LTI system,

$$e^{At} = L^{-1}\left\{(sI-A)^{-1}\right\}$$

(see Exercise 6.6). Therefore, demonstration of the boundedness of (7.59) can be achieved by considering the frequency-domain counterpart,

$$CL^{-1}\left\{(sI-A)^{-1}\right\}B \tag{7.62}$$

Boundedness of (7.62) is therefore equivalent to the condition of having the poles of the expression in (7.62) in the closed left-half plane.* However, providing for the existence of a bounded input, which might have characteristic frequencies on the imaginary axis, we must restrict the poles of (7.62) to the *open* left-half plane. Otherwise, the characteristic frequencies of the output

* We are assuming here that the reader is familiar with the relationship between the characteristic frequencies of a signal and the qualitative behavior of its time-domain representation.

$y(t)$ might be repeated on the imaginary axis and therefore produce an unbounded result. By taking the LaPlace transform of the LTI system equations, it is relatively easy to demonstrate that the transfer function of a system (i.e., the LaPlace transform of its impulse-response) is given by

$$H(s) = C(sI - A)^{-1} B + D \tag{7.63}$$

(see Section 1.1.5). Therefore, if the matrix D is bounded, our condition for BIBO stability can be stated as follows: *An LTI continuous-time system is BIBO stable if and only if the poles of its (proper) transfer function are located in the open left-half complex plane.* In the case of an LTI discrete-time system, this statement should be adjusted so that the poles must lie inside the open unit circle. Of course, by speaking of the transfer function (singular), we are implying a SISO system. However, the above statements apply if we add the further condition that a MIMO system is BIBO stable if and only if each individual input-to-output subsystem transfer function is BIBO stable.

Readers familiar with the Routh-Hurwitz method for testing stability of a system may now consider this test in the context of the stability properties discussed in this chapter [4]. Conventional Routh-Hurwitz testing determines BIBO stability. Extended Routh-Hurwitz testing, i.e., with additional analysis of the "auxiliary equation" in the case that a row of the Routh-Hurwitz array is entirely zero, has the further ability to determine so-called "marginal stability." Marginal stability corresponds to stability that results when the output is guaranteed to be bounded whenever the input asymptotically approaches zero, i.e., when the input has no characteristic frequencies with zero real parts.

7.4 Relationship Between Stability Types

Given the discussions of the various kinds of stability in this chapter, the relationship between different types of stability can be summarized by the Venn diagram in Figure 7.4. Although it is easy to envision systems that are Lyapunov stable but not BIBO stable, e.g., systems with poles on the imaginary axis, it is sometimes not obvious how a system can be BIBO stable, but not asymptotically stable, or even Lyapunov stable. Such a system is given in the following example.

Example 7.7: A BIBO, but Not Asymptotically Stable System

Consider the system given by the differential equation

$$\ddot{y} + \dot{y} - 2y = \dot{u} - u \tag{7.64}$$

Show that this system is BIBO but not asymptotically stable.

Solution:

First we will put the system into state variable form. From the simulation diagram given in Figure 1.7, we can derive a state variable model as follows:

$$\dot{x} = \begin{bmatrix} 0 & 1 \\ 2 & -1 \end{bmatrix} x + \begin{bmatrix} 0 \\ 1 \end{bmatrix} u$$
$$y = \begin{bmatrix} -1 & 1 \end{bmatrix} x \tag{7.65}$$

Finding the eigenvalues of this system, we determine that $\sigma(A) = \{1, -2\}$. From these eigenvalues, we see that this system is neither asymptotically nor Lyapunov stable. We will now check to see whether it is BIBO stable. It is an easy exercise to check that

$$e^{At} = \begin{bmatrix} \frac{2}{3}e^{t} + \frac{1}{3}e^{-2t} & \frac{1}{3}e^{t} - \frac{1}{3}e^{-2t} \\ \frac{2}{3}e^{t} - \frac{2}{3}e^{-2t} & \frac{1}{3}e^{t} + \frac{2}{3}e^{-2t} \end{bmatrix}$$

and that

$$Ce^{At}B = e^{-2t}$$

Therefore,

$$\int_{-\infty}^{t} \left\| Ce^{A(t-\tau)}B \right\| d\tau = \int_{-\infty}^{t} \left\| e^{-2(t-\tau)} \right\| d\tau = \frac{1}{2}$$

which is obviously bounded. The system is therefore BIBO stable.

As a further exercise, consider finding the transfer function from the original differential equation in (7.64). If the initial conditions are zero, we have

$$(s^2 + s - 2)Y(s) = (s-1)U(s)$$

giving

$$\frac{Y(s)}{U(s)} = \frac{(s-1)}{(s^2 + s - 2)} = \frac{(s-1)}{(s+2)(s-1)} = \frac{1}{(s+2)}$$

The transfer function shows poles in the open left-half of the complex plane, so again we have demonstrated that the system is BIBO stable.

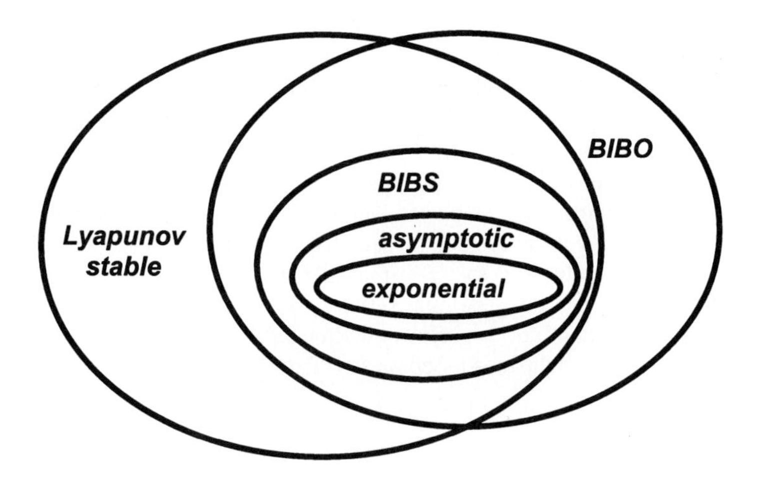

Figure 7.4 Venn diagram showing the relationship between stability types.

7.5 Summary

Before reading this chapter, one may already have formed an intuitive concept of system stability: A system is generally regarded as stable if its response remains finite. Although this interpretation often suffices in the sense that unstable systems are easy to recognize when they are encountered (we know one when we see one), the mathematically rigorous definitions for stability are more complex. Particularly, in the case of nonlinear or time-varying systems, we can encounter systems that are stable in the absence of inputs but unstable with them (Lyapunov but not BIBO stable); systems that are unstable when the internal dynamics (i.e., the states) are examined, but stable if only the output is of concern (BIBO but not Lyapunov stable); or systems that are stable for some initial conditions and/or initial times, but not from other initial states.

These distinctions are important for the study of newly modeled systems, for it is now clear that observation of one characteristic of a system or of one example of its behavior is not sufficient to characterize the stability of the system in general. A linear system that is designed and placed into a practical application should be assured of stability under all of its expected operating conditions. External stability, while usually desirable, is often insufficient for practical purposes. Usually, if internal states exceed their normal levels, the linear model upon which the analysis is performed will begin to fail, and all predictions of system performance will become invalid. Rarely, when internal states are conceptual abstractions, the fact that a system is internally not stable may not be important so long as the desired external behavior is stable.

The job of the engineer in system design is often predicated upon an analysis of the system's stability requirements. While undergraduate texts are fond of providing the many classical methods for achieving various performance specifications, real-life applications can be much more difficult. In complex systems, the goal of the designer is simply to achieve stability.

Tools for this purpose introduced in this chapter include the following:

- Stability in the sense of Lyapunov characterizes the system's state response in the absence of inputs, i.e., to initial conditions alone. For LTI systems, this type of stability is tested by examining the eigenvalues of a system to see that they are in the left-half plane (continuous-time systems) or inside the unit circle (discrete-time systems). For time-varying systems, as usual, the test is neither as simple to state nor to perform, and one should be careful not to reach erroneous conclusions based on the "frozen-time" eigenvalues.

- BIBO stability describes the behavior of the system in response to bounded inputs. Thus, BIBO stability is fundamentally different from Lyapunov stability, which neglects input. We have seen an example in which a system that is BIBO stable results from a state variable model whose zero-input counterpart is not Lyapunov stable. We will further explore this phenomenon after the following chapter on controllability and observability. As with Lyapunov stability, LTI systems prove to be much easier to test for BIBO stability than time-varying systems.

- Perhaps the most powerful tool introduced in the chapter is Lyapunov's direct, or *second* method.* This tool is particularly helpful in the study of nonlinear systems, for which there are relatively few other general stability analysis methods available. Lyapunov's method, wherein we construct so-called "Lyapunov functions," can be adapted to assist us in designing controllers, placing bounds on system parameters, and checking the sensitivity of a system to its initial conditions. We will provide more examples of its uses in the exercise problems.

7.6 Problems

7.1 Prove that the equilibrium of the LTI system $\dot{x} = Ax$ is asymptotically stable if and only if all the real parts of the eigenvalues of A are negative. Hint: Use Jordan forms.

7.2 Draw (or program a computer to draw) a phase portrait for the system in Example 7.3, which has the following nonlinear dynamic equations

---

* Lyapunov's *first* method is known as the method of exponents. It and *Lyapunov's linearization method* [18] are used exclusively for nonlinear systems and thus are not explored in detail here.

$$\dot{x}_1 = -x_1 - 2x_2^{\ 2}$$
$$\dot{x}_2 = x_1 x_2 - x_2^{\ 3}$$

7.3 A certain nonlinear system is given by the equations

$$\dot{x}_1 = -x_1 + x_2 + x_1(x_1^2 + x_2^2)$$
$$\dot{x}_2 = -x_1 - x_2 + x_2(x_1^2 + x_2^2)$$

a) Determine the equilibrium points of the system.

b) Choose a candidate Lyapunov function and use it to determine the bounds of stability for the system. Sketch the region in the $x_1 - x_2$ plane in which stability is guaranteed by the Lyapunov function.

c) Draw a phase portrait for the system.

7.4 Determine whether the following system is Lyapunov stable: $\dot{x} = tx$.

7.5 An LTI system is described by the equations

$$\dot{x} = \begin{bmatrix} a & 0 \\ 1 & -1 \end{bmatrix} x$$

Use Lyapunov's direct method to determine the range of variable a for which the system is asymptotically stable.

7.6 An LTI system is described by the equations

$$x(k+1) = \begin{bmatrix} 1-a & -1 \\ 0 & -1 \end{bmatrix} x(k)$$

Use Lyapunov's direct method to determine the range of variable a for which the system is asymptotically stable.

7.7 Show that a sufficient condition for a linear time-varying system to be asymptotically stable is for the matrix $A^{\mathrm{T}}(t) + A(t)$ to have all its eigenvalues in the open left-half of the complex plane.

7.8 Find a Lyapunov function that shows asymptotic stability for the system $\dot{x} = -(1+t)x$.

7.9 Show that matrix A has eigenvalues whose real parts are all less than $-\mu$ if and only if given an arbitrary positive definite matrix Q, there exists a unique, positive definite solution P to the matrix equation:

$$A^{\mathrm{T}}P + PA + 2\mu P = -Q$$

7.10 Determine whether the following system is BIBO stable: $\dot{x} = -x/t + u$.

7.11 Prove that an asymptotically stable system is by necessity also BIBO stable.

7.12 Prove the sufficiency of the condition given in Equation (7.45) for BIBS stability, i.e., that there exists a finite constant N_s such that

$$\int_{t_0}^{t} \|\Phi(t,\tau)B(\tau)\| \, d\tau \le N_s$$

7.13 Determine whether the system shown in Figure P7.13 is BIBO stable. Consider the input to be the current $u(t)$, and the output to be the voltage $v(t)$.

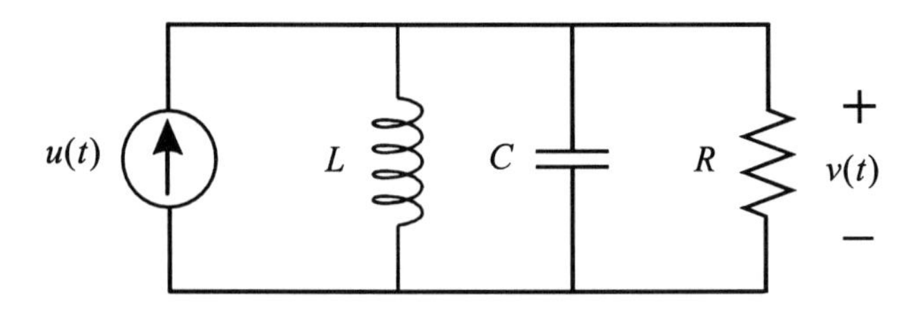

P7.13

Repeat if $R = \infty$ (i.e., remove the resistor).

7.14 Consider a system given by the matrix

$$A = \begin{bmatrix} 0 & 0 \\ 0 & 0 \end{bmatrix}$$

Determine all of the equilibrium points. Are they Lyapunov stable? If $b = [1 \quad 1]^T$, $c = [1 \quad 0]$ and $d = 0$, is the system BIBO stable? If instead $c = [0 \quad 0]$, is the system BIBO stable?

7.15 Consider the system given by

$$\dot{x} = \begin{bmatrix} -2 & 0 \\ 0 & 0 \end{bmatrix} x + \begin{bmatrix} 1 \\ 1 \end{bmatrix} u$$
$$y = \begin{bmatrix} -1 & 2 \end{bmatrix} x$$

a) Find all equilibrium solutions x_e.

b) Determine which equilibria are asymptotically stable.

c) Determine if the equilibrium solutions are Lyapunov stable.

d) Determine if the system is BIBO stable.

e) Let $z_1 = x_1$, $z_2 = -x_1 + x_2$, and $u(t) = 0$. If we denote $z \triangleq [z_1 \quad z_2]^T$ and $\dot{z} = \hat{A}z$, find the equilibrium solutions z_e and sketch them on the $z_1 - z_2$ plane.

f) Draw, on the same plane, a phase portrait.

7.16 For each of the following system matrices for systems of the form $\dot{x} = Ax + Bu$, determine a bounded input that *might* excite an unbounded output.

a) $A = \begin{bmatrix} 1 & 3 & 9 & 0 & 0 \\ 2 & 3 & 0 & 0 & 0 \\ 0 & 2 & 0 & 0 & 0 \\ 0 & 0 & 0 & 0 & 3 \\ 0 & 0 & 0 & -3 & 0 \end{bmatrix}$ b) $A = \begin{bmatrix} 1 & 2 & 3 \\ 4 & 5 & 6 \\ 4 & 5 & 6 \end{bmatrix}$

(Hint: Very little computation is necessary. Note that the B-matrix is undetermined in each case.)

7.17 For the three systems given below, determine the stability (i.e., Lyapunov, asymptotic, or BIBO).

a)
$$x(k+1) = \begin{bmatrix} 1 & 0 \\ -0.5 & 0.5 \end{bmatrix} x(k) + \begin{bmatrix} 1 \\ -1 \end{bmatrix} u(k)$$
$$y(k) = \begin{bmatrix} 5 & 5 \end{bmatrix} x(k)$$

b)
$$\dot{x} = \begin{bmatrix} -7 & -2 & 6 \\ 2 & -3 & -2 \\ -2 & -2 & 1 \end{bmatrix} x + \begin{bmatrix} 1 & 1 \\ 1 & -1 \\ 1 & 0 \end{bmatrix} u$$
$$y = \begin{bmatrix} -1 & -1 & 2 \\ 1 & 1 & -1 \end{bmatrix} x$$

c) $A = \begin{bmatrix} 2 & -5 \\ -4 & 0 \end{bmatrix}$ $b = \begin{bmatrix} 1 \\ -1 \end{bmatrix}$ $c = \begin{bmatrix} 1 & 1 \end{bmatrix}$

7.18 Given the system

$$\dot{x} = \begin{bmatrix} -25 & -45 & -55 \\ -5 & -15 & -15 \\ 15 & 35 & 35 \end{bmatrix} x + \begin{bmatrix} 2 \\ 0.2 \\ -0.8 \end{bmatrix} u$$
$$y = \begin{bmatrix} 1 & 0 & 0 \end{bmatrix} x$$

determine whether the system is Lyapunov stable, asymptotically stable, and/or BIBO stable. If it is not BIBO stable, give an example of a bounded input that will result in an unbounded output.

7.7 References and Further Reading

Although stability for time-invariant systems is relatively easy to test, determining the stability of time-varying and nonlinear systems can be very difficult. General texts used to distill the material presented here include [3], [7], and [15]. An interesting analysis of stability in the sense of Lyapunov is given in [14], where the authors establish the equivalence between Lyapunov stability and the stability determined by the more familiar Routh-Hurwitz test. Further ties between stability types are established in [1] and [16], where the relationships between BIBO, BIBS, and asymptotic stability are established for time-varying systems, although the controllability and observability concepts introduced in Chapter 8 are required. BIBO stability tests based on the impulse-response matrix were introduced in [16].

The issue of slowly varying systems, as briefly mentioned in Section 7.1.4, is addressed in [5], [11], and [17].

Lyapunov's direct method is the topic of much study, owing partially to the difficulty with which Lyapunov functions are found for time-varying and nonlinear systems. A good introductory discussion is included in [7], [9], [12], [13], [18], and [19], with [10] and [20] concentrating on discrete-time systems. The issue of *decrescence*, i.e., the importance of the γ_i-functions described in the theorems of Section 7.2.1, is explicitly discussed in [7] and [18]. These functions are critical for generating Lyapunov functions for time-varying systems, but were not discussed in depth here.

Also useful for nonlinear and time-varying systems are other types of stability tests, such as the small gain theorems used to determine input/output stability, as discussed in [6], [12], and [19].

[1] Anderson, Brian D. O., and J. B. Moore, "New Results in Linear System Stability," *SIAM Journal on Control and Optimization*, vol. 7, no. 3, 1969, pp. 398-414.

[2] Brogan, William L., *Modern Control Theory*, 3rd edition, Prentice-Hall, 1991.

[3] Chen, Chi-Tsong, *Linear System Theory and Design*, Holt, Rinehart, and Winston, 1984.

[4] D'Azzo, John D., and Constantine H. Houpis, *Linear Control System Analysis and Design: Conventional and Modern*, 4th edition, McGraw-Hill, 1995.

[5] Desoer, Charles A., "Slowly Varying System $\dot{x} = A(t)x$," *IEEE Transactions on Automatic Control*, vol. AC-14, 1969, pp. 780-781.

[6] Desoer, Charles A., and M. Vidyasagar, *Feedback Systems: Input-Output Properties*, Academic Press, 1975.

[7] Hahn, Wolfgang, *Stability of Motion*, Springer-Verlag, 1967.

[8] Kailath, Thomas, *Linear Systems*, Prentice-Hall, 1980.

[9] Kalman, Rudolph E., and J. E. Bertram, "Control System Analysis and Design Via the Second Method of Lyapunov: I. Continuous-Time Systems," *Transactions of the ASME, Journal of Basic Engineering*, vol. 82D, June 1960, pp. 371-393.

[10] Kalman, Rudolph E., and J. E. Bertram, "Control System Analysis and Design Via the Second Method of Lyapunov: II. Discrete-Time Systems," *Transactions of the ASME, Journal of Basic Engineering*, vol. 82D, June 1960, pp. 394-400.

[11] Kamen, Edward W., P. P. Khargonekar, and A. Tannenbaum, "Control of Slowly-Varying Linear Systems," *IEEE Transactions on Automatic Control*, vol. 34, no. 12, 1989, pp. 1283-1285.

[12] Khalil, Hassan, *Nonlinear Systems*, Macmillan, 1992.

[13] LaSalle, J., and S. Lefschetz, *Stability by Liapunov's Direct Method, with Applications*, Academic Press, 1961.

[14] Puri, N. N., and C. N. Weygandt, "Second Method of Liapunov and Routh's Canonical Form," *Journal of The Franklin Institute*, vol. 276, no. 5, 1963, pp. 365-384.

[15] Rugh, Wilson, *Linear System Theory*, 2nd edition, Prentice Hall, 1996.

[16] Silverman, Leonard M., and Brian D. O. Anderson, "Controllability, Observability, and Stability of Linear Systems," *SIAM Journal on Control and Optimization*, vol. 6, no. 1, 1968, pp. 121-130.

[17] Skoog, Ronald A., and Clifford G. Y. Lau, "Instability of Slowly Time-Varying Systems," *IEEE Transactions on Automatic Control*, vol. AC-17, no. 1, 1972, pp. 86-92.

[18] Slotine, Jean-Jacques, and Weiping Li, *Applied Nonlinear Control*, Prentice Hall, 1991.

[19] Vidyasagar, Mathukumalli, *Nonlinear Systems Analysis*, 2nd edition, Prentice Hall, 1993.

[20] Weiss, Leonard, "Controllability, Realization, and Stability of Discrete-Time Systems," *SIAM Journal on Control and Optimization*, vol. 10, no. 2, 1972, pp. 230-251.

8

Controllability and Observability

We first encountered the concepts of controllability and observability in Examples 3.10 and 3.11. In those examples, we demonstrated the rank conditions for solving systems of simultaneous linear algebraic equations. At that time, the controllability problem was to ascertain the existence of a sequence of inputs to a discrete-time system that would transfer the system's initial state to the zero vector. The observability problem was to determine the existence of the solution of the state equations such that, if the input and output sequences were known for a certain duration, the system's initial state could be deduced. In those examples, we derived a test for such conditions, indicating whether the system considered was controllable and/or observable.

In this chapter, we will continue this discussion, investigating the controllability and observability properties of linear systems in more detail, and extending the results to continuous-time systems. This will mark the first step in the study of linear system controller *design*. We will see in Chapter 10 that the first stage of the design of a linear controller is often the investigation of controllability and observability, and that the process of testing a system for these properties can sometimes dictate the subsequent procedures for the design process.

8.1 Definitions

The descriptions of controllability and observability given in Examples 3.10 and 3.11 will serve as the foundation of the definitions we will use in this chapter. In these definitions, and throughout most of this entire chapter, we make no distinction between SISO and MIMO systems.

> ***Controllability:*** A linear system is *controllable* in an interval $[t_0, t_1]$ if there exists an input $u(t)$ that, when applied to the system from an initial state $x(t_0)$, transfers the system to the state $x(t_1) = 0$. If this property holds regardless of the initial time t_0 or the initial state $x(t_0)$, the system is said to be *completely controllable.* (8.1)

> ***Observability:*** A linear system is *observable* in an interval $[t_0, t_1]$ if, for an initial state $x(t_0)$, knowing two functions $u(t)$ and $y(t)$ over the same interval is sufficient information to uniquely solve for $x(t_0)$. If this property holds regardless of the initial time t_0 or the initial state $x(t_0)$, the system is said to be *completely observable.* (8.2)

We have seen from the examples in Chapter 3 that the property of controllability depends (at least for the discrete-time case) on the matrices A and B, and the property of observability depends on the matrices A and C. This is because controllability is dependent only on the input and the state, and observability is dependent only on the state and the output (provided that the input signal, whatever it may be, is known).

As with most issues concerning linear systems, the time-invariant situation is usually easier to develop than the time-varying case. In order to understand controllability and observability in their simplest contexts, we will at first restrict ourselves to LTI systems.

8.2 Controllability Tests for LTI Systems

The results of Examples 3.10 and 3.11 were derived only for the discrete-time case. However, we shall see that the same matrix "tests" hold for continuous-time systems as well.

8.2.1 The Fundamental Tests for Controllability and Observability

In Example 3.10, we derived the result that a discrete-time LTI system is controllable if and only if the matrix

$$P = \left[B \mid AB \mid \cdots \mid A^{k-1}B \right] \tag{8.3}$$

had rank greater than or equal to n, the order of the system. At the time, we made the note that because this P matrix is $n \times k$, this rank condition will in general require that $k \geq n$. We have since encountered the Cayley-Hamilton theorem, which eliminates the need to consider any power of matrix A greater than $n-1$. We can therefore restate the condition for controllability as follows: An n-dimensional discrete-time LTI system is controllable if and only if the matrix[M]

ctrb(A,B)

$$P \triangleq \left[B_d \;\vdots\; A_d B_d \;\vdots\; \cdots \;\vdots\; A_d^{\,n-1} B_d \right] \tag{8.4}$$

has rank n.

Exactly analogous reasoning leads to the additional statement: An n-dimensional discrete-time LTI system is observable if and only if the matrix[M]

obsv(A,C)

$$Q \triangleq \begin{bmatrix} C_d \\ \hdashline C_d A_d \\ \hdashline \vdots \\ \hdashline C_d A_d^{\,n-1} \end{bmatrix} \tag{8.5}$$

has rank n. The subscript d in each case refers to the fact that the system matrices come from a *discrete-time* system.

We will call these two matrix tests the *fundamental* tests for controllability and observability. They are by far the most common tests and are preprogrammed into many control system analysis and design software packages. Matrix P is often referred to as the *controllability matrix*, and Q is referred to as the *observability matrix*. We will now proceed to show that the exact same tests can be used to determine the controllability and observability properties of continuous-time LTI systems. Only the derivation and proof differ. Actually, we will present the (fairly lengthy) constructive proof only for controllability. Observability is a dual concept, and the proof of the observability test can be reconstructed by analogy to controllability.

THEOREM: A n-dimensional continuous-time LTI system is completely controllable if and only if the matrix

$$P \triangleq \left[B \;\vdots\; AB \;\vdots\; \cdots \;\vdots\; A^{n-1} B \right] \tag{8.6}$$

has rank n. It is observable if and only if the matrix

$$Q \triangleq \begin{bmatrix} C \\ CA \\ \vdots \\ CA^{n-1} \end{bmatrix} \tag{8.7}$$

has rank n.

PROOF: We begin by recalling the solution of the LTI state equations

$$x(t)=e^{A(t-t_0)}x(t_0)+\int_{t_0}^{t} e^{A(t-\tau)}Bu(\tau)\,d\tau \tag{8.8}$$

Given an arbitrary initial condition, we would like to reach the zero state at some future (finite) time t_1. We desire $x(t_1)=0$, so

$$\int_{t_0}^{t_1} e^{A(t_1-\tau)}Bu(\tau)\,d\tau=-e^{A(t_1-t_0)}x(t_0)\stackrel{\Delta}{=}\xi_1$$

where ξ_1 is an arbitrary $n\times 1$ vector, because we are allowing arbitrary initial conditions $x(t_0)$. Using the Cayley-Hamilton expansion

$$e^{A(t_1-\tau)}=\sum_{i=1}^{n}\gamma_i(\tau)A^{n-i}$$

we can express

$$\begin{aligned}
\xi_1 &= \int_{t_0}^{t_1}\left[\sum_{i=1}^{n}\gamma_i(\tau)A^{n-i}\right]Bu(\tau)\,d\tau \\
&= \int_{t_0}^{t_1}\left[\sum_{i=1}^{n}A^{n-i}B\gamma_i(\tau)\right]u(\tau)\,d\tau \\
&= \int_{t_0}^{t_1}\Big[A^{n-1}B\gamma_1(\tau)u(\tau)+A^{n-2}B\gamma_2(\tau)u(\tau)+\cdots \\
&\qquad \cdots+A^{0}B\gamma_n(\tau)u(\tau)\Big]\,d\tau
\end{aligned}$$

$$= A^{n-1}B\int_{t_0}^{t_1}\gamma_1(\tau)u(\tau)\,d\tau + A^{n-2}B\int_{t_0}^{t_1}\gamma_2(\tau)u(\tau)\,d\tau + \cdots$$
$$\cdots + B\int_{t_0}^{t_1}\gamma_n(\tau)u(\tau)\,d\tau \tag{8.9}$$

Defining

$$\Gamma_i \stackrel{\Delta}{=} \int_{t_0}^{t_1}\gamma_i(\tau)u(\tau)\,d\tau \tag{8.10}$$

This gives

$$\xi_1 = \left[B \mid AB \mid \cdots \mid A^{n-1}B\right]\begin{bmatrix}\Gamma_n \\ \hline \vdots \\ \hline \Gamma_1\end{bmatrix} = P\begin{bmatrix}\Gamma_n \\ \hline \vdots \\ \hline \Gamma_1\end{bmatrix} \tag{8.11}$$

Again, since ξ_1 is arbitrary, we must guarantee that this matrix equation always has a solution by requiring matrix P to have full rank. In the most general multivariable case where the system has m inputs, this matrix P will be $n \times nm$, so full rank means $r(P) = n$.

As mentioned, the observability test follows dual reasoning and results in the requirement that $r(Q) = n$, where Q is defined in (8.5).

Example 8.1: Controllability and Observability of a Circuit

For the circuit shown in Figure 8.1, formulate a state variable description using v_C and i_L as the state variables, with source voltage v_s as the input and v_x as the output. Then determine the conditions on the resistor that would make the system uncontrollable and unobservable.

Solution.

First, we notice that

$$v_C(t) = v_s(t) - v_x(t) = v_s(t) - L\frac{di_L(t)}{dt}$$

so

$$\frac{di_L(t)}{dt} = \frac{1}{L}v_s(t) - \frac{1}{L}v_C(t) \tag{8.12}$$

which is one of the necessary state equations. Furthermore, for the capacitor voltage,

$$\begin{aligned}\frac{dv_C(t)}{dt} &= \frac{1}{C}i_C(t) \\ &= \frac{1}{C}\left(i_L(t) + i_R(t) - \frac{v_C(t)}{R}\right) \\ &= \frac{1}{C}\left(i_L(t) + \frac{v_s(t) - v_C(t)}{R} - \frac{v_C(t)}{R}\right) \\ &= \frac{i_L(t)}{C} - \frac{2v_C(t)}{RC} + \frac{v_s(t)}{RC}\end{aligned} \tag{8.13}$$

Therefore the state equations are:

$$\begin{aligned}\begin{bmatrix}\dot{v}_C(t) \\ \dot{i}_L(t)\end{bmatrix} &= \begin{bmatrix}-\frac{2}{RC} & \frac{1}{C} \\ -\frac{1}{L} & 0\end{bmatrix}\begin{bmatrix}v_C(t) \\ i_L(t)\end{bmatrix} + \begin{bmatrix}\frac{1}{RC} \\ \frac{1}{L}\end{bmatrix}v_s(t) \\ v_x(t) &= v_s(t) - v_C(t) \\ &= \begin{bmatrix}-1 & 0\end{bmatrix}\begin{bmatrix}v_C(t) \\ i_L(t)\end{bmatrix} + v_s(t)\end{aligned} \tag{8.14}$$

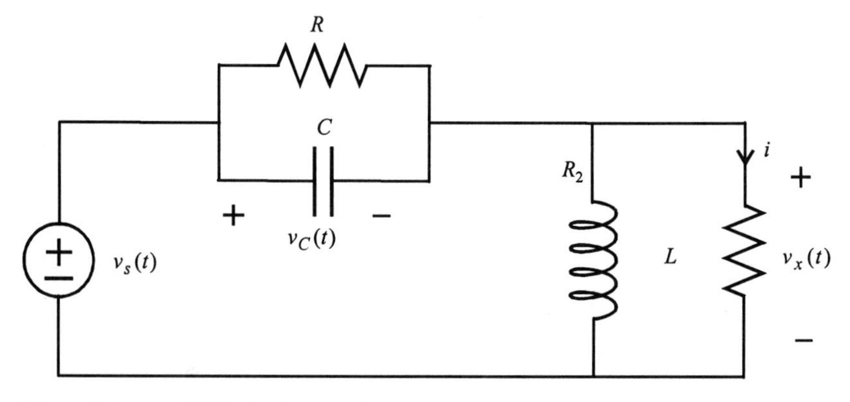

Figure 8.1 RLC circuit example.

Testing controllability using the fundamental controllability test,

$$P = \begin{bmatrix} B & AB \end{bmatrix} = \begin{bmatrix} \dfrac{1}{RC} & -\dfrac{2}{R^2C^2} + \dfrac{1}{LC} \\ \dfrac{1}{L} & -\dfrac{1}{RLC} \end{bmatrix} \tag{8.15}$$

The rank of this matrix can be checked by determining the determinant:

$$\begin{aligned} \begin{vmatrix} \dfrac{1}{RC} & -\dfrac{2}{R^2C^2} + \dfrac{1}{LC} \\ \dfrac{1}{L} & -\dfrac{1}{RLC} \end{vmatrix} &= -\frac{1}{R^2LC^2} + \frac{2}{R^2LC^2} - \frac{1}{L^2C} \\ &= \frac{1}{R^2LC^2} - \frac{1}{L^2C} \end{aligned} \tag{8.16}$$

If we want the conditions under which this determinant is zero, we solve the equation

$$\frac{1}{R^2LC^2} - \frac{1}{L^2C} = 0 \tag{8.17}$$

which gives

$$R = \sqrt{\frac{L}{C}} \tag{8.18}$$

Similarly, the observability matrix is:

$$Q = \begin{bmatrix} -1 & 0 \\ \dfrac{2}{RC} & -\dfrac{1}{C} \end{bmatrix} \tag{8.19}$$

which obviously is always of full rank. Hence, the system is always observable, but becomes uncontrollable whenever $R = \sqrt{L/C}$.

It is useful to think in classical terms in this example to see the significance of controllability. If we were to calculate the transfer function between the output $V_x(s)$ and the input $V_s(s)$ (using conventional circuit analysis techniques), we might first find the differential equation:

$$\frac{d^2 v_x(t)}{dt^2} + \frac{2}{RC}\frac{dv_x(t)}{dt} + \frac{1}{LC}v_x(t) = \frac{d^2 v_s(t)}{dt^2} + \frac{1}{RC}\frac{dv_s(t)}{dt} \tag{8.20}$$

from which it should be apparent that the transfer function is:

$$\frac{V_x(s)}{V_s(s)} = \frac{s\left(s + \frac{1}{RC}\right)}{s^2 + \frac{2}{RC}s + \frac{1}{LC}} \tag{8.21}$$

If we wish to express this function in pole-zero form, we will need to compute the roots of the denominator quadratic:

$$s_{1,2} = \frac{-\frac{2}{RC} \pm \sqrt{\frac{4}{R^2C^2} - \frac{4}{LC}}}{2} = -\frac{1}{RC} \pm \sqrt{\frac{1}{R^2C^2} - \frac{1}{LC}} \tag{8.22}$$

If $R = \sqrt{L/C}$ as computed from (8.18) above, then this becomes

$$s_{1,2} = -\frac{1}{RC} \pm \sqrt{\frac{1}{R^2C^2} - \frac{1}{LC}} = -\frac{1}{RC} \pm \sqrt{\frac{1}{LC} - \frac{1}{LC}} = -\frac{1}{RC} \tag{8.23}$$

giving a system with repeated roots at $s = -1/RC$. Thus, in pole-zero form, (8.21) is

$$\frac{V_x(s)}{V_s(s)} = \frac{s\left(s + \frac{1}{RC}\right)}{\left(s + \frac{1}{RC}\right)\left(s + \frac{1}{RC}\right)} = \frac{s}{\left(s + \frac{1}{RC}\right)} \tag{8.24}$$

which is now a *first*-order system. What has apparently happened is that the energy storage elements, in conjunction with this particular resistance value, are interacting in such a way that their effects *combine*, giving an equivalent first-order system. In this case, the time constants due to these two elements are the same, making them equivalent to a single element. One might contrast this to a circuit with two capacitors in series or parallel, which can be combined by adding to give a single equivalent capacitance. In this case, the capacitor and inductor together result in an equivalent first-order system, but only for a specific value of R.

8.2.2 Popov-Belevitch-Hautus Tests

The Popov-Belevitch-Hautus (PBH) tests, also commonly known as simply the Hautus tests, are not nearly as common as the fundamental test above nor the Jordan form test below. However, they have some interesting geometric interpretations and help us prove the Jordan form test.

PBH Eigenvector Test

Before presenting this Hautus test, we define a vector known as the left eigenvector. The eigenvectors we have been considering up until now have been *right* eigenvectors. A *left* eigenvector corresponding to eigenvalue λ is a nonzero vector v such that

$$vA = \lambda v \tag{8.25}$$

where obviously, v must be a row vector. Given this definition, a new test for controllability may be stated as follows:

> LEMMA: The LTI system is *not* controllable if and only if there exists a nonzero left eigenvector v of A such that
>
> $$vB = 0 \tag{8.26}$$
>
> PROOF: (Sufficiency) Suppose there exists a nonzero left eigenvector v such that $vB = 0$. Then we could multiply (8.25) from the right by B to achieve
>
> $$vAB = \lambda vB = 0$$
>
> We could also multiply (8.25) from the right by AB to achieve
>
> $$vA^2B = \lambda vAB = \lambda^2 vB = 0$$
>
> and so on, until we demonstrate that
>
> $$vP = v\left[B \mid AB \mid \cdots \mid A^{n-1}B\right] = 0$$
>
> implying that $r(P) < n$ and the system is not controllable.
>
> (Necessity) Conversely, suppose that the system is not controllable. Then we know that

$$r\left(\left[B \;\vdots\; AB \;\vdots\; \cdots \;\vdots\; A^{n-1}B\right]\right) < n$$

Then there exists a nonzero vector v such that

$$\begin{aligned} 0 &= v\left[B \;\vdots\; AB \;\vdots\; \cdots \;\vdots\; A^{n-1}B\right] \\ &= \left[vB \;\vdots\; vAB \;\vdots\; \cdots \;\vdots\; vA^{n-1}B\right] \end{aligned}$$

Restricting our attention to vectors v that are left eigenvectors, that is, for which (8.25) holds, then this implies

$$\begin{aligned} 0 &= \left[vB \;\vdots\; vAB \;\vdots\; \cdots \;\vdots\; vA^{n-1}B\right] \\ &= \left[vB \;\vdots\; \lambda vB \;\vdots\; \cdots \;\vdots\; \lambda^{n-1}vB\right] \end{aligned}$$

in turn implying that $vB = 0$ for the left eigenvector v.

A similar statement may be made about observability: A system that has a right eigenvector q ($Aq = \lambda q$) such that $Cq = 0$ will *not* be observable.

PBH Rank Test (Hautus Test)

Now we may state and prove the main PBH controllability test:

THEOREM (PBH rank test): An LTI system will be controllable if and only if $r\left(\left[sI - A \;\vdots\; B\right]\right) = n$ for all s. (8.27)

PROOF: (Sufficiency) If $r\left(\left[sI - A \;\vdots\; B\right]\right) = n$, then there can exist no nonzero vector v such that

$$v\left[sI - A \;\vdots\; B\right] = \left[v(sI - A) \quad vB\right] = 0$$

Consequently, there is *no* vector v such that $vs = vA$ and $vB = 0$. By the above lemma (i.e., the PBH eigenvector test), the system will therefore be controllable.

(Necessity) Now suppose the system is controllable. By reversing the above statements, we can demonstrate that $r\left(\left[sI - A \;\vdots\; B\right]\right) = n$.

Of course, the dual result for observability may also be stated:

> THEOREM (PBH rank test): An LTI system will be observable if and only if
>
> $$r\left(\begin{bmatrix} sI - A \\ \hdashline C \end{bmatrix}\right) = n \tag{8.28}$$
>
> for all s.

8.2.3 Controllability and Observability of Jordan Forms

When we first introduced the concept of the canonical form (e.g., the diagonal and Jordan forms), we indicated that canonical forms were useful for the purpose of exposing more saliently certain important properties of systems. Controllability and observability are foremost among these properties. By transforming systems to their Jordan forms, it is usually possible to ascertain controllability and observability by inspection [5].

Consider the standard-form of the linear LTI system:

$$\begin{aligned} \dot{x} &= Ax + Bu \\ y &= Cx + Du \end{aligned} \tag{8.29}$$

If we use the modal matrix M to perform the change of variables $x = M\overline{x}$ and thus compute the similarity transformation

$$\begin{aligned} \dot{\overline{x}} &= \overline{A}\overline{x} + \overline{B}u \\ y &= \overline{C}\overline{x} + \overline{D}u \end{aligned} \tag{8.30}$$

where $\overline{A} = M^{-1}AM$, $\overline{B} = M^{-1}B$, $\overline{C} = CM$, and $\overline{D} = D$, we arrive at the Jordan form. In the Jordan form, a simple test for controllability can be constructed.

Without loss of generality, we can arrange the Jordan form so that all Jordan blocks corresponding to the same eigenvalues are adjacent:

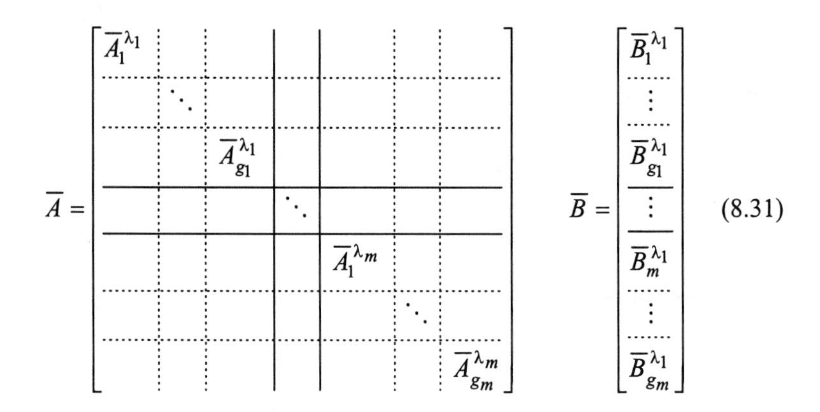

(8.31)

where $\overline{A}_j^{\lambda_i}$ stands for the jth Jordan block corresponding to eigenvalue λ_i. As we know, there will be a number of these blocks equal to the geometric multiplicity g_i of the eigenvalue. In this notation, m is the number of distinct eigenvalues. In the matrix $\overline{B}$, we have simply partitioned the rows to correspond to the partition of rows in the matrix $\overline{A}$.

Now consider performing the Hautus test with the matrices $\overline{A}$ and $\overline{B}$, i.e., determine the rank of the matrix $[sI-\overline{A} \quad \overline{B}]$. We will do this by counting the number of linearly independent rows in this matrix. First, if s is *not* an eigenvalue of $\overline{A}$, then each subblock of $sI-\overline{A}$ will itself be of full rank, so there will automatically be n linearly independent rows, and $r([sI-\overline{A} \quad \overline{B}])=n$. If however s *is* an eigenvalue, say $s=\lambda_1$, then the Jordan blocks corresponding to eigenvalues other than λ_1, i.e., $\overline{A}_j^{\lambda_i}$ for $i \neq 1$, will each have full rank, but the blocks corresponding to $s=\lambda_1$ will be rank deficient. Suppose for the sake of argument that there are only two such blocks for λ_1, so that the rows of $\left[sI-\overline{A} \quad \overline{B}\right]$ corresponding to $s=\lambda_1$ appear as

$$
\left[\begin{array}{ccccc:ccccc:c}
0 & -1 & 0 & \cdots & 0 & & & & & & \bar{b}_{11}^{\lambda_1} \\
 & 0 & & \ddots & \vdots & & & & & & \\
\vdots & & & \ddots & 0 & & & 0 & & & \vdots \\
 & & & \ddots & -1 & & & & & & \\
0 & & \cdots & & 0 & & & & & & \bar{b}_{1\ell_1}^{\lambda_1} \\
\hdashline
 & & & & & 0 & -1 & 0 & \cdots & 0 & \bar{b}_{21}^{\lambda_1} \\
 & & & & & & 0 & & \ddots & \vdots & \\
 & & 0 & & & \vdots & & & \ddots & 0 & \vdots \\
 & & & & & & & & \ddots & -1 & \\
 & & & & & 0 & & \cdots & & 0 & \bar{b}_{2\ell_2}^{\lambda_1}
\end{array}\right] \tag{8.32}
$$

where the last column represents the appropriate rows of the blocks $\bar{B}_1^{\lambda_1}$ and $\bar{B}_2^{\lambda_1}$ and where $\bar{b}_{ij}^{\lambda_1}$ is the jth row of block $\bar{B}_i^{\lambda_1}$. We are assuming that these two blocks have ℓ_1 and ℓ_2 rows, respectively. It is clear from this structure that the first $\ell_1 - 1$ rows of the first block are linearly independent and that the first $\ell_2 - 1$ rows of the second block are also linearly independent (of each other and of the first block). Therefore, if and only if the two rows $\bar{b}_{1\ell_1}^{\lambda_1}$ and $\bar{b}_{2\ell_2}^{\lambda_1}$ are independent will we have a complete set of n linearly independent rows. Of course, this applies to any number of Jordan blocks, for any number of repeated eigenvalues.

To state the results succinctly: Denote by $\bar{B}^{\lambda_i}$ the matrix consisting of the rows of the $\bar{B}$-matrix that each corresponds to the last row of a different Jordan block for eigenvalue λ_i. That is,

$$
\bar{B}^{\lambda_i} \triangleq \begin{bmatrix} \bar{b}_{1\ell_1}^{\lambda_i} \\ \vdots \\ \bar{b}_{g_i \ell_{g_i}}^{\lambda_i} \end{bmatrix} \tag{8.33}
$$

A system in Jordan form is *controllable if and only if all the rows of* $\bar{B}^{\lambda_i}$ *are linearly independent for every* λ_i. Rows of the $\bar{B}$-matrix that correspond to blocks from *different* eigenvalues do not necessarily have to be linearly independent of one another.

This statement has some obvious implications for simpler systems. For example, consider single-input systems. If a single-input system in Jordan form has a $\bar{B}$-matrix whose rows that correspond to the last rows of the Jordan blocks for any eigenvalue are linearly independent, then there must be only a single Jordan block for each eigenvalue. This is because $\bar{B}$ will consist of a single column. Therefore, if we gather all the rows as in $\bar{B}^{\lambda_i}$ above, we will again have a single column, which has at most rank 1. Two scalars are never linearly independent of one another. This extends to the case where the $\bar{A}$-matrix is diagonal, even if there are repeated eigenvalues (as is the case with symmetric matrices). There must be no more than one Jordan block for any eigenvalue, so any diagonal single-input system with repeated eigenvalues will inevitably be uncontrollable. In general, a system must have at least as many inputs (columns of B) as there are Jordan blocks for the eigenvalue that has the most Jordan blocks in order for the system to be controllable.

Furthermore, suppose a Jordan form consists of a diagonal $\bar{A}$-matrix with distinct eigenvalues. Then the diagonal form can be considered a collection of 1×1 blocks, each corresponding to a different eigenvalue. Therefore, for controllability, each row of the $\bar{B}$-matrix must simply be *nonzero*.

To extend these arguments to tests for observability, we change each reference from the *row*(s) of the $\bar{B}$-*matrix* corresponding to *last row*(s) of the Jordan form to the *column*(s) of the $\bar{C}$-*matrix* corresponding to *first column*(s) of the Jordan form. An example to illustrate this concept is presented here.

Example 8.2: Controllability and Observability of Multivariable Jordan Forms

A system is represented by the matrices given in Figure 8.2. Determine whether it is controllable and/or observable.

Solution:

Rather than redraw the system equations, we will mark them up as indicated to show the relevant rows and columns. Note that there are two Jordan blocks for the eigenvalues –5 and 0 while the other eigenvalues correspond to only a single block each. There are two inputs and one output, providing a two-column $\bar{B}$-matrix, and a one-row $\bar{C}$-matrix.

The arrows indicate the relevant rows and columns, with the arrows tied together for the rows that correspond to the Jordan blocks of a single eigenvalue. Examination of the $\bar{B}$-matrix reveals that the rows corresponding to Jordan blocks for $\lambda = -5$ are linearly independent of each other. However, the two rows corresponding to the two blocks for $\lambda = 0$ are not linearly independent. Therefore the system is not controllable. As for observability, we know immediately that because the $\bar{C}$-matrix is a single row and some eigenvalues are providing multiple Jordan blocks, the system cannot possibly be observable.

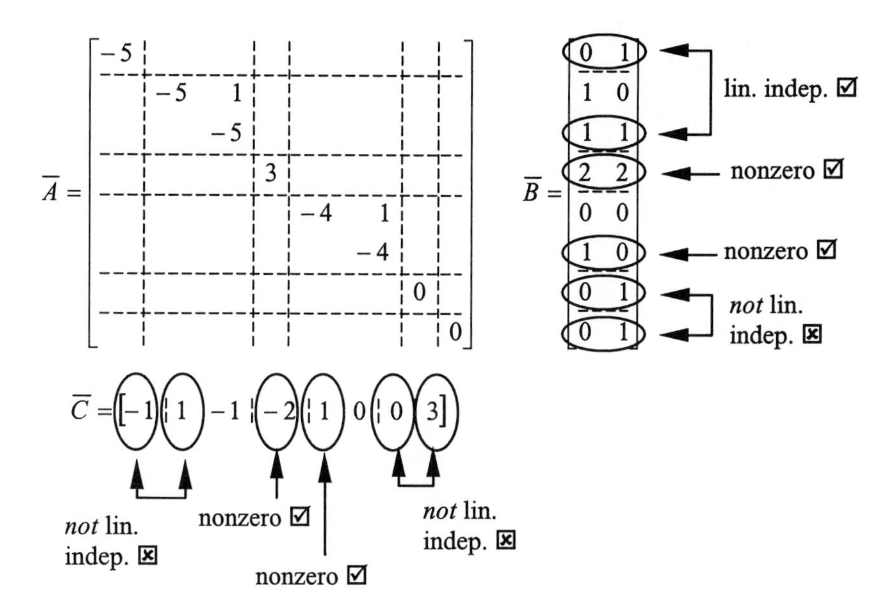

Figure 8.2 Marked-up Jordan form matrices showing criteria for controllability and observability.

One may reasonably ask the question "How can a system consisting of a single Jordan block be controllable if all the state variables must be controlled with a single input?" For example, the system

$$\dot{x} = \begin{bmatrix} -2 & 1 \\ 0 & -2 \end{bmatrix} x + \begin{bmatrix} 0 \\ 1 \end{bmatrix} u \triangleq Ax + Bu$$

can easily be shown to be controllable, because

$$P = \begin{bmatrix} B & AB \end{bmatrix} = \begin{bmatrix} 0 & 1 \\ 1 & -2 \end{bmatrix}$$

but this demonstration does not make clear how the first state variable, $x_1(t)$, can be made to go to zero when there is no direct input to the first differential equation of the system, $\dot{x}_1(t) = -2x_1(t) + x_2(t)$. The answer lies in the fact that the equations deriving from a Jordan block are *coupled*. Clearly, $x_2(t)$ is affected by the input $u(t)$ because it appears directly in the second equation, $\dot{x}_2(t) = -2x_2(t) + u(t)$. Essentially, $u(t)$ can make $x_2(t)$ behave as desired (i.e., approach zero), while at the same time $u(t)$ can force $x_2(t)$ to act as an input to the first equation, forcing $x_1(t)$ to zero. We say that $x_1(t)$ is controlled "through" $x_2(t)$.

8.2.4 Controllable and Observable Canonical Forms

`canon(sys, type)`

There are two other canonical forms[M] associated with controllability and observability. They are not strictly useful for testing for controllability and observability, because a system cannot be transformed into these canonical forms unless controllability and observability are already known. However, these forms will be important for future sections on controller and observer design, and the necessary transformations are derived from the controllability and observability matrices P and Q. It therefore seems appropriate to introduce these forms here.

Controllable Form

A system in controllable canonical form will appear as follows:

$$A_c = \begin{bmatrix} 0 & 1 & 0 & \cdots & 0 \\ 0 & 0 & 1 & \ddots & \vdots \\ \vdots & & \ddots & \ddots & 0 \\ 0 & & & 0 & 1 \\ -a_0 & -a_1 & \cdots & -a_{n-2} & -a_{n-1} \end{bmatrix} \qquad B_c = \begin{bmatrix} 0 \\ 0 \\ \vdots \\ 0 \\ 1 \end{bmatrix} \tag{8.34}$$
$$C_c = [\text{arbitrary; no special structure}]$$

A system in observable canonical form has a similar structure:

$$A_o = \begin{bmatrix} -a_{n-1} & 1 & 0 & \cdots & 0 \\ -a_{n-2} & 0 & 1 & \ddots & \vdots \\ \vdots & & \ddots & \ddots & 0 \\ -a_1 & & & 0 & 1 \\ -a_0 & 0 & \cdots & 0 & 0 \end{bmatrix} \qquad B_o = \begin{bmatrix} \text{arbitrary;} \\ \text{no special} \\ \text{structure} \end{bmatrix} \tag{8.35}$$
$$C_o = \begin{bmatrix} 1 & 0 & 0 & \cdots & 0 \end{bmatrix}$$

The similarity in these forms is readily apparent. The common structure of the A-matrix, i.e., the superdiagonal of 1's and the single row or column of nonzero entries (in either the last row or the first column, respectively) is known as a *companion form*.[M] These two forms, the controllable canonical form and the observable canonical form, will be especially useful in the design of controllers and observers that we will introduce in Chapter 10. For now, we will show an algorithmic method for computing the transformation that changes the system into these forms.

`compan(P)`

Consider a single-input system

$$\dot{x} = Ax + bu$$

and denote the controllability matrix by

$$P = \left[b \;\vdots\; Ab \;\vdots\; \cdots \;\vdots\; A^{n-1}b \right]$$

If this matrix is full rank, it will be invertible (for a single-input system). Compute this inverse and call its last row p:

$$P^{-1} = \begin{bmatrix} \vdots \\ \cdots\cdots\cdots \\ p \end{bmatrix}$$

Knowing that $P^{-1}P = I$, we have

$$P^{-1}P = \begin{bmatrix} \vdots \\ \cdots \\ p \end{bmatrix} \begin{bmatrix} b & Ab & \cdots & A^2 b \end{bmatrix} = I \tag{8.36}$$

Considering only the bottom row of this matrix product, we see that

$$\begin{aligned} pb &= 0 \\ pAb &= 0 \\ &\vdots \\ pA^{n-2}b &= 0 \\ pA^{n-1}b &= 1 \end{aligned} \tag{8.37}$$

so that if we define the matrix transformation

$$U^{-1} \triangleq \begin{bmatrix} p \\ pA \\ \vdots \\ pA^{n-1} \end{bmatrix} \tag{8.38}$$

then it is clear from (8.37) that $U^{-1}b = b_c$, demonstrating that the state space transformation $x = Ux_c$ will accomplish the transformation necessary for the b-matrix.

Now consider what this transformation does for the A-matrix. Given U^{-1} in (8.38), define the columns of U as follows:

$$U \triangleq \begin{bmatrix} u_1 & u_2 & \cdots & u_n \end{bmatrix}$$

Now, as we did before, we know that $U^{-1}U = I$, so that

$$U^{-1}U = \begin{bmatrix} p \\ pA \\ \vdots \\ pA^{n-1} \end{bmatrix} \begin{bmatrix} u_1 & u_2 & \cdots & u_n \end{bmatrix} = \begin{bmatrix} 1 & 0 & \cdots & 0 \\ 0 & 1 & \ddots & \vdots \\ \vdots & \ddots & \ddots & 0 \\ 0 & \cdots & 0 & 1 \end{bmatrix}$$

implying that

$$pA^i u_j = \begin{cases} 0 & \text{if } \; i \neq j-1 \\ 1 & \text{if } \; i = j-1 \end{cases}$$

This allows us to compute

$$\begin{aligned} U^{-1}AU &= \begin{bmatrix} p \\ pA \\ \vdots \\ pA^{n-1} \end{bmatrix} A \begin{bmatrix} u_1 & u_2 & \cdots & u_n \end{bmatrix} \\ &= \begin{bmatrix} pA \\ pA^2 \\ \vdots \\ pA^n \end{bmatrix} \begin{bmatrix} u_1 & u_2 & \cdots & u_n \end{bmatrix} \end{aligned}$$

$$
= \begin{bmatrix} 0 & 1 & 0 & \cdots & 0 \\ & 0 & 1 & \ddots & \vdots \\ \vdots & & \ddots & \ddots & 0 \\ 0 & \cdots & & 0 & 1 \\ pA^n u_1 & pA^n u_2 & & \cdots & pA^n u_n \end{bmatrix} \tag{8.39}
$$

In order to simplify the bottom row of this matrix, recall that the Cayley-Hamilton theorem tells us that $A^n + a_{n-1}A^{n-1} + \cdots + a_1 A + a_0 I = 0$, or

$$
A^n = -a_{n-1}A^{n-1} - \cdots - a_1 A - a_0 I
$$

where the coefficients a_i, $i = 0, \ldots, n-1$ come from the characteristic polynomial of the system. Therefore (8.39) becomes

$$
U^{-1}AU = \begin{bmatrix} 0 & 1 & 0 & \cdots & 0 \\ & 0 & 1 & \ddots & \vdots \\ \vdots & & \ddots & \ddots & 0 \\ 0 & \cdots & & 0 & 1 \\ -a_0 & -a_1 & & \cdots & -a_{n-1} \end{bmatrix} \tag{8.40}
$$

We see therefore that the transformation $x(t) = Ux_c(t)$ does indeed perform the conversion to controllable canonical form. This results in

$$
\begin{aligned}
\dot{x}_c(t) &= U^{-1}AUx_c(t) + U^{-1}bu(t) \\
&\stackrel{\Delta}{=} A_c x_c + b_c u \\
y(t) &= cUx_c(t) + du(t) \\
&\stackrel{\Delta}{=} c_c x_c(t) + d_c u(t)
\end{aligned} \tag{8.41}
$$

where A_c and b_c have the forms given in (8.34).

Equation (8.40) also shows another interesting property of the controllable (and observable) canonical forms: For a system that is observable and controllable, the matrix elements $a_0, \ldots, a_{n-1}$ seen in A_c [or A_o, Equation (8.35)], are exactly the same scalar coefficients of the system's characteristic polynomial, i.e.,

$$s^n + a_{n-1}s^{n-1} + \cdots + a_1 s + a_0 = 0 \tag{8.42}$$

In fact, if the c_c-matrix is given by $c_c = \begin{bmatrix} c_1 & c_2 & \cdots & c_n \end{bmatrix}$, then the transfer function of the system in (8.34) will be

$$G(s) = \frac{c_n s^{n-1} + \cdots + c_2 s + c_1}{s^n + a_{n-1}s^{n-1} + \cdots + a_1 s + a_0} + d \tag{8.43}$$

The reason that the system must be both observable and controllable for this to work will become apparent in Chapter 9.

One may correctly guess that an analogous procedure is used for the observable canonical form, using the *first column* of the inverse observability matrix, etc. The result in Equation (8.43) pertaining to the transfer function also follows the analogy, using the b_c-matrix entries as the numerator of the transfer function instead of the c_c-matrix.

8.2.5 Similarity Transformations and Controllability

Before considering other aspects of controllability and observability, we should make an observation about the similarity transformations necessary to change the basis of a system, for example, to transform a system to a canonical form. It is perhaps not obvious that a system's controllability and observability properties are invariant under a full-rank transformation, but this is indeed the case. Consider a linear system with controllability matrix

$$P = \left[B \mid AB \mid \cdots \mid A^{n-1}B \right] \tag{8.44}$$

Under a similarity transformation, say $x = M\tilde{x}$, we will have transformed matrices

$$\tilde{A} = M^{-1}AM \qquad \tilde{B} = M^{-1}B$$

Therefore, in this new basis, the controllability test will give

$$\begin{aligned} \tilde{P} &= \begin{bmatrix} \tilde{B} & \tilde{A}\tilde{B} & \tilde{A}^2\tilde{B} & \cdots \end{bmatrix} \\ &= \begin{bmatrix} M^{-1}B & M^{-1}AMM^{-1}B & M^{-1}AMM^{-1}AMM^{-1}B & \cdots \end{bmatrix} \\ &= \begin{bmatrix} M^{-1}B & M^{-1}AB & M^{-1}A^2B & \cdots \end{bmatrix} \end{aligned}$$

$$
\begin{aligned}
&= \left[M^{-1}B \quad M^{-1}AB \quad \cdots \quad M^{-1}A^{n-1}B \right] \\
&= M^{-1}\left[B \quad AB \quad \cdots \quad A^{n-1}B \right] \\
&= M^{-1}P
\end{aligned}
\tag{8.45}
$$

Because M is nonsingular, $r(\tilde{P}) = r(P)$. An analogous process shows that $r(\tilde{Q}) = r(Q)$. Thus, a system that is controllable (observable) will remain so after the application of a similarity transformation.

So far, we have spoken of systems as being only controllable (observable) or not. In the next section, we will consider uncontrollable (unobservable) systems and investigate *degrees* of controllability (observability). We will consider an uncontrollable (unobservable) system and ask: Is the system controllable (observable) *in part*? If so, what part?

8.3 Modal Controllability and Observability

Often control systems engineers are interested in the controllability and observability of the individual modes of a system, as defined and discussed in Section 6.3. Because in Jordan form the modes corresponding to the different eigenvalues are readily apparent and the system appears as a collection of decoupled blocks, it is sometimes useful to speak of controllable modes and observable modes (as well as uncontrollable modes and unobservable modes). In Example 8.2, we see that only the modes corresponding to eigenvalue $\lambda = 0$ are uncontrollable; all the other modes are controllable. Likewise, inspection of the columns of $\overline{C}$ shows that the modes corresponding to eigenvalues $\lambda = 3$ and $\lambda = -4$ are observable, but the modes corresponding to eigenvalues $\lambda = -5$ and $\lambda = 0$ are unobservable.

8.3.1 Geometric Interpretation of Modal Controllability

To understand how one part of a system can be controllable and/or observable and another part uncontrollable and/or unobservable, we consider in this section the concept of a controllable subspace and an observable subspace. It is helpful to first consider systems defined in discrete-time. From Examples 3.10 and 3.11, we may interpret the controllability matrix $P = [B \quad AB \quad \cdots \quad A^{n-1}B]$ as a mapping (operator) that takes the input sequence into the state space, by Equation (3.55). We have pictured this feedback relationship in Figure 8.3. From the feedback configuration, it can be seen that the state $x(k+1)$ is affected not only by $Bu(k)$, but, tracing through the feedback loop, it is also affected by $ABu(k)$, etc. By the Cayley-Hamilton theorem, there is no need to consider more than $n-1$ such passes around the feedback path.

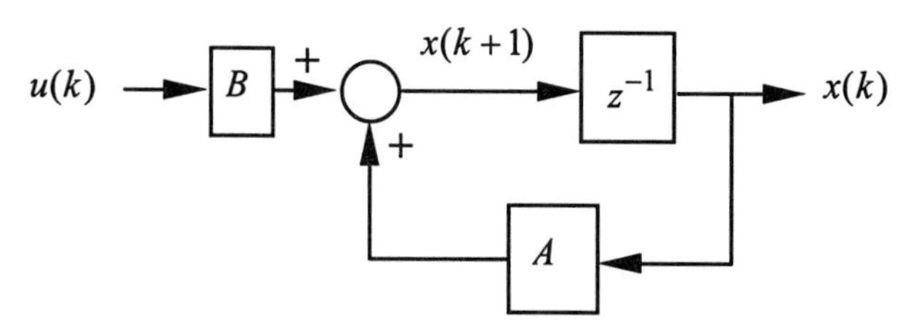

Figure 8.3 Feedback mechanism for discrete-time systems.

If we wish the system to be controllable and we do not know the initial state, then we surmise that the input signal might need to have some components, when mapped by B and traversing around the feedback loop up to $n-1$ passes, in every possible direction of the state space. Requiring that $\{u, Bu(t), ABu(t), \ldots, A^{n-1}Bu(t)\}$ span the entire state space is synonymous to the condition that $r(P)=n$. One might picture this on a phase plane as the ability to "push" the state vector around in all possible directions.

Now thinking about a similar system in continuous-time, we no longer have a matrix equation like (3.55) that maps discrete samples of $u(t)$ into the state space. Rather, the input feeds continuously into the state space and is integrated by the convolution integral (6.6) before we arrive at an expression for the state. This situation is pictured in Figure 8.4.

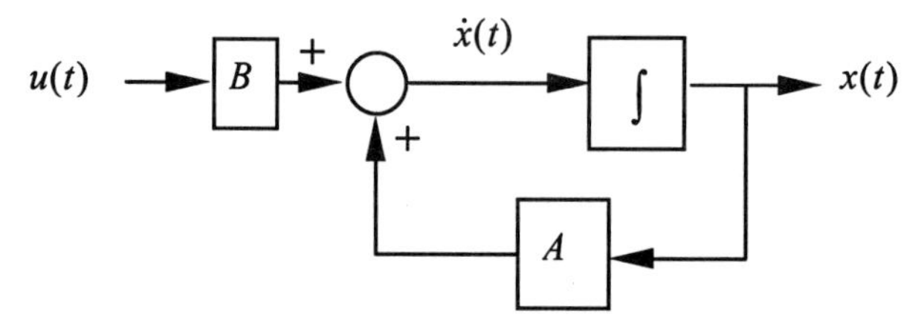

Figure 8.4 Feedback mechanism for continuous-time systems.

However, as we showed in the proof of the fundamental controllability test, the Cayley-Hamilton theorem may be used to reduce this continuous dependence on time into a dependence on the n values Γ_i, Equation (8.10), as mapped into the state space by the same controllability matrix P.

To state this interpretation in terms of subspaces, denote by β the range space of B: $\beta = \mathrm{R}(B)$. Defining the sum of two spaces X and Y as

$$\mathrm{X}+\mathrm{Y}=\{\boldsymbol{x}+\boldsymbol{y} \mid \boldsymbol{x} \in \mathrm{X}, \boldsymbol{y} \in \mathrm{Y}\}$$

we can define the controllable subspace of a system as

$$\mathrm{X}_c \overset{\Delta}{=} \beta + A\beta + \cdots + A^{n-1}\beta \tag{8.46}$$

Therefore, if the system is completely controllable, then $\mathrm{X}_c = \mathrm{X}$, or, the controllable subspace is the entire state space. This subspace X_c can also be interpreted as the smallest A-invariant subspace of X that contains β. This also leaves room to define the *uncontrollable* subspace as $\mathrm{X}_{\bar{c}} = \mathrm{X} - \mathrm{X}_c$. For systems that are not controllable, the span of the set of vectors $\{B, AB, \ldots, A^{n-1}B\}$ has a particular dimension, which we will call n_c. Obviously, $n_c = r([B \quad AB \quad \cdots \quad A^{n-1}B])$, and $n_{\bar{c}} = \dim(\mathrm{X}_{\bar{c}}) = n - n_c$.

We consider observable subspaces a little differently. We will define the *un*observable subspace as

$$\mathrm{X}_{\bar{o}} \overset{\Delta}{=} \bigcap_{i=1}^{n} \mathrm{N}(CA^{i-1}) \tag{8.47}$$

Therefore, the entire system will be observable if and only if

$$\bigcap_{i=1}^{n} \mathrm{N}(CA^{i-1}) = \varnothing \tag{8.48}$$

As with the controllable and uncontrollable subspaces, we define $\mathrm{X}_o \overset{\Delta}{=} \mathrm{X} - \mathrm{X}_{\bar{o}}$ and give the appropriate subspace dimensions as

$$n_{\bar{o}} \overset{\Delta}{=} \dim(\mathrm{X}_{\bar{o}}) = \dim\left(\bigcap_{i=1}^{n} \mathrm{N}(CA^{i-1})\right) \qquad n_o \overset{\Delta}{=} n - n_{\bar{o}}$$

It can be shown that $\mathrm{X}_{\bar{o}} = \mathrm{N}(Q)$, and naturally, $n_o = r(Q)$, where Q is the observability matrix. For a physical interpretation of this result, we envision the null spaces of C, CA, etc. If a state vector cannot be "observed" through the relationship $y = Cx + Du$ because it is in the null space of C, then, by the feedback diagrams above, it is perhaps possible to observe it through CA, etc. If a state vector cannot be observed through *any* of these mappings, then of course it cannot be observed at all. Geometrically, $\mathrm{X}_{\bar{o}}$ is the largest A-invariant subspace of X that is entirely contained in $\mathrm{N}(C)$.

8.3.2 Kalman Decompositions

In cases wherein some modes are controllable and/or observable and others are not, we may wish to group the controllable and observable modes together. Of course, there will be four possible groupings of modes: controllable and observable; controllable and unobservable; uncontrollable and observable; and uncontrollable and unobservable. It is possible to find a transformation matrix that accomplishes such groupings. We will present such a transformation based on the geometric interpretations introduced above.

Suppose it is known that for the SISO* LTI system (continuous or discrete)

$$\begin{aligned} \dot{x} &= Ax + bu \\ y &= cx + du \end{aligned} \tag{8.49}$$

that $r(P) = n_c < n$ from the controllability analysis and that $r(Q) = n_o < n$ from observability analysis, i.e., the system is neither controllable nor observable. What we seek in this section is a transformation that rearranges the state variables in such a way that the modes that are both controllable and observable are grouped together, the modes that are neither controllable nor observable are grouped together, the modes that are controllable but not observable are grouped together, and the modes that are observable but not controllable are grouped together. Such a transformation is called a *Kalman decomposition* [6].

`ctrbf(A,B,C)`

First we consider separating only the controllable subspace from the rest of the state space.^{M} To do this, we construct a change of basis matrix whose first n_c columns are a maximal set of n_c linearly independent columns from controllability matrix P. The remaining $n - n_c$ columns do not matter, except, of course, that they need to be linearly independent of the first n_c columns that span the controllable subspace and linearly independent themselves. We therefore construct the change of basis as the nonsingular $n \times n$ operator

$$V = \begin{bmatrix} v_1 & \cdots & v_{n_c} & | & v_{n_c+1} & \cdots & v_n \end{bmatrix} \tag{8.50}$$

By creating a new state vector $x = Vw$, we will get the transformed system with the form

$$\begin{aligned} \dot{w} &= V^{-1}AVw + V^{-1}bu \\ y &= CVw + du \end{aligned}$$

which has the decomposed structure

---

* MIMO systems present special cases, because different inputs (outputs) may be more or less useful in controlling (observing) the system. Multivariable systems are not discussed here, but are in Chapter 10.

$$\begin{bmatrix} \dot{w}_1 \\ \dot{w}_2 \end{bmatrix} = \begin{bmatrix} A_c & A_{12} \\ 0 & A_{\bar{c}} \end{bmatrix} \begin{bmatrix} w_1 \\ w_2 \end{bmatrix} + \begin{bmatrix} b_c \\ 0 \end{bmatrix} u$$
$$y = \begin{bmatrix} c_c & c_{\bar{c}} \end{bmatrix} \begin{bmatrix} w_1 \\ w_2 \end{bmatrix} + du \tag{8.51}$$

In this form, we use the notation (A_c, b_c) to denote the $n_c \times n_c$ subsystem consisting of the first n_c state variables w_1. Because of the way we constructed the transformation V, Equation (8.50), we know that this subsystem (A_c, b_c) is controllable, and the remainder of the system is not controllable. The uncontrollability of the second subsystem, corresponding to the state variables w_2, is obvious from the structures of the system and input matrices in (8.51). Not only is there no direct mapping of the input $u(t)$ into the state space, but there is also no coupling into this second subsystem from the controllable subsystem.

A similar procedure may be used to decompose a system into observable and unobservable parts[M]: First find the n_o-dimensional observability matrix Q. Gather n_o linearly independent *rows* from Q and append to these any $n - n_o$ other rows that are independent of the first n_o in order to create a nonsingular operator:

`obsvf(A,B,C)`

$$U = \begin{bmatrix} u_1 \\ \vdots \\ u_{n_o} \\ \hline u_{n_o+1} \\ \vdots \\ u_n \end{bmatrix}$$

With the change of basis $z = Ux$ (note the distinction between this transformation and the previous one $x = Vw$), a new system may be derived with the form

$$\begin{bmatrix} \dot{z}_1 \\ \dot{z}_2 \end{bmatrix} = \begin{bmatrix} A_o & 0 \\ A_{21} & A_{\bar{o}} \end{bmatrix} \begin{bmatrix} \dot{z}_1 \\ \dot{z}_2 \end{bmatrix} + \begin{bmatrix} b_o \\ b_{\bar{o}} \end{bmatrix} u$$
$$y = \begin{bmatrix} c_o & 0 \end{bmatrix} \begin{bmatrix} z_1 \\ z_2 \end{bmatrix} + du \tag{8.52}$$

Here again, it is clear that the subsystem consisting of only the state variable components z_2 is in the null space of the observability matrix in the new basis and will therefore be unobservable.

By carefully applying a sequence of such controllability and observability transformations,* the complete Kalman decomposition will result in a system of the block-form:

$$\begin{bmatrix} \dot{x}_{co} \\ \dot{x}_{c\bar{o}} \\ \dot{x}_{\bar{c}o} \\ \dot{x}_{\bar{c}\bar{o}} \end{bmatrix} = \begin{bmatrix} A_{co} & 0 & A_{13} & 0 \\ A_{21} & A_{c\bar{o}} & A_{23} & A_{24} \\ 0 & 0 & A_{\bar{c}o} & 0 \\ 0 & 0 & A_{43} & A_{\bar{c}\bar{o}} \end{bmatrix} \begin{bmatrix} x_{co} \\ x_{c\bar{o}} \\ x_{\bar{c}o} \\ x_{\bar{c}\bar{o}} \end{bmatrix} + \begin{bmatrix} b_{co} \\ b_{c\bar{o}} \\ 0 \\ 0 \end{bmatrix} u$$

$$y = \begin{bmatrix} c_{co} & 0 & c_{\bar{c}o} & 0 \end{bmatrix} \begin{bmatrix} x_{co} \\ x_{c\bar{o}} \\ x_{\bar{c}o} \\ x_{\bar{c}\bar{o}} \end{bmatrix} + du$$

In this form, the four distinct subsystems are appropriately grouped and denoted, e.g., $(A_{\bar{c}o}, c_{\bar{c}o})$ represents the subsystem that is not controllable but is observable. There are also coupling terms, e.g., A_{43}, but these do not affect the controllability and observability properties of the subsystems. These dependencies are illustrated in the block diagram of Figure 8.5. In this figure, the dependencies of the subsystems on the input and the dependency of the output on the subsystems are all apparent. We will see the further significance of the Kalman decomposition in Chapter 9.

Example 8.3: Decomposition of an Uncontrollable System

Given the system

$$\dot{x} = \begin{bmatrix} 2 & 1 & 1 \\ 5 & 3 & 6 \\ -5 & -1 & -4 \end{bmatrix} x + \begin{bmatrix} 1 \\ 0 \\ 0 \end{bmatrix} u$$
$$y = \begin{bmatrix} 1 & 1 & 2 \end{bmatrix} x$$

determine whether the system is controllable and/or observable. If it is not controllable, perform a Kalman decomposition to separate controllable and uncontrollable subsystems.

---

* It is not simply a matter of applying one transformation after another. The full procedure is left as an exercise.

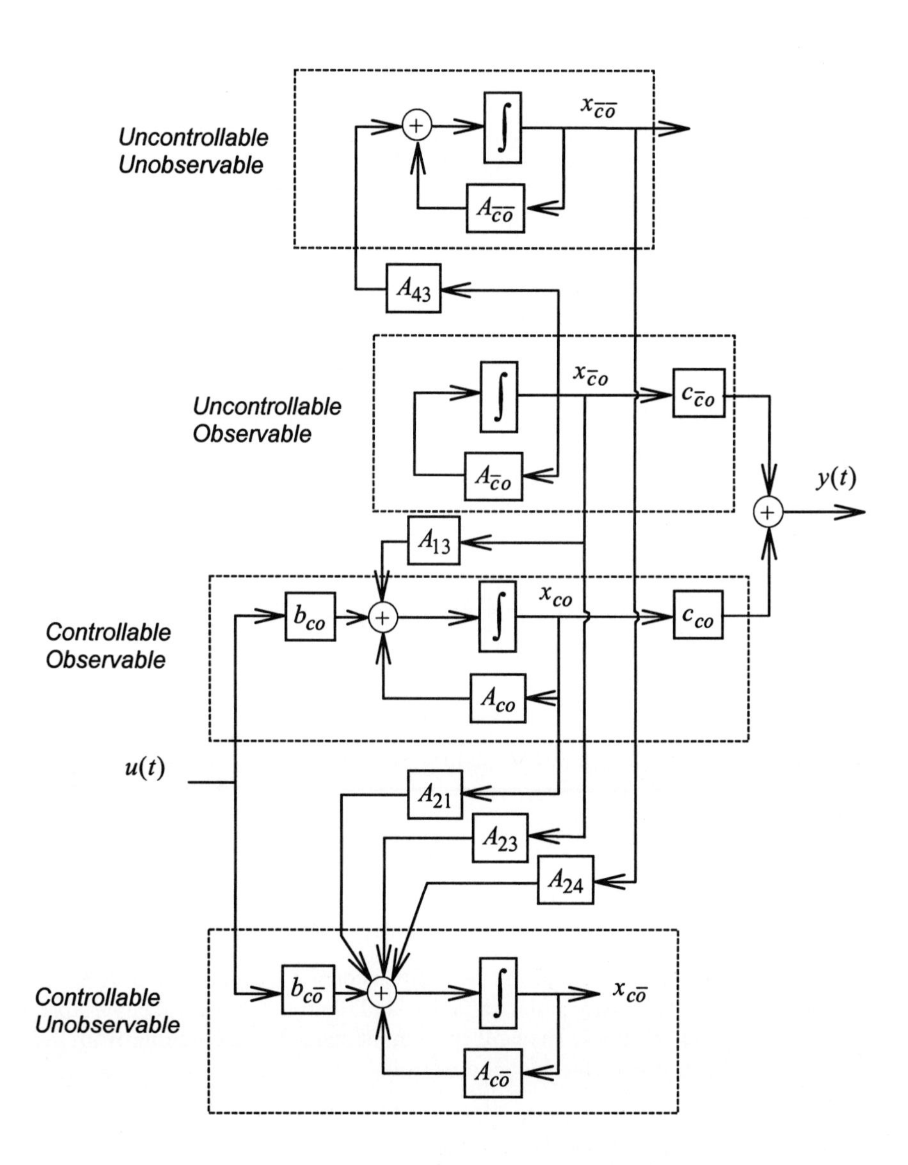

Figure 8.5 Block diagram of a system in Kalman decomposition.

Solution:

To first determine controllability and observability, we compute

$$P = \begin{bmatrix} b & Ab & A^2b \end{bmatrix} = \begin{bmatrix} 1 & 2 & -4 \\ 0 & 5 & -5 \\ 0 & -5 & 5 \end{bmatrix}$$

and

$$Q = \begin{bmatrix} c \\ cA \\ cA^2 \end{bmatrix} = \begin{bmatrix} 1 & 1 & 2 \\ -3 & 2 & -1 \\ 9 & 4 & 13 \end{bmatrix}$$

It is then a simple matter to determine that $r(P) = 2$ and $r(Q) = 2$, so this system is neither controllable nor observable.

To decompose into controllable and uncontrollable parts, we note that the first two columns of P are linearly independent, so by appending a third linearly independent column, we can construct a suitable similarity transformation:

$$V = \left[\begin{array}{cc|c} 1 & 2 & 0 \\ 0 & 5 & 0 \\ 0 & -5 & 1 \end{array}\right]$$

By now performing the similarity transformation $x = Vw$, we compute

$$A_w = V^{-1}AV = \left[\begin{array}{cc|c} 0 & 6 & -1.4 \\ 1 & -1 & 1.2 \\ \hline 0 & 0 & 2 \end{array}\right] \qquad b_w = V^{-1}b = \left[\begin{array}{c} 1 \\ 0 \\ \hline 0 \end{array}\right] \qquad c_w = cV = \left[\begin{array}{cc|c} 1 & -3 & 2 \end{array}\right]$$

From this structure, we see that the two-dimensional controllable subsystem is grouped into the first two state variables and that the third state variable, represented by the '2' block, has neither an input signal nor a coupling from the other subsystem. It is therefore not controllable.

8.3.3 Stabilizability and Detectability

Although we have not yet considered the problems of controller and observer design, we have noted in Chapter 7 that the primary goal in control system design is stability. We will see in Chapter 10, as we might suspect from the terminology, that controllability of a mode is a prerequisite for successful controller design and observability of a mode is a prerequisite for observer design. If indeed the designer's primary design criterion is to achieve system stability, however, it is sometimes not necessary for all modes to be controllable and/or observable. This may be the case if the uncontrollable modes are already

Lyapunov stable and the unobservable modes are also Lyapunov stable, thus leading to the following two simple definitions:

> ***Stabilizability:*** A system is *stabilizable* if its uncontrollable modes, if any, are stable. Its controllable modes may be stable or unstable. (8.53)

> ***Detectability:*** A system is *detectable* if its unobservable modes, if any, are stable. Its observable modes may be stable or unstable. (8.54)

8.4 Controllability and Observability of Time-Varying Systems

Controllability and observability studies of time-varying systems are sufficiently different from those of time-invariant systems that they are treated separately in this section. Although the definitions of controllability and observability are basically the same as for time-invariant systems, the tests are considerably different. As was the case for stability, we must be careful not to conclude anything about controllability and observability properties from the "frozen-time" equations of the system. The "duality" of controllability and observability is preserved for the time-varying case, so again we will omit explicit derivations and proofs for observability when they are sufficiently parallel to those for controllability.

8.4.1 Continuous-Time Systems

Consider the time-varying system

$$\begin{aligned} \dot{x}(t) &= A(t)x(t) + B(t)u(t) \\ y(t) &= C(t)x(t) \end{aligned} \tag{8.55}$$

(Because feedthrough affects neither controllability nor observability, we will simplify our equations by dropping the Du term.) We have seen that in the time-invariant case, the fundamental controllability test was derived from a linear independence test on the columns of a matrix that was a function of A and B. For our first test of controllability for time-varying systems, we draw from the same concept; however, the similarities end soon thereafter.

> THEOREM: The state equations in (8.55) are *controllable* in an interval $[t_0, t_1]$ if the rows of the matrix product

$\Phi(t_0,\tau)B(\tau)$, where $\Phi(t_0,\tau)$ is the state-transition matrix, are linearly independent. (8.56)

PROOF: We know from the results of Chapter 2 that linear independence of time functions is established through the testing of a grammian matrix. In this case, the statement that the rows of $\Phi(t_0,\tau)B(\tau)$ are linearly independent is equivalent to saying that the so-called "controllability grammian"[M]

```
gram(sys,'c')
```

$$G_c(t_0,t_1) \triangleq \int_{t_0}^{t_1} \Phi(t_0,\tau)B(\tau)B^{\mathrm{T}}(\tau)\Phi^{\mathrm{T}}(t_0,\tau)\,d\tau \qquad (8.57)$$

is nonsingular, i.e., invertible. To begin the proof, we suppose (8.57) is invertible. Then we may consider an input to the system of the form

$$u(t) = -B^{\mathrm{T}}(t)\Phi^{\mathrm{T}}(t_0,t)G_c^{-1}(t_0,t_1)x(t_0) \qquad (8.58)$$

assuming, as we must, that $x(t_0)$ is known. This being the input, we can use the convolution integral solution to determine the state at the end of the interval, $x(t_1)$:

$$\begin{aligned} x(t_1) &= \Phi(t_1,t_0)x(t_0) \\ &\quad -\int_{t_0}^{t_1} \Phi(t_1,\tau)B(\tau)B^{\mathrm{T}}(\tau)\Phi^{\mathrm{T}}(t_0,\tau)\,G_c^{-1}(t_0,t_1)x(t_0)\,d\tau \end{aligned}$$

Factoring $\Phi(t_1,t_0)$ from the left side of this expression (recalling the properties of state-transition matrices),

$$\begin{aligned} x(t_1) &= \Phi(t_1,t_0)\left[x(t_0) - \int_{t_0}^{t_1} \Phi(t_0,\tau)B(\tau)B^{\mathrm{T}}(\tau)\Phi^{\mathrm{T}}(t_0,\tau)\,d\tau\, G_c^{-1}(t_0,t_1)x(t_0)\right] \\ &= \Phi(t_1,t_0)\left[x(t_0) - G_c(t_0,t_1)G_c^{-1}(t_0,t_1)x(t_0)\right] \\ &= \Phi(t_1,t_0)\left[x(t_0) - x(t_0)\right] \\ &= 0 \end{aligned}$$

This implies that the (cleverly chosen) control signal in (8.58) can force the system to reach the zero state in finite time.

Now we suppose that the system is controllable and show that the controllability grammian $G_c(t_0,t_1)$ is invertible. We do this by contradiction. If $G_c(t_0,t_1)$ were *not* invertible, then it would, by definition, be rank-deficient, and would therefore have a nontrivial null space. Therefore, there would exist a nonzero vector z such that $G_c(t_0,t_1)z=0$ and likewise $z^{\mathrm{T}}G_c(t_0,t_1)z=0$. Explicitly,

$$\begin{aligned} z^{\mathrm{T}}G_c(t_0,t_1)z &= \int_{t_0}^{t_1} z^{\mathrm{T}}\Phi(t_0,\tau)B(\tau)B^{\mathrm{T}}(\tau)\Phi^{\mathrm{T}}(t_0,\tau)z\,d\tau \\ &= \left\|B^{\mathrm{T}}(\tau)\Phi^{\mathrm{T}}(t_0,\tau)z\right\|^2 \\ &= 0 \end{aligned}$$

implying

$$z^{\mathrm{T}}\Phi(t_0,\tau)B(\tau)=0 \tag{8.59}$$

If the system is, as we assumed, controllable, then it is possible to find an input $u_z(t)$ that, when applied from an arbitrary initial state, transfers the system to the zero state at time t_1. Selecting $x(t_0)=z$, we have

$$0=\Phi(t_1,t_0)z+\int_{t_0}^{t_1}\Phi(t_1,\tau)B(\tau)u_z(\tau)\,d\tau$$

Solving this equation for z:

$$\begin{aligned} z &= -\Phi^{-1}(t_1,t_0)\int_{t_0}^{t_1}\Phi(t_1,\tau)B(\tau)u_z(\tau)\,d\tau \\ &= -\int_{t_0}^{t_1}\Phi(t_0,\tau)B(\tau)u_z(\tau)\,d\tau \end{aligned} \tag{8.60}$$

Multiplying this equation through from the left by z^{T} and applying Equation (8.59) results in

$$z^{\mathrm{T}}z=-\int_{t_0}^{t_1}z^{\mathrm{T}}\Phi(t_0,\tau)B(\tau)u_z(\tau)\,d\tau=0$$

However, this can only be true if $z \equiv 0$. Therefore a nonzero z such that $G_c(t_0,t_1)z = 0$ does not exist, and it follows that $G_c(t_0,t_1)$ must be invertible.

As has been repeatedly noted, any operation involving the state-transition matrix $\Phi(t_1,t_0)$ is difficult. Therefore, the computation of the controllability grammian $G_c(t_0,t_1)$ from its definition in (8.57) is not recommended and in fact, is not always possible. Instead, it is a straightforward application of calculus to find a differential equation for which $G_c(t_0,t_1)$ is a solution:

$$\begin{aligned}
\frac{d}{dt}G_c(t,t_1) &= \frac{d}{dt}\left[\int_t^{t_1} \Phi(t,\tau)B(\tau)B^{\mathrm{T}}(\tau)\Phi^{\mathrm{T}}(t,\tau)\,d\tau\right] \\
&= \int_t^{t_1} \dot{\Phi}(t,\tau)B(\tau)B^{\mathrm{T}}(\tau)\Phi^{\mathrm{T}}(t,\tau)\,d\tau \\
&\quad + \int_t^{t_1} \Phi(t,\tau)B(\tau)B^{\mathrm{T}}(\tau)\dot{\Phi}^{\mathrm{T}}(t,\tau)\,d\tau - \Phi(t,\tau)B(\tau)B^{\mathrm{T}}(\tau)\Phi^{\mathrm{T}}(t,\tau)\Big|_{\tau=t} \\
&= A(t)\int_t^{t_1} \Phi(t,\tau)B(\tau)B^{\mathrm{T}}(\tau)\Phi^{\mathrm{T}}(t,\tau)\,d\tau \\
&\quad + \int_t^{t_1} \Phi(t,\tau)B(\tau)B^{\mathrm{T}}(\tau)\Phi^{\mathrm{T}}(t,\tau)\,d\tau\, A^{\mathrm{T}}(t) - B(t)B^{\mathrm{T}}(t) \\
&= A(t)G_c(t,t_1) + G_c(t,t_1)A^{\mathrm{T}}(t) - B(t)B^{\mathrm{T}}(t)
\end{aligned}$$

so

$$\frac{d}{dt}G_c(t,t_1) = A(t)G_c(t,t_1) + G_c(t,t_1)A^{\mathrm{T}}(t) - B(t)B^{\mathrm{T}}(t) \tag{8.61}$$

for which numerical solutions may be sought.

Clearly, Equation (8.61) is more convenient for the computation of $G_c(t_0,t_1)$ than is (8.57) because (8.61) does not require knowledge of the state-transition matrix. However, solution of (8.61), especially in closed form, is not easy either. Another method for testing controllability of time-varying systems is based on the following definitions [8]: Let $M_0(t) \triangleq B(t)$ and

$$M_j(t) \triangleq -A(t)M_{j-1}(t) + \dot{M}_{j-1}(t) \tag{8.62}$$

for $j = 1, \ldots, n$. [This definition requires that $A(t)$ and $B(t)$ be continuously differentiable at least $n-1$ times.] Then we can state the following result:

THEOREM: The system (8.55) is controllable in the interval $[t_0, t_1]$ if there exists a $\tau \in [t_0, t_1]$ such that

$$r\left(\left[M_0(\tau) \mid M_1(\tau) \mid \cdots \mid M_{n-1}(\tau)\right]\right) = n \tag{8.63}$$

PROOF: (By contradiction) Suppose (8.63) holds for some τ yet the system is not controllable. Then the controllability grammian $G_c(t_0, t_1)$ is not invertible, and there will exist as a consequence a nonzero vector z such that

$$z^{\mathrm{T}} \Phi(t_0, t) B(t) = 0 \tag{8.64}$$

for $t \in [t_0, t_1]$. Define another vector $\bar{z} \stackrel{\Delta}{=} \Phi^{\mathrm{T}}(t_0, \tau) z$. Then (8.64) gives

$$\begin{aligned} z^{\mathrm{T}} \Phi(t_0, t) B(t) &= \bar{z}^{\mathrm{T}} \Phi(\tau, t_0) \Phi(t_0, t) B(t) \\ &= \bar{z}^{\mathrm{T}} \Phi(\tau, t) B(t) \\ &= 0 \end{aligned}$$

for any $t \in [t_0, t_1]$. Supposing that $t = \tau$, this gives

$$\begin{aligned} \bar{z}^{\mathrm{T}} \Phi(\tau, \tau) B(\tau) &= \bar{z}^{\mathrm{T}} B(\tau) \\ &= \bar{z}^{\mathrm{T}} M_0(\tau) \\ &= 0 \end{aligned} \tag{8.65}$$

Now differentiating (8.64) and evaluating at $t = \tau$,

$$\begin{aligned} \frac{d}{dt}\left(z^{\mathrm{T}} \Phi(t_0, t) B(t)\right)\Big|_{t=\tau} &= z^{\mathrm{T}}\left[\dot{\Phi}(t_0, t) B(t) + \Phi(t_0, t) \dot{B}(t)\right]\Big|_{t=\tau} \\ &= z^{\mathrm{T}}\left[-\Phi(t_0, t) A(t) B(t) + \Phi(t_0, t) \dot{B}(t)\right]\Big|_{t=\tau} \\ &= z^{\mathrm{T}} \Phi(t_0, t)\left[-A(t) B(t) + \dot{B}(t)\right]\Big|_{t=\tau} \\ &= z^{\mathrm{T}} \Phi(t_0, \tau) M_1(\tau) \\ &= \bar{z}^{\mathrm{T}} M_1(\tau) \\ &= 0 \end{aligned} \tag{8.66}$$

where the fact that $\dot{\Phi}(t_0,t) = -\Phi(t_0,t)A(t)$ is a result of applying what we learned in Exercise 3.2 to find $\frac{d}{dt}\left(\Phi^{-1}(t,t_0)\right)$. We can take a second derivative of (8.64), which is most easily accomplished by taking the first derivative of the fourth line of (8.66) above. This will lead to

$$\begin{aligned}\frac{d^2}{dt^2}\left(z^{\mathrm{T}}\Phi(t_0,t)B(t)\right)\Big|_{t=\tau} &= \frac{d}{dt}\left(z^{\mathrm{T}}\Phi(t_0,t)M_1(t)\right)\Big|_{t=\tau}\\ &= \bar{z}^{\mathrm{T}}M_2(\tau)\\ &= 0\end{aligned}$$

by a similar process. Continuing a total of $n-1$ times, we will develop the expression

$$\bar{z}^{\mathrm{T}}\left[M_0(\tau) \quad M_1(\tau) \quad \cdots \quad M_{n-1}(\tau)\right] = 0$$

which contradicts the linear independence implied by the assertion of (8.63). Therefore, if (8.63) holds, then the system is controllable.

We might note at this point the parallel alternative route this theorem provides to the fundamental controllability test. If the system in question is time-invariant, then in (8.62), $\dot{M}_j(t) \equiv 0$, and the rank test will reduce to

$$r\left[B \mid -AB \mid A^2B \mid \cdots \mid (-1)^{n-1}A^{n-1}B\right] \stackrel{?}{=} n \tag{8.67}$$

However, if the matrices A and B are constant, then the rank of the above matrix is not affected by the alternating negative signs, and the controllability matrix P is apparent. Likewise, there is a route to the fundamental controllability test from the controllability grammian condition, which exploits the series expansion for

$$\Phi(t,t_0) = e^{A(t-t_0)}$$

but we will not belabor the point by presenting that derivation here.

Example 8.4: Controllability for a Time-Varying System

Test the system below for controllability over the interval $t \in [0,1]$:

$$\dot{x}(t)=\begin{bmatrix}0 & t\\ 1 & t\end{bmatrix}x(t)+\begin{bmatrix}2\\ t\end{bmatrix}u(t)$$

Solution:

Using the test in (8.63), we generate the matrix

$$\begin{bmatrix}M_0(t) & M_1(t)\end{bmatrix}=\begin{bmatrix}2 & -t^2\\ t & -(t^2+1)\end{bmatrix}$$

To test the rank of this matrix, we can check the determinant:

$$\det\begin{bmatrix}2 & -t^2\\ t & -(t^2+1)\end{bmatrix}=t^3-2t^2-2$$

Because the equation $t^3-2t^2-2=0$ does not hold identically in $t\in[0,1]$, we can conclude that the system is controllable in that interval.

Analogous Observability Results

In the development of the controllability results for time-varying systems above, we neglected to point out the corresponding tests for observability. This was so that the sometimes algebraically involved derivations and proofs did not get interrupted by such dual issues. We have instead gathered them here, where with one exception, we present them without detailed derivation or proof. Once again, the definition of observability is the same for time-varying systems as it is for time-invariant systems, and the results generated here apply equally well for MIMO systems as they do for SISO systems.

That exception is in the definition and use of the *observability grammian*[M] to show observability. In our proof that an invertible controllability grammian implies controllability, we relied on a cleverly chosen input, Equation (8.58), that drives the system to the origin. Naturally, in the case of observability, the control input is irrelevant, so the system's output is not a variable that we are free to specify. Therefore, the analogous observability result is obtained a little differently.

```
gram(sys,'o')
```

THEOREM: The linear, time-varying system in Equation (8.55) is observable in the interval $t\in[t_0,t_1]$ if the columns of the matrix $C(t)\Phi(t,t_0)$ are linearly independent. (8.68)

PROOF: As before, this linear independence condition is equivalent to the invertibility of the matrix we call the observability Grammian:

$$G_o(t_0,t_1) \stackrel{\Delta}{=} \int_{t_0}^{t_1} \Phi^{\mathrm{T}}(\tau,t_0)C^{\mathrm{T}}(\tau)C(\tau)\Phi(\tau,t_0)\, d\tau \qquad (8.69)$$

In the absence of an input, the output of a system is given by

$$y(t) = C(t)\Phi(t,t_0)x(t_0)$$

Multiplying both sides of this equation by $\Phi^{\mathrm{T}}(t,t_0)C^{\mathrm{T}}(t)$ and integrating from $t = t_0$ to $t = t_1$ give

$$\begin{aligned}\int_{t_0}^{t_1} \Phi^{\mathrm{T}}(t,t_0)C^{\mathrm{T}}(t)y(t)\, dt &= \int_{t_0}^{t_1} \Phi^{\mathrm{T}}(t,t_0)C^{\mathrm{T}}(t)C(t)\Phi(t,t_0)x(t_0)\, dt \\ &= G_o(t_0,t_1)x(t_0)\end{aligned} \qquad (8.70)$$

Considering the left-hand side of (8.70) to be an arbitrary vector ξ in the state space, we are left with the linear algebraic system to solve:

$$\xi = G_o(t_0,t_1)x(t_0)$$

This equation is familiar to us, being quite similar to the condition for observability of the discrete-time systems given originally in Chapter 3. For a unique solution to exist in general, we therefore require the observability grammian $G_o(t_0,t_1)$ to be invertible. This proves the sufficiency of the grammian condition. Proof of necessity is by contradiction, and is omitted here.

The observability grammian $G_o(t_0,t_1)$ satisfies the matrix differential equation

$$\frac{d}{dt}G_o(t,t_1) = -A^{\mathrm{T}}(t)G_o(t,t_1) - G_o(t,t_1)A(t) - C^{\mathrm{T}}(t)B^{\mathrm{T}}(t) \qquad (8.71)$$

For an observability test analogous to that culminating in the rank test of Equation (8.63), define $N_0(t) \triangleq C(t)$, and

$$N_j(t) \triangleq N_{j-1}(t)A(t) + \dot{N}_{j-1}(t) \tag{8.72}$$

for $j = 1, \ldots, n$.

THEOREM: The system (8.55) is observable in the interval $[t_0, t_1]$ if there exists a $\tau \in [t_0, t_1]$ such that

$$r\begin{bmatrix} N_0(\tau) \\ \text{------} \\ N_1(\tau) \\ \text{------} \\ \vdots \\ \text{------} \\ N_{n-1}(\tau) \end{bmatrix} = n \tag{8.73}$$

As usual, the duality between controllability and observability is readily apparent. As a practical matter, one can get away with only a single set of tests, i.e., for controllability or observability. Observability of the pair (A, C) is equivalent to controllability of the pair $(A^{\mathrm{T}}, C^{\mathrm{T}})$, as direct substitution into any of the above formulas will reveal. For example, if we wish to write a set of subroutines that test for controllability, a single routine suffices for both observability and controllability.

8.4.2 Reachability and Reconstructibility

Depending on the nature of the system to be controlled, it may be desirable that the final state of the system be something other than zero. For example, in nonzero set-point control, the goal is to take a system *from* the zero state *to* a nonzero state. Note that this is the opposite of the problem implied by the definition of controllability, which says that a system can be transferred from a nonzero state to the zero state. If a system can be moved from the zero initial state to an arbitrary nonzero final state, the system is said to be *reachable*. For continuous-time systems, a system is reachable if and only if it is controllable, so there is no pressing need for this distinction. However, this is not the case for discrete-time systems. We will consider discrete-time systems in the next section of this chapter.

Likewise, it may sometimes be desired not to deduce the starting state from a sequence of inputs and outputs, as specified in the observability problem, but rather to "reconstruct" the final state from a knowledge of only the output. That is, we may want to deduce the final state rather than the initial state. A system for which this is possible is called *reconstructible*. As with reachability, reconstructibility is equivalent to observability in continuous-time systems but is somewhat different in discrete-time systems. We will revisit this distinction in the next section as well.

8.5 Discrete-Time Systems

Throughout this part of the book, we have often pointed to the dualities between controllability and observability, and used these similarities as excuses for omitting some of the detailed derivations and proofs of certain properties. At times, such as in Chapter 7 on stability, we have made similar observations about the analogies between discrete-time and continuous-time systems. For most situations, this is sufficient, as the distinctions between the two domains do not warrant the time spent illustrating algebraic differences in the derivations. However, we have devoted this final section of this chapter to just such discrepancies because, in the case of controllability and observability, they are fundamentally different.

We should first point out that the examples we presented in Chapter 3 were from a somewhat simplified viewpoint because their purpose at the time was to illustrate an example of solving simultaneous equations. Here, we will revisit the discrete-time controllability and observability properties in more depth.

8.5.1 Controllability and Reachability

Recall that in Chapter 6, the solution for the state of the set of discrete-time state equations

$$\begin{aligned} x(k+1) &= A_d(k)x(k) + B_d(k)u(k) \\ y(k) &= C_d(k)x(k) + D_d(k)u(k) \end{aligned} \tag{8.74}$$

beginning with initial state $x(j_0) = x_0$ was

$$x(j) = \Psi(j, j_0)x_0 + \sum_{i=j_0+1}^{j} \Psi(j,i)B(i-1)u(i-1) \tag{8.75}$$

which has a slight change in notation from Equation (6.52). Also recall that

$$\Psi(j,k) \stackrel{\Delta}{=} \prod_{i=k}^{j-1} A_d(i))$$

To first consider the problem of *controllability* on the time interval $[j_0, j_1]$, we take the final value to be $x(j_1) = 0$ and seek the circumstances under which the equations

$$
\begin{aligned}
-\Psi(j_1, j_0)x_0 &= \sum_{i=j_0+1}^{j} \Psi(j_1, i)B(i-1)u(i-1) \\
&= \left[B(j_1-1) \;\vdots\; \Psi(j_1, j_1-1)B(j_1-2) \;\vdots\; \cdots \right. \\
&\qquad \left. \cdots \;\vdots\; \Psi(j_1, j_0+1)B(j_0)\right] \begin{bmatrix} u(j_1-1) \\ u(j_1-2) \\ \vdots \\ u(j_0) \end{bmatrix} \\
&\triangleq R_c(j_0, j_1) \begin{bmatrix} u(j_1-1) \\ u(j_1-2) \\ \vdots \\ u(j_0) \end{bmatrix}
\end{aligned}
\tag{8.76}
$$

have a solution for the vector of inputs (which is not necessarily unique). We call the matrix $R_c(j_0, j_1)$ defined in (8.76) the *reachability* matrix. In the case of a time-invariant system, as we have seen, the matrix becomes

$$R_c(j_0, j_1) = \begin{bmatrix} B_d & A_d B_d & \cdots & A_d^{j_1-1} B_d \end{bmatrix}$$

The first interesting point to illustrate is that in Chapter 3, we assumed that the left-hand side of Equation (8.76) is a generic vector in $\Re^n$, so that the requirement we reached was that $r(R_c(j_0, j_1)) = n$. In the sense that we must allow for an arbitrary initial condition x_0, this is true. However, in the discrete-time case we find situations in which $r(\Psi(j_1, j_0)) < n$, implying that the left-hand side of (8.76) may be restricted to only a proper subspace of the entire state space. This means that in such cases, the system may be controllable, in its strict definition, with $r(R_c(j_0, j_1)) < n$. In such a situation, controllability would result if $\mathrm{R}(R_c(j_0, j_1)) = \mathrm{R}(\Psi(j_1, j_0))$. Physically, this situation is possible because a rank-deficient matrix $A_d(k)$ might result in a system that is controllable to the zero state with zero input. For example, in the extreme case wherein $A_d(k) = [0]_{n \times n}$, then the state of the system will go immediately to zero with a zero input regardless of the matrix $B_d(k)$. Such a situation does not happen with rank deficient A-matrices in continuous-time because rank deficiency in continuous-time does not take the state itself to zero; rather, it makes the state's *derivative* zero, resulting in an initial state that does not change. Thus, controllability conditions in discrete-time are fundamentally different from those in continuous-time.

Another distinction between discrete-time and continuous-time controllability lies in the fact that the number of terms in the matrix $R_c(j_0, j_1)$ depends on the number of time steps in the interval $[j_0, j_1]$. Suppose that $r(\Psi(j_1, j_0)) = n_1$ so that we require $\mathrm{R}(R_c(j_0, j_1)) = \mathrm{R}(\Psi(j_1, j_0))$ for controllability. Suppose also that the system is single-input. Then the number of columns of $R_c(j_0, j_1)$ in (8.76) is equal to the number of time steps $j_1 - j_0$. Naturally, if $j_1 - j_0 < n_1$, we have no hope for controllability whatsoever. The Cayley-Hamilton theorem tells us that there is no reason to ever consider *more* than n columns for $R_c(j_0, j_1)$ [or, more generally, powers of $A_d(k)$ higher than $n-1$], just as it did in continuous-time. However, in continuous-time, the controllability matrix never has *fewer* than n columns.

Consider now the discrete-time reachability problem. In this problem, we consider that $x(j_0) = 0$ and the desired final state is $x(j_1) = x_f$. When that is the case, Equation (8.75) reduces to

$$\begin{aligned} x_f &= \sum_{i=j_0+1}^{j} \Psi(j,i)B(i-1)u(i-1) \\ &= R_c(j_0, j_1)\begin{bmatrix} u(j_1-1) \\ u(j_1-2) \\ \vdots \\ u(j_0) \end{bmatrix} \end{aligned} \tag{8.77}$$

It is clear that in this case, regarding x_f as an arbitrary vector in $\Re^n$, we *must* have

$$r(R_c(j_0, j_1)) = n \tag{8.78}$$

for reachability, hence the name *reachability matrix*. This is true regardless of the number of system inputs. [For multi-input systems, the solution of (8.76) may require a pseudoinverse, as discussed in Chapter 3.] For this reason, the property of reachability is more desirable than controllability for discrete-time systems. *A discrete-time system can be controllable without being reachable.* Note that the observation about the minimum number of time steps required still applies, just as it did for controllability. Thus, although singularity of the state-transition matrix is no longer an issue, the length of the control interval still is.

For the time-invariant case, with $A_d(k) = A_d$ and $B_d(k) = B_d$ the reachability matrix condition is

$$r(R_c) \overset{?}{=} r\left(\begin{bmatrix} B_d & A_d B_d & \cdots & A_d^{n-1} B_d \end{bmatrix}\right) = n \tag{8.79}$$

which would imply that for time-invariant systems, reachability and controllability are equivalent. Even though the concept of reachability is fundamentally distinct from controllability, this condition indicates that the test is much the same.

Reachability Grammians

As with controllability, there are alternate tests for reachability although, as in continuous-time, (8.6) or (8.7) is by far the most common. Fortunately, the discrete-time reachability grammian test is considerably easier to prove than it was in continuous-time.

Define the *reachability grammian* as the matrix[M]

`gram(sys,'c')`

$$H_c(j_0, j_1) \overset{\Delta}{=} \sum_{i=j_0}^{j_1-1} \Psi(j_1, i+1) B(i) B^{\mathrm{T}}(i) \Psi^{\mathrm{T}}(j_1, i+1) \tag{8.80}$$

We will state the reachability grammian test in the form of a theorem.

> THEOREM: The system in (8.74) is reachable on the interval $[j_0, j_1]$ if and only if the reachability grammian in (8.80) is nonsingular. (8.81)

Rather than presenting a detailed proof of this theorem, which can be constructed parallel to the proof given for continuous-time, we need merely point out that $H_c(j_0, j_1) = R_c(j_0, j_1) R_c^{\mathrm{T}}(j_0, j_1)$. Our knowledge of the rank of operators makes it immediately clear that $r(H_c(j_0, j_1)) = n$ if and only if $r(R_c(j_0, j_1)) = n$, which has already been proven a valid reachability test [7].

Example 8.5: A Discrete-Time Control Sequence

Determine whether the state equation given below is controllable. Then determine the *minimum* number of time steps necessary to take the system from

$x(0)=\begin{bmatrix}-2 & -2 & 2 & -4 & -3\end{bmatrix}^T$ to the zero state, and find an input sequence that does so.

$$x(k+1)=\begin{bmatrix}1 & 0 & 0 & 0 & 0\\ 0 & 0 & 1 & 0 & 0\\ 0 & -0.5 & -1 & 0 & 0\\ 0 & 0 & 0 & 0 & 1\\ 0 & 0 & 0 & -0.5 & 1\end{bmatrix}x(k)+\begin{bmatrix}1\\ 0\\ 1\\ 1\\ 0\end{bmatrix}u(k) \tag{8.82}$$

Solution:

First, it is a trivial computation (on a computer) to determine that

$$r(P)=r\left(\begin{bmatrix}b_d & A_d b_d & A_d^2 b_d & A_d^3 b_d & A_d^4 b_d\end{bmatrix}\right)=5$$

so that the system is clearly controllable (and reachable). This guarantees that after five time steps, the equation

$$-A_d^j x(0)=P\begin{bmatrix}u(j-1)\\ \vdots\\ u(1)\\ u(0)\end{bmatrix}$$

with $j=5$ has a unique solution for any possible $x(0)$.

However, for the given $x(0)$ in particular, we need only satisfy the requirement that

$$-A_d^j x(0)\in \mathrm{R}\left(\begin{bmatrix}b_d & A_d b_d & \cdots & A_d^{j-1} b_d\end{bmatrix}\right)$$

which might occur even if $j<n$. If we try at first *one* time step, we have the equation

$$-A_d x(0)=b_d u(0) \tag{8.83}$$

from which we find that $r([b_d])\neq r([-A_d x(0)\quad b_d])$, so (8.83) cannot have a solution. After two time steps, we must solve

$$-A_d^2x(0)=\begin{bmatrix} b_d & A_db_d \end{bmatrix}\begin{bmatrix} u(1) \\ u(0) \end{bmatrix} \tag{8.84}$$

This time, we can find that

$$r([b_d \quad A_db_d])=r\left(\begin{bmatrix} -A_d^2x(0) & b_d & A_db_d \end{bmatrix}\right) \quad (=2) \tag{8.85}$$

so that (8.85) must have a solution. That solution is easily found as

$$\begin{bmatrix} u(1) \\ u(0) \end{bmatrix}=\begin{bmatrix} b_d & A_db_d \end{bmatrix}^+\left(-A_d^2x(0)\right)=\begin{bmatrix} 1 \\ 1 \end{bmatrix}$$

using the pseudoinverse. If we apply this sequence to the original equations in (8.82) starting at $x(0)=\begin{bmatrix} -2 & -2 & 2 & -4 & -3 \end{bmatrix}^{\mathrm{T}}$, we get

$$x(1)=A_dx(0)+b_du(0)=\begin{bmatrix} -1 & 2 & 0 & 2 & 1 \end{bmatrix}^{\mathrm{T}}$$
$$x(2)=A_dx(1)+b_du(1)=\begin{bmatrix} 0 & 0 & 0 & 0 & 0 \end{bmatrix}^{\mathrm{T}}$$

Thus, although the system requires at least five steps to ensure a solution for an *arbitrary* initial condition, there are certain initial conditions which can be "controlled" with fewer steps.

8.5.2 Observability and Reconstructibility

We will see in this section that although the concepts of observability and reconstructibility are closely related, there is an important difference between the two, just as we found with controllability and reachability. First, we must clarify the definition of observability to suit the discrete-time interval:

> ***Observability:*** The linear discrete-time system in (8.74) is said to be *observable* in the time interval $[j_0, j_1]$ if an arbitrary initial state $x(j_0)$ can be uniquely determined given knowledge of the sequence $y(j)$ for $j=j_0,\cdots,j_1-1$. (8.86)

The distinction between this definition and that for continuous-time is primarily the specification of the output sequence at the discrete-time instants rather than in a continuous interval. This definition is sometimes given with the requirement that the input sequence $u(j)$ be known as well. While this is true, we can show that the conditions for the existence of the actual solution for the initial state $x(j_0)$ is the same for the zero-state response as is it is for the complete response. We will thereafter simplify our algebra by considering only the zero-state response.

First we consider the general solution for output $y(j)$ as derived in Chapter 6:

$$\begin{aligned} y(j) &= C_d(j)\Psi(j, j_0)x(j_0) \\ &\quad + C_d(j) \sum_{i=j_0+1}^{j} \Psi(j,i)B_d(i-1)u(i-1) + D_d(j_0)u(j_0) \end{aligned} \tag{8.87}$$

Beginning with the initial condition $x(j_0) = x_0$, we can write the first few terms of this solution, similar to what we did for the time-invariant case in Chapter 3:

$$\begin{aligned} y(j_0) &= C_d(j_0)x_0 + D_d(j_0)u(j_0) \\ y(j_0+1) &= C_d(j_0+1)\Psi(j_0+1, j_0)x_0 \\ &\qquad + C(j_0)B_d(j_0)u(j_0) + D_d(j_0+1)u(j_0+1) \\ &\vdots \end{aligned} \tag{8.88}$$

To justify considering the zero-input response only, consider that in order to solve the above equations in (8.88) for the required x_0, we rewrite the sequence of equations as

$$\begin{bmatrix} y(j_0) \\ y(j_0+1) \\ \vdots \\ y(j_1-1) \end{bmatrix} = \begin{bmatrix} C_d(j_0) \\ C_d(j_0+1)\Psi(j_0+1, j_0) \\ \vdots \\ C_d(j_1-1)\Psi(j_1-1, j_0) \end{bmatrix} x_0 + \begin{bmatrix} y_{zs}(j_0) \\ y_{zs}(j_0+1) \\ \vdots \\ y_{zs}(j_0-1) \end{bmatrix} \tag{8.89}$$

where $y_{zs}(j)$ stands for the zero-state response at time j, which includes all the terms of (8.88) that include input $u(j)$. In order to determine x_0, we can subtract this zero-state response vector from the left-hand side, but this will not change the rank condition that then becomes apparent for the coefficient matrix of x_0. It is therefore not important that we consider the zero-input portion of (8.88) at all, and for simplicity we simply omit it.

Therefore, because $x_0 \in \Re^n$, and we require a unique solution, (8.89) implies the observability condition that

$$r\left(\begin{bmatrix} C_d(j_0) \\ C_d(j_0+1)\Psi(j_0+1,j_0) \\ \vdots \\ C_d(j_1-1)\Psi(j_1-1,j_0) \end{bmatrix}\right) \triangleq r\big(R_0(j_0,j_1)\big) = n \tag{8.90}$$

We will refer to the matrix $R_0(j_0, j_1)$ as the *observability matrix*. If the system has p outputs, this matrix has dimensions $j_1 p \times n$. As with controllability and reachability, observability is possible only if $j_1 p \geq n$, implying a minimum number of observations (specifically, n of them for a single-input system). When the system is time-invariant, the condition in (8.90) reduces to the fundamental observability condition, Equation (8.5). In both cases, the likelihood of observability of a specific state is improved if more time instants are considered.

As for reconstructibility, we first give the formal definition from which we will derive a test.

> ***Reconstructibility:*** A linear discrete-time system, Equation (8.74), is *reconstructible* in the time interval $[j_0, j_1]$ if given an arbitrary initial condition $x(j_0)$, the final state $x(j_1)$ can be uniquely determined from the output response sequence $y(j)$ for $j = j_0, \cdots, j_1 - 1$. (8.91)

Again, we will work with (8.89) in such a way that the zero-state responses are irrelevant and will therefore be neglected. To determine our reconstructibility test, we note that without this zero-input part, (8.89) becomes

$$\begin{bmatrix} y(j_0) \\ y(j_0+1) \\ \vdots \\ y(j_1-1) \end{bmatrix} = \begin{bmatrix} C_d(j_0) \\ C_d(j_0+1)\Psi(j_0+1,j_0) \\ \vdots \\ C_d(j_1-1)\Psi(j_1-1,j_0) \end{bmatrix} x_0 \tag{8.92}$$

However, we also know from the inversion property of the state-transition matrix that if $\Psi^{-1}(j_1, j_0)$ exists [it will exist only if *all* of its component factors $A_d^{-1}(k)$, for $k = j_0, \ldots, j_1 - 1$, exist], then $\Psi^{-1}(j_1, j_0) = \Psi(j_0, j_1)$ and $x(j_0) = \Psi(j_0, j_1)x(j_1)$. If this is the case, (8.92) can be rewritten as

$$\begin{bmatrix} y(j_0) \\ y(j_0+1) \\ \vdots \\ y(j_1-1) \end{bmatrix} = \begin{bmatrix} C_d(j_0)\Psi(j_0,j_1) \\ C_d(j_0+1)\Psi(j_0+1,j_0)\Psi(j_0,j_1) \\ \vdots \\ C_d(j_1-1)\Psi(j_1-1,j_0)\Psi(j_0,j_1) \end{bmatrix} x(j_1) = \begin{bmatrix} C_d(j_0)\Psi(j_0,j_1) \\ C_d(j_0+1)\Psi(j_0+1,j_1) \\ \vdots \\ C_d(j_1-1)\Psi(j_1-1,j_1) \end{bmatrix} x(j_1) \tag{8.93}$$

From (8.93), the obvious condition for the solution for $x(j_1)$ is the *reconstructibility* test:

$$r\left(\begin{bmatrix} C_d(j_0)\Psi(j_0,j_1) \\ C_d(j_0+1)\Psi(j_0+1,j_1) \\ \vdots \\ C_d(j_1-1)\Psi(j_1-1,j_1) \end{bmatrix}\right) \triangleq r\left(\overline{R}_0(j_0,j_1)\right) = n \tag{8.94}$$

We call the matrix $\overline{R}_0(j_0,j_1)$ thus defined the *reconstructibility matrix.*

Recall now that controllability of a discrete-time system did not imply reachability. Can we make an analogous statement about observability and reconstructibility? The answer is yes but in the opposite sense. *In discrete-time, observability implies reconstructibility, but not vice versa.* This is apparent if we consider the relationship between the observability and reconstructibility matrices. Although $\Psi^{-1}(j_1,j_0)$ may not exist because of singularity of one of the $A_d(k)$ factors, we are assured from its construction that $\Psi(j_1,j_0)$ itself exists. Therefore, the relationship between (8.92) and (8.93) can be expressed as

$$\overline{R}_0(j_0,j_1)\Psi(j_1,j_0) = R_0(j_0,j_1)$$

Therefore, we have

$$r\left(\overline{R}_0(j_0,j_1)\right) \geq r\left(R_0(j_0,j_1)\right)$$

If the system is observable and $r\left(R_0(j_0,j_1)\right) = n$, then certainly $r\left(\overline{R}_0(j_0,j_1)\right) = n$. For this reason, we normally desire to ascertain the *observability* of a system more often than the reconstructibility.

Observability Grammians

With the observability test in (8.90), we can construct a grammian test by defining the *observability grammian* as follows:

$$H_o(j_0,j_1) \stackrel{\Delta}{=} \sum_{i=j_0}^{j_1-1} \Psi^{\mathrm{T}}(i,j_0)C^{\mathrm{T}}(i)C(i)\Psi(i,j_0) \tag{8.95}$$

Now, as we did with reachability, a grammian test can be given in the form of a theorem:

> THEOREM: The system (8.74) is observable on the interval $[j_0, j_1]$ if and only if the observability grammian in (8.95) is nonsingular. (8.96)

Again, the detailed proof of this test is not provided since once again it is clear that $H_o(j_0,j_1) = R_o^{\mathrm{T}}(j_0,j_1)R_o(j_0,j_1)$ and the full-rank test, Equation (8.90), for $R_o(j_0,j_1)$ suffices to establish invertibility of (8.95).

8.6 Controllability and Observability Under Sampling

As a final note, we should remember that many discrete-time systems are derived as discretizations of continuous-time systems. When considering stability, we discovered that the stability of the discretization of a system is not necessarily preserved. It is an important fact that the same phenomenon can occur with the controllability (or reachability) and observability (or reconstructibility) properties. When a controllable and observable continuous-time system is sampled and discretized, such as shown in Section 6.5.1, it may not remain controllable or observable [6]. Usually this degradation occurs at specific sample periods *T*. Nevertheless, such occurrences should not be allowed to go unnoticed.

8.7 Summary

It is unfortunate that the material presented in this chapter must necessarily come before Chapter 10 on controller and observer design. The concepts of controllability and observability provide answers to the questions "Can this system be controlled?" and "Can this system be observed?" Until we know exactly what a "controller" is and what an "observer" is, it may be difficult to appreciate the implications of controllability and observability. However, we will soon find that the processes of designing a controller and observer will be completely precluded if a system is not controllable or observable, so we must investigate these properties first.

As with any good modeling paradigm, such as the state space system, we hope to find that a single technique will suffice for the most commonly encountered problems and that more sophisticated methods will be needed only for exceptional situations. This is true with controllability and observability. We have already seen in previous chapters that time-invariant systems, whether continuous-time or discrete-time, are much easier to analyze than time-varying systems. We therefore apply the time-invariant state space model whenever it sufficiently captures the physical phenomena we seek to study. If this is the case, the two tests we have called the fundamental controllability test and the fundamental observability test [Equations (8.6) and (8.7)] will suffice to ascertain controllability and observability. It is these tests and the matrices P and Q that will immediately spring to the control engineer's mind upon hearing the words "controllability" and "observability." Other tests are rarely used for time-invariant systems, and the average controls engineer will steer away from time-varying systems as well as nonlinear systems.

Nevertheless, systems of various characteristics will inevitably arise so we have presented here a relatively comprehensive study of controllability and observability determination. Among our findings:

- As mentioned above, most controllability and observability tests are performed with the matrices P and Q. If these matrices have rank n, the systems are controllable and observable, respectively.

- Various other tests, such as the Hautus eigenvector and rank tests, are useful mostly for their value in proofs or in generating still further tests, such as the Jordan form test. The Jordan form test is convenient because it can usually be applied by inspection, provided that the system already appears in Jordan form.

- The controllable and observable canonical forms were derived in preparation for their use in controller and observer design, respectively. Their special structure will facilitate the computation of the gain matrices that are integral components of controllers and observers. Thus, they are not *tests* per se because we cannot even find them unless the system is already known to be controllable or observable.

- The concept of modal controllability was used as a vehicle to introduce the geometric interpretation of controllability and observability. These interpretations are the topic of a large body of literature (for example, [10]), and are particularly helpful when generalizations to nonlinear systems are developed. Furthermore, by accepting the fact that *parts* of systems may be controllable and/or observable while other parts are not, we were able to derive the Kalman decomposition, which groups subsystems with comparable properties. Such decompositions are used mostly to aid in the understanding of system behaviors, as depicted in Figure 8.5. We will refer back to the Kalman decomposition in the next chapter as well.

- We have shown that although time-varying systems are more difficult to analyze than time-invariant systems, controllability and observability tests are nevertheless available. Many texts will derive these tests at the outset and later treat time-invariant systems as a special case, for which the time-varying results simplify. We have not taken this approach, preferring instead to build on the two linear equation solving examples from Chapter 3.
- In the special case of time-varying discrete-time systems, we distinguished the property of reachability from controllability, and the property of reconstructibility from observability. Of these two new properties, we concentrated on reachability over controllability because reachability implies controllability. We concentrate on observability over reconstructibility for the same reason.

Although we can lament the sometimes tedious study of controllability and observability before the study of controllers and observers, we will nevertheless not directly proceed at this point with their introduction. That will occur in Chapter 10. The next chapter presents yet another property of state space systems, realizability. This is also an important property that will allow us to better understand the relationship between controllability, observability, and stability. It will also lead to a practical understanding of issues relevant to *implementing* state space systems. Finally, in Chapter 10 we begin to *design* in earnest.

8.8 Problems

8.1 Determine whether each of the systems below is controllable and/or observable

a)
$$x(k+1)=\begin{bmatrix} 1 & 0 \\ -\frac{1}{2} & \frac{1}{2} \end{bmatrix} x(k)+\begin{bmatrix} 1 \\ -1 \end{bmatrix} u(k)$$
$$y(k)=\begin{bmatrix} 5 & 1 \end{bmatrix} x(k)$$

b)
$$\dot{x}=\begin{bmatrix} -7 & -2 & 6 \\ 2 & -3 & -2 \\ -2 & -2 & 1 \end{bmatrix} x+\begin{bmatrix} 1 & 1 \\ 1 & -1 \\ 1 & 0 \end{bmatrix} u$$
$$y=\begin{bmatrix} -1 & -1 & 2 \\ 1 & 1 & -1 \end{bmatrix} x$$

c)
$$A=\begin{bmatrix} 2 & -5 \\ -4 & 0 \end{bmatrix} \quad b=\begin{bmatrix} 1 \\ -1 \end{bmatrix} \quad c=\begin{bmatrix} 1 & 1 \end{bmatrix}$$

d)
$$\dot{x} = \begin{bmatrix} 0 & 1 \\ -4 & -4 \end{bmatrix} x + \begin{bmatrix} 0 \\ 1 \end{bmatrix} u$$
$$y = \begin{bmatrix} 1 & 1 \end{bmatrix} x$$

8.2 For the system given by the state equations

$$\dot{x} = \begin{bmatrix} 2 & 1 & 1 \\ 5 & 3 & 6 \\ -5 & -1 & -4 \end{bmatrix} x + \begin{bmatrix} 1 \\ 0 \\ 0 \end{bmatrix} u$$
$$y = \begin{bmatrix} 1 & 1 & 2 \end{bmatrix} x$$

a) Determine whether the system is controllable and/or observable by using the fundamental controllability and observability tests.

b) Transform the system into its Jordan canonical form, then determine which individual modes are controllable and/or observable.

c) Find the transfer function of this system and observe which modes appear as system poles.

8.3 Consider the single-input system $\dot{x} = Ax + bu$. Suppose that the system is controllable. Suppose we select a new basis as the n linearly independent columns of the controllability matrix $P = \begin{bmatrix} b & Ab & \cdots & A^{n-1}b \end{bmatrix}$. Find the structure of the system in this new basis.

8.4 Consider the p-input state equations $\dot{x} = Ax + Bu$. Suppose there exists a transformation matrix T such that

$$TB = \begin{bmatrix} B_1 \\ \hdashline 0 \end{bmatrix} \quad \text{and} \quad TAT^{-1} = \begin{bmatrix} A_{11} & \vdots & A_{12} \\ \hdashline A_{21} & \vdots & A_{22} \end{bmatrix}$$

where matrix B_1 has n_1 rows and $rank(B_1) = n_1$, and matrix A_{11} is $n_1 \times n_1$. The other submatrices are correspondingly dimensioned. Suppose also that $r(B) = r(B_1)$. Show that the pair (A, B) is controllable if and only if the pair (A_{22}, A_{21}) is controllable.

8.5 Determine whether the system described by the following matrices is controllable and/or observable.

$$A=\begin{bmatrix}2&1&0&0&0&0&0\\0&2&0&0&0&0&0\\0&0&-3&0&0&0&0\\0&0&0&-3&0&0&0\\0&0&0&0&2&0&0\\0&0&0&0&0&4&1\\0&0&0&0&0&0&4\end{bmatrix}\qquad B=\begin{bmatrix}1&-1\\0&1\\1&0\\1&1\\-1&1\\0&0\\1&0\end{bmatrix}$$

$$C=\begin{bmatrix}0&1&-1&-1&1&0&1\\0&0&-1&-1&-1&1&0\end{bmatrix}$$

8.6 Determine whether the system in Problem 8.5 is stabilizable and/or detectable.

8.7 For the system shown in Figure P8.7, use the indicated state variables to write complete state equations and determine which modes are controllable and which are observable.

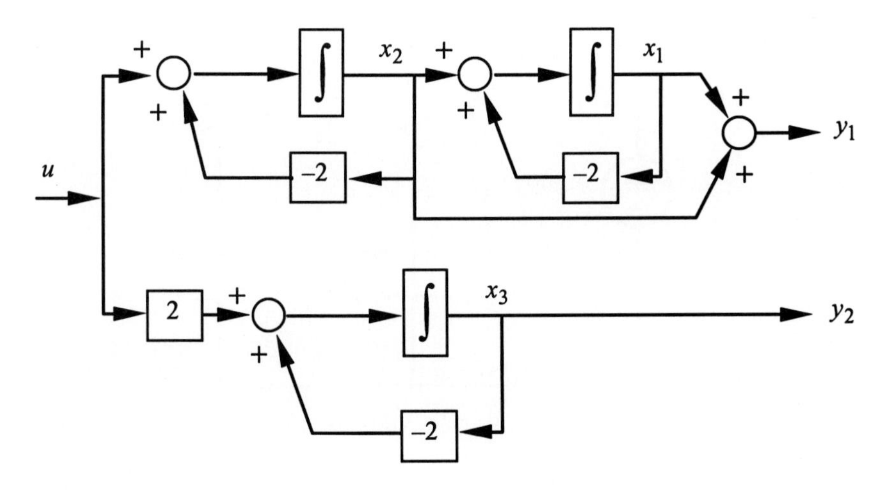

P8.7

8.8 For the electrical circuit shown in Figure P8.8, find conditions on C_1 and C_2 that will make the system uncontrollable. Consider v_g to be the input, and v_1 and v_2 to be the state variables.

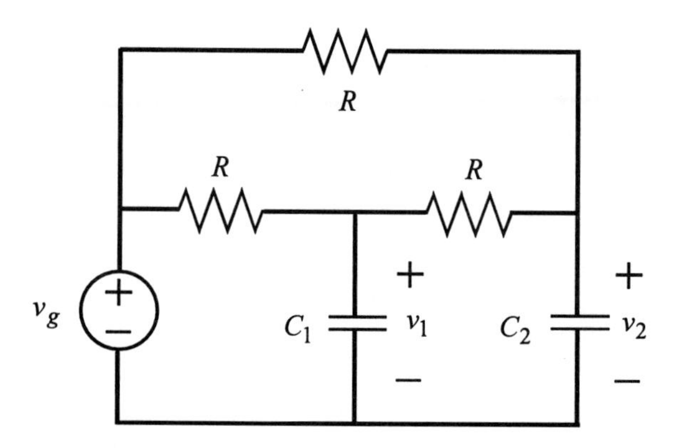

P8.8

8.9 For the circuit shown in Figure 8.9, find conditions on the system components $R_1, R_2, C_1,$ and C_2 that result in an uncontrollable system. Consider v_g to be the input, and v_1 and v_2 to be the state variables.

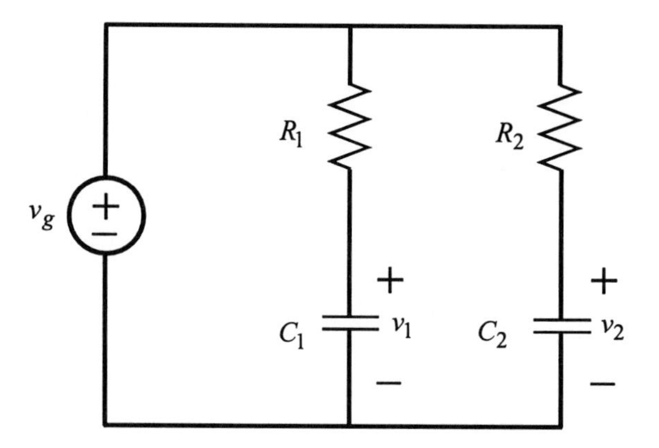

P8.9

8.10 For the system given in the block diagram in Figure 8.10, find necessary and sufficient conditions for the values of α, β, k_1, and k_2 such that the system will be both controllable and observable.

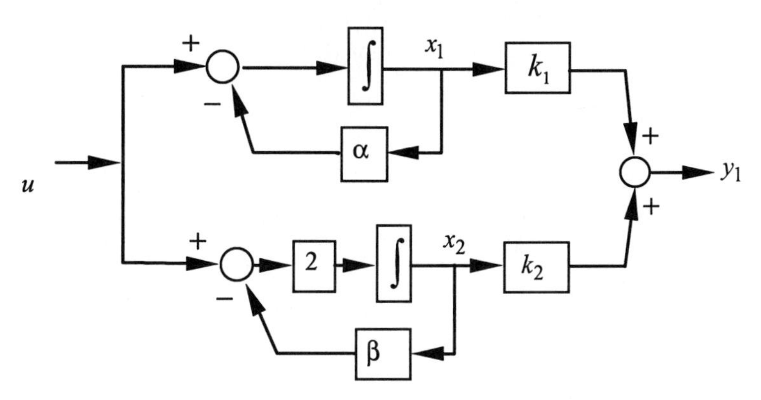

P8.10

8.11 Consider a linear, time-invariant system $\dot{x} = Ax + Bu$, $y = Cx$, that is also asymptotically stable. Define a matrix as

$$G_o = \int_0^\infty e^{A^{\mathrm{T}}t} C^{\mathrm{T}} C e^{At}\, dt$$

Show that the matrix G_o is positive definite if and only if the system is observable.

8.12 Consider a linear, time-invariant system $\dot{x} = Ax + Bu$, $y = Cx$. Suppose it is known that this system is controllable. Show that the system is asymptotically stable if and only if the equation

$$AN + NA^{\mathrm{T}} = -BB^{\mathrm{T}}$$

has a positive-definite solution N.

8.13 Consider a linear, time-invariant system $x(k+1) = Ax(k) + Bu(k)$, $y(k) = Cx(k)$. Suppose it is known that this system is controllable. Show that the system is asymptotically stable if and only if the equation

$$ANA^{\mathrm{T}} - N = -BB^{\mathrm{T}}$$

has a positive-definite solution N.

8.14 Devise a "reconstructibility grammian" test corresponding to the observability grammian test [Equation (8.69)].

8.15 Determine if the following systems are controllable.

a) $$\dot{x} = \begin{bmatrix} 0 & 0 \\ 0 & 1 \end{bmatrix} x + \begin{bmatrix} 1 \\ e^{-t} \end{bmatrix} u$$

b) $$\dot{x} = \begin{bmatrix} 0 & 0 \\ 0 & 1 \end{bmatrix} x + \begin{bmatrix} 0 \\ e^{-2t} \end{bmatrix} u$$

c) $$\dot{x} = \begin{bmatrix} 0 & 1 \\ 0 & t \end{bmatrix} x + \begin{bmatrix} 0 \\ 1 \end{bmatrix} u$$
$$y = \begin{bmatrix} 0 & 1 \end{bmatrix} x$$

8.16 A p input, q output system $\dot{x} = A(t)x + B(t)u$, $y = C(t)x$ is said to be *output controllable* at an initial time t_0 if there exists an input $u(t)$ defined over $[t_0, t_1]$ such that with the arbitrary initial state $x(t_0) = x_0$, the output can be transferred to value $y(t_1) = 0$, with t_1 finite.

a) Show that a system with impulse-response matrix $H(t,\tau) = C(t)\Phi(t,\tau)B(\tau)$ (see *Testing for BIBO Stability* in Section 7.3.1) is output controllable if the rows of $H(t_1,\tau)$ are linearly independent (over the field of complex numbers).

b) Show that if the system is time-invariant, then it is output controllable if and only if

$$r\left(\begin{bmatrix} CB & CAB & \cdots & CA^{n-1}B \end{bmatrix}\right) = q$$

c) Is there a condition on matrix C such that state controllability can then imply output controllability?

8.9 References and Further Reading

The topics of controllability and observability can be relatively simple, as for time-invariant linear systems, or they can be very difficult to understand and apply, such as with time-varying and nonlinear systems. The treatments given here originate with the work of Kalman [6] (including, of course, the Kalman decomposition), and are completely discussed in [2], [3], [4], and [7]. The PBH tests can be found in a number of sources, but are applied to circuit theory by Belevitch in [1].

For time-varying systems, the tests for controllability and observability are due to [8], and are also specifically discussed in [4], [9], the latter concentrating on discrete-time systems.

The geometric interpretation is, for controllability and observability, one of the best conceptual tools for understanding the basic concepts, and is best explained in by Wonham in [10].

[1] Belevitch, V., *Classical Network Theory*, Holden-Day, 1968.

[2] Brockett, Roger W., *Finite Dimensional Linear Systems*, John Wiley & Sons, 1970.

[3] Casti, John L., *Linear Dynamical Systems*, Academic Press, 1987.

[4] Chen, Chi-Tsong, *Linear System Theory and Design*, Holt, Rinehart, and Winston, 1984.

[5] Chen, Chi-Tsong, and Charles A. Desoer, "A Proof of Controllability of Jordan Form State Equations," *IEEE Transactions on Automatic Control*, vol. AC-13, no. 2, April 1968, pp. 195-196.

[6] Kalman, Rudolph E., "Mathematical Description of Linear Dynamical Systems," *Journal of the Society of Industrial and Applied Mathematics-Control Series, Series A*, vol., 1, no. 2., 1963, pp. 152-192.

[7] Rugh, Wilson, *Linear System Theory*, 2nd edition, Prentice Hall, 1996.

[8] Silverman, Leonard M., and H. E. Meadows, "Controllability and Observability in Time-Variable Linear Systems," *SIAM Journal on Control and Optimization*, vol. 5, no. 1, 1967, pp. 64-73.

[9] Weiss, Leonard, "Controllability, Realization, and Stability of Discrete-Time Systems," *SIAM Journal on Control and Optimization*, vol. 10, no. 2, 1972, pp. 230-251.

[10] Wonham, W. Murray, *Linear Multivariable Control: A Geometric Approach*, 3rd edition, Springer-Verlag, 1985.

9

System Realizations

In Chapter 1, we gave an introduction to the relationship between transfer functions and state equations. In this chapter, we will continue this investigation with added attention to the complications introduced by controllability and observability of systems. Also in Chapter 1, we presented what we called at the time *simulation diagrams*. A simulation diagram, such as Figures 1.7 or 1.8, is sometimes also known as an *integrator realization*, because it shows how a differential equation or transfer function can be *realized* with physical components, in this case, integrators. Note that once a transfer function has been represented by such a diagram, the state equations can be written directly, with state variables taken as the outputs of the integrators. In this chapter, we will explore the details of constructing such realizations of transfer functions. We will show a number of alternatives to the realizations of Figures 1.7 or 1.8, along with the mathematical methods by which they are obtained. In the course of this presentation, we will discover the relationship among controllability, observability, and the relative size of the transfer function and associated state matrices. One such realization method, which uses quantities called Markov parameters, will lead to a method for *system identification*, which allows us to estimate the state matrices from input and output data.

Throughout this chapter, we will concentrate on time-invariant systems. The main reason for this is that time-varying systems are not described by transfer functions, so conversion from transfer function to state space is not necessary. Some results can be obtained by "realizing" the impulse-response matrix for time-varying systems [16], but because this matrix is not often available in closed form, that process is considered an academic exercise.

Unless otherwise indicated by our standard notation, the results shown here apply equally well to SISO and MIMO systems. They also apply to continuous- and discrete-time systems alike, with the necessary change from s-domain notation to z-domain notation.

9.1 Minimal Realizations

As shown in Chapter 1, we can derive the transfer function representation for the state space system

$$\begin{aligned} \dot{x} &= Ax + Bu \\ y &= Cx + Du \end{aligned} \tag{9.1}$$

ss2tf(A,B,C,D)

by first assuming zero initial conditions, then taking the LaPlace transform of both sides of the equation (9.1). The result is[M]

$$H(s) = C(sI - A)^{-1}B + D \tag{9.2}$$

For the discrete-time system

$$\begin{aligned} x(k+1) &= A_d x(k) + B_d u(k) \\ y(k) &= C_d x(k) + D_d u(k) \end{aligned} \tag{9.3}$$

the result is much the same,

$$H(z) = C_d(zI - A_d)^{-1}B_d + D_d \tag{9.4}$$

The relationship between a state space system, which we will denote by the quadruple $\{A, B, C, D\}$, and the transfer function $H(s)$ is described by the terminology given in the following definition:

> ***Realization:*** A realization of transfer function $H(s)$ is any state space quadruple $\{A, B, C, D\}$ such that $H(s) = C(sI - A)^{-1}B + D$. If such a set $\{A, B, C, D\}$ exists, then $H(s)$ is said to be *realizable*. (9.5)

Because of the existence of the methods shown in Chapter 1, we can immediately say that a transfer function is realizable if it is *proper* (or, in the multivariable case, if each component transfer function is proper). Also as seen in Chapter 1, it is possible for two different state space descriptions to give the same transfer function. For example, consider the system in (9.1) and an equivalent system given by the similarity transformation provided by nonsingular change of basis matrix M:

$$\begin{aligned} \dot{\hat{x}} &= M^{-1}AM\hat{x} + M^{-1}Bu \\ y &= CMx + Du \end{aligned} \tag{9.6}$$

Forming the transfer function of this matrix, we get

$$\begin{aligned}\hat{H}(s) &= CM(sI - M^{-1}AM)^{-1}M^{-1}B + D \\ &= C(sMM^{-1} - MM^{-1}AMM^{-1})B + D \\ &= C(sI - A)^{-1}B + D \\ &= H(s)\end{aligned}$$

Therefore, we have the result that any two systems related by a similarity transformation have the same transfer function.

Now we wish to consider the *order* of these transfer functions relative to the dimension of the state space description, i.e., relative to the value n, which is the size of the matrix A. To motivate this discussion, we will find the transfer function for the differential equation $\ddot{y} + 4\dot{y} + 3y = \dot{u} + 3u$. With zero initial conditions, a LaPlace transform of both sides gives the result

$$H(s) = \frac{Y(s)}{U(s)} = \frac{s+3}{s^2 + 4s + 3} = \frac{s+3}{(s+1)(s+3)} = \frac{1}{s+1} \tag{9.7}$$

This shows the curious fact that the transfer function is of lower order (i.e., the degree of the characteristic polynomial in the denominator) than the order of the differential equation, which would also be the size of the state space description of the system derived from the realization methods of Chapter 1. We could also realize this transfer function with the first-order state space system

$$\begin{aligned}\dot{x} &= -x + u \\ y &= x\end{aligned} \tag{9.8}$$

To discover the reason for this occurrence, consider putting the differential equation into state space form, according to the realization from Figure 1.8:

$$\begin{aligned}\begin{bmatrix}\dot{x}_1 \\ \dot{x}_2\end{bmatrix} &= \begin{bmatrix}-4 & 1 \\ -3 & 0\end{bmatrix}\begin{bmatrix}x_1 \\ x_2\end{bmatrix} + \begin{bmatrix}1 \\ 3\end{bmatrix}u \\ y &= \begin{bmatrix}1 & 0\end{bmatrix}\begin{bmatrix}x_1 \\ x_2\end{bmatrix}\end{aligned} \tag{9.9}$$

The first thing we notice about this particular realization is that it is in the observable canonical form. This means that the realization in (9.9) is observable by default. However, if we examine the controllability matrix[M] for (9.9), we get

`ctrb(A,B)`

$$P = \begin{bmatrix} 1 & -1 \\ 3 & -3 \end{bmatrix}$$

Clearly this is a rank-deficient matrix, and we see that this realization is not controllable. If we had chosen the controllable canonical realization from Figure 1.7, we would have found that

$$\begin{bmatrix} \dot{x}_1 \\ \dot{x}_2 \end{bmatrix} = \begin{bmatrix} 0 & 1 \\ -3 & -4 \end{bmatrix} \begin{bmatrix} x_1 \\ x_2 \end{bmatrix} + \begin{bmatrix} 0 \\ 1 \end{bmatrix} u$$
$$y = \begin{bmatrix} 3 & 1 \end{bmatrix} \begin{bmatrix} x_1 \\ x_2 \end{bmatrix} \tag{9.10}$$

which is easily found to be controllable but not observable.

From these examples we make two observations. The first is that for a given transfer function, it is possible to find realizations of different order. This necessitates the following definition:

> ***Minimal Realization:*** A realization $\{A, B, C, D\}$ is called a *minimal realization* (also called an *irreducible realization*) of a transfer function if there is no other realization of smaller size. (9.11)

The second observation we make from the examples above results in the following theorem, which we will prove somewhat less than rigorously:

> THEOREM: A minimal realization is both controllable and observable. (9.12)

To demonstrate this theorem, we note from the Kalman decomposition of Chapter 8 that if a system is not controllable, then it is possible to decompose it into controllable and uncontrollable parts:[M]

`ctrbf(A,B,C)`

$$\begin{bmatrix} \dot{x}_c \\ \dot{x}_{\bar{c}} \end{bmatrix} = \begin{bmatrix} A_c & A_{12} \\ 0 & A_{\bar{c}} \end{bmatrix} \begin{bmatrix} x_c \\ x_{\bar{c}} \end{bmatrix} + \begin{bmatrix} B_c \\ 0 \end{bmatrix} u$$
$$y = \begin{bmatrix} C_1 & C_2 \end{bmatrix} \begin{bmatrix} x_c \\ x_{\bar{c}} \end{bmatrix} + Du \tag{9.13}$$

Deriving the transfer function from this form,

$$
\begin{aligned}
H(s) &= \begin{bmatrix} C_c & C_{\bar{c}} \end{bmatrix} \begin{bmatrix} sI - A_c & -A_{12} \\ 0 & sI - A_{\bar{c}} \end{bmatrix}^{-1} \begin{bmatrix} B_c \\ 0 \end{bmatrix} + D \\
&= \begin{bmatrix} C_c & C_{\bar{c}} \end{bmatrix} \begin{bmatrix} (sI - A_c)^{-1} & (sI - A_c)^{-1} A_{12} (sI - A_{\bar{c}})^{-1} \\ 0 & (sI - A_{\bar{c}})^{-1} \end{bmatrix} \begin{bmatrix} B_c \\ 0 \end{bmatrix} + D \\
&= \begin{bmatrix} C_c & C_{\bar{c}} \end{bmatrix} \begin{bmatrix} (sI - A_c)^{-1} B_c \\ 0 \end{bmatrix} + D \\
&= C_c (sI - A_c)^{-1} B_c + D
\end{aligned} \tag{9.14}
$$

This demonstrates the interesting fact that the transfer function does not depend on the uncontrollable part of the realization. A similar exercise shows that the transfer function is independent of the unobservable part of the realization as well. Together, the statements can be generalized to say that *the transfer function of the system depends on only the controllable and observable part of the system. Conversely, a nonminimal realization is either uncontrollable, unobservable, or both,* i.e.,

$$
H(s) = C_{co} (sI - A_{co})^{-1} B_{co} + D \tag{9.15}
$$

The even more general statement can be made:

> THEOREM: (Minimal realization theorem) If $\{A, B, C, D\}$ is a minimal realization and $\{\bar{A}, \bar{B}, \bar{C}, \bar{D}\}$ is another minimal realization, then the two realizations are similar to one another (i.e., they are related by a similarity transformation). (9.16)

The proof of this theorem will be deferred until we investigate Markov parameters in Section 9.3.

In the case of SISO systems, wherein we can denote the transfer function as

$$
h(s) = \frac{n(s)}{d(s)} \tag{9.17}
$$

we identify the degree of the denominator polynomial $d(s)$ as the highest power of s. If a minimal realization is found for (9.17) of size less than n, then the two polynomials $n(s)$ and $d(s)$ will have a common (polynomial) factor, which could be canceled in order to arrive at a transfer function of smaller degree, not altering the input-output behavior. Note that this form of pole-zero cancellation is *exact* in the sense that it happens internally and identically within the system. It is not a violation of the admonishment given in classical control theory that poles and zeros never exactly cancel each other.

If the polynomials $n(s)$ and $d(s)$ have no common factor, they are said to be *coprime*, and no such pole-zero cancellation will occur. If $n(s)$ and $d(s)$ are coprime, and $d(s)$ is of degree n, then the minimal realization will have size n as well. This is often stated as a theorem whose proof simply follows the explanation of minimal realizations above. The definition of coprimeness and this statement of the size of the minimal realization also extends to matrix-fraction descriptions for MIMO systems, although we do not pursue such descriptions here.

As a final note for this section, we might remark on the implication of modal controllability and observability on BIBO stability. We learned in Chapter 7 that BIBO stability depends on the poles of the transfer function, not on the eigenvalues of the state space representation. In Chapter 8 we discovered that the controllability and observability of the individual modes of the system can be determined by inspection of either the Jordan form or the Kalman decomposition. We have seen in this section that any uncontrollable and or unobservable modes will not appear in the transfer function. Therefore, combining these facts, we can arrive at a method by which we can determine BIBO stability by inspection: *If all of the unstable modes of a system (as determined by their eigenvalues) are found to be uncontrollable, unobservable, or both, by inspection of the canonical form, then the system will be BIBO stable.*

9.2 Specific Realizations

In Chapter 1, we presented two basic "integrator" realizations, which we referred to as simulation diagrams, because they can be used directly to simulate systems with analog integrators. They apply equally well to continuous- and discrete-time systems, with only the substitution of unit delay blocks in the place of the integrators. We see now that these simulations are in fact realizations in the sense they are used here: a technique for converting from transfer function to state space notation. In this section, we will discuss these realizations from a new perspective and add some additional realizations.

We start with the differential equation

$$\begin{aligned}\frac{d^n y(t)}{dt^n}+a_{n-1}\frac{d^{n-1}y(t)}{dt^{n-1}}+\cdots+a_1\frac{dy(t)}{dt}+a_0 y(t)=\\ b_n\frac{d^n u(t)}{dt^n}+\cdots+b_0 u(t)\end{aligned} \tag{9.18}$$

We begin with the frequency-domain expression that is obtained by finding the LaPlace transform of both sides of (9.18) while assuming zero initial conditions:

$$h(s) = \frac{y(s)}{u(s)} = \frac{b_n s^n + \cdots + b_1 s + b_0}{s^n + a_{n-1}s^{n-1} + \cdots + a_1 s + a_0} \tag{9.19}$$

9.2.1 Controllable Canonical Realization

In Chapter 1, this realization, shown in Figure 9.1, was referred to as the "phase variable" realization. From the perspective of Chapter 6, this term refers to the fact that the state variables defined as the outputs of the integrators are the variables depicted in phase plots, i.e., an output variable and its successive derivatives (or time delays).

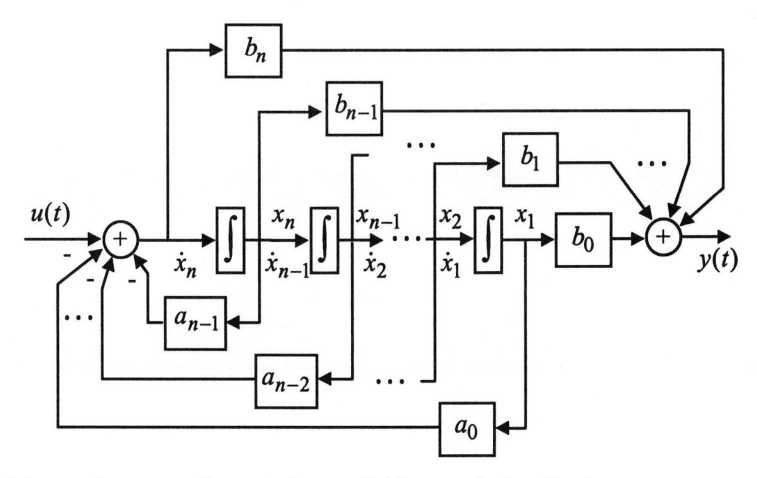

Figure 9.1 Integrator diagram of controllable canonical realization.

From the perspective of Chapter 8, the phase variable realization, written as the set of state equations

$$\begin{bmatrix} \dot{x}_1 \\ \vdots \\ \dot{x}_{n-1} \\ \dot{x}_n \end{bmatrix} = \begin{bmatrix} 0 & 1 & 0 & 0 \\ \vdots & \ddots & \ddots & 0 \\ 0 & \cdots & 0 & 1 \\ -a_0 & -a_1 & \cdots & -a_{n-1} \end{bmatrix} \begin{bmatrix} x_1 \\ \vdots \\ x_{n-1} \\ x_n \end{bmatrix} + \begin{bmatrix} 0 \\ \vdots \\ 0 \\ 1 \end{bmatrix} u(t)$$

$$y(t) = \begin{bmatrix} b_0 - b_n a_0 & b_1 - b_n a_0 & \cdots & b_{n-1} - b_n a_0 \end{bmatrix} \begin{bmatrix} x_1 \\ \vdots \\ x_{n-1} \\ x_n \end{bmatrix} + b_n u(t) \tag{9.20}$$

is seen to yield the controllable canonical form. Note that this need not yield a *minimal* realization, only a controllable one. If the minimal realization is of size less than n, then (9.20), while being controllable by construction (see Chapter 8), will not be observable because we know that nonminimal realizations must be either uncontrollable, unobservable, or both.

9.2.2 Observable Canonical Realization

Also in Chapter 1 we presented the realization shown in Figure 9.2.

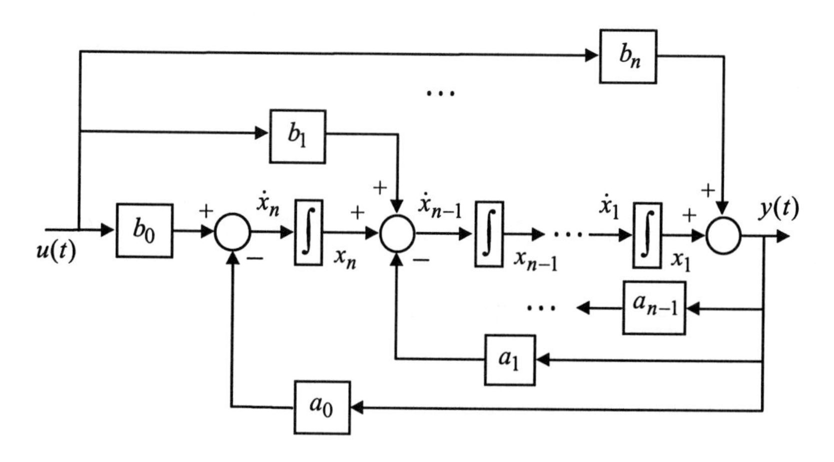

Figure 9.2 Integrator diagram of observable canonical realization.

Here, we need only remark that this realization results in the state equations of the form

$$\begin{bmatrix} \dot{x}_1 \\ \vdots \\ \dot{x}_{n-1} \\ \dot{x}_n \end{bmatrix} = \begin{bmatrix} -a_{n-1} & 1 & 0 & 0 \\ \vdots & \ddots & \ddots & 0 \\ -a_1 & \cdots & 0 & 1 \\ -a_0 & 0 & \cdots & 0 \end{bmatrix} \begin{bmatrix} x_1 \\ \vdots \\ x_{n-1} \\ x_n \end{bmatrix} + \begin{bmatrix} b_{n-1} - a_{n-1}b_n \\ \vdots \\ b_1 - a_1 b_n \\ b_0 - a_0 b_n \end{bmatrix} u(t)$$
$$y(t) = \begin{bmatrix} 1 & 0 & \cdots & 0 \end{bmatrix} \begin{bmatrix} x_1 \\ \vdots \\ x_{n-1} \\ x_n \end{bmatrix} + b_n u(t) \tag{9.21}$$

We refer to this realization as the observable canonical realization, because of its special companion form. (In some texts, the term "observable canonical realization" is given as a companion form with the $-a_i$ coefficients appearing in the *last* column of the A-matrix instead of the first and with a somewhat different b-matrix [4].)

Again, we note that these realizations (controllable and observable canonical) can be constructed just as easily in discrete-time, with unit delay blocks instead. Also, we should remark that there is a multivariable counterpart to each, which we will discuss in Chapter 10.

9.2.3 Jordan Canonical Realizations

By this time, the reader may have noticed that for every canonical form of the state equations, we have identified a corresponding realization. This is true for the Jordan canonical form as well. To see how the Jordan canonical form becomes the Jordan canonical realization, we will consider the specific example of the strictly proper transfer function:

$$h(s) = \frac{4s^4 + 54s^3 + 247s^2 + 500s + 363}{s^5 + 11s^4 + 63s^3 + 177s^2 + 224s + 100} \tag{9.22}$$

[If we have a non-strictly proper transfer function, it should first be manipulated so that it appears in the form:

$$h(s) = \frac{n(s)}{d(s)} + k \tag{9.23}$$

where the ratio $n(s)/d(s)$ is strictly proper. The following discussion applies to the ratio $n(s)/d(s)$, and the k term is simply a feedthrough constant.] The transfer function (9.22) can be expanded into the partial-fraction expansion[M]

`residue(n,d)`

$$h(s) = \frac{1-j2}{s-(-3+j4)} + \frac{1+j2}{s-(-3-j4)} + \frac{-1}{s+2} + \frac{1}{(s+2)^2} + \frac{3}{s+1} \tag{9.24}$$

which is then realized with the block-diagram realization shown in Figure 9.3.

The first observation we make about the Jordan canonical realization, given Figure 9.3, is that it is not an "integrator" realization since the dynamic blocks contain the form $1/(s-\lambda)$ rather than $1/s$. If it is desired that this realization be constructed entirely from pure integrators, then each $1/(s-\lambda)$ can be individually realized by the subdiagram shown in Figure 9.4. However, this decomposition into pure integrators is usually not necessary, since the operational amplifier circuit of the ideal integrator $1/s$ can be easily converted to the "nonideal integrator" with the addition of a single feedback resistor in parallel with the feedback capacitor.

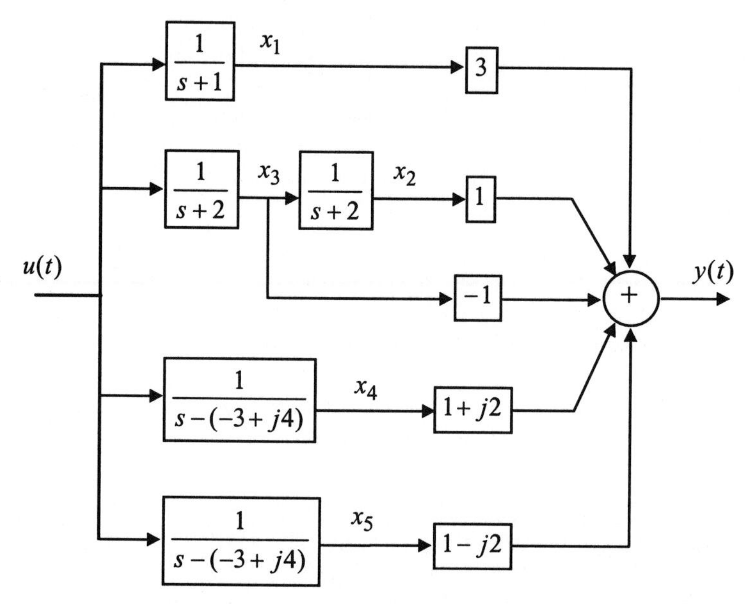

Figure 9.3 Complex-gain version of the Jordan form realization of the transfer function of Equation (9.22).

By defining the state variables as the output of the dynamic blocks, as we have in Figure 9.3, and keeping in mind Figure 9.4 as the internal realization of these blocks, we arrive at the state equations

$$\begin{bmatrix} \dot{x}_1 \\ \dot{x}_2 \\ \dot{x}_3 \\ \dot{x}_4 \\ \dot{x}_5 \end{bmatrix} = \left[\begin{array}{c|cc|c|c} -1 & 0 & 0 & 0 & 0 \\ \hline 0 & -2 & 1 & 0 & 0 \\ 0 & 0 & -2 & 0 & 0 \\ \hline 0 & 0 & 0 & -3-j4 & 0 \\ \hline 0 & 0 & 0 & 0 & -3+j4 \end{array}\right] \begin{bmatrix} x_1 \\ x_2 \\ x_3 \\ x_4 \\ x_5 \end{bmatrix} + \begin{bmatrix} 1 \\ 0 \\ 1 \\ 1 \\ 1 \end{bmatrix} u$$

$$y = \begin{bmatrix} 3 & 1 & -1 & 1-j2 & 1+j2 \end{bmatrix} \begin{bmatrix} x_1 \\ x_2 \\ x_3 \\ x_4 \\ x_5 \end{bmatrix} \tag{9.25}$$

which is easily recognizable as being in the Jordan form.

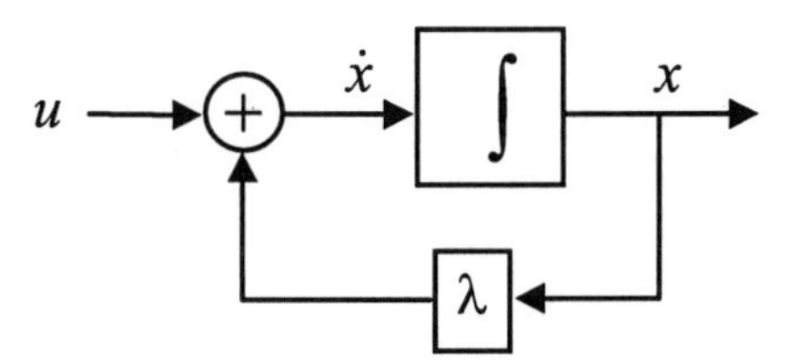

Figure 9.4 Integrator realization of $1/(s-\lambda)$ blocks in Figure 9.3.

The second observation we wish to make about the Jordan form realization of Figure 9.3 and Equation (9.25) is that it contains complex gains. If the need for a realization is physical, i.e., if it is desired that the system be constructed in hardware, then we cannot allow complex numbers. To remedy this, each pair of complex conjugate poles in (9.24) should be left as a quadratic, i.e.,

$$h(s)=\frac{2s+22}{(s+3)^2+4^2}+\frac{-1}{s+2}+\frac{1}{(s+2)^2}+\frac{3}{s+1} \tag{9.26}$$

which can be obtained simply by adding the parallel paths in Figure 9.3. The quadratic term can be realized with any of the other realization methods (e.g., the controllable or observable), and is then considered as a block. The quadratic term then becomes a block in a *block* Jordan form such as[M]

`cdf2rdf(V,D)`

$$\dot{x}=\left[\begin{array}{c|cc|c} -1 & & & \\ \hline & -2 & 1 & \\ & & -2 & \\ \hline & & & \hat{A}_{2\times 2} \end{array}\right]x+\begin{bmatrix} 1 \\ 0 \\ 1 \\ \hat{b}_{2\times 1} \end{bmatrix}u \tag{9.27}$$
$$y=\begin{bmatrix} 3 & 1 & -1 & \hat{c}_{1\times 2} \end{bmatrix}x$$

For example, the block corresponding to the quadratic term might be given as

$$\hat{A}_{2\times 2}=\begin{bmatrix} -3 & 4 \\ -4 & -3 \end{bmatrix} \qquad \hat{b}_{2\times 1}=\begin{bmatrix} 2 \\ 4 \end{bmatrix} \qquad \hat{c}_{1\times 2}=\begin{bmatrix} 1 & 0 \end{bmatrix}$$

We should also remark on the controllability and/or observability of such a realization. It is guaranteed to be neither. Note that in Figure 9.3, each eigenvalue is associated with a parallel path from input to output. In the example given, each path has a nonzero gain connecting it to the output and a unity gain connection to the input. We know from Chapter 8 that for this SISO system, which has only a single Jordan block for each distinct eigenvalue, the row (column) of the b-matrix (c-matrix) that corresponds to the last row (first column) of the corresponding Jordan block must be nonzero. In this system, that is the case. A zero in such a row (column) would correspond to one of the parallel paths in Figure 9.3 being "disconnected" from the input, the output, or perhaps both. The special structure of Jordan forms makes them "decoupled," so that the modes do not interact; therefore, any such disconnected path results in a nonminimal realization.

If there were more than one Jordan block for any particular eigenvalue, we would have two separate paths in Figure 9.3 for that eigenvalue. If this were to happen in our system, we already know that the system would be either uncontrollable, unobservable, or both, because there are an insufficient number of inputs and outputs to allow for linearly independent rows (columns) of the b-matrix (c-matrix).

In the case that the system has more than one input and/or output, the Jordan form can still be computed and realized in a form similar to Figure 9.3. Of course, the diagram will include p input channels and q output channels, each with a separate summing junction. The other interconnections will look very similar to those in Figure 9.3. The procedure will also require finding the partial fraction expansion of the $(q \times p)$ *matrix* transfer function $H(s)$, which entails a partial-fraction expansion whose residues (numerator coefficients) are all $(q \times p)$ constant matrices. This, in turn, requires the definition of the characteristic polynomial of the matrix $H(s)$ as the least common denominator of all its minors [4]. Rather than pursue this method in detail, we will proceed to a more common technique for realizing a multivariable system, with *Markov parameters*.

9.3 Markov Parameters

An interesting way to realize a transfer function that has some beneficial side effects is through the power series expansion. Thus, if we have the (SISO) transfer function for a causal LTI system (implying a proper transfer function),

$$g(s) = \frac{b_n s^n + b_{n-1} s^{n-1} + \cdots + b_1 s + b_0}{s^n + a_{n-1} s^{n-1} + \cdots + a_1 s + a_0} \tag{9.28}$$

we seek an expansion in the form:

$$g(s) = h(0) + h(1)s^{-1} + h(2)s^{-2} + \ldots \tag{9.29}$$

By simple long division, it is clear that $h(0) = b_n$. The values $h(i)$, with $i = 0,1,\ldots$, are known as *Markov parameters*. From (9.29), it is apparent that $g(s)$ could be realized with an infinite chain of integrators, although this would clearly not be a good idea. We suspect instead that the realization should require no more than n integrators and hence, the transfer function should be realizable with a system of order n or less.

To generate a more suitable expression for Markov parameters, rewrite the transfer function of the system in terms of the matrices in the realization:

$$g(s) = C(sI - A)^{-1}B + D \tag{9.30}$$

[For generality, we allow for the possibility of a MIMO system, although Equation (9.28) cannot be expressed as such in that case, but rather as a matrix of transfer functions.] Manipulating the term that must be inverted by using the *resolvent identity* given in Appendix A, i.e., Equation (A.42),

$$\begin{aligned} g(s) &= D + C\left[\frac{1}{s}\left(I - \frac{A}{s}\right)^{-1}\right]B \\ &= D + C\left[\frac{1}{s}\left(I + \frac{A}{s} + \frac{A^2}{s^2} + \cdots\right)\right]B \\ &= D + \frac{CB}{s} + \frac{CAB}{s^2} + \frac{CA^2B}{s^3} + \cdots \end{aligned} \tag{9.31}$$

(Convergence of the series generated may be ensured by scaling the A-matrix if necessary.) This shows that the Markov parameters are $h(0) = D$ and $h(i) = CA^{i-1}B$ for $i = 1, 2, \ldots$. One might also observe that if the inverse LaPlace transform were performed on the last line of (9.31), then the impulse response would result:

$$\begin{aligned} g(t) &= D\delta(t) + CB + CABt + \frac{CA^2Bt^2}{2!} + \cdots \\ &= D\delta(t) + C\left[1 + At + \frac{A^2t^2}{2!} + \cdots\right]B \\ &= D\delta(t) + Ce^{At}B \end{aligned} \tag{9.32}$$

just as we would expect. Equivalently, we can see that for $i = 1, 2, \ldots$,

$$h(i) = CA^{i-1}B = \left.\frac{d^{i-1}g(t)}{dt^{i-1}}\right|_{t=0} \tag{9.33}$$

Again, this expression takes the form of a *matrix* in the case of a MIMO system.

An alternative method to derive the Markov parameters (for SISO systems) is to realize that from (9.28) and (9.29),

$$\begin{aligned} & b_n s^n + b_{n-1}s^{n-1} + \cdots + b_1 s + b_0 \\ & \quad = \left(s^n + a_{n-1}s^{n-1} + \cdots + a_1 s + a_0\right)\left(h(0) + h(1)s^{-1} + h(2)s^{-2} + \cdots\right) \end{aligned}$$

Equating coefficients of powers of s^{-1}, we have

$$\begin{aligned} b_n &= h(0) \\ b_{n-1} &= h(1) + h(0)a_{n-1} \\ b_{n-2} &= h(2) + h(1)a_{n-1} + h(0)a_{n-2} \\ &\vdots \\ b_i &= h(n-i) + h(n-i-1)a_{n-1} + \cdots + h(0)a_i \end{aligned}$$

so the recursive relationship can be given as:

$$\begin{aligned} h(0) &= b_n \\ h(1) &= b_{n-1} - h(0)a_{n-1} \\ h(2) &= b_{n-2} - h(1)a_{n-1} - h(0)a_{n-2} \\ &\vdots \\ h(n) &= b_0 - h(n-1)a_{n-1} - \cdots - h(0)a_0 \end{aligned} \tag{9.34}$$

An important observation suggested by this result is that the Markov parameters $h(i)$, $i = 0,1,\ldots$, ought to be invariant to the particular realization $\{A, B, C, D\}$, i.e., they are unique to the transfer function. We will use this fact in the proof of the minimal realization theorem (on page 371) that follows.

9.3.1 Proof of the Minimal Realization Theorem

At this point we will reconsider the statement made in the minimal realization theorem. In that theorem it was claimed that any two minimal realizations of the same transfer function were similar to one another. We can now demonstrate this using the newly defined Markov parameters. We will first demonstrate some interesting properties of Markov parameters.

First, note that if $\{A, B, C, D\}$ and $\{\overline{A}, \overline{B}, \overline{C}, \overline{D}\}$ are two different realizations of the same transfer function $g(s)$, then (even if they are of different size), $D = \overline{D}$ [or $h(0) = \overline{h}(0)$], and, by equating powers of s^{-1} in the respective series expansions (9.31),

$$\bar{h}(i) = \overline{CA}^{i-1}\bar{B} = h(i) = CA^{i-1}B \tag{9.35}$$

Thus, Markov parameters themselves are unique to a transfer function, even though the matrices from which they may be computed might differ. This same result is even more apparent from direct application of the similarity transformation:

$$\overline{CA}^{i-1}\bar{B} = (CT)(T^{-1}A^{i-1}T)(T^{-1}B) = CA^{i-1}B$$

Next, consider two minimal realizations $\{A, B, C, D\}$ and $\{\bar{A}, \bar{B}, \bar{C}, \bar{D}\}$ so that we are sure each is of the same size. Also neglect $h(0)$ and consider only $h(i)$, $i = 1, 2, \ldots$. If we form the product of the observability and controllability matrices

$$\begin{aligned} QP &= \begin{bmatrix} C \\ CA \\ \vdots \\ CA^{n-1} \end{bmatrix} \begin{bmatrix} B & AB & \cdots & A^{n-1}B \end{bmatrix} \\ &= \begin{bmatrix} CB & CAB & \cdots & CA^{n-1}B \\ CAB & CA^2B & & \\ \vdots & & \ddots & \vdots \\ CA^{n-1}B & & \cdots & CA^{2(n-1)}B \end{bmatrix} \end{aligned} \tag{9.36}$$

so of course for $\{\bar{A}, \bar{B}, \bar{C}, \bar{D}\}$,

$$\overline{QP} = \begin{bmatrix} \overline{CB} & \overline{CAB} & \cdots & \overline{CA}^{n-1}\bar{B} \\ \overline{CAB} & \overline{CA}^2\bar{B} & & \\ \vdots & & \ddots & \vdots \\ \overline{CA}^{n-1}\bar{B} & & \cdots & \overline{CA}^{2(n-1)}\bar{B} \end{bmatrix} \tag{9.37}$$

Now because of (9.35), we have

$$QP = \overline{QP} \tag{9.38}$$

Because both realizations are minimal, they are known to be controllable and observable, so that the $(n \times qn)$ matrix $\bar{Q}$ is full rank n. Then multiplying both side of (9.38) by $\bar{Q}^{\mathrm{T}}$ gives

$$\overline{Q}^{\mathrm{T}}QP=\overline{Q}^{\mathrm{T}}\overline{Q}\overline{P} \tag{9.39}$$

Being observable, the $(n \times n)$ matrix $\overline{Q}^{\mathrm{T}}\overline{Q}$ is invertible, so (9.39) can be solved to get

$$\begin{aligned}\overline{P} &= (\overline{Q}^{\mathrm{T}}\overline{Q})^{-1}\overline{Q}^{\mathrm{T}}QP \\ &\overset{\Delta}{=} TP\end{aligned} \tag{9.40}$$

Thus we have a relationship between the controllability matrices P and $\overline{P}$.

Performing a similar procedure to solve (9.38) for Q,

$$QPP^{\mathrm{T}} = \overline{Q}\overline{P}P^{\mathrm{T}} \tag{9.41}$$

$$Q=\overline{Q}\overline{P}P^{\mathrm{T}}\left(PP^{\mathrm{T}}\right)^{-1} \tag{9.42}$$

Now substituting (9.40) into (9.42),

$$\begin{aligned}Q &= \overline{Q}TPP^{\mathrm{T}}\left(PP^{\mathrm{T}}\right)^{-1} \\ &= \overline{Q}T\end{aligned} \tag{9.43}$$

Together, (9.40) and (9.43) demonstrate that

$$\begin{aligned}P = T^{-1}\overline{P} &= T^{-1}\left[\overline{B} \quad \overline{A}\overline{B} \quad \cdots \quad \overline{A}^{n-1}\overline{B}\right] \\ &= \left[T^{-1}\overline{B} \quad T^{-1}\overline{A}\overline{B} \quad \cdots \quad T^{-1}\overline{A}^{n-1}\overline{B}\right] \\ &= \left[T^{-1}\overline{B} \quad \left(T^{-1}\overline{A}T\right)T^{-1}\overline{B} \quad \cdots \quad \left(T^{-1}\overline{A}^{n-1}T\right)T^{-1}\overline{B}\right] \\ &= \left[B \quad AB \quad \cdots \quad A^{n-1}B\right]\end{aligned} \tag{9.44}$$

and

$$\begin{aligned}Q &= \overline{Q}T \\ &= \begin{bmatrix}\overline{C} \\ \overline{C}\overline{A} \\ \vdots \\ \overline{C}\overline{A}^{n-1}\end{bmatrix}T = \begin{bmatrix}\overline{C}T \\ \overline{C}\overline{A}T \\ \vdots \\ \overline{C}\overline{A}^{n-1}T\end{bmatrix} = \begin{bmatrix}\overline{C}T \\ \overline{C}T\left(T^{-1}\overline{A}T\right) \\ \vdots \\ \overline{C}T\left(T^{-1}\overline{A}^{n-1}T\right)\end{bmatrix}\end{aligned}$$

$$= \begin{bmatrix} C \\ CA \\ \vdots \\ CA^{n-1} \end{bmatrix} \tag{9.45}$$

These relationships imply that $A = T^{-1}\overline{A}T$, $B = T^{-1}\overline{B}$, and $C = \overline{C}T$, meaning that $\{A, B, C, D\}$ and $\{\overline{A}, \overline{B}, \overline{C}, \overline{D}\}$ are indeed related by the similarity transformation T. This proves the minimal realization theorem.

9.3.2 Hankel Matrices and System Order

We have shown that for transfer function $g(s)$, a minimal realization is one that is of minimal order, and we know that this realization will be both controllable and observable. However, we have not yet shown that this order will be, as we have been implicitly assuming since Chapter 1, the same as the degree of the denominator in the SISO case, i.e., n. This can be shown with Markov parameters. We will do this explicitly only for the SISO situation.

We form an $(n_r \times n_c)$ matrix from the Markov parameters, called a *Hankel* matrix,[M] as follows:

`hankel(C)`

$$H(n_r, n_c) = \begin{bmatrix} h(1) & h(2) & \cdots & h(n_c) \\ h(2) & h(3) & \cdots & h(n_c+1) \\ \vdots & \vdots & \ddots & \vdots \\ h(n_r) & h(n_r+1) & \cdots & h(n_r+n_c-1) \end{bmatrix}$$

$$= \begin{bmatrix} cb & cAb & \cdots & cA^{n_c-1}b \\ cAb & cA^2b & \cdots & cA^{n_c}b \\ \vdots & \vdots & \ddots & \vdots \\ cA^{n_r-1}b & cA^{n_r}b & \cdots & cA^{n_r+n_c-2}b \end{bmatrix} \tag{9.46}$$

The construction of this matrix is easier to justify for discrete-time systems than for continuous-time systems and will be presented in Section 9.5. For now, we assume that we can perform a power series expansion on the transfer function to obtain the Markov parameters analytically. The main result of this section can now be stated in terms of this Hankel matrix as the following theorem:

> THEOREM: Given a sequence of Markov parameters in terms of the system matrices in (9.33), the order n of the

transfer function that is realized by these matrices (and hence, the size of the minimal realization) is

$$n = \max_{n_r, n_c} r\left(H(n_r, n_c)\right) \tag{9.47}$$

i.e., the maximal rank of the Hankel matrix as more rows and columns are considered.

This theorem implies that if we begin constructing the so-called "infinite Hankel matrix,"

$$H(n_r, n_c) = \begin{bmatrix} h(1) & h(2) & h(3) & \cdots \\ h(2) & h(3) & & \\ h(3) & & \ddots & \\ \vdots & & & \end{bmatrix} \tag{9.48}$$

its rank will grow up to a maximum value of n, which will be the size of the minimal realization. (Why we would happen to have such a sequence of Markov parameters for a system as opposed to some other information will become clear in Section 9.5.) The theorem is easily proven by observing that if indeed we have a minimal realization (of order n), then by (9.36),

$$H(n, n) = QP \tag{9.49}$$

where Q and P are, respectively, the rank n observability and controllability matrices constructed from A, B, and C. If an additional row or column is added to (9.49), then by the Cayley-Hamilton theorem, that row or column must be linearly dependent on a previous row or column, so the rank of (9.48) cannot exceed n. Thus, $r\left(H(n_r, n_c)\right) = n$ for all n_r and $n_c > n$.

Realizations from Hankel Matrices

In addition to indicating the size of the minimal realization of a system, the Hankel matrix may be used to give a realization similar to those seen in Section 9.2. We will present this method as an algorithm applicable to MIMO as well as SISO systems, because it is considerably easier to treat single-variable systems as a special case than it would be to generalize to multivariable systems had we given a presentation that applied only to single-variable systems.

We therefore begin by assuming that we have available a Hankel matrix $H(n_r, n_c)$, with $n_r > n+1$, $n_c > n$, and n being the size of the minimal realization. If n is not known, we can successively test the rank of the Hankel matrix as we add rows and columns to it, until it appears to stop growing in rank. This might result in a realization that is *smaller* than the actual minimal realization, but it would probably qualify as a reasonable approximation, since it would lack only in high-order Markov parameters, i.e., the ones most likely to be negligible in the power series expansion in (9.31). It is quite reasonable to seek an mth-order realization that approximates an nth-order realization, $m < n$, especially if the transfer functions are similar. This is known as the *partial realization* problem and will be addressed again in Section 9.4. The algorithm is as follows:

1. Search the rows of the $(n_r p \times n_c q)$ (p inputs, q outputs) matrix $H(n_r, n_c)$, from top to bottom, for the first n linearly independent rows. (Again, if we do not know n, we can choose an integer likely to approximate n.) Denote these rows as rows $r_1, r_2, \cdots, r_n$. Gather these rows into a single $(n \times n_c q)$ matrix called H_r.
2. Gather into another $(n \times n_c q)$ matrix the rows $r_1 + q, r_2 + q, \cdots, r_n + q$ of $H(n_r, n_c)$ and call this matrix H_{r+q}. We asked that $n_r > n+1$ in order to have sufficient rows available to perform this step.
3. From matrix H_r, search from left to right for the first n linearly independent columns and refer to these as columns $c_1, c_2, \cdots, c_n$. Gather these columns together and call the resulting invertible $(n \times n)$ matrix M.
4. Create the $(n \times n)$ matrix M_A as columns $c_1, c_2, \cdots, c_n$ from the matrix H_{r+q}.
5. Create the $(n \times p)$ matrix M_B as the first p columns of matrix H_r.
6. Create the $(q \times n)$ matrix M_C as columns $c_1, c_2, \cdots, c_n$ from the matrix $H(1, n_c)$, i.e., from the first *block* row of the Hankel matrix.

From the matrices thus defined, the realization can be given as:

$$A \stackrel{\Delta}{=} M_A M^{-1} \qquad B \stackrel{\Delta}{=} M_B \qquad C \stackrel{\Delta}{=} M_C M^{-1} \tag{9.50}$$

Example 9.1: Realization from a Hankel Matrix

The SISO transfer function

$$g(s) = \frac{s+2}{s^2 + 4s + 3} \tag{9.51}$$

can be expanded into a power series that yields the following sequence of Markov parameters:

$$\begin{aligned} h(0) &= 0 \\ h(1) &= 2 \\ h(2) &= -7 \\ h(3) &= 22 \\ h(4) &= -67 \\ h(5) &= 202 \\ &\vdots \end{aligned} \tag{9.52}$$

Find a realization from this set of Markov parameters.

Solution:

The fact that the Markov parameter $h(0) = 0$ indicates that the system will have no feedthrough term, i.e., $d = 0$. If we construct the Hankel matrices

$$H(1,1) = [2] \quad H(2,2) = \begin{bmatrix} 2 & -7 \\ -7 & 22 \end{bmatrix} \quad H(3,3) = \begin{bmatrix} 2 & -7 & 22 \\ -7 & 22 & -67 \\ 22 & -67 & 202 \end{bmatrix} \dots \tag{9.53}$$

we find that $r(H(3,3)) = r(H(2,2)) = 2$. We assume that a realization of order 2 is therefore desired. (In fact, it is easy to show from rank arguments on the expression in (9.46) that, for noiseless data, when the rank first stops growing, it is sufficient to stop; no further rank increase can be expected.) We will therefore seek a second-order realization. Using a Hankel matrix with an extra row, we will start with

$$H(2,3) = \begin{bmatrix} 2 & -7 \\ -7 & 22 \\ 22 & -67 \end{bmatrix} \tag{9.54}$$

Searching for the $n = 2$ linearly independent rows of this matrix gives the result that $r_1 = 1$, $r_2 = 2$, and

$$H_r = \begin{bmatrix} 2 & -7 \\ -7 & 22 \end{bmatrix}$$

Correspondingly, because $q = 1$,

$$H_{r+q} = \begin{bmatrix} -7 & 22 \\ 22 & -67 \end{bmatrix}$$

The first two linearly dependent columns of H_r are columns $c_1 = 1$ and $c_2 = 2$. Steps 3, 4, and 6 in the algorithm give, respectively,

$$M = \begin{bmatrix} 2 & -7 \\ -7 & 22 \end{bmatrix} \qquad M_A = \begin{bmatrix} -7 & 22 \\ 22 & -67 \end{bmatrix} \qquad M_C = \begin{bmatrix} 2 & -7 \end{bmatrix}$$

and the $p = 1^{st}$ column of H_r is

$$M_B = \begin{bmatrix} 2 \\ -7 \end{bmatrix}$$

From all these values, we have the system matrices

$$A = M_A M^{-1} = \begin{bmatrix} 0 & 1 \\ -3 & -4 \end{bmatrix} \qquad b = M_B = \begin{bmatrix} 2 \\ -7 \end{bmatrix}$$

$$c = M_C M^{-1} = \begin{bmatrix} 1 & 0 \end{bmatrix} \qquad d = h(0) = 0, \tag{9.55}$$

These matrices give a second-order realization for the transfer function in (9.51).

9.4 Balanced Realizations

We now know that uncontrollable or unobservable modes do not contribute to the transfer function of a system, and that realizations of minimal order therefore result in completely controllable and observable models. We could, in principle, reduce the order of our nonminimal state space model by performing the Kalman decomposition and retaining only the fully controllable and observable subsystems, thus achieving *model reduction* without altering the input/output behavior of the system (i.e., the transfer function). However, it has been found that the controllability and observability of a system (and in particular, a mode) can depend on arbitrarily small variations in the parameters of the system matrices. That is, arbitrarily small perturbations of a parameter might change a controllable system to an uncontrollable system, thereby changing a minimal realization to a nonminimal one, without significantly changing the transfer function. However, up to this point, we have no way of determining which mode is *more* or *less* controllable and/or observable.

balreal(sys)

We now investigate further this notion of model reduction with a special realization known as a *balanced realization*[M] [13]. The balanced realization identifies the modes that are *almost* unobservable and uncontrollable, making them candidates for deletion. However, as revealed by the next section, we must be careful to delete only the modes that are almost *both* uncontrollable and unobservable.

9.4.1 Grammians and Signal Energy

gram(sys,'c')

gram(sys,'o')

Restricting our discussion to continuous-time LTI systems, we will use the controllability and observability grammians[M] discussed in Chapter 8. The continuous-time grammians, Equations (8.57) and (8.69), are defined as

$$G_c(t_0,t_1)=\int_{t_0}^{t_1} e^{A(t_0-t)}BB^{\mathrm{T}}e^{A^{\mathrm{T}}(t_0-t)}\,dt \tag{9.56}$$

and

$$G_o(t_0,t_1)=\int_{t_0}^{t_1} e^{A^{\mathrm{T}}(t-t_0)}C^{\mathrm{T}}Ce^{A(t-t_0)}\,dt \tag{9.57}$$

respectively. From Chapter 8, the system is controllable and observable in the interval $[t_0,t_1]$ if and only if these two symmetric matrices are invertible.

Suppose we are given an arbitrary initial state $x(t_0)=x_0$ and wish to find a control that drives the system from this state to any other state $x(t_1)=x_1$. It is easily verified by substitution into the state equation solution, i.e., Equation (6.6), that a control signal that accomplishes this can be written as:

$$u(t)=B^{\mathrm{T}}e^{A^{\mathrm{T}}(t_0-t)}G_c^{-1}(t_0,t_1)\left[e^{A(t_0-t_1)}x_1-x_0\right] \tag{9.58}$$

[a similar control was used in Equation (8.58)]. If we define the *energy* content of this control signal as

$$E_u=\int_{t_0}^{t_1}\|u(t)\|^2\,dt=\int_{t_0}^{t_1}u^{\mathrm{T}}(t)u(t)\,dt \tag{9.59}$$

then it can be shown that the energy of the particular control signal in (9.58) is smaller than or equal to the energy of any other control signal that accomplishes the same transfer from x_0 to x_1 in the same time [4]. This energy becomes

$$
\begin{aligned}
E_u &= \int_{t_0}^{t_1} u^{\mathrm{T}}(t)u(t)\,dt \\
&= \int_{t_0}^{t_1} \left[[e^{A(t_0-t_1)}x_1 - x_0]^{\mathrm{T}} G_c^{-1}(t_0,t_1) e^{A(t_0-t)} B \right] \\
&\qquad \cdot \left[B^{\mathrm{T}} e^{A^{\mathrm{T}}(t_0-t)} G_c^{-1}(t_0,t_1) [e^{A(t_0-t_1)}x_1 - x_0] \right] dt \\
&= [e^{A(t_0-t_1)}x_1 - x_0]^{\mathrm{T}} G_c^{-1}(t_0,t_1) \left[\int_{t_0}^{t_1} e^{A(t_0-t)} B B^{\mathrm{T}} e^{A^{\mathrm{T}}(t_0-t)}\, dt \right] \\
&\qquad \cdot G_c^{-1}(t_0,t_1) [e^{A(t_0-t_1)}x_1 - x_0] \\
&= [e^{A(t_0-t_1)}x_1 - x_0]^{\mathrm{T}} G_c^{-1}(t_0,t_1) [e^{A(t_0-t_1)}x_1 - x_0]
\end{aligned}
$$

Taking $t_0 = 0$ and $x_1 = 0$, then the energy to reach the origin in time t_1 becomes simply

$$
E_u = x_0^{\mathrm{T}} G_c^{-1}(0,t_1) x_0 \tag{9.60}
$$

In a similar fashion, suppose we wish to compute the energy in the output as the system decays from the initial condition x_0 to zero in the absence of any input. This energy would be:

$$
\begin{aligned}
E_y &= \int_0^{t_1} \|y(t)\|^2\, dt \\
&= \int_0^{t_1} x_0^{\mathrm{T}} e^{A^{\mathrm{T}}t} C^{\mathrm{T}} C e^{At} x_0\, dt \\
&= x_0^{\mathrm{T}} G_o(0,t_1) x_0
\end{aligned} \tag{9.61}
$$

Thus, the control energy and the "observation" energy are quadratic forms on the initial condition. Because the grammians are both symmetric positive-definite, we can interpret the expressions in (9.60) and (9.61) according the geometric interpretation of quadratic forms as discussed in Chapters 4 and 5. Recall that a symmetric positive-definite quadratic form defines the surface of an ellipsoid whose major axes lie along the eigenvectors, with lengths along those axes equal to the corresponding eigenvalues. Equation (9.60) implies that certain initial conditions require more energy to control than others, namely, those at the long axes of the ellipsoid defined by the *inverse* controllability grammian. Initial conditions at such locations are "harder" to control than others. Equation (9.61) implies that certain initial conditions provide more observation energy or are "easier" to observe. These conditions are at the long end of the ellipsoid defined by the observability grammian.

In general, though, the controllability and observability grammians are independent of each other, so a mode that is easy to control might be difficult to observe (i.e., it might take little control energy, but also contributes little to the output) or vice versa. This is why we cannot simply use grammians to deduce that modes that are almost uncontrollable *or* almost unobservable do not contribute to the transfer function; on the contrary, their corresponding observability or controllability properties may give just the opposite effect.

9.4.2 Internal Balancing

The definitions for the grammians, Equations (9.56) and (9.57), are often given with limits of integration of 0 and ∞, which results in no loss of generality for time-invariant systems. If this is the case and if the system is asymptotically stable, then the grammians can also be found as the positive-definite solutions to the two Lyapunov equations[M]

`lyap(A,Q)`

$$AG_c + G_c A^{\mathrm{T}} = -BB^{\mathrm{T}} \tag{9.62}$$

and

$$A^{\mathrm{T}} G_o + G_o A = -C^{\mathrm{T}} C \tag{9.63}$$

Using these equations, we will investigate the effect of a (not necessarily orthonormal) similarity transformation, $\tilde{x} = Tx$. We already know that the state matrices transform as

$$\tilde{A} = T^{-1}AT \qquad \tilde{B} = T^{-1}B \qquad \tilde{C} = CT \tag{9.64}$$

In the new basis, the Lyapunov equation that gives the controllability grammian becomes

$$\begin{gathered}\tilde{A}\tilde{G}_c + \tilde{G}_c \tilde{A}^{\mathrm{T}} = -\tilde{B}\tilde{B}^{\mathrm{T}} \\ T^{-1}AT\tilde{G}_c + \tilde{G}_c T^{\mathrm{T}} A^{\mathrm{T}} \left(T^{-1}\right)^{\mathrm{T}} = -T^{-1}BB^{\mathrm{T}}\left(T^{-1}\right)^{\mathrm{T}}\end{gathered} \tag{9.65}$$

Multiplying this equation from the left by T and from the right by T^{T}, we get

$$A\left(T\tilde{G}_c T^{\mathrm{T}}\right)+\left(T\tilde{G}_c T^{\mathrm{T}}\right)A^{\mathrm{T}}=-BB^{\mathrm{T}}$$

implying that

$$\begin{aligned} G_c &= T\tilde{G}_c T^{\mathrm{T}} \\ \tilde{G}_c &= T^{-1}G_c\left(T^{-1}\right)^{\mathrm{T}} \end{aligned} \tag{9.66}$$

Performing analogous steps on the observability grammian and on Equation (9.63) results in

$$\begin{aligned} G_o &= \left(T^{-1}\right)^{\mathrm{T}}\tilde{G}_o T^{-1} \\ \tilde{G}_o &= T^{\mathrm{T}}G_o T \end{aligned} \tag{9.67}$$

From (9.66) and (9.67), it is apparent that the controllability and observability grammians can be scaled by a similarity transformation, so their eigenvalues can be changed by appropriate selection of transformation T. Note however that their product

$$\begin{aligned} \tilde{G}_c\tilde{G}_o &= T^{-1}G_c\left(T^{-1}\right)^{\mathrm{T}}T^{\mathrm{T}}G_o T \\ &= T^{-1}G_c G_o T \end{aligned} \tag{9.68}$$

which implies that the eigenvalues of the product G_cG_o are the same as the eigenvalues of the product $\tilde{G}_c\tilde{G}_o$. There therefore exists a trade-off between the controllability ellipsoid and the observability ellipsoid.

In order to exploit this trade-off, we seek a transformation T that results in

$$\tilde{G}_c=\tilde{G}_o=\Sigma$$

where Σ is a diagonal matrix containing the square roots of the eigenvalues of the product G_cG_o. This will accomplish two things. First, because the new grammians will be diagonal, their major axes will be aligned with the coordinate axes in the new basis. Second, because the grammians will be equal, each mode (i.e., along each coordinate axis) will be just *as controllable as it is observable*. Modes with large control energies will have small observation energies, and vice versa. Modes with large control energies, being *almost uncontrollable*, and with small observation energies, being *almost unobservable*, will thus be identified. They can then be assumed to have a relatively small effect on the transfer function and might be deleted from a realization. Such a realization is said to be *internally balanced.* It is important to note that we are restricted to stable systems, so that we are safe from discarding unstable modes.

In [13], an algorithm to accomplish this balancing is given as follows:

1. Determine the singular value decomposition of the positive-definite matrix G_c. Because of its symmetry, the result will be of the form $G_c = M_c \Sigma_c M_c^T$, where Σ_c is the diagonal matrix of eigenvalues of G_c.

2. Define a preliminary transformation as $T_1 = M_c \Sigma_c^{1/2}$ and apply this transformation[M] to the original realization $\{A, B, C\}$ to get $\{A_1, B_1, C_1\} = \{T_1^{-1} A T_1, T_1^{-1} B, C T_1\}$.

ss2ss(sys,T)

3. Compute the new observability grammian $\hat{G}_o$ from the transformed system and determine its singular value decomposition $\hat{G}_o = M_o \hat{\Sigma}_o M_o^T$, where $\hat{\Sigma}_o$ is a diagonal matrix of the eigenvalues of $\hat{G}_o$.

4. Define the transformation $T_2 = M_o \Sigma_o^{-1/4}$. Apply this new transformation to the previously transformed system to result in a composite transformation $T = T_1 T_2$. The balanced realization is then

$$
\begin{aligned}
\tilde{A} &= T^{-1} A T = T_2^{-1} T_1^{-1} A T_1 T_2 \\
\tilde{B} &= T^{-1} B = T_2^{-1} T_1^{-1} B \\
\tilde{C} &= C T = C T_1 T_2
\end{aligned}
\tag{9.69}
$$

Suppose that when a system is thus balanced, the transformation matrices are arranged such that $\tilde{G}_c = \tilde{G}_o = \Sigma = diag(\sigma_1, \ldots, \sigma_j, \sigma_{j+1}, \ldots, \sigma_n)$, with $\sigma_1 \geq \sigma_2 \geq \cdots \geq \sigma_n$. Suppose now that $\sigma_j \gg \sigma_{j+1}$. Then it can be concluded that modes corresponding to state variables $\tilde{x}_{j+1}$ through $\tilde{x}_n$ are much less controllable *and* observable than are modes corresponding to variables $\tilde{x}_1$ through $\tilde{x}_j$. Then in the partitioned state equations

$$
\begin{bmatrix} \dot{\tilde{x}}_1 \\ \vdots \\ \dot{\tilde{x}}_j \\ \hline \dot{\tilde{x}}_{j+1} \\ \vdots \\ \dot{\tilde{x}}_n \end{bmatrix} = \left[\begin{array}{c|c} \tilde{A}_{11} & \tilde{A}_{12} \\ \hline \tilde{A}_{12} & \tilde{A}_{22} \end{array} \right] \begin{bmatrix} \tilde{x}_1 \\ \vdots \\ \tilde{x}_j \\ \hline \tilde{x}_{j+1} \\ \vdots \\ \tilde{x}_n \end{bmatrix} + \begin{bmatrix} \tilde{B}_1 \\ \hline \tilde{B}_2 \end{bmatrix} u
$$

$$y = \begin{bmatrix} \tilde{C}_1 & \tilde{C}_2 \end{bmatrix} \begin{bmatrix} \tilde{x}_1 \\ \vdots \\ \tilde{x}_j \\ \hline \tilde{x}_{j+1} \\ \vdots \\ \tilde{x}_n \end{bmatrix} \tag{9.70}$$

the true, exact transfer function $H(s) = \tilde{C}(sI - \tilde{A})^{-1}\tilde{B}$ can be approximated by the transfer function of the simplified, reduced system:

$$\tilde{H}(s) = \tilde{C}_1(sI - \tilde{A}_{11})^{-1}\tilde{B}_1 \tag{9.71}$$

This system can also be shown to be asymptotically stable. (Note that the same concepts apply to discrete-time systems, but the computations are different because of the different Lyapunov equations. See [14].)

9.5 Discrete-Time System Identification

We close this chapter with a discussion of one of the most useful applications of Markov parameters: discrete-time system identification. System identification is the name given to the process of determining a system description (e.g., the matrices A, B, and C) from input/output data (the feedthrough matrix D is trivially determined by direct measurement). There are many approaches to this problem, including both frequency-domain (e.g., curve-fitting frequency responses) and time-domain (i.e., finding a realization). System identification has become an important topic for controls engineers, because often in practical problems the description of the plant is not analytically available. Without a mathematical model for the plant, most control strategies are useless.

The methods we use here follow directly from the definitions and uses of Markov parameters for realizations as discussed above. In fact, the algorithm given here can be used interchangeably with the algorithm given in Section 9.3.2; they will however result in different realizations. In Section 9.3.2, we assumed that the Markov parameters were available as a result of a series expansion on the transfer function. Here, we assume that we do not know the transfer function but that we have access to input and output data. We consider discrete-time systems because in most situations, empirical input and output measurements are obtained at discrete-time instants. It is therefore natural to seek a discrete-time model, even if the plant itself is continuous.

9.5.1 Eigensystem Realization Algorithm

For the first method we will use a sequence of noiseless measurements of input/output pairs of a discrete-time system [7]. As in Example 3.11, we can assume a zero-state response and generate, in response to the known input sequence $u(0), u(1), \cdots$, the following output sequence:

$$\begin{aligned} y(0) &= Du(0) \\ y(1) &= CBu(0) + Du(1) \\ y(2) &= CABu(0) + CBu(1) + Du(0) \\ &\vdots \end{aligned} \tag{9.72}$$

or

$$\begin{bmatrix} y(0) \\ y(1) \\ y(2) \\ \vdots \\ y(j-1) \end{bmatrix} = \begin{bmatrix} D & 0 & 0 & 0 & 0 \\ CB & D & 0 & 0 & 0 \\ CAB & CB & D & 0 & 0 \\ \vdots & \ddots & \ddots & D & 0 \\ CA^{j-1}B & \cdots & CAB & CB & D \end{bmatrix} \begin{bmatrix} u(0) \\ u(1) \\ u(2) \\ \vdots \\ u(j-1) \end{bmatrix} \tag{9.73}$$

Clearly, in this case, the natural form of the input/output data in discrete-time is the Markov parameter. Given such pairs of input/output data, we can rearrange (9.73) to give the formula:

$$\begin{aligned} &\begin{bmatrix} y(0) & y(1) & \cdots & y(j-1) \end{bmatrix} \\ &= \begin{bmatrix} h(0) & h(1) & \cdots & h(j-1) \end{bmatrix} \begin{bmatrix} u(0) & u(1) & \cdots & u(j-1) \\ 0 & u(0) & \ddots & \vdots \\ \vdots & \ddots & \ddots & u(1) \\ 0 & \cdots & 0 & u(0) \end{bmatrix} \end{aligned}$$

Defining this equation symbolically as $Y \triangleq H_1 U$, then if the matrix U is invertible, we can solve for the Markov parameters:

$$H_1 = \begin{bmatrix} h(0) & h(1) & \cdots & h(j-1) \end{bmatrix} = YU^{-1} \tag{9.74}$$

provided, of course, that matrix U is indeed invertible. The invertibility of U is related to the question of whether the input is *sufficiently exciting* for the equation to be solved. If our simple system is single input, then we clearly would require the property $u(0) \neq 0$.

Even if the initial condition on the state is nonzero, the input/output sequence is still sufficient for identifying the system. This is because we will be computing a minimal realization, which is therefore controllable and observable. For a controllable and observable system, we will always be able to *make* the initial state zero by choosing appropriate input signals in negative time. Then because the starting time for a time-invariant system is irrelevant for identification purposes, we can include the negative-time input/output data into (9.73) above while estimating the system matrices A, B, and C [4].

Thus, assuming that the Markov parameters for a MIMO system are known, we can begin constructing, after k samples have been acquired, the $(kq \times kp)$ block Hankel matrix as in Equation (9.46):

$$H_k = H(k,k) = \begin{bmatrix} CB & CAB & \cdots & CA^{k-1}B \\ CAB & CA^2B & & \\ \vdots & & \ddots & \vdots \\ CA^{k-1}B & & \cdots & CA^{2(k-1)}B \end{bmatrix} = \begin{bmatrix} H(1) & H(2) & \cdots & H(k) \\ H(2) & H(3) & & \\ \vdots & & \ddots & \vdots \\ H(k) & & \cdots & H(2k-1) \end{bmatrix} \tag{9.75}$$

As mentioned before, we will increase the size of this matrix by acquiring new data until the rank of (9.75) stops growing. When this occurs, the order of the minimal realization will be equal to the order of the block Hankel matrix when it stops growing, i.e., $n = k$ such that $n = r(H_k) = r(H_{k+1}) = r(H_{k+2}) = \cdots$. According to the eigensystem realization algorithm (ERA) [7], the next step is to perform a singular value decomposition on H_k:

$$H_k = U\Sigma V^{\mathrm{T}} \tag{9.76}$$

To carry out the rest of the algorithm, we compute the following matrices:

- U_1 $(kq \times n)$ is defined as the first n columns from the $(kq \times kq)$ matrix U.
- V_1 $(kp \times n)$ is defined as the first n columns from the $(kp \times kp)$ matrix V.
- U_{11} $(q \times n)$ is defined as the first q rows of the matrix U_1 (where q is the number of outputs).
- V_{11} $(p \times n)$ is defined as the first p rows of the matrix V_1 (where p is the number of inputs).

- S is defined as the $(n \times n)$ invertible, diagonal matrix as in Equation (4.34):

$$\Sigma = \begin{bmatrix} S & 0 \\ 0 & 0 \end{bmatrix}$$

where $S = diag\{\sigma_1, \sigma_2, \ldots, \sigma_n\}$.

- $\widetilde{H}_k$ is defined as the modified Hankel matrix

$$\widetilde{H}_k = \begin{bmatrix} H(2) & H(3) & \cdots & H(k+1) \\ H(3) & H(4) & & \\ \vdots & & \ddots & \vdots \\ H(k+1) & & \cdots & H(2k) \end{bmatrix}$$

Given these computations, the system matrices may be derived as

$$\begin{aligned} A &= S^{-\frac{1}{2}} U_1^T \widetilde{H}_k V_1 S^{-\frac{1}{2}} \\ B &= S^{\frac{1}{2}} V_{11}^T \\ C &= U_{11} S^{\frac{1}{2}} \end{aligned} \tag{9.77}$$

Again, if there is a feedthrough term, it is measured directly, just like the Markov parameters.

9.5.2 Simplified ERA Identification

The basic procedure spelled out in the algorithm above can be altered to give different (minimal) realizations for the same set of Markov parameters [5]. The only modification to the ERA algorithm above is that instead of first performing the SVD on the block Hankel matrix, only selected rows and columns of the Hankel matrix are used; this step is similar to the procedure used for the realization based on the Hankel matrix in Section 9.3.2. The algorithm then offers some flexibility in the form of the resulting realization. The steps in this modified algorithm are:

1. Choose n linearly independent columns of H_k. Arrange them into a matrix defined as H_1. Denote the chosen columns as $v_1, v_2, \ldots, v_n$.

2. Select from H_k the columns $v_1 + p, v_2 + p, \ldots, v_n + p$ and arrange them into a matrix called H_2.

3. The system matrices in a special canonical form (the "pseudo-controllable canonical form") are computed as follows:

$$A = \left(H_1^{\mathrm{T}} H_1\right)^{-1} H_1^{\mathrm{T}} H_2$$
$$B = \begin{bmatrix} I_{p\times p} \\ 0_{(n-p)\times p} \end{bmatrix} \tag{9.78}$$

   where A is the least-squares solution to the equation $H_1 A = H_2$ and C is the first q rows of H_1.

The flexibility offered by this approach is that the choice of the particular columns of H_k can be guided by special integers known as *(pseudo)controllability indices*. These numbers pertain to the number of modes that can be controlled by each individual input, i.e., each column of the B-matrix. There may be several sets of controllability indices so the designer will have certain freedoms in selecting the form of the resulting system matrices. In fact, an analogous procedure can be written in terms of the *(pseudo)observability indices* and the *rows* of the Hankel matrix. However, because we have not yet encountered controllability and observability indices, we will not delve further into these options. Although this algorithm does indeed offer such flexibilities, the computational complexity is comparable to the original ERA, because the best way to accurately compute (9.78) is through the SVD.

9.6 Summary

In this chapter we have tied together the two most common representations for linear systems: the transfer function and the state space description. In essence, the material presented in this chapter represents the more rigorous theory behind the material we presented in Chapter 1. In doing so, we have answered some questions posed or suggested in that chapter. For example, we have discovered why transfer functions derived from state matrices do not necessarily have degree equal to the order of the A-matrix. We also learned that there are as many different realizations as there are similarity transformations (infinite).

Perhaps more importantly, this chapter introduces the practical techniques necessary for the actual physical construction of linear systems (and, as we shall see, their controllers). By constructing the realizations herein, time-domain systems can be modeled from frequency-domain or empirical data. (Usually, these are implemented on a computer, although analog state variable controllers are becoming more common.) This led us naturally into the subject of system identification, which is itself the topic of many books and papers, e.g., [10] and [11]. The small subset of system identification procedures mentioned here is the construction of a minimal realization from the Markov parameters that naturally result from the measurement of input and output data in a discrete-time system.

Summarizing the important results of this chapter:

- It was found that uncontrollable and/or unobservable modes do not contribute to the transfer function of a system, so that if it is only the input/output behavior that is of interest, such modes (provided they are stable) can be discarded from the state space model. Thus, we have a *minimal* realization when the model is both controllable and observable.

- Several specific realizations were shown, including the controllable canonical, observable canonical, and Jordan canonical forms. We see from these realizations that there is a direct correspondence between similarity transformations and integrator realizations (i.e., simulation diagrams). The further significance of the controllable and observable canonical realizations will be revealed in the next chapter.

- Markov parameters were introduced as the coefficients of a power series expansion of the (causal) transfer function. It was seen that Markov parameters are invariant to similarity transformations. Arranged into the Hankel matrix, Markov parameters provide a means of generating a realization either from this series expansion or from empirical input/output data from discrete-time systems.

- The Hankel matrix itself is also shown to be the product of the observability and controllability matrices for a realization. Its rank, as more rows and columns are added to it, grows to a maximal size that is equal to the order of the minimal realization. An algorithm is given to generate state matrices based on the Hankel matrix.

- Balanced realizations were introduced to tie together the ideas of controllability, observability, and system order. If a partial realization, i.e., one with smaller order than the minimal realization, is desired, a balanced realization can be computed that illustrates the extent to which individual modes are almost uncontrollable and unobservable. Balanced realizations are particularly important to the study of large-scale systems, where model reduction is sometimes the first order of business.

- As we have mentioned, system identification is sometimes the first step in the design of a controller for a practical physical system. Many times, systems are either too complex or too poorly modeled to provide an analytical description of a plant. In these situations, it is first necessary to perform system identification so that a working model is available as the basis for controller design. By acquiring input/output data in discrete-time, a sequence of Markov parameters can be used to generate a model for "black-box" data. Although this technique nicely illustrates some of the contents of this chapter, it is neither the most efficient nor common method for performing realistic system identification.

In the next chapter, we will at last be able to use the tools we have developed so far to design controllers and observers. The concepts of control via state feedback and observation are fundamental to time-domain linear system studies, and we will explore them in Chapter 10.

9.7 Problems

9.1 Find one minimal and one nonminimal realization for the transfer function

$$g(s)=\frac{s^3+6s^2+3s-10}{s^3+4s^2+s-6}$$

9.2 Find a minimal realization for the transfer function

$$g(z)=\frac{z+3}{z^2+7z+10}$$

9.3 Show using the Kalman observability decomposition that the unobservable part of a realization does not contribute to the transfer function for that system. [Similar to the construction of Equation (9.14).]

9.4 Given below are two different realizations.

a) Show that they are each minimal realizations of the same system.

b) Find the similarity transformation matrix T that relates the two.

$$A_1=\begin{bmatrix}0 & 1 & 0\\ 0 & 0 & 1\\ -12 & -19 & -8\end{bmatrix}\qquad b_1=\begin{bmatrix}0 & 1\\ 1 & 1\\ 0 & -1\end{bmatrix}\qquad c_1=\begin{bmatrix}1 & 0 & 0\\ -1 & 0 & 0\end{bmatrix}$$

$$A_2=\begin{bmatrix}-90 & 29 & -55\\ -3 & 1 & -2\\ 133 & -42 & 81\end{bmatrix}\qquad b_2=\begin{bmatrix}-1 & -2\\ 0 & 1\\ 1 & 3\end{bmatrix}\qquad c_2=\begin{bmatrix}0 & 1 & 0\\ 0 & -1 & 0\end{bmatrix}$$

9.5 Compute a minimal realization for the sequence of Markov parameters $h(i) = 0, 1, -4, 16, -64, 256, \ldots$, for $i = 0, 1, 2, \ldots$.

9.6 A Fibonacci sequence is a sequence of numbers, each of which is the sum of the previous two, for example: $0, 1, 1, 2, 3, 5, 8, 13, 21, \ldots$. Consider these numbers as a sequence of Markov parameters and determine a discrete-time state variable system that produces such a sequence. Express the result as a difference equation.

9.7 The *Padé approximation* used to approximate a time-delay with a transfer function [5] can be viewed as a partial realization problem that uses the Hankel matrix methods outlined in this chapter. Use these methods to determine a second-order Padé approximation to the time delay operator e^{-Ts}. (Hint: expand e^{-Ts} into a McClaurin series in terms of the parameter $\sigma = 1/s$.)

9.8 Write a MATLAB routine (or code in another language) to perform the realization based on a Hankel matrix as in Section 9.3.2.

9.9 Write a MATLAB routine (or code in another language) to perform the ERA as in Section 9.5.1.

`deconv(p,q)`

9.10 Write MATLAB code (or code in another language) that can be used to determine the Markov parameters from an arbitrary proper transfer function. (Hint: Use the DECONV[M] command.)

9.11 For a SISO time-invariant system, show that a realization based on Markov parameters can be written as

$$A = \begin{bmatrix} h(2) & h(3) & \cdots & h(n+1) \\ h(3) & h(4) & & \\ \vdots & & \ddots & \vdots \\ h(n+1) & & \cdots & h(2n-1) \end{bmatrix} \begin{bmatrix} h(1) & h(2) & \cdots & h(n) \\ h(2) & h(3) & & \\ \vdots & & \ddots & \vdots \\ h(n) & & \cdots & h(2n) \end{bmatrix}^{-1}$$

$$b = \begin{bmatrix} h(1) \\ h(2) \\ \vdots \\ h(n) \end{bmatrix} \qquad c = \begin{bmatrix} 1 & 0 & \cdots & 0 \end{bmatrix} \qquad d = h(0)$$

9.12 Show that if the LTI system $\dot{x} = Ax + Bu$ is controllable and asymptotically stable, then there exists a symmetric, positive-definite solution G to the Lyapunov equation

$$AG + GA^{\mathrm{T}} = -BB^{\mathrm{T}}$$

9.13 Show that if the LTI system $x(k+1) = A_d x(k) + B_d u(k)$ is reachable and asymptotically stable, then there exists a symmetric, positive-definite solution G to the Lyapunov equation

$$AGA^{\mathrm{T}} - G = -BB^{\mathrm{T}}$$

9.14 Use Hankel matrix methods to derive realizations for the following transfer functions and draw the integrator realizations for each.

$$G_1(s) = \begin{bmatrix} \dfrac{s}{(s+1)^2} \\ \dfrac{(s+3)}{(s+1)(s+2)} \end{bmatrix} \qquad G_2(s) = \begin{bmatrix} \dfrac{s}{(s+1)^2} & \dfrac{1}{(s+1)^2(s+2)} \\ \dfrac{(s+3)}{(s+1)(s+2)} & \dfrac{(s-2)}{(s+2)} \end{bmatrix}$$

9.15 Show that step 2 in the model-balancing algorithm results in a transformed controllability grammian

$$\hat{G}_c = T_1^{-1} G_c \left(T_1^{-1}\right)^{\mathrm{T}} = I$$

9.16 Perform a model balancing on the following system:

$$\dot{x} = \begin{bmatrix} -3 & 5 & 4 & 0 & 0 \\ 0 & -4 & 2 & 0 & 1 \\ 2 & 0 & -5 & 1 & 0 \\ -2 & 3 & 0 & -8 & 1 \\ 2 & 0 & 3 & 0 & -9 \end{bmatrix} x + \begin{bmatrix} 1 \\ 1 \\ 1 \\ 1 \\ 1 \end{bmatrix} u$$

$$y = \begin{bmatrix} 3 & 1 & 0 & 1 & 0 \end{bmatrix}$$

9.17 The following set of input/output data is collected from a single-input, two-output discrete-time system. Determine a realization and a transfer function that identify the system.

k	0	1	2	3	4	5	6	7	8	9
$u(k)$	1	0.5	0	−0.5	−1	−0.5	0	.5	1	0.5
$y_1(k)$	1	1.5	0.5	−1	−1.25	−1.5	−0.375	0.875	1.3125	1.5
$y_2(k)$	0	−1	1.5	−0.5	0.25	0.5	−1.125	0.375	−0.3125	−0.375

9.8 References and Further Reading

The realization theory we present here can be found in its original form in Kalman [9]. Our discussion of the minimal realization and the significance of controllability and observability are based on that work. It is also discussed in similar terms in the texts [4], [8], and [16].

For a good introduction to Markov parameters and Hankel matrix methods, see [4], [8], [16], and [17]. In addition to obtaining the realizations themselves, we have applied these methods to system identification problems (using the ERA algorithm). However, system identification is a much broader science, and the interested reader should consult [10] and [11]. The ERA algorithm itself can be pursued in further detail, in its original description [7] and in some of the variations and enhancements of it in [1], [2], and [5]. Model balancing and balanced realizations were introduced by Moore in [13].

As was briefly mentioned in Chapter 1, we do not discuss most of the details of frequency-domain representations in this book. However, considerable literature is available, including introductory but comprehensive treatments in [3], [4], [8], [12], and [15].

[1] Bingulac, Stanoje, and N. K. Sinha, "On the Identification of Continuous-Time Multivariable Systems from Input-Output Data," *Proc. 7th International Conference on Mathematical and Computer Modelling*, vol. 14, 1990, pp. 203-208.

[2] Bingulac, Stanoje, and Hugh F. VanLandingham, *Algorithms for Computer-Aided Design of Multivariable Control Systems*, Dekker, 1993.

[3] Callier, Frank M., and Charles A. Desoer, *Multivariable Feedback Systems*, Springer-Verlag, 1982.

[4] Chen, Chi-Tsong, *Linear System Theory and Design*, Holt, Rinehart, and Winston, 1984.

[5] D'Azzo, John D., and Constantine H. Houpis, *Linear Control System Analysis and Design: Conventional and Modern*, 4th edition, McGraw-Hill, 1995.

[6] Gorti, Bhaskar M., S. Bingulac, and H. F. VanLandingham, "Computational Simplification in ERA (Eigensystem Realization Algorithm)," *Proc. 28th Annual Allerton Conference on Communication, Control, and Computing*, October 1990, pp. 529-530.

[7] Ho, B. L., and Rudolph E. Kalman, "Effective Construction of Linear State variable Models from Input/Output Data," *Proc. 3rd Annual Allerton Conference on Circuit and System Theory*, 1966, pp. 449-459.

[8] Kailath, Thomas, *Linear Systems*, Prentice Hall, 1980.

[9] Kalman, Rudolph E. , "Mathematical Description of Linear Dynamical Systems," *Journal of the Society of Industrial and Applied Mathematics-Control Series, Series A*, vol. 1, no. 2, 1963, pp. 152-192.

[10] Ljung, Lennart, *System Identification: Theory for the User*, Prentice Hall, 1987.

[11] Ljung, Lennart, and Torsten Söderström, *Theory and Practice of Recursive Identification*, MIT Press, 1986.

[12] MacFarlane, A. G. J., "A Survey of Some Results in Linear Multivariable Feedback Theory," *Automatica*, vol. 8, no. 4, July 1972, pp. 455-492.

[13] Moore, Bruce C., "Principal Component Analysis in Linear Systems: Controllability, Observability, and Model Reduction," *IEEE Transactions on Automatic Control*, vol. AC-26, no. 1, February 1981, pp. 17-32.

[14] Pernebo, Lars, and Leonard M. Silverman, "Model Reduction via Balanced State Space Representations," *IEEE Transactions on Automatic Control*, vol. AC-27, no. 2, April 1982, pp. 382-387.

[15] Rosenbrock, H. H., *State Space and Multivariable Theory*, John Wiley & Sons, 1970.

[16] Rugh, Wilson, *Linear System Theory*, 2nd edition, Prentice-Hall, 1996.

[17] Silverman, Leonard M., "Realization of Linear Dynamical Systems," *IEEE Transactions on Automatic Control*, vol. AC-16, 1971, pp. 554-567.

10

State Feedback and Observers

We are now ready to consider the *control* of state space systems. In an overly general but realistic way, we can define the term "control" as the alteration of a dynamic system such that it behaves as if it were a different dynamic system. Often, the motivation for control results from more specific criteria, such as stability, the tuning of a transient response, the reduction of error in the output, or improvement in the system's tolerance to disturbance and unmodeled dynamics. Nevertheless, control is achieved through the introduction of a controller, or compensator, which changes the equations to achieve a desired behavior.

In this chapter, we introduce the most fundamental form of control for state space systems: *state feedback*. Using state feedback, we will be able to change the A-matrix of a system, under certain conditions. This will require access to the state variables, which sometimes are not available in physical systems. However, we can reconstruct the state variables using a construct known as an *observer*. As might well be guessed from these introductory remarks, the conditions of controllability and observability will be necessary for the design of controllers and observers.

We will consider in this chapter only time-invariant systems, both continuous-time and discrete-time. At the end of the chapter, we will have discussed all the necessary tools for the design of state-feedback controllers for SISO and MIMO systems. Throughout, we should keep in mind that the concept of observers and controllers is not necessarily an exercise in electronics or mechanical design, but is applicable to mathematical models that are quite broad. For example, state space models (as well as observers and controllers) have been developed for physiological systems, economic models, and social behaviors. It is interesting to consider what the consequences of *control* and *observations* are in those domains [2].

10.1 State Feedback for SISO Systems

The concepts and procedures for state feedback are the same for both continuous-time and discrete-time systems, although there are some distinctions in the results that might be achieved with each. We will therefore present state feedback in continuous-time, as the notation is more compact.

To begin, we resort to an example that will be familiar to readers with any classical controls systems experience. Consider the control configuration in Figure 10.1.

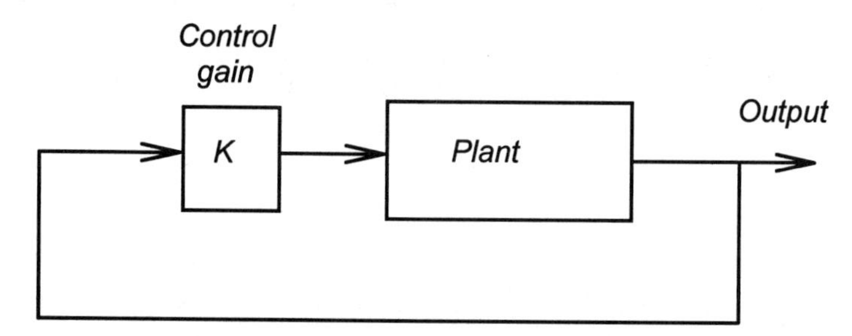

Figure 10.1 Basic regulator configuration for scalar systems.

This is the so-called "regulator system," wherein there is no input to track; there is only a plant output that we wish to maintain at zero. In this type of control, the output is fed back to the input and multiplied by a gain factor. By changing the value of the feedback gain K, the poles of the plant can be moved to different locations in the s- (or z-)plane. Techniques such as root locus may be used to determine the sets on which the poles may move.

If the system in Figure 10.1 is scalar, i.e., one-dimensional, then the output can be made equal to the single state variable, and the location of the single pole can be moved arbitrarily along the real line by choice of gain K. Continuing this idea to state space notation, consider the SISO first-order system

$$\begin{aligned} \dot{x} &= ax + bu \\ y &= cx + du \end{aligned} \tag{10.1}$$

This system can be depicted in the simple block diagram shown in Figure 10.2. We cannot directly access the summing junction at the input, but we can connect the input signal u to a feedback path. That is, we can feed the output y back to the input u, or, under certain circumstances, we may have access to the state variable x and the ability to feed it back to the input. Feeding y back to u is known as *output feedback*, and feeding x back to u is known as *state feedback*. This is shown in the dotted connection in Figure 10.2.

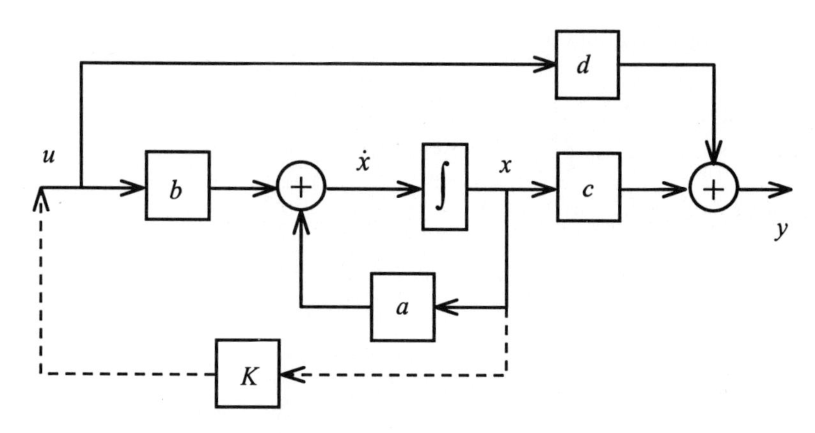

Figure 10.2 Block diagram for a scalar state space system.

State feedback is expressed by the assignment $u = Kx$, and by rewriting the state equations in the presence of such feedback, we get the "closed-loop" system

$$\begin{aligned}\dot{x} &= ax + bKx = (a + bK)x \\ y &= cx + dKx = (c + dK)x\end{aligned} \tag{10.2}$$

It is clear from (10.2) that the behavior of the system will be altered by the introduction of state feedback, and that the single eigenvalue (pole) is dependent on the feedback gain *K*. It is apparent that by appropriately choosing $K = (a^{\text{des}} - a)/b$, where a^{des} is the *desired* eigenvalue, feedback can be used to give the system any pole at all, provided that $b \neq 0$, which is equivalent to asking that the system be controllable.

Extending this idea further, we can consider nonscalar systems. In general time-invariant state space systems, we propose to use feedback of the form

$$u(t) = Kx(t) \tag{10.3}$$

where, as usual, $x \in \Re^n$, so *K* will be a $1 \times n$ matrix of *feedback gains*. More generally, because we may want to preserve the possibility of an exogenous input, which, like $u(t)$ in the original system, might be multiplied by a system matrix, we model the state feedback input as

$$u(t) = Kx(t) + Ev(t) \tag{10.4}$$

where $v(t)$ is the exogenous (i.e., externally applied) input and *E* is its input matrix, just as *B* is the input matrix to signal $u(t)$.

Substituting (10.4) into the general state equations

$$\begin{aligned} \dot{x} &= Ax + bu \\ y &= cx + du \end{aligned} \tag{10.5}$$

we arrive at

$$\begin{aligned} \dot{x} &= Ax + b(Kx + Ev) = (A + bK)x + bEv \\ y &= cx + d(Kx + Ev) = (c + dK)x + dEv \end{aligned} \tag{10.6}$$

whose transfer function is consequently

$$H(s) = (c + dK)(sI - A - bK)^{-1} bE + dE \tag{10.7}$$

As is usually the case, K is assumed here to be constant (i.e., "static" state feedback). The block diagram of the new system is pictured in Figure 10.3.

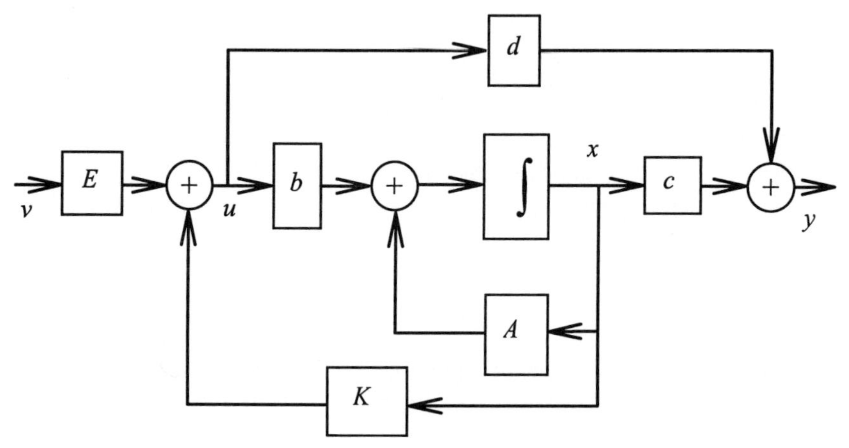

Figure 10.3 Block diagram of state feedback in a state space system.

We know from our stability studies of Chapter 7 that asymptotic stability depends only on the eigenvalues of a time-invariant system, and we also know from the behavior of differential equations that a system's transient response is also dependent on the eigenvalues (or poles, if the realization is minimal). Therefore, we need to study the procedure by which gain matrix K can be found such that the matrix $A + bK$ has a desired set of eigenvalues. Because this matrix does not appear in the output equation for y, we can concentrate on the state equation alone: $\dot{x} = (A + bK)x + bEv$.

10.1.1 Choosing Gain Matrices

The task of selecting an appropriate gain matrix K is facilitated by the controllable canonical form. Recall that the state matrices in this canonical form are:

$$\begin{bmatrix} \dot{x}_1 \\ \vdots \\ \dot{x}_{n-1} \\ \dot{x}_n \end{bmatrix} = \begin{bmatrix} 0 & 1 & 0 & 0 \\ \vdots & \ddots & \ddots & 0 \\ 0 & \cdots & 0 & 1 \\ -a_0 & -a_1 & \cdots & -a_{n-1} \end{bmatrix} \begin{bmatrix} x_1 \\ \vdots \\ x_{n-1} \\ x_n \end{bmatrix} + \begin{bmatrix} 0 \\ \vdots \\ 0 \\ 1 \end{bmatrix} u(t)$$

$$y = \begin{bmatrix} c_1 & c_2 & \cdots & c_n \end{bmatrix} \begin{bmatrix} x_1 \\ \vdots \\ x_{n-1} \\ x_n \end{bmatrix} + du(t) \tag{10.8}$$

Using state feedback expanded explicitly as

$$u = Kx = \begin{bmatrix} k_0 & k_1 & \cdots & k_{n-1} \end{bmatrix} \begin{bmatrix} x_1 \\ x_2 \\ \vdots \\ x_n \end{bmatrix} \tag{10.9}$$

(and neglecting temporarily the exogenous input term and the output equation), we can substitute (10.9) into (10.8) to achieve

$$\begin{aligned} \begin{bmatrix} \dot{x}_1 \\ \vdots \\ \dot{x}_{n-1} \\ \dot{x}_n \end{bmatrix} &= \begin{bmatrix} 0 & 1 & 0 & 0 \\ \vdots & \ddots & \ddots & 0 \\ 0 & \cdots & 0 & 1 \\ -a_0 & -a_1 & \cdots & -a_{n-1} \end{bmatrix} \begin{bmatrix} x_1 \\ \vdots \\ x_{n-1} \\ x_n \end{bmatrix} + \begin{bmatrix} 0 \\ \vdots \\ 0 \\ 1 \end{bmatrix} \begin{bmatrix} k_0 & k_1 & \cdots & k_{n-1} \end{bmatrix} \begin{bmatrix} x_1 \\ \vdots \\ x_{n-1} \\ x_n \end{bmatrix} \\ &= \begin{bmatrix} 0 & 1 & 0 & 0 \\ \vdots & \ddots & \ddots & 0 \\ 0 & \cdots & 0 & 1 \\ -a_0 & -a_1 & \cdots & -a_{n-1} \end{bmatrix} \begin{bmatrix} x_1 \\ \vdots \\ x_{n-1} \\ x_n \end{bmatrix} + \begin{bmatrix} 0 & 0 & \cdots & 0 \\ \vdots & \vdots & \ddots & 0 \\ 0 & 0 & \cdots & 0 \\ k_0 & k_1 & \cdots & k_{n-1} \end{bmatrix} \begin{bmatrix} x_1 \\ \vdots \\ x_{n-1} \\ x_n \end{bmatrix} \\ &= \begin{bmatrix} 0 & 1 & 0 & 0 \\ \vdots & \ddots & \ddots & 0 \\ 0 & \cdots & 0 & 1 \\ -a_0 + k_0 & -a_1 + k_1 & \cdots & -a_{n-1} + k_{n-1} \end{bmatrix} \begin{bmatrix} x_1 \\ \vdots \\ x_{n-1} \\ x_n \end{bmatrix} \end{aligned} \tag{10.10}$$

Knowing that the bottom row of constant coefficients in this matrix is the reverse-ordered set of negative coefficients in the characteristic polynomial for the system, it is clear from (10.10) that the gain matrix K allows us to completely specify the terms in the characteristic polynomial

$$\phi(s) = s^n + (a_{n-1} - k_{n-1})s^{n-1} + \cdots + (a_1 - k_1)s + (a_0 - k_0) \tag{10.11}$$

place(A,B,P)

and hence, allows us to arbitrarily select the poles (eigenvalues) of the system. State feedback is thus a method by which we can achieve *pole placement*.[M]

If the system (10.5) is not originally in the controllable canonical form, then the transformation technique presented in Chapter 8 can be used to transform it to controllable form. Specifically, transformation matrix U – see Equation (8.38) – will be used to transform the general system (10.5) into the controllable canonical form

$$\dot{\bar{x}} = \bar{A}\bar{x} + \bar{b}u \tag{10.12}$$

where, as usual, $x = U\bar{x}$, $\bar{A} = U^{-1}AU$, etc. The feedback applied to this system in order to place the poles will then be denoted by $\bar{K}$, i.e., $u = \bar{K}\bar{x}$, resulting in a system matrix $\bar{A} + \bar{b}\bar{K}$ with the desired eigenvalues. However, it is important to remember that, as in most cases, the change of basis just performed is a mathematical convenience, and the state variables $\bar{x}$ might not be accessible. Rather, only the state variables x will be accessible. In this case, we must finish the feedback design by undoing the similarity transformation:

$$\begin{aligned} u &= \bar{K}\bar{x} = \bar{K}U^{-1}x \\ &\triangleq Kx \end{aligned} \tag{10.13}$$

Thus, feedback $K = \bar{K}U^{-1}$ is used instead of the feedback $\bar{K}$ directly.

The convenience of the controllable form will be illustrated in the following example, which shows how difficult it might be to compute feedback, even for a simple system, if the matrices are not first converted to controllable canonical form.

Example 10.1: SISO State Feedback

Find a state feedback gain matrix that will place the eigenvalues of the following system at the new locations $\lambda_1 = -5$ and $\lambda_2 = -6$.

$$\dot{x} = \begin{bmatrix} 1 & 3 \\ 4 & 2 \end{bmatrix} x + \begin{bmatrix} 1 \\ 1 \end{bmatrix} u \tag{10.14}$$

Solution:

As an exercise, we will first attempt to place the poles of this system at the desired location *without* first transforming to controllable form. Defining $K = \begin{bmatrix} k_0 & k_1 \end{bmatrix}$, we have by direct substitution into (10.14),

$$\begin{aligned} A + bK &= \begin{bmatrix} 1 & 3 \\ 4 & 2 \end{bmatrix} + \begin{bmatrix} 1 \\ 1 \end{bmatrix} \begin{bmatrix} k_0 & k_1 \end{bmatrix} \\ &= \begin{bmatrix} 1 & 3 \\ 4 & 2 \end{bmatrix} + \begin{bmatrix} k_0 & k_1 \\ k_0 & k_1 \end{bmatrix} \\ &= \begin{bmatrix} 1 + k_0 & 3 + k_1 \\ 4 + k_0 & 2 + k_1 \end{bmatrix} \end{aligned} \tag{10.15}$$

Computing the eigenvalues of (10.15) above in the usual way:

$$\begin{aligned} \left| \lambda I - (A + bK) \right| &= \det\left(\begin{bmatrix} \lambda - 1 - k_0 & -3 - k_1 \\ -4 - k_0 & \lambda - 2 - k_1 \end{bmatrix} \right) \\ &= \lambda^2 + \lambda(-k_0 - k_1 - 3) - (10 + k_0 + 3k_1) \end{aligned} \tag{10.16}$$

Knowing that the desired eigenvalues are to be at -5 and -6, we must set (10.16) equal to the *desired* characteristic polynomial,

$$\begin{aligned} \phi^{des}(\lambda) &= (\lambda + 5)(\lambda + 6) \\ &= \lambda^2 + 11\lambda + 30 \\ &= \lambda^2 + \lambda(-k_0 - k_1 - 3) - (10 + k_0 + 3k_1) \end{aligned}$$

By matching coefficients in this equation, we arrive at the necessary conditions for the two gains:

$$\begin{aligned} 11 &= -k_0 - k_1 - 3 \\ 30 &= -10 - k_0 - 3k_1 \end{aligned} \tag{10.17}$$

We will not solve these equations because as they appear in (10.17), they illustrate the difficulty of not using the controllable form. The result in (10.17) is a coupled set of algebraic equations for the feedback gains, which we do know how to solve. However, the solution is not apparent by inspection, and it gets computationally more intensive as the system size increases.

Instead, we will compute the transformation matrix U that converts (10.14) to controllable canonical form. Using the results of Section 8.2.4, we find that

$$U = \begin{bmatrix} 1 & 1 \\ 3 & 1 \end{bmatrix}$$

resulting in

$$\bar{A} = \begin{bmatrix} 0 & 1 \\ 10 & 3 \end{bmatrix} \qquad \bar{b} = \begin{bmatrix} 0 \\ 1 \end{bmatrix} \tag{10.18}$$

Applying feedback $\bar{K} = [\bar{k}_0 \quad \bar{k}_1]$ to this system,

$$\begin{aligned} \bar{A} + \bar{b}\bar{K} &= \begin{bmatrix} 0 & 1 \\ 10 & 3 \end{bmatrix} + \begin{bmatrix} 0 \\ 1 \end{bmatrix} \begin{bmatrix} \bar{k}_0 & \bar{k}_1 \end{bmatrix} \\ &= \begin{bmatrix} 0 & 1 \\ 10 + \bar{k}_0 & 3 + \bar{k}_1 \end{bmatrix} \end{aligned} \tag{10.19}$$

The characteristic equation of this matrix is, by inspection, $\phi(\lambda) = \lambda^2 + \lambda(-3 - \bar{k}_1) + (-10 - \bar{k}_0)$. It is now easy to set this $\phi(\lambda)$ equal to $\phi^{des}(\lambda) = \lambda^2 + 11\lambda + 30$ to give the much simpler system of linear equations:

$$\begin{aligned} 11 &= -3 - \bar{k}_1 \\ 30 &= -10 - \bar{k}_0 \end{aligned} \tag{10.20}$$

Equations (10.20) may be solved by inspection, giving $\bar{K} = [\bar{k}_0 \quad \bar{k}_1] = \begin{bmatrix} -40 & -14 \end{bmatrix}$. Now we must remember to undo the similarity transform via (10.13):

$$\begin{aligned} K &= \begin{bmatrix} k_0 & k_1 \end{bmatrix} = \bar{K} U^{-1} \\ &= \begin{bmatrix} -1 & -13 \end{bmatrix} \end{aligned} \tag{10.21}$$

This value for K can be seen to satisfy the original design equations in (10.17). It is also easy to verify that the eigenvalues of $A + bK$ are indeed -5 and -6 as desired.

Formulas for State Feedback Gains

In addition to the above procedure, which requires the designer to actually transform the system into the controllable form, there are two other formulas for finding the gain K for a SISO system that do not require the complete transformation process. They are similar in that they both require knowledge of the original and the desired characteristic polynomials, but if the desired eigenvalues are known, then these will be available. In addition, they each require knowledge of the original controllability matrix[M] and, as we might expect, assurance of its invertibility.

`ctrb(A,B)`

Ackermann's Formula

To derive the first formula, known as *Ackermann's formula,*[M] we denote the *desired* characteristic polynomial as

`acker(A,B,P)`

$$\phi^{\text{des}}(\lambda) = \lambda^n + a_{n-1}^{\text{des}}\lambda^{n-1} + \cdots + a_1^{\text{des}}\lambda + a_0^{\text{des}} \tag{10.22}$$

while the *original* characteristic polynomial is denoted, as usual, as

$$\phi(\lambda) = \lambda^n + a_{n-1}\lambda^{n-1} + \cdots + a_1\lambda + a_0 \tag{10.23}$$

We know from the Cayley-Hamilton theorem that the controllable-form A-matrix, i.e., $\overline{A}$, must satisfy its own characteristic polynomial, so it must satisfy (10.23):

$$\phi(\overline{A}) = \overline{A}^n + a_{n-1}\overline{A}^{n-1} + \cdots + a_1\overline{A} + a_0 I = 0$$

or

$$\overline{A}^n = -a_{n-1}\overline{A}^{n-1} - \cdots - a_1\overline{A} - a_0 I \tag{10.24}$$

Although $\overline{A}$ does not satisfy the *desired* characteristic polynomial in (10.22), we can nevertheless evaluate (10.22) at $\overline{A}$:

$$\begin{aligned}\phi^{\text{des}}(\overline{A}) &= \overline{A}^n + a_{n-1}^{\text{des}}\overline{A}^{n-1} + \cdots + a_1^{\text{des}}\overline{A} + a_0^{\text{des}} I \\ &= \left(-a_{n-1}\overline{A}^{n-1} - \cdots - a_1\overline{A} - a_0 I\right) + a_{n-1}^{\text{des}}\overline{A}^{n-1} + \\ &\qquad \cdots + a_1^{\text{des}}\overline{A} + a_0^{\text{des}} I \\ &= \left(a_{n-1}^{\text{des}} - a_{n-1}\right)\overline{A}^{n-1} + \cdots + \left(a_1^{\text{des}} - a_1\right)\overline{A} + \left(a_0^{\text{des}} - a_0\right) I\end{aligned} \tag{10.25}$$

Now we note that *if* we were to use the controllable form for our computations, as above, we would exploit the form of (10.11) to write the equalities

$$
\begin{aligned}
-a_{n-1}^{\text{des}} &= -a_{n-1} + \bar{k}_{n-1} \\
&\vdots \\
-a_1^{\text{des}} &= -a_1 + \bar{k}_1 \\
-a_0^{\text{des}} &= -a_0 + \bar{k}_0
\end{aligned}
$$

which in turn imply

$$
\begin{aligned}
a_{n-1}^{\text{des}} - a_{n-1} &= -\bar{k}_{n-1} \\
&\vdots \\
a_1^{\text{des}} - a_1 &= -\bar{k}_1 \\
a_0^{\text{des}} - a_0 &= -\bar{k}_0
\end{aligned}
\tag{10.26}
$$

This allows us to express (10.25) as

$$
\phi^{\text{des}}(\bar{A}) = -\bar{k}_{n-1}\bar{A}^{n-1} - \cdots - \bar{k}_1\bar{A} - \bar{k}_0 I \tag{10.27}
$$

A clever use of the special structure of controllable form A-matrices allows us to recognize that if we define the *selection* vectors as

$$
\begin{aligned}
e_1 &= \begin{bmatrix} 1 & 0 & 0 & \cdots & 0 \end{bmatrix}^{\text{T}} \\
e_2 &= \begin{bmatrix} 0 & 1 & 0 & \cdots & 0 \end{bmatrix}^{\text{T}} \\
&\vdots \\
e_n &= \begin{bmatrix} 0 & 0 & \cdots & 0 & 1 \end{bmatrix}^{\text{T}}
\end{aligned}
$$

then

$$
e_1^{\text{T}}\bar{A} = e_2^{\text{T}} \qquad e_2^{\text{T}}\bar{A} = e_3^{\text{T}} \qquad \cdots \qquad e_{n-1}^{\text{T}}\bar{A} = e_n^{\text{T}} \tag{10.28}
$$

By repeatedly multiplying the equations in (10.28) from the right by $\bar{A}$, we generate the identities

$$
e_1^{\text{T}}\bar{A}^{n-1} = e_2^{\text{T}}\bar{A}^{n-2} = e_3^{\text{T}}\bar{A}^{n-3} = \cdots = e_n^{\text{T}} \tag{10.29}
$$

Multiplying (10.27) from the left by e_1^{T} therefore results in

$$\begin{aligned}
e_1^{\mathrm{T}}\phi^{\mathrm{des}}(\bar{A}) &= -\bar{k}_{n-1}e_1^{\mathrm{T}}\bar{A}^{n-1} - \bar{k}_{n-2}e_1^{\mathrm{T}}\bar{A}^{n-2} - \cdots - \bar{k}_1 e_1^{\mathrm{T}}\bar{A} - \bar{k}_0 e_1^{\mathrm{T}} \\
&= -\bar{k}_{n-1}e_n^{\mathrm{T}} - \bar{k}_{n-2}e_{n-1}^{\mathrm{T}} - \cdots - \bar{k}_0 e_1^{\mathrm{T}} \\
&= -\begin{bmatrix} \bar{k}_0 & \bar{k}_1 & \cdots & \bar{k}_{n-1} \end{bmatrix} \begin{bmatrix} 1 & 0 & 0 & \cdots & 0 \\ 0 & 1 & 0 & & \vdots \\ 0 & 0 & 1 & \ddots & 0 \\ \vdots & & \ddots & \ddots & 0 \\ 0 & 0 & \cdots & 0 & 1 \end{bmatrix} \\
&= -\begin{bmatrix} \bar{k}_0 & \bar{k}_1 & \cdots & \bar{k}_{n-1} \end{bmatrix} \\
&= -\bar{K}
\end{aligned} \tag{10.30}$$

While (10.30) may at first seem to be a suitable formula for feedback gain, we must realize that we have again found $\bar{K}$ and not K, so that the reverse similarity transformation in (10.13) must be applied. However, the entire derivation was motivated by the desire to not have to compute $\bar{A}$ or the similarity transformation matrix U. Fortunately, a few more identities will simplify Equation (10.30) even more.

Knowing from (10.13) that $K = \bar{K}U^{-1}$, and that $\bar{A} = U^{-1}AU$, we can rewrite (10.30) as

$$e_1^{\mathrm{T}}\phi^{\mathrm{des}}(U^{-1}AU) = -KU$$

or

$$\begin{aligned}
e_1^{\mathrm{T}}U^{-1}\phi^{\mathrm{des}}(A)U &= -KU \\
e_1^{\mathrm{T}}U^{-1}\phi^{\mathrm{des}}(A)UU^{-1} &= -K \\
e_1^{\mathrm{T}}U^{-1}\phi^{\mathrm{des}}(A) &= -K
\end{aligned} \tag{10.31}$$

In Equation (8.45), we found that $U^{-1} = \bar{P}P^{-1}$, where $\bar{P}$ is the controllability matrix of the controllable canonical form and P is the original controllability matrix. Therefore, (10.31) gives

$$K = -e_1^{\mathrm{T}}U^{-1}\phi^{\mathrm{des}}(A) = -e_1^{\mathrm{T}}\bar{P}P^{-1}\phi^{\mathrm{des}}(A)$$

Within this expression, the term $e_1^{\mathrm{T}}\bar{P}$ will be equal to the first row of the controllable form controllability matrix $\bar{P}$, which will be

$$e_1^{\mathrm{T}}\bar{P} = \begin{bmatrix} 0 & \cdots & 0 & 1 \end{bmatrix}^{\mathrm{T}} = e_n^{\mathrm{T}}$$

So finally, we have the final version of Ackermann's formula:

$$K = -e_n^{\mathrm{T}} P^{-1} \phi^{\mathrm{des}}(A) \tag{10.32}$$

where, as derived, $\phi^{\mathrm{des}}(A)$ is the *desired* characteristic polynomial (i.e., with the *desired* poles), evaluated at the *original* A-matrix. This matrix polynomial evaluation is relatively easy to perform with the aid of a computer and an analysis tool such as MATLAB.[M]

`polyvalm(V,X)`

We should remark on the use of (10.32) for feedback computation. For "manual" calculations, (10.32) is convenient and accurate provided that the controllability matrix P is well-conditioned, so that its inverse may be accurately computed. In cases where numerical accuracy is questionable because of the inversion operation, more numerically stable algorithms are available,[M] such as treating (10.32) as a simultaneous equation problem and solving it through the use of SVDs.

`place(A,B,P)`

Bass-Gura Formula

A second but similar formula for the gain matrix results from equally clever uses of matrix identities, but without the need to even consider the transformation U that converts the system to controllable form. This formula is known as the *Bass-Gura formula* [7] and again starts with the expression for the *desired* characteristic polynomial:

$$\phi^{\mathrm{des}}(\lambda) = \det[\lambda I - (A + bK)] \tag{10.33}$$

Appealing to the identity that $\det(AB) = \det(A)\det(B)$, we can factor (10.33) to get

$$\begin{aligned} \phi^{\mathrm{des}}(\lambda) &= \det\left[(\lambda I - A)\left(I - (\lambda I - A)^{-1} bK\right)\right] \\ &= \det[\lambda I - A]\det\left[I - (\lambda I - A)^{-1} bK\right] \\ &= \phi(\lambda)\det\left[I - (\lambda I - A)^{-1} bK\right] \end{aligned} \tag{10.34}$$

where $\phi(\lambda)$ is the original characteristic polynomial. Another identity from Appendix A, Equation (A.18), and the fact that a scalar quantity is also its own determinant allow us to express (10.34) as

$$\begin{aligned}\phi^{\text{des}}(\lambda) &= \phi(\lambda)\det\left[1 - K(\lambda I - A)^{-1}b\right] \\ &= \phi(\lambda)\left(1 - K(\lambda I - A)^{-1}b\right) \\ &= \phi(\lambda) - \phi(\lambda)K(\lambda I - A)^{-1}b\end{aligned}$$

or

$$\phi^{\text{des}}(\lambda) - \phi(\lambda) = -\phi(\lambda)K(\lambda I - A)^{-1}b \tag{10.35}$$

Another identity, known as a *resolvent* identity, Equation (A.42), gives the expansion of (10.35) as

$$\begin{aligned}\phi^{\text{des}}(\lambda) - \phi(\lambda) &= -\phi(\lambda)K\frac{1}{\phi(\lambda)}\Big[\lambda^{n-1}I + (A + a_{n-1}I)\lambda^{n-2} + \\ &\qquad \cdots + (A^{n-1} + a_{n-1}A^{n-2} + \cdots + a_0 I)\Big]b \\ &= -K\Big[\lambda^{n-1}I + (A + a_{n-1}I)\lambda^{n-2} + \\ &\qquad \cdots + (A^{n-1} + a_{n-1}A^{n-2} + \cdots + a_0 I)\Big]b\end{aligned} \tag{10.36}$$

Recalling from (10.22) and (10.23) that

$$\begin{aligned}\phi^{\text{des}}(\lambda) - \phi(\lambda) &= \lambda^n + \left(a_{n-1}^{\text{des}} - a_{n-1}\right)\lambda^{n-1} + \\ &\qquad \cdots + \left(a_1^{\text{des}} - a_1\right)\lambda + \left(a_0^{\text{des}} - a_0\right)\end{aligned} \tag{10.37}$$

we can match coefficient of like powers of λ in (10.36) and (10.37) to get

$$\begin{aligned}a_{n-1}^{\text{des}} - a_{n-1} &= -Kb \\ a_{n-2}^{\text{des}} - a_{n-2} &= -K(A + a_{n-1}I)b = -KAb - Ka_{n-1}b \\ a_{n-3}^{\text{des}} - a_{n-3} &= -K(A^2 + a_{n-1}A + a_{n-2}I)b \\ &= -KA^2b - a_{n-1}KAb - a_{n-2}Kb \\ &\vdots\end{aligned} \tag{10.38}$$

Rearranging (10.38) into a matrix equation,

$$\begin{bmatrix} a_{n-1}^{\text{des}} - a_{n-1} \\ \vdots \\ a_1^{\text{des}} - a_1 \\ a_0^{\text{des}} - a_0 \end{bmatrix}^{\text{T}} = -K\left[b \quad Ab \quad \cdots \quad A^{n-1}b\right] \begin{bmatrix} 1 & a_{n-1} & a_{n-2} & \cdots & a_0 \\ 0 & 1 & a_{n-1} & \ddots & \vdots \\ 0 & 0 & 1 & \ddots & a_{n-2} \\ \vdots & \ddots & \ddots & \ddots & a_{n-1} \\ 0 & \cdots & 0 & 0 & 1 \end{bmatrix} \quad (10.39)$$

`toeplitz(C,Z)`

The matrix on the right of Equation (10.39) is known as an upper triangular Toeplitz[M] matrix and is obviously invertible. The controllability matrix appearing in (10.39) is clearly invertible as well, so that Equation (10.39) can be rewritten more compactly as

$$K = -\begin{bmatrix} a_{n-1}^{\text{des}} - a_{n-1} \\ \vdots \\ a_1^{\text{des}} - a_1 \\ a_0^{\text{des}} - a_0 \end{bmatrix}^{\text{T}} \begin{bmatrix} 1 & a_{n-1} & a_{n-2} & \cdots & a_0 \\ 0 & 1 & a_{n-1} & \ddots & \vdots \\ 0 & 0 & 1 & \ddots & a_{n-2} \\ \vdots & \ddots & \ddots & \ddots & a_{n-1} \\ 0 & \cdots & 0 & 0 & 1 \end{bmatrix}^{-1} P^{-1} \quad (10.40)$$

which is the Bass-Gura formula.

Although we have taken some pains to present the gain computation matrices for SISO systems, it should be clear that from (10.32) and (10.40) (in particular, the fact that the inverse controllability matrix appears in each), these formulas do not extend to MIMO systems. For MIMO systems, there are a number of numerical techniques, as well as the methods we present in Section 10.2, which essentially reduce the p-input state feedback problem into p single-input feedback problems. First, though, we will make some important observations about the properties of state feedback in general. These remarks *do* apply to single-input and multi-input systems alike.

10.1.2 Properties of State Feedback

Obviously, once we have introduced state feedback into a system, we have fundamentally altered it. Primarily, its stability and transient response properties are altered because the eigenvalues have been changed. This fact raises a number of questions that we can easily address.

Equivalence of Controllability and Pole Placement

From the form of the system in (10.10) that includes state feedback gains, we have already concluded that when state feedback is applied to a controllable system, each individual coefficient of the characteristic polynomial can be specified. If the locations of the new poles are chosen such that complex poles appear in conjugate pairs, then it is clear that *controllability implies the ability to arbitrarily place poles*.

Through a simple example, we can illustrate what typically happens when state feedback is applied to a system that is *not* controllable.

Example 10.2: State Feedback for an Uncontrollable System

Consider the system

$$\dot{x} = \begin{bmatrix} -2 & 0 \\ 0 & -2 \end{bmatrix} x + \begin{bmatrix} 1 \\ 1 \end{bmatrix} u \tag{10.41}$$

Apply state feedback and determine the location of the resulting eigenvalues.

Solution:

The system is in Jordan form, with two Jordan blocks corresponding to the same eigenvalue. However, because there is only a single input, we know from Chapter 8 that the system must be uncontrollable. Taking a state feedback gain matrix of the usual structure, $K = \begin{bmatrix} k_0 & k_1 \end{bmatrix}$, and substituting $u = Kx$ into (10.41) give

$$\dot{x} = (A + bK)x = \begin{bmatrix} -2 + k_0 & k_1 \\ k_0 & -2 + k_1 \end{bmatrix} x \tag{10.42}$$

Finding the eigenvalues:

$$\begin{aligned} \det(\lambda I - (A + bK)) &= \begin{vmatrix} \lambda + 2 - k_0 & -k_1 \\ -k_0 & \lambda + 2 - k_1 \end{vmatrix} \\ &= (\lambda + 2 - k_0)(\lambda + 2 - k_1) - k_0 k_1 \\ &= \lambda^2 + \lambda(4 - k_0 - k_1) + (4 - 2k_0 - 2k_1) \\ &= (\lambda + 2)(\lambda + (2 - k_0 - k_1)) \end{aligned} \tag{10.43}$$

By setting (10.43) equal to zero, it is apparent that one eigenvalue may be arbitrarily located via state feedback, but the other one is fixed at $\lambda = -2$. Thus, when state feedback is applied to an uncontrollable system, one or more of the eigenvalues will be unaffected by the state feedback gains. (In fact, it has been proposed that an easy way to test for controllability is to apply a random feedback gain matrix and check to see if all the eigenvalues have changed. If one or more have not changed, it is very likely that the system is uncontrollable.

If those that do *not* move are in the left half-plane already, then the system is probably stabilizable.)

Controllability and Observability After Feedback

Unless there is a compelling physical reason to do otherwise, the exogenous input applied to a system after state feedback is often applied with $E = I$. Considering Equation (10.10), then, it is clear that if the system is assumed to be in controllable canonical form before feedback, then it will be in controllable canonical form after feedback. If indeed $E = I$, as we will assume hereafter, then the input matrix (i.e., the matrix through which input v is applied) is again the same as the original input matrix:

$$\bar{b} = \begin{bmatrix} 0 \\ \vdots \\ 0 \\ 1 \end{bmatrix}$$

We conclude, therefore, that *a system that is controllable before state feedback will be controllable after state feedback.* This will be true as well if $E \neq I$ provided that E is full rank.

A similar question can be asked of the observability of the system, i.e., is an observable system still observable after the introduction of state feedback? This question can be answered through the related question: Does state feedback alter the *zeros* of a system? To answer this question, we recall that the relationship between the transfer function and the state space matrices is given Equations (9.20) and (9.21). Therefore, *without* feedback, the transfer function of a system (10.8) is

$$\begin{aligned} g(s) &= \frac{c_n s^{n-1} + c_{n-1}s^{n-2} + \cdots + c_2 s + c_1}{s^n + a_{n-1}s^{n-1} + \cdots + a_1 s + a_0} + d \\ &= \frac{d\left(s^n + a_{n-1}s^{n-1} + \cdots + a_1 s + a_0\right) + c_n s^{n-1} + c_{n-1}s^{n-2} + \cdots + c_2 s + c_1}{s^n + a_{n-1}s^{n-1} + \cdots + a_1 s + a_0} \\ &= \frac{ds^n + \left(da_{n-1} + c_n\right)s^{n-1} + \cdots + \left(da_0 + c_1\right)}{s^n + a_{n-1}s^{n-1} + \cdots + a_1 s + a_0} \end{aligned} \tag{10.44}$$

After the introduction of state feedback, the new system equations are (assuming again that $E = I$),

$$\begin{aligned} \dot{x} &= (A+bK)x + bv \\ y &= (c+dK)x + dv \end{aligned} \tag{10.45}$$

Since the "new" matrix $A+bK$ has the same form as (10.10), the system in (10.45) will give a transfer function analogous to (10.44):

$$\begin{aligned} g(s) &= \frac{(c_n + dk_{n-1})s^{n-1} + (c_{n-1} + dk_{n-2})s^{n-2} + \cdots + (c_2 + dk_1)s + (c_1 + dk_0)}{s^n + (a_{n-1} - k_{n-1})s^{n-1} + \cdots + (a_1 - k_1)s + (a_0 - k_0)} + d \\ &= \frac{ds^n + (da_{n-1} - dk_{n-1} + c_n + dk_{n-1})s^{n-1} + \cdots + (da_0 - dk_0 + c_1 + dk_0)}{s^n + (a_{n-1} - k_{n-1})s^{n-1} + \cdots + (a_1 - k_1)s + (a_0 - k_0)} \\ &= \frac{ds^n + (da_{n-1} + c_n)s^{n-1} + \cdots + (da_0 + c_1)}{s^n + (a_{n-1} - k_{n-1})s^{n-1} + \cdots + (a_1 - k_1)s + (a_0 - k_0)} \end{aligned} \tag{10.46}$$

The numerator of this transfer function is clearly equal to that of (10.44). We conclude therefore that *the zeros of a SISO transfer function are not affected by state feedback*. This is true for MIMO system as well, although we have not shown this here.

With this fact, we can immediately conclude that *observability is* <u>*not*</u> *preserved by state feedback*. Why is this obvious? If we start with a controllable and observable system (i.e., a minimal realization), the feedback gain can be used to place the poles but the zeros remain unaffected. Suppose therefore that the poles move to locations that coincide with one or more zeros. Then the system is no longer minimal. Because we have already determined that controllability cannot have changed because of the introduction of state feedback, then observability must have been lost.

The final remark we make regarding state feedback is that it follows exactly the same procedure in continuous-time as in discrete-time. Of course, the design criteria will change because the stability region in the *z*-plane is inside the unit circle as opposed to being in the left-half of the *s*-plane. Other than that, the process for converting to controllable form, computing state feedback matrices, and for converting back again is exactly the same. It is only the *desired* characteristic polynomial that will likely be different and that is at the choice of the designer.

10.2 Multivariable Canonical Forms and Feedback

In both of the feedback formulas above (Ackermann and Bass-Gura), it is necessary to invert the controllability matrix P in order to compute appropriate gain matrices. This implies both that P is full rank (which it must be if the system is controllable) *and* that it is square. If the system has $p > 1$ inputs, then matrix P will be $n \times pn$, so it will of course not be invertible. Furthermore, the matrix structures found in the controllable canonical form are particular to single-input systems, i.e., those with a single-column b-matrix. The concept of state feedback is equally applicable to multiple-input systems, and Figure 10.3 still represents an accurate diagram of the feedback structure. However, the implementation and design are somewhat more complex.

10.2.1 Controllability Indices and Canonical Forms

To accommodate such difficulties with multi-input systems and enable us to compute state-feedback gains, we use *multivariable canonical forms*. A multivariable canonical form takes into account the differences in the different inputs, and the extent to which each input contributes to the control of the entire system. We can think of each input as having a certain capacity to control the system as a whole such that the controllable subspace is the sum of the spaces controllable by the individual inputs. This will enable us to design control signals one input at a time. The mechanism by which we perform this analysis is the *controllability index*. In this section we will define controllability indices and learn to use them to transform a multi-input system to multivariable canonical form.

Consider the $n \times pn$ controllability matrix for a multi-input system:

$$P = \left[B \mid AB \mid \cdots \mid A^{n-1}B \right] \tag{10.47}$$

Assume that the system is controllable. Then this matrix must have rank n or, equivalently, n linearly independent columns. If we wish to derive a similarity transformation from this matrix, we will need n such linearly independent columns, but of course there will be many ways that we might select this set from (10.47). Denote the b-matrix column-wise as

$$b = \left[b_1 \mid b_2 \mid \cdots \mid b_p \right] \tag{10.48}$$

Then (10.47) becomes

$$P = \left[b_1 \quad \cdots \quad b_p \mid Ab_1 \quad \cdots \quad Ab_p \mid \cdots \mid A^{n-1}b_1 \quad \cdots \quad A^{n-1}b_p \right] \tag{10.49}$$

Because we can select n linearly independent columns from this matrix in many ways, the canonical form we produce might have an equal number of variations in structures. We will present one common method for seeking out our n columns and discuss possible alternatives later.

We proceed by examining the columns of this matrix from left to right, noting how many columns in each partition of the matrix are linearly dependent on columns appearing to their left. For example, in the partition of (10.49) containing the columns $Ab_1, \dots, Ab_p$, we will probably find that at least one column is linearly dependent on those to its left, which includes the columns in the partition $b_1 \cdots b_p$. We will denote the number of these dependent columns we find in the ith partition as q_i, $i = 0, \dots, n-1$. (The q is reminiscent of the notation for nullity, and the subscript indicates that this is the ith partition of P.) Thus, if $r(B) = p$, then $q_0 = 0$, but q_i for $i > 0$ depends not only on the rank of $A^i B$, but also on the partitions to the left as well.

Continuing this procedure, we will generate a set of integers $\{q_0, q_1, \dots, q_{n-1}\}$. Note that if a single column in particular, say Ab_2, is found to be linearly dependent on the columns to its left, then of course $A^2 b_2$ will also be dependent on the columns to its left. Therefore it must be true that

$$0 \le q_0 \le q_1 \le \cdots \le q_{n-1} \le p \tag{10.50}$$

Because there can be at most n linearly independent columns among the np total columns, there must be some partition at which we stop encountering any more *in*dependent columns in P. We denote by μ the number of partitions that have at least one independent column. Then,

$$0 \le q_0 \le q_1 \le \cdots \le q_{\mu-1} < q_\mu = q_{\mu+1} = \cdots = q_{n-1} = p \tag{10.51}$$

Because we have searched the P matrix from left to right, the linearly dependent partitions are going to be the rightmost, so it would be possible to eliminate partitions $\mu, \mu+1, \cdots, n$, and consider only the reduced controllability matrix

$$P = \left[B \mid AB \mid \cdots \mid A^{\mu-1} B \right]$$

when continuing to search for columns to use in the similarity transformation. This integer μ is known as the overall *controllability index*.

Next, having identified all the linearly independent columns, from left to right, we will gather them into a separate matrix and rearrange them according to the input (i.e., the column of B) with which they are associated. Of course, there will not be n columns for each input. The number of columns associated with input i will be denoted μ_i. Therefore, the collection of *in*dependent columns is expressed as the matrix

$$\begin{aligned} M = \Big[b_1 \quad Ab_1 \quad \cdots \quad A^{\mu_1 - 1} b_1 \mid b_2 \quad Ab_2 \quad \cdots \quad A^{\mu_2 - 1} b_2 \mid \cdots \\ \cdots \mid b_p \quad Ab_p \quad \cdots \quad A^{\mu_p - 1} b_p \Big] \end{aligned} \tag{10.52}$$

The integers $\{\mu_1, \cdots, \mu_p\}$ are known as the *individual controllability indices*. With a little thought, it can be observed that

$$\mu = \max\{\mu_1, \cdots, \mu_p\}$$

and

$$\mu_1 + \mu_2 + \cdots + \mu_p \le n$$

where

$$\mu_1 + \mu_2 + \cdots + \mu_p = n$$

only for a controllable system. These controllability indices serve as our measure of "how controllable" the system is from each input acting alone. For example, if $\mu_1 = n$, then the controllability matrix would contain n linearly independent columns resulting from the *first* input alone, i.e., the entire system would be controllable even if the first input were the only one present. The remaining inputs would not contribute to the controllability of the system. The higher the controllability index of an input, the larger is the controllability subspace corresponding to that input.

Following the procedure set forth in [3] and [9], the next step in the derivation of the transformation to multi-input canonical form is to compute the inverse of the matrix M [Equation (10.52)]. Symbolically, this matrix can be written row-by-row with the notation

$$M^{-1} = \begin{bmatrix} m_{11} \\ m_{12} \\ \vdots \\ m_{1\mu_1} \\ \hline \vdots \\ \hline m_{p1} \\ \vdots \\ m_{p\mu_p} \end{bmatrix} \tag{10.53}$$

Notice that this matrix is arranged into partitions, where the ith partition has μ_i rows. By extracting the last row from each partition of this matrix (i.e., the rows denoted $m_{i\mu_i}$, for $i = 1, 2, \ldots, p$), we construct the matrix

$$T = \begin{bmatrix} m_{1\mu_1} \\ m_{1\mu_1}A \\ \vdots \\ m_{1\mu_1}A^{\mu_1-1} \\ \hdashline \vdots \\ \hdashline m_{p\mu_p} \\ \vdots \\ m_{p\mu_p}A^{\mu_p-1} \end{bmatrix} \tag{10.54}$$

Using this transformation matrix to change the basis[M] in the sense that $\bar{x} = Tx$ will result in transformed system matrices $\bar{A} = TAT^{-1}$, $\bar{B} = TB$, and $\bar{C} = CT^{-1}$ such that $\bar{A}$ and $\bar{B}$ have the following multi-input canonical form:

```
ss2ss(sys,T)
```

$$\bar{A} = \left[\begin{array}{c:c:c:c} \bar{A}_{\mu_1\times\mu_1} & \begin{matrix} \varnothing \\ \# \cdots \# \end{matrix} & \cdots & \begin{matrix} \varnothing \\ \# \cdots \# \end{matrix} \\ \hdashline \begin{matrix} \varnothing \\ \# \cdots \# \end{matrix} & \bar{A}_{\mu_2\times\mu_2} & \ddots & \vdots \\ \hdashline \vdots & \ddots & \ddots & \begin{matrix} \varnothing \\ \# \cdots \# \end{matrix} \\ \hdashline \begin{matrix} \varnothing \\ \# \cdots \# \end{matrix} & \cdots & \begin{matrix} \varnothing \\ \# \cdots \# \end{matrix} & \bar{A}_{\mu_p\times\mu_p} \end{array}\right]$$

$$\bar{B} = \begin{bmatrix} 0 & 0 & \cdots & 0 \\ \vdots & \vdots & \ddots & \vdots \\ 0 & 0 & \cdots & 0 \\ 1 & \# & \cdots & \# \\ \hline 0 & 0 & \cdots & 0 \\ \vdots & \vdots & \ddots & \vdots \\ 0 & 0 & \cdots & 0 \\ 0 & 1 & \# & \cdots \\ \hdashline & & \vdots & \\ \hdashline 0 & 0 & \cdots & 0 \\ \vdots & \vdots & \ddots & \vdots \\ 0 & 0 & \cdots & 0 \\ 0 & \cdots & 0 & 1 \end{bmatrix} \tag{10.55}$$

In this notation, $\overline{A}_{\mu_i \times \mu_i}$ is a $(\mu_i \times \mu_i)$ companion form block matrix, i.e., in the exact same structure with which we are familiar for the single-input system, Equation (10.8). Its bottom row, as usual, is filled with essentially arbitrary numbers. Along each such row of matrix $\overline{A}$ is an entire row of arbitrary numbers, indicated by the nonspecific indicator # (this does not imply that these numbers are all the same). The remaining rows of all the off-diagonal blocks are equal to zero, as indicated by the symbol $\varnothing$. The $(i, j)^{\text{th}}$ block has dimension $(\mu_i \times \mu_j)$.

In the $\overline{B}$-matrix, all the elements are zero except for the rows corresponding to the rows with #-elements in $\overline{A}$, i.e., rows $\mu_1, \mu_1 + \mu_2, \ldots$. The nonzero row that appears in the ith block of $\overline{B}$ will be zeros up to its ith column, which will be a one. The remaining columns of each such row can again be any number (again denoted by #). (In fact, the jth column in the ith block will be zero if $\mu_j > \mu_i$ [3].) The reader may want to relate this procedure to the analogous procedure for a single-input system as given in Section 8.2.4.

The benefit of this canonical form can be seen by considering the rows and columns of $\overline{B}$ one at a time. The first row to have a nonzero element will be the μ_1 st row. Multiplying a $p \times n$ matrix $\overline{K}$ by the $n \times p$ matrix $\overline{B}$ will therefore give a product $\overline{B}\overline{K}$ that has all zero rows until the μ_1 st row, which will be filled with gains. These gains will be added to the μ_1 st row of $\overline{A}$. This will be true thereafter, as well: the next nonzero row of $\overline{B}\overline{K}$ will be the $(\mu_1 + \mu_2)$ st row, which can again be used to alter the # values in the $(\mu_1 + \mu_2)$ st row of $\overline{A}$, etc. By judicious choice of the elements in $\overline{K}$, these special rows of $\overline{A}$ can be arbitrarily specified. Usually, this specification is such that overall, $\overline{A}$ becomes block-diagonal and subsets of eigenvalues are independently determined within the distinct blocks.

At this stage, a numerical example is the best illustration of how the entries of $\overline{K}$ are chosen, given a multivariable canonical form.

Example 10.3: Multi-input Pole Placement

Find at least two different gain matrices that will place the eigenvalues of the following controllable system at the set $\{-5, -5, -3 \pm j3\}$.

$$\dot{x} = \begin{bmatrix} 2 & -11 & 12 & 31 \\ -3 & 13 & -11 & -33 \\ 4 & -25 & 14 & 51 \\ -3 & 16 & -11 & -36 \end{bmatrix} x + \begin{bmatrix} -1 & -4 \\ 1 & 6 \\ -1 & -11 \\ 1 & 7 \end{bmatrix} u \tag{10.56}$$

Solution:

The first computation performed in any state-space controller design is likely to be the controllability matrix:[M]

`ctrb(A,B)`

$$P = \left[\begin{array}{cc|cc|cc|cc} -1 & -4 & 6 & 11 & -12 & -27 & 16 & 73 \\ 1 & 6 & -6 & -20 & 14 & 59 & -24 & -175 \\ -1 & -11 & 8 & 37 & -20 & -111 & 36 & 331 \\ 1 & 7 & -6 & -23 & 14 & 68 & -24 & -202 \end{array}\right] \tag{10.57}$$

It is easily verified by computer[M] that $r(P) = n = 4$. Using the independent column counting procedure, we find (by checking the rank of progressively more columns in a collection beginning at the left) that the first four columns of this P matrix are linearly independent. Therefore, $\mu_1 = \mu_2 = 2$ and $\mu = 2$. The matrix M is constructed according to (10.52) as the first, third, second, and fourth columns of P, in that order. Then,

`rank(P)`

$$M^{-1} = \left[\begin{array}{cccc} 4/3 & -1 & 3 & 19/3 \\ 1/3 & -1 & 1/2 & 11/6 \\ \hline -1 & -4 & 0 & 3 \\ -1/3 & -1 & 0 & 2/3 \end{array}\right] \tag{10.58}$$

Taking the second (i.e., μ_1) and fourth (i.e., $\mu_1 + \mu_2$) rows of this matrix as m_{12} and m_{22}, respectively, the following transformation matrix is found:

$$T = \begin{bmatrix} m_{12} \\ m_{12}A \\ m_{22} \\ m_{22}A \end{bmatrix} = \left[\begin{array}{cc|cc} 1/3 & -1 & 1/2 & 11/6 \\ 1/6 & 1/6 & 11/6 & 17/6 \\ \hline -1/3 & -1 & 0 & 2/3 \\ 1/3 & 4/3 & -1/3 & -4/3 \end{array}\right] \tag{10.59}$$

which produces the transformed system:

$$\bar{A} = TAT^{-1} = \left[\begin{array}{cc|cc} 0 & 1 & 0 & 0 \\ -\frac{4}{3} & -\frac{7}{3} & \frac{8}{3} & \frac{7}{6} \\ \hline 0 & 0 & 0 & 1 \\ -2 & -\frac{2}{3} & -5 & -\frac{14}{3} \end{array}\right] \qquad \bar{B} = TB = \left[\begin{array}{cc} 0 & 0 \\ 1 & 0 \\ \hline 0 & 0 \\ 0 & 1 \end{array}\right]$$

At this point we are ready to design the control. With an $n = 4$ and $p = 2$ system, the state-feedback gain matrix is written (remembering to retain the "bar" basis notation):

$$\bar{K} = \begin{bmatrix} \bar{k}_{11} & \bar{k}_{12} & \bar{k}_{13} & \bar{k}_{14} \\ \bar{k}_{21} & \bar{k}_{22} & \bar{k}_{23} & \bar{k}_{24} \end{bmatrix} \tag{10.60}$$

Directly applying this feedback,

$$\begin{aligned} \bar{A} + \bar{B}\bar{K} &= \left[\begin{array}{cc|cc} 0 & 1 & 0 & 0 \\ -\frac{4}{3} & -\frac{7}{3} & \frac{8}{3} & \frac{7}{6} \\ \hline 0 & 0 & 0 & 1 \\ -2 & -\frac{2}{3} & -5 & -\frac{14}{3} \end{array}\right] + \left[\begin{array}{cc} 0 & 0 \\ 1 & 0 \\ \hline 0 & 0 \\ 0 & 1 \end{array}\right] \begin{bmatrix} \bar{k}_{11} & \bar{k}_{12} & \bar{k}_{13} & \bar{k}_{14} \\ \bar{k}_{21} & \bar{k}_{22} & \bar{k}_{23} & \bar{k}_{24} \end{bmatrix} \\ &= \left[\begin{array}{cc|cc} 0 & 1 & 0 & 0 \\ -\frac{4}{3} & -\frac{7}{3} & \frac{8}{3} & \frac{7}{6} \\ \hline 0 & 0 & 0 & 1 \\ -2 & -\frac{2}{3} & -5 & -\frac{14}{3} \end{array}\right] + \begin{bmatrix} 0 & 0 & 0 & 0 \\ \bar{k}_{11} & \bar{k}_{12} & \bar{k}_{13} & \bar{k}_{14} \\ 0 & 0 & 0 & 0 \\ \bar{k}_{21} & \bar{k}_{22} & \bar{k}_{23} & \bar{k}_{24} \end{bmatrix} \\ &= \left[\begin{array}{cc|cc} 0 & 1 & 0 & 0 \\ -\frac{4}{3} + \bar{k}_{11} & -\frac{7}{3} + \bar{k}_{12} & \frac{8}{3} + \bar{k}_{13} & \frac{7}{6} + \bar{k}_{14} \\ \hline 0 & 0 & 0 & 1 \\ -2 + \bar{k}_{21} & -\frac{2}{3} + \bar{k}_{22} & -5 + \bar{k}_{23} & -\frac{14}{3} + \bar{k}_{24} \end{array}\right] \end{aligned} \tag{10.61}$$

The matrix in (10.61) should be examined closely before deciding on any feedback gains. The locations of the gain terms at our disposal determine the flexibility we have with the structure of the "closed-loop" system. In particular, it should be noted that we can easily transform this system into two different structures:

$$\left[\begin{array}{cc|cc} 0 & 1 & 0 & 0 \\ -\alpha_0 & -\alpha_1 & 0 & 0 \\ \hline 0 & 0 & 0 & 1 \\ 0 & 0 & -\beta_0 & -\beta_1 \end{array}\right] \quad \text{or} \quad \begin{bmatrix} 0 & 1 & 0 & 0 \\ 0 & 0 & 1 & 0 \\ 0 & 0 & 0 & 1 \\ -\alpha_0 & -\alpha_1 & -\alpha_2 & -\alpha_3 \end{bmatrix}$$

That is, we can specify that the system consist of two (2×2) block-diagonal parts, each in companion form, or as a single (4×4) block, also in companion form. Each option, of course, dictates different sets of $\bar{k}_{ij}$'s. To determine the desired values for the bottom row of each companion block, we should consider the desired eigenvalues two at a time, or all four at a time, i.e.,

$$\left\{ \begin{array}{c} (s+5)(s+5) = s^2 + 10s + 25 \stackrel{\Delta}{=} s^2 + \alpha_1 s + \alpha_0 \\ \text{and} \\ (s+3+j3)(s+3-j3) = s^2 + 6s + 18 \stackrel{\Delta}{=} s^2 + \beta_1 s + \beta_0 \end{array} \right\} \tag{10.62}$$

or

$$\begin{aligned} (s+5)(s+5)(s+3+j3)(s+3-j3) &= s^4 + 16s^3 + 103s^2 + 330s + 450 \\ &= s^4 + \alpha_3 s^3 + \alpha_2 s^2 + \alpha_1 s + \alpha_0 \end{aligned} \tag{10.63}$$

In order to achieve the two-block structure using the two separate characteristic polynomials in (10.62), the gain matrix $\bar{K}$ should be chosen such that

$$\bar{A} + \bar{B}\bar{K} = \left[\begin{array}{cc|cc} 0 & 1 & 0 & 0 \\ -25 & -10 & 0 & 0 \\ \hline 0 & 0 & 0 & 1 \\ 0 & 0 & -18 & -6 \end{array} \right] \tag{10.64}$$

By inspection and by simple term-by-term matching, such a gain matrix is seen to be:

$$\bar{K}_1 = \begin{bmatrix} -23\frac{2}{3} & -7\frac{2}{3} & -\frac{8}{3} & -\frac{7}{6} \\ 2 & \frac{2}{3} & -13 & -\frac{4}{3} \end{bmatrix} \tag{10.65}$$

On the other hand, creating a single block requires that (10.63) be matched term-by-term with

$$\bar{A} + \bar{B}\bar{K}_2 = \begin{bmatrix} 0 & 1 & 0 & 0 \\ 0 & 0 & 1 & 0 \\ 0 & 0 & 0 & 1 \\ -450 & -330 & -103 & -16 \end{bmatrix} \tag{10.66}$$

This matching results in a different gain matrix that gives the closed-loop system the same eigenvalues:

$$\bar{K}_2 = \begin{bmatrix} 4/3 & 7/3 & -8/3 & -1/6 \\ -448 & -329\,1/3 & -98 & -34/3 \end{bmatrix} \tag{10.67}$$

The reader can verify that in each case, the eigenvalues of $\bar{A} + \bar{B}\bar{K}$ are in the desired locations.

A few remarks are in order. First, we must remember that this system was transformed before the state feedback gains were found. Therefore the problem is not finished with the calculation of the gain; we must undo the transformation with $K = \bar{K}T$. Second, we observe from (10.65) and (10.67) that of the two feedback matrices we generated, $\bar{K}_1$ will be much better conditioned[M] (i.e., have a smaller condition number) than $\bar{K}_2$. This can be guessed because the numbers in $\bar{K}_1$ are much more similar to one another than are the numbers in $\bar{K}_2$. This may have practical implications in physical circuits, wherein it is desirable to have amplifier gains all of the same order of magnitude. Also, the transient response of a system with smaller gains is likely to be better than the transient response of a system with larger gains. Further considerations of the factors that lead one to prefer one feedback gain over another can be addressed with the concept of *optimal control*, which we introduce only briefly in the next chapter. Finally, we point out that although our method for computing these gains resulted in only two "obvious" options, *as long as* $\mu > 1$, *there are an infinite number of ways in which different gain matrices might be found.* (If $\mu = 1$, the system has essentially a single input.) This is simply an easy "manual" method for finding a few possibilities. The opposite can be said about single-input systems: *In a single-input system, the feedback that places the eigenvalues at a specific set of locations is unique.*

`cond(K)`

10.3 Observers

A significant assumption taken for granted in all of the above development is the availability of the state vector x. The implementation of state feedback very obviously relies on the physical presence of the complete state. However, we must not forget the *output* equation. In particular, the output equation models the way in which the system presents its internal workings to the outside world. More often than not, it is the output alone that is physically measurable in the physical system. Then how can we perform state feedback without the presence of the state vector? The answer is in a separate system called an *observer*.

Sometimes called an *estimator*,* an observer is itself a linear system whose task is to accept as inputs the original system's (i.e., the plant's) input and output signals, and produce as its output an estimate of the plant's state vector. This state vector estimate should asymptotically track the exact state vector. In this way, the output of the observer, rather than the true state, can be used to compute state feedback. The observer and state feedback combination is often thought of by controls engineers as the most fundamental control system available in state space.

10.3.1 Continuous-Time Observers

We begin by considering a continuous-time LTI system. Discrete-time system observers, like state feedback, follow analogous equations but also admit variations on the design of the observer. They will therefore be treated separately later. For the system

$$\begin{aligned} \dot{x} &= Ax + Bu \\ y &= Cx + Du \end{aligned} \tag{10.68}$$

we seek an estimate of $x(t)$ from knowledge of $u(t)$ and $y(t)$. A naive approach, the *open-loop observer*, is to assume knowledge of the system's initial state, $x(t_0)$, and use the known system matrices to create an exact copy of the plant in (10.68). If we call the state estimate $\hat{x}(t)$, then such an open-loop observer might be written as

$$\dot{\hat{x}} = A\hat{x} + Bu \qquad \hat{x}(t_0) = x(t_0) \tag{10.69}$$

The difficulty with an estimator such as this (i.e., one that does not take into account the performance y or the output of the plant assuming the estimated state, $\hat{y} = C\hat{x} + Du$) is that any errors it produces are never measured nor compensated. For example, the initial condition guess $\hat{x}(t_0)$ will invariably be inaccurate and the plant might not be asymptotically stable. This would create a signal $\hat{x}(t)$ that diverges from $x(t)$ such that they very soon have no resemblance at all. Creating a feedback $u = K\hat{x}$ would have unpredictable, and very likely disastrous, results.

Instead we will design a *closed-loop* observer.[M] A closed-loop observer continually compares its estimate (as measured via the output equation) with the true plant in order to create an error signal. The error signal is used to drive the

`estim(sys,L)`

---

* We will reserve the term *estimator* for observers that operate in systems that also include noise models. Observation of such systems is more literally an *estimation* process.

observer toward asymptotic convergence with the plant. Only then will we be able to use the estimate for state feedback $u = K\hat{x}$.

We have already stated that the observer is a linear system whose inputs are the plant input and the plant output. But because the observer's output should track the plant and the error in the outputs should drive the observer, it is also possible to think of the observer as having the same basic structure as the plant, but with the additional input $y - \hat{y}$:

$$\begin{aligned} \dot{\hat{x}} &= A\hat{x} + Bu + L(y - \hat{y}) \\ &= A\hat{x} + Bu + L(y - C\hat{x} - Du) \\ &= (A - LC)\hat{x} + (B - LD)u + Ly \end{aligned} \tag{10.70}$$

where the matrix L is a matrix of gains that remain to be determined.

How do we enforce convergence of this system's state $\hat{x}$ to the plant state x? Suppose the state *error* could be measured: $\tilde{x} \overset{\Delta}{=} x - \hat{x}$. Then, of course, $\dot{\tilde{x}} = \dot{x} - \dot{\hat{x}}$ as well. Substituting the state matrices into this relationship gives

$$\begin{aligned} \dot{\tilde{x}} &= \dot{x} - \dot{\hat{x}} \\ &= Ax + Bu - (A - LC)\hat{x} - (B - LD)u - L(Cx + Du) \\ &= A(x - \hat{x}) - LC(x - \hat{x}) \\ &= (A - LC)(x - \hat{x}) \\ &= (A - LC)\tilde{x} \end{aligned} \tag{10.71}$$

If the autonomous system in (10.71), with system matrix $A - LC$, is asymptotically stable, then $\tilde{x} = x - \hat{x} \to 0$, or equivalently, $\hat{x} \to x$ as $t \to \infty$. In order to guarantee the asymptotic stability of (10.71), the only choice we have is the selection of the gain matrix L. Matrix L must be chosen such that the eigenvalues of $A - LC$ lie in the open left half of the complex plane. To facilitate this, notice the resemblance between the familiar quantity $A + BK$ and the quantity $A - LC$, or more usefully, $(A - LC)^{\mathrm{T}} = A^{\mathrm{T}} - C^{\mathrm{T}} L^{\mathrm{T}}$. (The transpose is immaterial because the eigenvalues of any matrix are same as those of its transpose.) Making the correspondences $C^{\mathrm{T}} \sim B$ and $-L^{\mathrm{T}} \sim K$, the problem of placing the eigenvalues of $A^{\mathrm{T}} - C^{\mathrm{T}} L^{\mathrm{T}}$ by choosing L^{T} is exactly the same (save for the minus sign) as the problem of placing the eigenvalues of $A + BK$ by choosing K. The matrix dimensions are analogous, and A (A^{T}) is available in each case. For the placement of the observer poles, it may be apparent by analogy that certain canonical forms for the A^{T}-matrix are more convenient than others. As might be guessed, it is the *observable canonical form* – see Equation (9.21) – that facilitates observer design, exactly as the controllable canonical form was appropriate for state-feedback design.

Naturally, it must first be established that the eigenvalues of the observer system *can* be arbitrarily placed, which, by analogy, matches the question of controllability of the pair $(A^{\mathrm{T}}, C^{\mathrm{T}})$. However, construction of a *controllability* matrix for $(A^{\mathrm{T}}, C^{\mathrm{T}})$ is identical to construction of the *observability* matrix for (A, C). Therefore, it can be said that *the eigenvalues of the observer error system in* (10.70) *can be arbitrarily placed in the left half plane if the system in* (10.68) *is observable.* Notice that the observer system in (10.70) must be given an initial condition. If indeed the error system is made asymptotically stable, then unlike the open-loop observer, even an inaccurate initial condition will cause only a transient error, because this error can be proven to asymptotically approach zero.

Full-Order Augmented Observer and Controller

The construction of the observer is motivated by the need for an estimated state vector that we can feed back to the input. However, feeding back the true state vector is undeniably not the same thing as feeding back the estimated state vector, because the two vectors are not identical. This raises the question of whether the observer and feedback together accomplish the goal of placing the plant eigenvalues and whether the two systems interact to produce unexpected results. While the estimates are still converging, we must ensure that the composite system (i.e., observer and controller together) behaves according to the K and L gains we have chosen. To address this concern, the plant, with state feedback of the form $u = K\hat{x} + v$, and the observer can be combined together into an *augmented* system. First, considering the plant itself, the closed-loop equation is:

$$\dot{x} = Ax + Bu = Ax + B(K\hat{x} + v) = Ax + BK\hat{x} + Bv \tag{10.72}$$

Then the observer equation begins with (10.70) and becomes:

$$\begin{aligned}\dot{\hat{x}} &= (A - LC)\hat{x} + (B - LD)u + Ly \\ &= (A - LC)\hat{x} + (B - LD)(K\hat{x} + v) + L(Cx + Du) \\ &= \left[A - LC + (B - LD)K\right]\hat{x} + (B - LD)v + LCx + LD(K\hat{x} + v) \\ &= \left[A - LC + (B - LD)K + LDK\right]\hat{x} + LCx + \left[B - LD + LD\right]v \\ &= \left[A - LC + BK\right]\hat{x} + LCx + Bv\end{aligned} \tag{10.73}$$

The two systems, (10.72) and (10.73), together are written in augmented state space form as:

$$\begin{bmatrix} \dot{x} \\ \dot{\hat{x}} \end{bmatrix} = \begin{bmatrix} A & BK \\ LC & A - LC + BK \end{bmatrix} \begin{bmatrix} x \\ \hat{x} \end{bmatrix} + \begin{bmatrix} B \\ B \end{bmatrix} v \tag{10.74}$$

Such a system can be depicted in the block diagram of Figure 10.4.

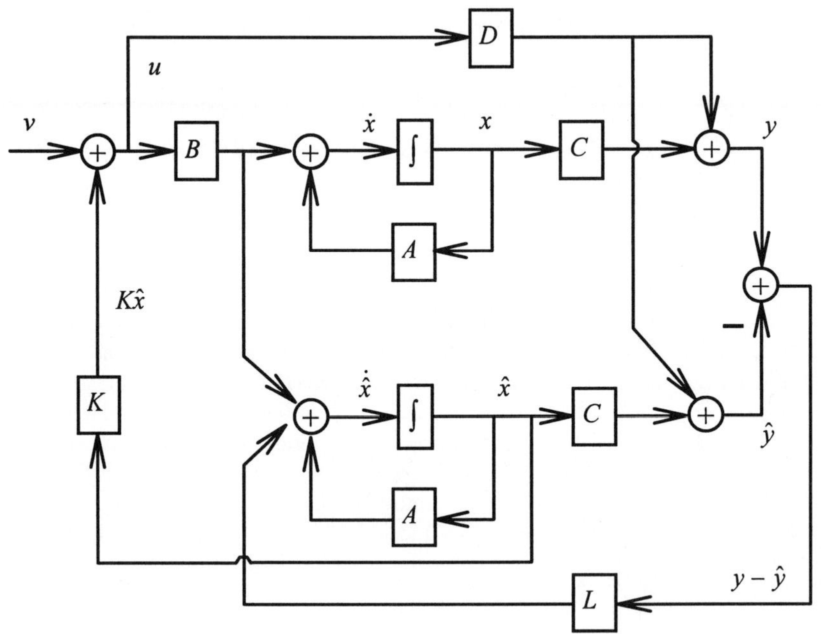

Figure 10.4 Block diagram of observer and state feedback combination.

To check whether the eigenvalues of this composite system are truly at the desired locations, we must determine the zeros of the determinant:

$$\det \begin{bmatrix} \lambda I_{n \times n} - A & -BK \\ -LC & \lambda I_{n \times n} - A + LC - BK \end{bmatrix} \tag{10.75}$$

To facilitate the computation of this determinant, we will perform elementary row and column operations, known to not alter the value of the determinant. First, we subtract the second row from the first, without altering the second row itself:

$$\det \begin{bmatrix} \lambda I_{n \times n} - A + LC & -\lambda I_{n \times n} + A - LC \\ -LC & \lambda I_{n \times n} - A + LC - BK \end{bmatrix}$$

Next, add the first column to the second, without altering the first column:

$$\det\begin{bmatrix} \lambda I_{n\times n} - A + LC & 0 \\ -LC & \lambda I_{n\times n} - A - BK \end{bmatrix}$$
$$= \det[\lambda I_{n\times n} - A + LC]\cdot\det[\lambda I_{n\times n} - A - BK] \tag{10.76}$$

where the last factorization of the determinant is made possible by the block of zeros in the upper right corner. Equation (10.76) indicates that the eigenvalues of the composite system are the union of the eigenvalues of a system with matrix $A - LC$ and those of a system with matrix $A + BK$. Therefore, even though the feedback control $u = K\hat{x} + v$ does not contain the *true* plant state, but rather the *observed* plant state, the closed-loop system still has the correctly placed eigenvalues. This result is known as the *separation principle.* A consequence of the separation principle is that the observer and the controller need not be designed simultaneously; the controller gain K can be computed independently of the observer gain L.

One must be careful, however, when thinking of the controller and observer gains as independent. It would make little sense for the closed-loop plant eigenvalues to be "faster" than the observer eigenvalues. If this were the case, the plant would "outrun" the observer. Rather, the observer eigenvalues should be faster than the plant eigenvalues, in order that the observer state converges to the plant state faster than the plant state converges to zero (or approaches infinity, if it is unstable). A workable rule of thumb is to place the observer eigenvalues two to five times farther left on the complex plane than the closed-loop plant eigenvalues. Eigenvalues with more disparity than that might result in poorly conditioned system matrices and difficulties with physical implementations. Fast eigenvalues also imply a large bandwidth, making the system susceptible to noise and poor transient performance.

Another consideration in the design of the closed-loop observer and controller system is the steady-state behavior. We can examine steady-state behavior by considering the transfer function of the system in (10.74), which presumes zero initial conditions. Computing the transfer function from the system equations in (10.74) would be extremely tedious. However, as we have already discovered, transfer functions are invariant to similarity transformations, which are simply changes of coordinates. Because $\tilde{x} = x - \hat{x}$ is a linear combination of x and $\hat{x}$, we can transform the system to use state variable x and $\tilde{x}$ instead of x and $\hat{x}$. Equation (10.72) will become

$$\begin{aligned}\dot{x} &= Ax + BK\hat{x} + Bv \\ &= Ax + BK\hat{x} - BKx + BKx + Bv \\ &= (A + BK)x - BK(x - \hat{x}) + Bv \\ &= (A + BK)x - BK\tilde{x} + Bv\end{aligned} \tag{10.77}$$

whereas (10.71) can be used directly. Together, their combination is the state space system

$$\begin{bmatrix} \dot{x} \\ \dot{\tilde{x}} \end{bmatrix} = \begin{bmatrix} A + BK & -BK \\ 0 & A - LC \end{bmatrix} \begin{bmatrix} x \\ \tilde{x} \end{bmatrix} + \begin{bmatrix} B \\ 0 \end{bmatrix} v \tag{10.78}$$

with output equation

$$\begin{aligned}y &= Cx + Du \\ &= Cx + D(K\hat{x} + v) \\ &= Cx + DK\hat{x} - DKx + DKx + Dv \\ &= (C + DK)x - DK\tilde{x} + Dv \\ &= \begin{bmatrix} (C + DK) & -DK \end{bmatrix} \begin{bmatrix} x \\ \tilde{x} \end{bmatrix} + Dv\end{aligned} \tag{10.79}$$

`ctrbf(A,B,C)`

Equations (10.78) and (10.79) can be recognized as being in the form of a Kalman controllability decomposition[M*] (see Chapter 8). The realization theory of Chapter 9 then indicates that the transfer function of the system is independent of the uncontrollable part. The transfer function resulting from (10.78) and (10.79) is therefore

$$H(s) = (C + DK)(sI - A - BK)^{-1} B + D \tag{10.80}$$

which is the multivariable version of the transfer function derived for the SISO case, Equation (10.7), assuming that $E = I$.

It might at first seem surprising that the transfer function for the closed-loop observer and controller system is completely independent of the observer. The reason that this is true is that transfer functions always assume zero initial conditions. If the initial conditions for (10.71) are zero, then $\tilde{x}(t_0) = x(t_0) - \hat{x}(t_0) = 0$, or $\hat{x}(t_0) = x(t_0)$. The asymptotically stable "error system" in (10.71), with this zero initial condition, will therefore remain in a zero state forever and will not affect the system output. This however does not

* It might also be observed that this form would have made the computation of the eigenvalues of (10.74) much easier.

prevent the observer from affecting the transient response, as we will illustrate in an example.

Example 10.4: An Observer and Controller With Simulation

Compute state feedback to place the eigenvalues of the system below at $-5 \pm j5$ and -10. Use an observer to estimate the state of the system and simulate the observer and controller composite system, comparing the true state to the estimated state.

$$\begin{aligned} \dot{x} &= \begin{bmatrix} 0 & 1 & 0 \\ 0 & 0 & 1 \\ 24 & 14 & -1 \end{bmatrix} x + \begin{bmatrix} 0 \\ 0 \\ 1 \end{bmatrix} u \\ y &= \begin{bmatrix} 2 & -1 & 1 \end{bmatrix} x \end{aligned} \tag{10.81}$$

Solution:

The system is already in controllable canonical form so it provides a shortcut in the computations. (It is not often that one will compute feedback gains "by hand" anyway, so there is little sense in going through the exercise of performing the similarity transformation.) MATLAB has two commands for placing the poles of a system, ACKER and PLACE. The ACKER command uses Ackermann's formula (although warning of its numerical difficulties). Using this command or some other program to compute feedback gains, we find that

$$K = \begin{bmatrix} -524 & -164 & -19 \end{bmatrix} \tag{10.82}$$

Using the same command but with A^{T} and C^{T} as arguments, the observer gain can be computed as well:

$$L = \begin{bmatrix} -47.9 & 186.6 & 341.4 \end{bmatrix}^{\mathrm{T}} \tag{10.83}$$

For the observer, we selected all eigenvalues to be located at -20, at least two times farther left on the complex plane than the desired plant eigenvalues.

Together, the augmented system may be written[M] as in (10.74) and numerically simulated. One can verify (again, by computer) that the eigenvalues of the system matrix in (10.74) are indeed the union of the sets of desired plant and desired observer eigenvalues: $\{-10, -5 \pm j5, -20, -20, -20\}$. In the simulation shown in Figure 10.5, a zero input is assumed and the initial conditions are given as $x(0) = \begin{bmatrix} -3 & -3 & -3 \end{bmatrix}^{\mathrm{T}}$ and $\hat{x}(0) = \begin{bmatrix} 0 & 0 & 0 \end{bmatrix}^{\mathrm{T}}$.

`reg(sys,K,L)`

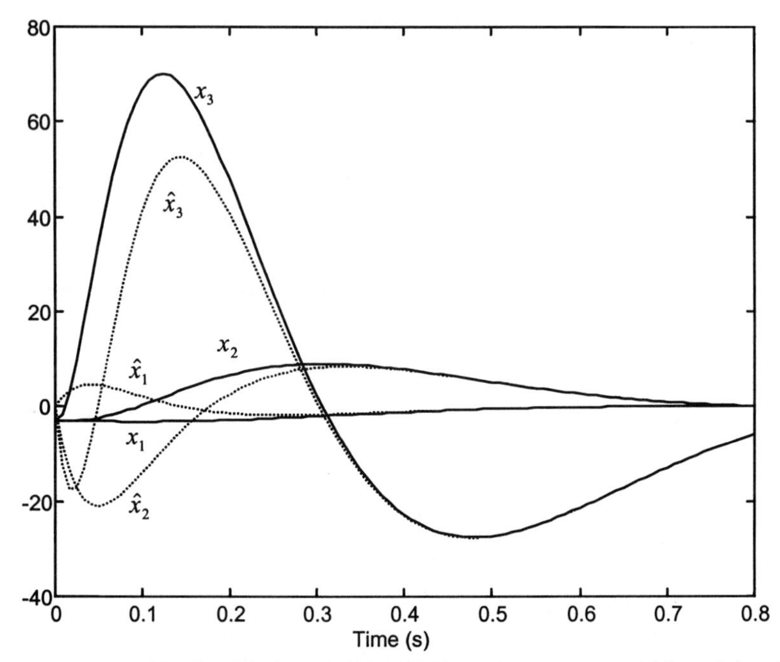

Figure 10.5 Simulated trajectories for the three (true) state variables of the plant, along with their three estimates, as provided by the observer.

The observers developed in this section are referred to as *full-order observers* because if $t=(k+1)T$, then $\hat{x} \in \Re^n$. That is, the observer estimates the full state of the system, not just a portion of it. In the event that only some of the state variables need to be estimated, a *reduced-order observer* can be designed, which will estimate only the subspace of the state space that is not already measured through the output equations.

Reduced-Order Observers

The output equation of a state space description can be viewed as a q-dimensional projection of the n-dimensional state vector. If the coordinate axes (i.e., basis vectors) of this reduced space are judiciously chosen, then the output y can be used to directly measure any q state variables. In this case, the state equations would have the form

$$\begin{bmatrix} \dot{x}_1 \\ \dot{x}_2 \end{bmatrix} = \begin{bmatrix} A_{11} & A_{12} \\ A_{21} & A_{22} \end{bmatrix} \begin{bmatrix} x_1 \\ x_2 \end{bmatrix} + \begin{bmatrix} B_1 \\ B_2 \end{bmatrix} u$$
$$y = \begin{bmatrix} I_{q \times q} & 0 \end{bmatrix} \begin{bmatrix} x_1 \\ x_2 \end{bmatrix} + Du = x_1 + Du \tag{10.84}$$

so that the output directly provides the q state variables $x_1(t)$. This portion of the state vector is readily available for feedback, but the rest of the vector, $x_2(t)$, must still be observed. The system that observes such a partial state is called a *reduced-order observer*, and we will describe it below.

First, however, we should note that not all systems will have the form seen in (10.84). If not, a simple transformation matrix may be obtained to transform any system into one resembling (10.84). To construct such a transformation, we will create a new basis from the linearly independent rows of the C-matrix. Usually, there will be q such rows (otherwise, some outputs would be redundant). To create a new basis, then, we will collect these q rows and augment them with $n-q$ additional rows, linearly independent of the others, in order to create a nonsingular transformation. That is, we create the transformation W as:

$$W = \begin{bmatrix} C \\ R \end{bmatrix} \begin{matrix} \text{original output matrix} \\ \text{any } n-q \text{ additional linearly indpendent rows} \end{matrix} \tag{10.85}$$

Then with the similarity transformation $\overline{A} = WAW^{-1}$, $\overline{B} = WB$, and $\overline{C} = CW^{-1}$, the new system will be of the form in (10.84). Hereafter, we will assume that the system starts in the form in (10.84).

The strategy for observing $x_2(t)$ will be to find a state space equation for it alone and to build an observer similar to the full-order observer. To find the equations for the $x_2(t)$ subsystem, we will take advantage of the fact that $x_1(t)$ can be measured directly from y in (10.84), so that it can be treated as a known signal. If this is the case, then the two subsystem equations in (10.84) can be rewritten as

$$\begin{aligned} A_{12}x_2 &= \dot{x}_1 - A_{11}x_1 - B_1 u \\ \dot{x}_2 &= A_{22}x_2 + \left[B_2 u + A_{21}x_1\right] \end{aligned} \tag{10.86}$$

Simplifying the notation, if we define the (known) signals

$$\bar{u} \triangleq A_{21}x_1 + B_2 u$$

$$\bar{y} \triangleq \dot{x}_1 - A_{11}x_1 - B_1 u$$

then (10.86) may be written as

$$\begin{aligned} \dot{x}_2 &= A_{22}x_2 + \bar{u} \\ \bar{y} &= A_{12}x_2 \end{aligned} \tag{10.87}$$

This set of equations should now be viewed as a state space system in and of itself. As such, the design of an observer for it begins with the specification of the observer equation as in (10.70). Writing this equation for the system in (10.87) and performing the necessary substitutions,

$$\begin{aligned} \dot{\hat{x}}_2 &= A_{22}\hat{x}_2 + \bar{u} + L(\bar{y} - \hat{\bar{y}}) \\ &= A_{22}\hat{x}_2 + \bar{u} + L(\bar{y} - A_{12}\hat{x}_2) \\ &= (A_{22} - LA_{12})\hat{x}_2 + \bar{u} + L\bar{y} \end{aligned} \tag{10.88}$$

The definition of $\bar{y}$ as a "known" signal, while true, is impractical. In practice, the value of $x_1(t)$ is indeed known, but because $\bar{y}$ depends on $\dot{x}_1(t)$, and pure derivatives are difficult to realize, we wish to avoid the use of $\bar{y}$ in (10.88). Toward this end, we will introduce a change of variables $z = \hat{x}_2 - Lx_1$. Now if we use the state vector $\dot{z}$ instead of $\dot{\hat{x}}_2$, the state equation for the observer in (10.88) will simplify:

$$\begin{aligned} \dot{z} &= \dot{\hat{x}}_2 - L\dot{x}_1 \\ &= (A_{22} - LA_{12})\hat{x}_2 + \bar{u} + L\bar{y} - L\dot{x}_1 \\ &= (A_{22} - LA_{12})\hat{x}_2 + \bar{u} - L(A_{11}x_1 + B_1 u) \\ &= (A_{22} - LA_{12})(\hat{x}_2 - Lx_1) + (A_{22} - LA_{12})Lx_1 + \bar{u} - L(A_{11}x_1 + B_1 u) \\ &= (A_{22} - LA_{12})z + \left[(A_{22} - LA_{12})L - LA_{11}\right]x_1 + \bar{u} - LB_1 u \\ &= (A_{22} - LA_{12})z + \left[(A_{22} - LA_{12})L + A_{21} - LA_{11}\right]x_1 + (B_2 - LB_1)u \end{aligned} \tag{10.89}$$

(The transition from the third line to the fourth is a result of adding and subtracting the term $(A_{22} - LA_{12})Lx_1$ from the equation, and the transition from the fifth line to the sixth is a result of substitution for $\bar{u}$.) This is the equation that can be simulated or realized in order to generate the signal $z(t)$, after which the estimated state vector is computed from $\hat{x}_2 = z + Lx_1$. (Recall that the vector signal $x_1(t)$ is directly measured from (10.84) as $x_1(t) = y(t) - Du(t)$.)

The reduced-order observer (10.89) has an observer gain L in it, as did the full-order observer. In order to find a suitable gain, the "error" system must again be constructed, which will ensure that the estimated vector approaches the true vector. In this case, the error system is $e = x_2 - \hat{x}_2$, so the "error dynamics" are [with substitutions from (10.87) and (10.88)]:

$$\begin{aligned}
\dot{e} &= \dot{x}_2 - \dot{\hat{x}}_2 \\
&= \left[A_{22}x_2 + \bar{u}\right] - \left[A_{22}\hat{x}_2 + \bar{u} + L(A_{12}x_2 - A_{12}\hat{x}_2)\right] \\
&= A_{22}(x_2 - \hat{x}_2) - LA_{12}(x_2 - \hat{x}_2) \\
&= (A_{22} - LA_{12})(x_2 - \hat{x}_2) \\
&= (A_{22} - LA_{12})e
\end{aligned} \tag{10.90}$$

Comparing (10.90) to (10.71), it is apparent that the gain L is again computed in a manner analogous to the computation of the state feedback gains; i.e., by using eigenvalue placement methods[M] on the transposed system matrix $A_{22}^{\mathrm{T}} - A_{12}^{\mathrm{T}}L^{\mathrm{T}}$, exactly as done for the full-order observer. This will ensure that the system (10.90) is made asymptotically stable, so that $\hat{x}_2 \to x_2$ as $t \to \infty$. The question that should arise with the observant reader is that of observability of the pair of matrices (A_{22}, A_{12}). Provided that the original system (A, C) is observable, this will always be the case, as is shown in Problem 10.7 at the end of this chapter.

`place(A,B,P)`

Summarizing, assuming that the state equations are in the form (10.84), the "observed" state vector is

$$\begin{bmatrix} x_1 \\ \hat{x}_2 \end{bmatrix} = \begin{bmatrix} y - Du \\ z + Lx_1 \end{bmatrix} \tag{10.91}$$

where the signal $y(t)$ is the output of the original system, and the signal $z(t)$ is the state of the system given by (10.89). One can then proceed to compute state feedback in the form

$$u = K\begin{bmatrix} x_1 \\ \hat{x}_2 \end{bmatrix} + v \tag{10.92}$$

Although the reduced-order observer has fewer state variables than the full-order observer, one might question whether the equations in (10.89) are actually more efficient than using the larger but simpler full-order observer. In fact, they might not be, but the comparison depends on a number of other factors, such as the sparsity of the system matrices and the size of the system relative to the number of output variables q. The choice of full- versus reduced-order observer should also be taken in consideration of engineering factors as well. Because the signal x_1 appears directly in the estimate for the state variables x_2 (or z) in (10.91), and x_1 is derived from the system output y, there is "direct feedthrough" of the output to the input in the reduced-order observer. Theoretically, this presents no difficulty at all. From a practical viewpoint, however, this can be a source of excess noise in the observed value $\hat{x}_2$. The output of many physical systems is measured with physical sensors, which almost always include some noise. In the reduced-order observer, this noise passes through directly to the estimate. In the full-order observer, the estimated state also relies on the output signal, but only as an input to the observer state equations. Thus, the observer (assuming it is stable) always has a smoothing effect on y. The noise in y does not pass directly to the estimate. Finally, although we do not provide an example to illustrate the point here, it is often found that the reduced-order observer has a better transient response. This is because some of the state variables (i.e., x_1) are directly measured, and have no transient responses at all: $\hat{x}_1 \equiv x_1$. Thus, there are fewer contributions to the error dynamics early in time, while the transients in $x_2 - \hat{x}_2$ are occurring. Because of all these factors, the selection of the reduced- versus full-order observer is not automatic.

10.3.2 Discrete-Time Observers

As is usually true, the discrete-time observer is different from the continuous-time observer in notation only. To demonstrate this, we will derive the full-order observer using the same sequence of steps and the same reasoning as above, but we will leave the reduced-order observer as an exercise.

Full- and Reduced-Order Observers

Given the discrete-time plant equations (assumed to be observable):

$$\begin{aligned} x(k+1) &= A_d x(k) + B_d u(k) \\ y(k) &= C_d x(k) + D_d u(k) \end{aligned} \tag{10.93}$$

a full-order observer can be modeled after (10.70):

$$\begin{aligned} \hat{x}(k+1) &= A_d \hat{x}(k) + B_d u(k) + L(y(k) - \hat{y}(k)) \\ \hat{y}(k) &= C_d \hat{x}(k) + D_d u(k) \end{aligned} \tag{10.94}$$

Substituting for $\hat{y}(k)$, this simplifies to the same form as (10.70):

$$\begin{aligned}\hat{x}(k+1) &= A_d\hat{x}(k) + B_d u(k) + L\left[y(k) - C_d\hat{x}(k) - D_d u(k)\right] \\ &= (A_d - LC_d)\hat{x}(k) + (B_d - LD_d)u(k) + Ly(k)\end{aligned} \tag{10.95}$$

To determine the proper choice of the observer gain *L*, we again form the error signal $e(k) = x(k) - \hat{x}(k)$ and the error-dynamics equations:

$$\begin{aligned}e(k+1) &= x(k+1) - \hat{x}(k+1) \\ &= A_d x(k) + B_d u(k) - A_d\hat{x}(k) - B_d u(k) - L\left[y(k) - \hat{y}(k)\right] \\ &= A_d\left[x(k) - \hat{x}(k)\right] - L\left[C_d x(k) - C_d\hat{x}(k)\right] \\ &= (A_d - LC_d)(x(k) - \hat{x}(k)) \\ &= (A_d - LC_d)e(k)\end{aligned} \tag{10.96}$$

Exactly as for the continuous-time case, the procedure for making this time-invariant system asymptotically stable is to place the eigenvalues of the matrix $(A_d - LC_d)$ using the gain matrix *L*. Of course, this implies that the eigenvalues will be inside the unit circle, not in the left half-plane. Also, the eigenvalues should be "faster" than the closed-loop eigenvalues of the plant, as specified with the feedback gain matrix *K*; faster implies a smaller *magnitude* for these eigenvalues, i.e., closer to the origin on the complex plane. Naturally, all the same gain matrix calculation tools will apply. Other properties of observers remain the same as well, such as the fact that the transfer function of the combined controller and observer contains only the closed loop plant poles, and not the observer poles, and the fact that *K* and *L* can be computed independently, i.e., the separation principle.

Current Estimators

The development of this discrete-time full-order observer might serve to convince the reader that the reduced-order discrete-time observer is analogous to the reduced-order continuous-time observer. It is. However, there is a unique peculiarity of discrete-time systems that we can use to our advantage when designing observers. This inherent difference is in the moment at which the feedback is computed. In particular, there is a single one-period time lag between the time that the state estimate $\hat{x}(k)$ is computed, and the time that it is applied to the system through the feedback equation $u(k) = K\hat{x}(k) + r(k)$. Thus, at the instant in time $t = kT$, the value $\hat{x}(k)$ is found, based on the known values of $\hat{x}(k-1)$, $u(k-1)$, and $y(k-1)$. This value is computed in a relatively short time, probably considerably less than the sample interval *T*.

Once $\hat{x}(k)$ is determined, the feedback $K\hat{x}(k)$ *could* be immediately determined. It is not used, however, until the *next* computation time, which occurs at the instant $t=(k+1)T$, at which time $\hat{x}(k+1)$ is computed. The result is that each computation of $\hat{x}$ is always based on data that is at least one sample period old.

An alternative formulation has been developed that incorporates an estimate of the state data at the "current" time: a *current estimator* [6]. In the current estimator, the process of observation is separated into two components: *prediction* of the *estimated* state at time $k+1$ and the correction of that estimate with the system's true and estimated outputs at time $k+1$. Explicitly, we create the predicted state estimate as:

$$\bar{x}(k+1) = A_d\hat{x}(k) + B_d u(k) \tag{10.97}$$

which can be computed at time *k*, not $k+1$. This can be interpreted as the "time update" portion of the observer equation. Then at the moment sampling instant $k+1$ occurs, we perform the "measurement update" of this estimate with the expression:

$$\begin{aligned}\hat{x}(k+1) &= \bar{x}(k+1) + L(y(k+1) - \bar{y}(k+1)) \\ &= \bar{x}(k+1) + L\left[y(k+1) - C_d\bar{x}(k+1) - D_d u(k+1)\right]\end{aligned} \tag{10.98}$$

Substituting (10.97) into (10.98) gives the observer equation:

$$\begin{aligned}\hat{x}(k+1) = {} & A_d\hat{x}(k) + B_d u(k) \\ & + L\left\{y(k+1) - C_d\left[A_d\hat{x}(k) + B_d u(k)\right] - D_d u(k+1)\right\}\end{aligned} \tag{10.99}$$

Because this is significantly different from the previous observer equations we have been using, we will again write the error equations to determine the appropriate conditions on gain matrix *L*:

$$\begin{aligned}e(k+1) &= x(k+1) - \hat{x}(k+1) \\ &= A_d x(k) + B_d u(k) - A_d\hat{x}(k) - B_d u(k) \\ &\quad - L\left\{C_d x(k+1) + D_d u(k+1)\right. \\ &\quad \left. - C_d\left[A_d\hat{x}(k) + B_d u(k)\right] - D_d u(k+1)\right\} \\ &= A_d x(k) - A_d\hat{x}(k) \\ &\quad - L\left\{C_d\left[A_d x(k) + B_d u(k)\right] + D_d u(k+1)\right. \\ &\quad \left. - C_d\left[A_d\hat{x}(k) + B_d u(k)\right] - D_d u(k+1)\right\}\end{aligned}$$

$$\begin{aligned} &= (A_d - LC_d A_d)[x(k) - \hat{x}(k)] \\ &= (A_d - LC_d A_d)e(k) \end{aligned} \tag{10.100}$$

Note that this error equation is somewhat different from (10.71). In this case, it is the eigenvalues of the matrix $A_d - LC_d A_d$ that must be placed using appropriate selection of gain L. This is not guaranteed by the observability of the system, as it was with the conventional observer in Section 10.3.1. However, if system matrix A_d is of full rank, then it will not change the space spanned by the rows of C_d. If that is the case, we can think of the product $C_d A_d$ as a "new" output matrix

$$\overline{C}_d \triangleq A_d C_d$$

Then the eigenvalues of $A_d - L\overline{C}_d$ can be placed as usual. If A_d is *not* of full rank, then arbitrary placement of all the eigenvalues of $A_d - LC_d A_d$ will not be possible.

10.3.3 Output Feedback and Functional Observers

Before closing this chapter, we will briefly consider a different type of feedback and observer. We have seen that full state feedback is able to arbitrarily place the eigenvalues of a controllable system, and that if an observer is constructed, the full state will either be available or estimated. However, what can be said about the capabilities of feedback if an observer is *not* used, i.e., if only the output is fed back, in the form $u = K_o y$?

If $u = K_o y$, then the state equation in (10.68) would become

$$\begin{aligned} \dot{x} &= Ax + Bu \\ &= Ax + B(K_o y) \\ &= Ax + B[K_o(Cx + Du)] \\ &= (A + BK_o C)x + BK_o Du \end{aligned} \tag{10.101}$$

Of course, from (10.101) it is apparent that the eigenvalues of the matrix $A + BK_o C$ must be selectable through proper choice of the gain K_o. Matrix K_o in this case will be $p \times q$ rather than $p \times n$ as was the state-feedback matrix K. In the situation that $q = 1$, for example, only a single feedback gain will be available, so of course we could not possibly independently specify the locations of n eigenvalues. This is the situation in classical control that is illustrated by the root locus technique: A change in gain will alter all the system poles, but not independently of one another. In general, output feedback will be able to assign *some* eigenvalues, but not all, unless the number of inputs (p) and outputs (q) are both large. If C happens to be square and invertible, then of course the output feedback problem is equivalent to the state feedback problem, because the output is simply a transformed version of the state. However, pole assignment using output feedback is not a completely solved problem (see [5] and [13]).

Another variation on the concepts of controllers and observers is the *functional observer* [11]. The design of the full- and reduced-order observers above represents a time investment to extract the n state variables of a plant, whereas the real goal of the designer is actually to find the control signal $K\hat{x}$ that has only p components. The functional observer combines these functions by directly observing not the state vector, but a linear combination of the state variables, i.e., $K\hat{x}$. While it is only a single computational step to multiply the estimated state $\hat{x}$ by the gain K, the functional estimator can do this with a system of order $\mu_o - 1$, where μ_o is the observability index of the system (defined analogously to the controllability index discussed above) [10]. Thus, some economy can be realized via the use of the functional observer. In fact, if a system with a large n has a small observability index, then the functional observer will guarantee the existence of a low-order observer and controller that can still achieve arbitrary pole placement.

10.4 Summary

In this chapter, we have presented the most fundamental of all state space control system tools: feedback and state observation. State feedback is the vector generalization of "position" feedback in classical systems, although it is understood that the state vector might contain variables quite different from the concept of position. The state vector might contain positions, velocities, accelerations, or any other abstract type of state variable. Because we have defined the state vector of a system as the vector of information sufficient to describe the complete behavior of the system, feedback of the state is a general type of control.

However, the feedback we have considered in this chapter is sometimes referred to as *static* state feedback. The term *static* refers to the fact that the state vector is not filtered by a dynamic system before reintroducing it as a component of the input. This is in contrast to *dynamic* state feedback, which is the type of control usually studied in frequency-domain. In dynamic state feedback, the state vector or state estimate is processed by a dynamic system (a compensator), which has a transfer function (and hence, a state space) of its own. More often, dynamic feedback is applied to the *output* of a system, which is the classical control problem. One can view the observer and controller combined system as dynamic output feedback, because the output is "filtered" by the observer dynamics before resulting in the control input.

In the same sense, it might be argued that classical dynamic compensators, represented by a controller transfer function in the forward path of the feedback loop, really represent a form of observer and gain combination. The dynamics of the compensator "prefilters" the error signal in order to generate more information about the underlying system, thereby providing more data on which the gains can act.

The important topics presented in this chapter include:

- Computational models and formulas were given for placing the eigenvalues of a system using full state feedback. Ackermann's formula and the Bass-Gura formula are the two most common "pencil-and-paper" techniques, but these are not often the most numerically desirable. Computer-aided engineering programs such as MATLAB include one-word commands that can efficiently compute feedback gains for single- and multiple-input systems.

- It was found that the use of state feedback does not change the zeros of a system. Thus, because the poles are free to move and, potentially, cancel the zeros, it is possible that observability is lost by the introduction of state feedback. Controllability, however, is preserved by state feedback.

- The placement of eigenvalues in multi-input systems is difficult if performed "by hand." In order to facilitate this process, the concept of controllability indices was introduced. Using a transformation based on controllability indices, a system can be transformed into a multivariable controllable canonical form in which the feedback gains can be determined by considering their effects one input at a time. Again, this is a manual process that may be dissimilar from the numerical algorithms written into modern computer-aided control system design (CACSD) programs. However, the definitions and properties of controllability indices are useful in broader applications in multivariable control.

- The observer is a dynamic system that produces an estimate of the state variable of the plant system when that state is not directly measured. The full-order observer produces estimates of all the state variables, while the reduced-order observer estimates only those variables that are not measured via output y. The reduced-order observer is therefore of lower dimension than the full-order observer, but this does not necessarily guarantee that it requires fewer computations because of the transformations involved in its construction. There are other factors to consider in the choice of full- versus reduced-order observer as well, such as transient response and noise levels.

- In discrete time, the state feedback and observer calculations are exactly the same as in continuous time; it is only the desired eigenvalue locations that change. However, discrete-time systems offer the additional possibility of the *current* observer. The current observer separates the so-called "time update" from the "measurement update." This distinction will appear again in the Kalman filter, which will be discussed again in the next and final chapter.

The next chapter will explore two concepts that were mentioned only in passing in this chapter. The first is the problem of choosing a state feedback gain K for multi-input systems. Although we showed in this chapter a technique to select a few possible gains, we noted that there are many choices. In order to select a unique gain, a secondary criterion can be introduced: the cost function. By minimizing this cost function rather than placing the system eigenvalues, a unique gain can be determined. This is the basis for the *linear-quadratic regulator* (discussed in Chapter 11).

The second concept is embedded in the idea of the current estimator in discrete time. For systems with noise, provided that plant noise and sensor (output) noise have known statistical properties, a special observer is available that finds (statistically) the best possible observer given the noise model. This type of observer will be referred to as an *estimator*, and in discrete time, it will have the property of separating the computations into "time-update" and "measurement update" components. This estimator is known as a *Kalman filter* and is the standard for designing observers in noisy environments.

Together, the linear-quadratic regulator and the Kalman filter form a controller and observer pair, similar in concept to those described in this chapter, except that the feedback gain and the observer gain are both designed according to more sophisticated criteria. The combination is known as the linear-quadratic-gaussian controller (discussed in Chapter 11). Whereas the state feedback and full-order observer were described here as the most fundamental of the state space control techniques, the linear-quadratic-gaussian controller is often regarded as the most useful.

10.5 Problems

10.1 An uncontrollable state-space equation is given below. If possible, determine a gain matrix that transforms the system such that its eigenvalues are $\{-3,-3,-3,-3\}$. Repeat for eigenvalues $\{-3,-3,-2,-2\}$ and for $\{-2,-2,-2,-2\}$. If the solution is impossible, explain why.

$$\dot{x} = \begin{bmatrix} 3 & 3 & 0 & 2 \\ 0 & 87 & 0 & 60 \\ 6 & 3 & -3 & 2 \\ 0 & -126 & 0 & -87 \end{bmatrix} x + \begin{bmatrix} 0 \\ 3 \\ -1 \\ 4 \end{bmatrix} u$$

10.2 A two-input linear system has the following matrices:

$$A = \begin{bmatrix} 0 & 1 \\ 7 & -4 \end{bmatrix} \quad B = \begin{bmatrix} 1 & 1 \\ 1 & 0 \end{bmatrix} \quad C = \begin{bmatrix} 1 & 3 \end{bmatrix} \quad D = 0$$

a) Find two different state-feedback gain matrices such that the poles of the closed-loop system are at $s = -4$ and $s = -6$.

b) Change the B-matrix as given above to $b = \begin{bmatrix} 1 & -1 \end{bmatrix}^{\mathrm{T}}$. Then apply *output* feedback $u = -ky$ to the system. Find a range of k such that the closed-loop system is asymptotically stable.

10.3 A discrete-time system with transfer function

$$\frac{y(z)}{u(z)} = \frac{5}{z-3}$$

is to be controlled through state feedback. The closed-loop pole should be placed at $z = 0.6$, and the observer should have pole $z = 0.3$. Find the state feedback gain, observer gain, and the closed-loop augmented state variable description of the controller and observer pair.

10.4 A permanent magnet DC motor with negligible armature inductance has transfer function

$$G(s) = \frac{\theta(s)}{v(s)} = \frac{-50}{s(s+5)}$$

where $\theta(s)$ is the shaft angle and $v(s)$ is the applied voltage.

a) Find a state variable representation of the system, where the two state variables are $x_1(t) = \theta(t)$ and $x_2(t) = \dot{\theta}(t)$.

b) Design a full-state feedback matrix K such that the closed-loop transfer function of the system has damping $\zeta = 0.707$ and undamped natural frequency $\omega_n = 10 \text{ rad s}^{-1}$ (i.e., the characteristic polynomial has the form $s^2 + 2\zeta\omega_n s + \omega_n^2$).

c) Find a full-order observer with a gain matrix L such that the observer itself has $\zeta_o = 0.5$ and $\omega_{n_o} = 20 \text{ rad s}^{-1}$.

d) Compute the transfer function of the combined system with the observer and controller.

10.5 The dynamics of a wind turbine drive train are given in [12] as:

$$\dot{x} = \begin{bmatrix} -0.94 & 0.43 & 7.14\times 10^{-6} \\ 0.98 & -0.98 & 1.64\times 10^{-5} \\ 1.2\times 10^{7} & -1.2\times 10^{7} & 0 \end{bmatrix} x + \begin{bmatrix} 0 \\ -1.64\times 10^{-5} \\ 0 \end{bmatrix} u$$
$$y = \begin{bmatrix} 0 & 1 & 0 \end{bmatrix} x$$

where the state vector $x = \begin{bmatrix} x_1 & x_2 & x_3 \end{bmatrix}^T$ consists of, respectively, the turbine angular velocity, the generator angular velocity, and the shaft torque due to torsional flexing. The input u represents the generator torque reference. Design a full-order observer and state-feedback controller combination that places the closed-loop poles at $s_1 = -10$ and $s_{2,3} = -5 \pm j5$. Notice how poorly conditioned this model is. How does this affect your procedure?

10.6 Find a state feedback gain and an observer such that the following system is stabilized.

$$\dot{x} = \begin{bmatrix} 2 & 1 & 0 & 0 & 0 \\ 0 & 2 & 0 & 0 & 0 \\ 0 & 0 & 2 & 0 & 0 \\ 0 & 0 & 0 & -3 & 0 \\ 0 & 0 & 0 & 0 & -3 \end{bmatrix} x + \begin{bmatrix} 1 & 0 \\ 1 & 1 \\ 2 & 1 \\ 0 & 1 \\ 0 & 1 \end{bmatrix} u$$
$$y = \begin{bmatrix} -1 & 0 & 0 & 0 & 0 \\ 0 & 0 & 1 & 1 & 0 \\ 1 & 0 & 0 & 1 & 1 \end{bmatrix} x$$

10.7 Show that for the system

$$\begin{bmatrix} \dot{x}_1 \\ \dot{x}_2 \end{bmatrix} = Ax + Bu = \begin{bmatrix} A_{11} & A_{12} \\ A_{21} & A_{22} \end{bmatrix} \begin{bmatrix} x_1 \\ x_2 \end{bmatrix} + \begin{bmatrix} B_1 \\ B_2 \end{bmatrix} u$$

$$y = Cx + Du = \begin{bmatrix} I_{q \times q} & 0 \end{bmatrix} \begin{bmatrix} x_1 \\ x_2 \end{bmatrix} + Du = x_1 + Du$$

where each block submatrix is of consistent dimensions, the system (A_{22}, A_{12}) is observable if and only if the system (A, C) is observable.

10.8 Define the *observability* index of a system analogous to the controllability index, and show that when a system is in the form of Equation (10.84), if the observability index of the system (A, C) is ν, then the observability index of the system (A_{22}, A_{12}) is $\nu - 1$. Show that when the *individual* observability indices of (A, C) are ν_i, then the observability indices of (A_{22}, A_{12}) are $\nu_i - 1$.

10.9 For the following SISO continuous-time system:

$$\dot{x} = \begin{bmatrix} 3 & 5 & 4 & 0 & 0 \\ 0 & 1 & 2 & 0 & 1 \\ 2 & 0 & 1 & 1 & 0 \\ 0 & 0 & 0 & 0 & 5 \\ 0 & 0 & 0 & -5 & 0 \end{bmatrix} x + \begin{bmatrix} 1 \\ 1 \\ 1 \\ 1 \\ 1 \end{bmatrix} u$$

$$y = \begin{bmatrix} 3 & 1 & 0 & 1 & 0 \end{bmatrix} x$$

a) Determine the controllability and observability.

b) Design a full-order observer and a reduced-order (fourth-order) observer. Give the system suitable initial conditions and simulate each observer. For each, compare to the true states. Place the observer poles in the region [–15,–10] in the complex plane. Compare each observer's output with the true states.

c) Design and implement full-state feedback $u = K\hat{x} + v$ such that the closed-loop poles are placed on the set $\{-2, -2, -3, -5, -5\}$. Verify by determining the closed-loop eigenvalues of the controller and observer (i.e., augmented) system.

10.10 An inverted pendulum on a driven cart is modeled in [15] as

$$\dot{x} = \begin{bmatrix} 0 & 1 & 0 & 0 \\ 30.66 & 0 & 0 & 20.27 \\ 0 & 0 & 0 & 1 \\ -1.63 & 0 & 0 & -7.56 \end{bmatrix} x + \begin{bmatrix} 0 \\ -24.1 \\ 0 \\ 8.99 \end{bmatrix} u$$

$$y = \begin{bmatrix} 37.57 & 0 & 0 & 0 \\ 0 & 0 & 10 & 0 \\ 0 & 0 & 0 & -4.5 \end{bmatrix} x$$

where the state vector $x = \begin{bmatrix} x_1 & x_2 & x_3 & x_4 \end{bmatrix}^T$ consists of, respectively, the pendulum angle, the pendulum's angular velocity, the cart position, and the cart's velocity. The input represents the voltage applied to the wheels on the cart.

a) Design a stabilizing state-feedback controller with a full-order observer and a reduced-order (first order) observer. Compare the performance of each.

b) Can the system be observed through only one of the outputs? Which one?

10.11 Design a state feedback controller to place the closed-loop eigenvalues of the system below at $z = 0.25 \pm 0.25j$:

$$x(k+1) = \begin{bmatrix} 0 & 1 \\ -0.5 & 1 \end{bmatrix} x(k) + \begin{bmatrix} 0 \\ 1 \end{bmatrix} u(k)$$

$$y(k) = \begin{bmatrix} 0.5 & 1 \end{bmatrix} x(k)$$

10.12 A discrete-time system is given as

$$x(k+1) = \begin{bmatrix} 0 & 1 \\ -1 & -2 \end{bmatrix} x(k) + \begin{bmatrix} 1 & 0 \\ 0 & 1 \end{bmatrix} u(k)$$

Determine a state feedback $u(k) = Kx(k)$ such that the closed-loop poles are located at $z = 0, 0$. Is it possible to do this with feedback to only $u_1(k)$? Is it possible to do this with feedback to only $u_2(k)$? In each case, find the feedback if possible.

10.13 The discrete-time system below is unstable. Design a state-feedback controller that places its poles at $z = 0.5 \pm 0.5j$. Also design both a full-order observer and a reduced-order (first-order) observer to estimate the system state vector. Specify suitable observer poles for each. Combine both types of observer with the feedback and simulate the zero-input response of each system. Compare and comment.

$$x(k+1) = \begin{bmatrix} 2 & 0 \\ 1 & 1 \end{bmatrix} x(k) + \begin{bmatrix} 1 \\ 0 \end{bmatrix} u(k)$$
$$y(k) = \begin{bmatrix} 0 & 1 \end{bmatrix} x(k)$$

10.14 For the system given in Problem 10.11, design and simulate a current estimator, and compare the results with that problem.

10.15 Show that a valid representation for a current estimator for a system with state feedback K is with the equations

$$\xi(k+1) = A_c \xi(k) + B_c y(k)$$
$$u(k) = C_c \xi(k) + D_c y(k)$$

where

$$\xi(k) \triangleq \hat{x}(k) - Ly(k)$$

$$A_c \triangleq (I - LC_d)(A_d - B_d K) \qquad B_c \triangleq A_d L$$
$$C_c \triangleq -K \qquad D_c \triangleq -KL$$

10.16 Show that if the output matrix C is $n \times n$ and invertible, then for every *state* feedback gain matrix K that results in a particular set of closed-loop eigenvalues, there exists an *output* feedback gain matrix K_o that results in the same set of eigenvalues.

10.6 References and Further Reading

Fundamental procedures and properties of static state feedback can be found in the texts [2], [3], and [7]. The basic principles of state feedback depend of course on controllability. However, it was not originally obvious that controllability and the ability to arbitrarily place a system's eigenvalues were synonymous. That result can be found in [4] and [16].

Although not discussed here, the concept of *output* feedback is perhaps more basic (in the sense that a student will encounter it first in classical controls courses), but is actually a more difficult problem. Results on the possibility of arbitrary eigenvalue placement through output feedback can be found in [5] and [13]. A discussion of *dynamic* output feedback is given in [1], which includes a discussion of graphical techniques such as Bode and Nyquist plots, with which classical controls students are probably familiar.

The concept of the observer is attributable to Luenberger, [10] and [11]. This is interestingly a result of his Ph.D. thesis of 1963 from Stanford University. This work includes most of the basic aspects of observers, including the reduced order observer, canonical transformations, and even functional observers. Observers have since become a mainstay of state-feedback control theory and can be used as the basis for the *optimal estimators* [8] and [14] that we will consider in the next chapter.

[1] Byrnes, Christopher I., "Pole Assignment by Output Feedback," in *Three Decades of Mathematical System Theory*, H. Nijmeijier and J. Schumacher, eds., Springer-Verlag, 1989, pp. 31-78.

[2] Casti, John L., *Linear Dynamical Systems*, Academic Press, 1987.

[3] Chen, Chi-Tsong, *Linear System Theory and Design*, Holt, Rinehart, and Winston, 1984.

[4] Davison, Edward J., "On Pole Assignment in Multivariable Linear Systems," *IEEE Transactions on Automatic Control*, vol. AC-13, no. 6, December 1968, pp. 747-748.

[5] Davison, Edward J., and S. H. Wang, "On Pole Assignments in Linear Multivariable Systems Using Output Feedback," *IEEE Transactions on Automatic Control*, vol. AC-20, no. 4, 1975, pp. 516-518.

[6] Franklin, Gene, and J. David Powell, *Digital Control of Dynamic Systems*, Addison-Wesley, 1981.

[7] Kailath, Thomas, *Linear Systems*, Prentice-Hall, 1980.

[8] Leondes, C. T., and L. M. Novak, "Optimal Minimal-Order Observers for Discrete-Time Systems – A Unified Theory," *Automatica*, vol. 8, no. 4, 1972, pp. 379-387.

[9] Luenberger, David G., "Canonical Forms for Linear Multivariable Systems," *IEEE Transactions on Automatic Control*, vol. AC-12, 1967, pp. 290-293.

[10] Luenberger, David G., "Observers for Multivariable Systems," *IEEE Transactions on Automatic Control*, vol. AC-11, December 1966, pp. 563-603.

[11] Luenberger, David G., "Observing the State of a Linear System," *IEEE Transactions on Military Electronics*, vol. MIL-8, 1964, pp. 74-80.

[12] Novak, P., T. Ekelund, I. Jovik, and B. Schmidtbauer, "Modeling and Control of Variable-Speed Wind Turbine Drive-System Dynamics," *IEEE Control Systems Magazine*, vol. 15, no. 4, August 1995, pp. 28-38.

[13] Sarachik, P. E., and E. Kriendler, "Controllability and Observability of Linear Discrete-Time Systems," *International Journal of Control*, vol. 1, no. 5, 1965, pp. 419-432.

[14] Tse, Edison, and Michael Athans, "Optimal Minimal-Order Observer-Estimators for Discrete Linear Time-Varying Systems," *IEEE Transactions on Automatic Control*, vol. AC-15, no. 4, 1970, pp. 416-426.

[15] Van der Linden, Gert-Wim, and Paul F. Lambrechts, "H_∞ Control of an Experimental Inverted Pendulum with Dry Friction," *IEEE Control Systems Magazine*, vol. 13, no. 4, August 1993, pp. 44-50.

[16] Wonham, W. Murray, "On Pole Assignment in Multi-Input, Controllable Linear Systems," *IEEE Transactions on Automatic Control*, vol. AC-12, no. 6, December 1967, pp. 660-665.

11

Introduction to Optimal Control and Estimation

The observer and state feedback combination was described in the last chapter as the most fundamental form of state space control system. However, as was mentioned in the summary of Chapter 10, it is not necessarily the most *useful* method. The distinction lies in the practicality of the feedback and observer gains. First, it was demonstrated that the feedback gains that place the eigenvalues of a multivariable system at specified locations are not unique. This difficulty is compounded because the designer does not actually know the best locations for closed-loop eigenvalues. Often, performance criteria are specified in terms of quantities that are of course related to, but not necessarily obvious from, the eigenvalue locations. How many of us can accurately predict the transient-response or frequency-response characteristics of a large system from a list of eigenvalues? Second, a similar problem occurs in the choice of the observer gain. Although we might agree that the observer eigenvalues should be "faster" than the controller eigenvalues, is there any other reason to prefer one location to another, and if there is, is any guidance available on the choice of observer gain which results in these eigenvalues?

These questions are addressed with the methods introduced in this chapter. First, the matter of feedback gains is approached by defining an optimization criterion that should be satisfied by the state feedback controller. By minimizing this criterion, the eigenvalues will be automatically placed and the feedback gains will be uniquely selected. The optimization criterion chosen is a functional of quadratic forms in the state vector and the input of the system. Such controller designs are known as linear-quadratic (LQ) controllers or linear-quadratic regulators (LQRs). We will see how such a criterion can be constructed, and how a choice of feedback gain can be made to minimize it.

In order to determine the gain of the observer, similar optimization criteria must be imposed. However, these criteria must be imposed not on the performance of the state trajectory but on the observation error of the system. In order to "optimize" the state observation of a system, we will include explicit statistical models of the noise that might affect a state space description. Noise might appear as an input to the system or as a disturbance introduced by the sensors, modeled in the output equation. By minimizing the effects of this noise in the observed state vector, we will have created an optimal statistical *estimate* of the state, which can then be fed back through the state feedback gains. The estimator thus created is known as the Kalman filter.

When a Kalman filter estimates the state vector of a noisy plant and an LQ controller is used to compute state feedback, the combination is known as a linear-quadratic-gaussian (LQG) controller. It is the LQG controller that is regarded by many as the most useful state space controller. It is the simplest of many forms of *optimal controllers*. Optimal control is a broad and sometimes complicated topic that is only briefly introduced by this chapter.

11.1 The Principle of Optimality

In order to understand the idea of the optimization criterion in state variable systems, we will consider a discrete state space. We can imagine the state trajectory of a system making discrete transitions from one state to another under the influence of a continuous-valued input that is also applied at discrete-times. Thus, we can imagine that the goal of the control system is to force the transition of the plant from an initial state to a final state, with one or more intermediate states in between.

In the process of making the transition from one state to another, the system incurs a *cost*. The cost of a transition can be thought of as a penalty. A system might be penalized for being too far away from its final state, for staying away from the final state for too long a time, for requiring a large control signal u in order to make the next transition, or for any other imaginable criterion. As the system moves from state to state, these costs add up until a total cost is accumulated at the end of the trajectory. If the final state of the trajectory is not at the desired goal state, further penalties can be assessed as well.

To simplify this concept, consider the set of states and possible transitions represented in the graph of Figure 11.1. In this figure, the initial state is denoted by "1", and the final, desired state is denoted by "8". The system jumps from state to state at each time k, as determined by the input $u(k)$ through the state equations $x(k+1) = Ax(k) + Bu(k)$. The possible transitions under the influence of the input are represented by the arcs that connect the initial state, through the intermediate states, to the final state. The cost associated with each transition is represented by the label on the arc, e.g., J_{47} represents the cost of making the transition from state 4 to state 7. In real systems, this cost may be a function of the state vector, the input, and the time, but we represent the cost in Figure 11.1 as simply a real number. If it is assumed that costs accumulate additively, then the total cost incurred in any path is the sum of the costs incurred in each step.

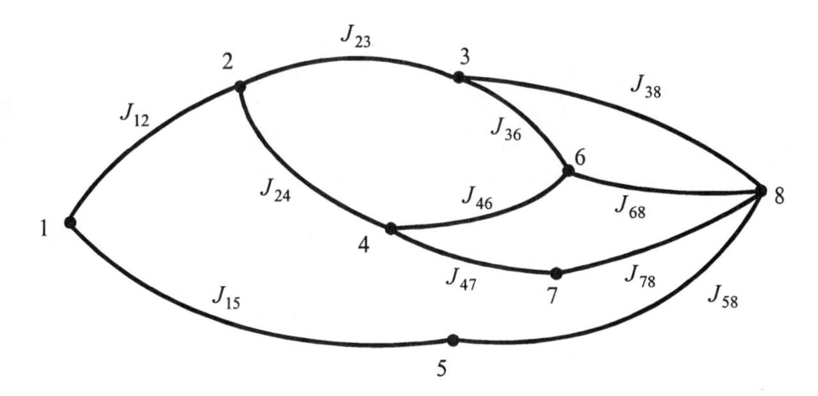

Figure 11.1 Possible paths from an initial state 1 to a final state 8.

Because there are several alternate routes from the initial state to the final state, the total cost in reaching state 8 from state 1 will vary according to the path chosen:

$$J_{18} = \begin{cases} J_{15} + J_{58} \\ \text{or} \\ J_{12} + J_{24} + J_{46} + J_{68} \\ \vdots \end{cases} \tag{11.1}$$

If each segment results in a different cost, then it is clear that the cost of the whole trajectory might be minimized by the judicious selection of the path taken. Of course, it should not be assumed that the path with the fewest individual segments will be the one with the minimal cost. The input sequence $u(k)$ applied such that J_{18} is minimized will be called the *optimal policy* $u^*(k)$, and the resulting minimal cost that results from this input will be referred to as J_{18}^*. The goal of an optimal controller is therefore to determine the input $u^*(k)$. In continuous-time, as we will discuss in detail later, the accumulation of costs is represented by integration rather than by summation.

The mathematical tool by which we will find the optimal policy is called *Bellman's principle of optimality*. This deceptively obvious principle states that *at any intermediate point x_i in an optimal path from x_o to x_f, the policy from x_i to the goal x_f must itself constitute optimal policy.* This statement may seem too obvious to be useful, but it will enable us to solve, in closed form, for the optimal control in our problems. It is also used in recursive calculations, in a procedure known as *dynamic programming*, to numerically compute optimal policy.

11.1.1 Discrete-Time LQ Control

Continuing with the discrete-time case so that the interpretation of incurred costs can be maintained, we will now consider the discrete-time state equation

$$x(k+1) = Ax(k) + Bu(k) \tag{11.2}$$

and the problem of forcing the system to reach a final state $x_f = x(N)$ from an initial state $x_o = x(i)$. We wish to accomplish this task through full-state feedback, and we will specify a cost function to minimize in the process. The cost function we will use has the quadratic form:

$$J_{i,N} = \tfrac{1}{2}x^{\mathrm{T}}(N)Sx(N) + \tfrac{1}{2}\sum_{k=i}^{N-1}\left[x^{\mathrm{T}}(k)Qx(k) + u^{\mathrm{T}}(k)Ru(k)\right] \tag{11.3}$$

which may be interpreted as the total cost associated with the transition from state $x(i)$ to the goal state $x(N)$. In this expression, the term $\tfrac{1}{2}x^{\mathrm{T}}(N)Sx(N)$ gives the penalty for "missing" a desired goal state [if the desired goal of $x(N)$ is zero]. The terms in the summation represent the penalties for excessive size of the state vector and for the size of the input. These terms are all quadratic forms, with weighting matrices *S*, *Q*, and *R* selected to penalize some state variables (inputs) more than others. It is necessary that the matrices *S* and *Q* be positive semidefinite, and *R* be positive definite, for reasons we will see later. As discussed in Section 5.3, this implies that they are, without loss of generality, considered to be symmetric. They can be functions of time *k* as well, but we will treat them as constants to simplify their notation. When selecting appropriate *S*, *Q*, and *R* matrices, it should be remembered that larger terms will penalize the corresponding variables more than smaller terms, and that it is only the *relative* size of the entries in this matrix that matter.

To find a control input function $u(k)$ that should be applied to (11.2) such that (11.3) is minimized, we will begin at the last step and apply the principle of optimality as the state transitions are traced backward. Thus, according to (11.3), the cost incurred in the transition from step $N-1$ to step *N* is

$$J_{N-1,N} = \tfrac{1}{2}\left[x^{\mathrm{T}}(N)Sx(N) + x^{\mathrm{T}}(N-1)Qx(N-1) + u^{\mathrm{T}}(N-1)Ru(N-1)\right] \tag{11.4}$$

We wish to minimize this function over all possible choices of $u(N-1)$, but we must first remember to express $x(N)$ as a function of $u(N-1)$ according to (11.2):

$$J_{N-1,N} = \tfrac{1}{2}\Big[\big(Ax(N-1)+Bu(N-1)\big)^{\mathrm{T}} S\big(Ax(N-1)+Bu(N-1)\big) \\ +x^{\mathrm{T}}(N-1)Qx(N-1)+u^{\mathrm{T}}(N-1)Ru(N-1)\Big] \tag{11.5}$$

Now, because J is continuous in the input u, this function can be minimized over all inputs by differentiation:

$$\begin{aligned}\frac{\partial^{\mathrm{T}} J_{N-1,N}}{\partial u(N-1)} &= 0 \\ &= B^{\mathrm{T}} S\big[Ax(N-1)+Bu(N-1)\big]+Ru(N-1) \\ &= \big[R+B^{\mathrm{T}}SB\big]u(N-1)+B^{\mathrm{T}}SAx(N-1)\end{aligned} \tag{11.6}$$

(We use the *transpose* derivative in order to express the results as a column rather than a row.) Therefore,

$$u^{*}(N-1) = -\big[R+B^{\mathrm{T}}SB\big]^{-1} B^{\mathrm{T}}SA\, x(N-1) \tag{11.7}$$

That this a minimizing solution rather than a maximizing solution may be checked with the second derivative:

$$\frac{\partial^2 J_{N-1,N}}{\partial u^2(N-1)} = \big[R+B^{\mathrm{T}}SB\big] > 0 \tag{11.8}$$

Obviously, the expression (11.7) is in the form of state feedback, as we wished. Using a more compact and familiar notation, this expression can be written as

$$u^{*}(N-1) = K_{N-1}\, x(N-1) \tag{11.9}$$

with

$$K_{N-1} \triangleq -\big[R+B^{\mathrm{T}}SB\big]^{-1} B^{\mathrm{T}}SA \tag{11.10}$$

Using (11.5) to express the value of $J^{*}{}_{N-1,N}$:

$$
\begin{aligned}
J^*_{N-1,N} &= \tfrac{1}{2}\Big[\big(Ax(N-1)+BK_{N-1}x(N-1)\big)^{\mathrm{T}} S\big(Ax(N-1)+BK_{N-1}x(N-1)\big) \\
&\qquad + x^{\mathrm{T}}(N-1)Qx(N-1) + x^{\mathrm{T}}(N-1)K^{\mathrm{T}}_{N-1}RK_{N-1}x(N-1)\Big] \\
&= \tfrac{1}{2}x^{\mathrm{T}}(N-1)\Big[\big(A+BK_{N-1}\big)^{\mathrm{T}} S\big(A+BK_{N-1}\big)+Q+K^{\mathrm{T}}_{N-1}RK_{N-1}\Big]x(N-1)
\end{aligned}
\tag{11.11}
$$

In order to simplify this notation, define

$$
S_{N-1} \triangleq \big(A+BK_{N-1}\big)^{\mathrm{T}} S\big(A+BK_{N-1}\big)+Q+K^{\mathrm{T}}_{N-1}RK_{N-1} \tag{11.12}
$$

so that (11.11) becomes

$$
J^*_{N-1,N} = \tfrac{1}{2}x^{\mathrm{T}}(N-1)S_{N-1}x(N-1) \tag{11.13}
$$

The choice of the notation in (11.12) is motivated by the fact that if we set $i = N$ in (11.3), then we get

$$
J_{N,N} = J^*_{N,N} = \tfrac{1}{2}x^{\mathrm{T}}(N)Sx(N) \tag{11.14}
$$

(which is the *optimal* $J_{N,N}$ because it is the *only* $J_{N,N}$). Therefore, comparing (11.14) and (11.13), we can make the notational identification that $S = S_N$.

Now we take another step backward and compute the cost $J_{N-2,N}$ in going from state $N-2$ to the goal state *N*. First, we realize that

$$
J_{N-2,N} = J_{N-2,N-1} + J_{N-1,N} \tag{11.15}
$$

as computed in (11.1) above. Therefore, to find the *optimal* policy in going from $N-2$ to *N*, we use the principle of optimality to infer that

$$
J^*_{N-2,N} = J_{N-2,N-1} + J^*_{N-1,N} \tag{11.16}
$$

So now step $N-1$ is the goal state, and we can find $J_{N-2,N-1}$ using the same Equation (11.5), but substituting $N-2$ for $N-1$ and $N-1$ for *N*:

$$
\begin{aligned}
J_{N-2,N-1} &= \tfrac{1}{2}\Big[\big(Ax(N-2)+Bu(N-2)\big)^{\mathrm{T}} S\big(Ax(N-2)+Bu(N-2)\big) \\
&\qquad + x^{\mathrm{T}}(N-2)Qx(N-2) + u^{\mathrm{T}}(N-2)Ru(N-2)\Big]
\end{aligned}
\tag{11.17}
$$

This can be minimized over all possible $u(N-2)$ by differentiation just as in (11.6) and (11.7), but it should be apparent that the result will be the same as (11.7) except, again, for the substitutions of $N-2$ for $N-1$ and $N-1$ for N.

$$u^*(N-2) = K_{N-2}\, x(N-2) \tag{11.18}$$

where

$$K_{N-2} \stackrel{\Delta}{=} -\left[R + B^{\mathrm{T}} S_{N-1} B\right]^{-1} B^{\mathrm{T}} S_{N-1} A \tag{11.19}$$

Continuing these backward steps, we will get similar expressions for each time k. We can therefore summarize the results at each step with the set of equations

$$u^*(k) = K_k\, x(k) \tag{11.20}$$

where

$$K_k = -\left[R + B^{\mathrm{T}} S_{k+1} B\right]^{-1} B^{\mathrm{T}} S_{k+1} A \tag{11.21}$$

and

$$S_k = \left(A + BK_k\right)^{\mathrm{T}} S_{k+1}\left(A + BK_k\right) + Q + K_k^{\mathrm{T}} R K_k \tag{11.22}$$

Note that (11.22) is a difference equation whose *starting* condition occurs at the *final* time and is computed toward earlier time. This starting condition is derived from (11.14) as

$$S_N = S \tag{11.23}$$

At such time k, one can also compute the "cost-to-go," i.e., the optimal cost to proceed from $x(k)$ to $x(N)$ as in (11.13):

$$J_{k,N}^* = \tfrac{1}{2} x^{\mathrm{T}}(k) S_k x(k) \tag{11.24}$$

The set of equations in (11.20) through (11.22) represents the complete LQ controller[M] for a given discrete-time system. Equation (11.22) is known as the discrete-time *matrix Riccati equation*. The fact that it is a "backward-time" equation gives the system's feedback gain K_k the peculiar behavior that its transients appear at the end of the time interval rather than at the beginning.

`reg(sys,K,L)`

Example 11.1: A Discrete-Time LQ Controller

Simulate a feedback controller for the system

$$x(k+1)=\begin{bmatrix}2 & 1\\ -1 & 1\end{bmatrix}x(k)+\begin{bmatrix}0\\ 1\end{bmatrix}u(k) \qquad x(0)=\begin{bmatrix}2\\ -3\end{bmatrix} \tag{11.25}$$

such that the cost criterion

$$J=\tfrac{1}{2}x^{\mathrm{T}}(10)\begin{bmatrix}5 & 0\\ 0 & 5\end{bmatrix}x^{\mathrm{T}}(10)+\tfrac{1}{2}\sum_{k=1}^{9}\left(x^{\mathrm{T}}(k)\begin{bmatrix}2 & 0\\ 0 & .1\end{bmatrix}x^{\mathrm{T}}(k)+2u^{2}(k)\right) \tag{11.26}$$

is minimized.

Solution:

Given the cost criterion in (11.26), the only work necessary to solve this problem is to iterate the equations in (11.21) and (11.22) in backward time, saving the S-matrix and gain matrix at each step. When these values are computed, then the plant equation in (11.25) is simulated in forward time using the stored values of gain.

Plotted in Figure 11.2 below is the state variable sequence versus time. Although it should be obvious that the original plant is unstable, it is apparent from these results that the controlled system asymptotically approaches the origin of the two-dimensional space.

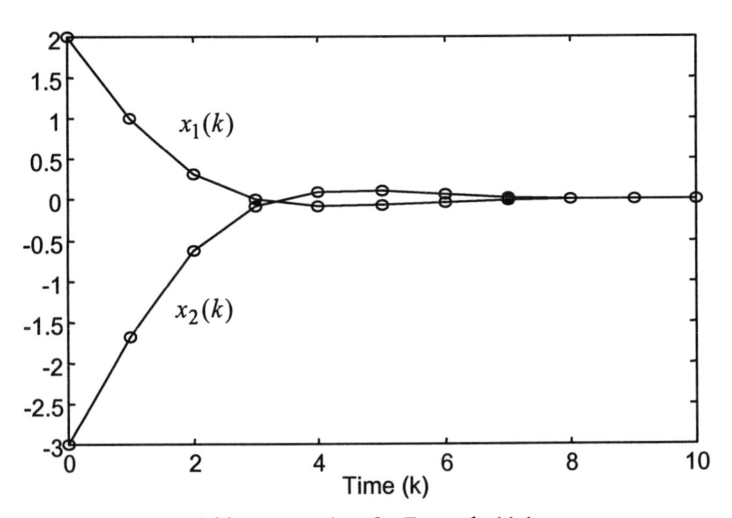

Figure 11.2 State variables versus time for Example 11.1.

Figure 11.3 is a plot of the feedback gains versus time. From this plot we note the interesting feature that the gains change at the *end* of the time interval rather than at the beginning. This is because of the backward-time nature of the matrix equation in 11.22. This observation will become important in our discussion of infinite-time optimal control in Section 11.1.3.

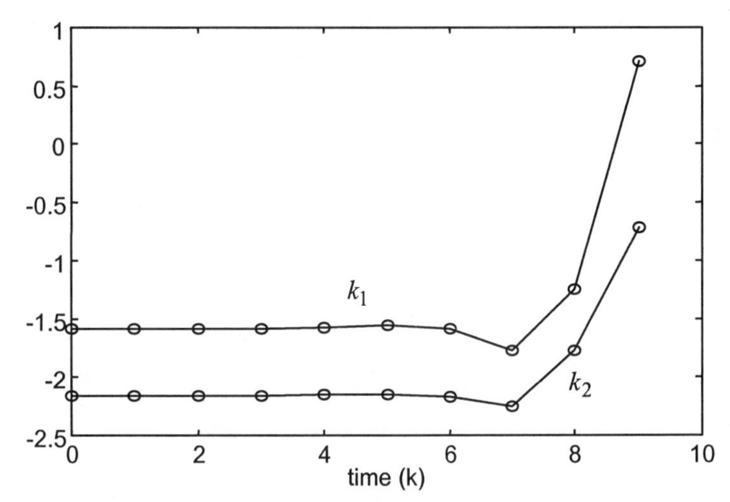

Figure 11.3 Feedback gains versus time for Example 11.1. Note that they are computed only up through $k = N - 1 = 9$ because $K(10)$ is not needed to compute $x(10)$.

Figure 11.4 is a plot of the matrix S versus the step number. Normally, this matrix itself, the feedback gain matrix K_k, need not be stored, but we wish to make the observation here that in the early stages of the time interval, the matrix appears to be constant or nearly constant. This observation will become important in Section 11.1.3.

11.1.2 Continuous-Time LQ Control

The development of the continuous-time LQ feedback controller follows the same principles as in the discrete-time situation, but the notion of "steps" is absent. Instead, we will first perform the derivation assuming small *increments*, then let these increments approach zero.

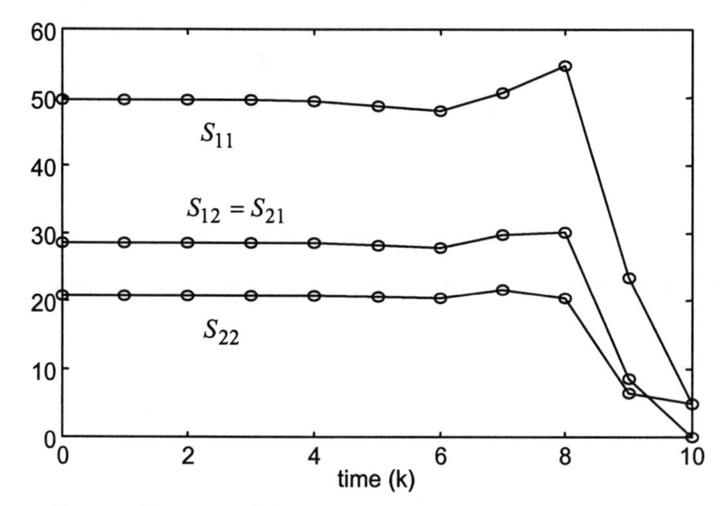

Figure 11.4 Elements of S-matrix versus step number. Because S is symmetric, $S_{12} = S_{21}$.

The continuous-time LQ cost function takes the form:

$$J(x(t_o),u(t_o),t_o) = \tfrac{1}{2}x^{\mathrm{T}}(t_f)Sx(t_f) + \tfrac{1}{2}\int_{t_o}^{t_f}\left[x^{\mathrm{T}}(t)Qx(t) + u^{\mathrm{T}}(t)Ru(t)\right]dt \qquad (11.27)$$

This criterion can be interpreted in the same manner as the discrete-time cost function in (11.3): it is the cost in making the transition from the initial state $x(t_o)$ to the final state $x(t_f)$ using the control function $u(t)$. It contains a final state cost and an accumulation term that penalizes deviation of the state and inputs from zero, as measured by the quadratic forms with user-defined weighting matrices. This accumulation of costs is now represented by the integration from initial time t_o to final time t_f. Once again, S and Q are symmetric, positive-semidefinite matrices and R is a symmetric, positive-definite matrix. It may be noted that in more general optimal control problems, J may also be an explicit function of time and of possible "cross terms" such as $x^{\mathrm{T}}Nu$. We will make use of only the quadratic functions seen in (11.27).

For the continuous-time system

$$\dot{x} = Ax + Bu \qquad (11.28)$$

the optimal control problem can thus be stated as the problem of finding the optimal control function $u^*(t)$, where $t \in [t_o, t_f]$, and applying it to (11.28), such that the function (11.27) is minimized. Thus the optimal cost will be denoted

$$J^*(x,t) = \min_{u(t)} \left\{ \tfrac{1}{2} x^{\mathrm{T}}(t_f) S x(t_f) + \tfrac{1}{2} \int_{t_o}^{t_f} \left[x^{\mathrm{T}}(t) Q x(t) + u^{\mathrm{T}}(t) R u(t) \right] dt \right\} \tag{11.29}$$

Note that strictly, this notation removes from J^* any explicit dependence on input $u(t)$. We will drop the arguments on J^* in order to economize on the notation.

Suppose that we decompose the computation of the cost of moving from t_o to t_f into two stages: one from t_o to an infinitesimal time δt later and then from $t_o + \delta t$ to t_f. At time $t_o + \delta t$, we will assume that we have reached state $x(t_o + \delta t)$. We can then write (11.27) as:

$$\begin{aligned} J(x,u,t) = {} & \tfrac{1}{2} x^{\mathrm{T}}(t_f) S x(t_f) + \tfrac{1}{2} \int_{t_o}^{t_o+\delta t} \left[x^{\mathrm{T}}(t) Q x(t) + u^{\mathrm{T}}(t) R u(t) \right] dt \\ & + \tfrac{1}{2} \int_{t_o+\delta t}^{t_f} \left[x^{\mathrm{T}}(t) Q x(t) + u^{\mathrm{T}}(t) R u(t) \right] dt \end{aligned} \tag{11.30}$$

Bellman's principle of optimality would then imply that for an optimal path,

$$J^*(x,t) = \min_{u(t)} \left\{ \tfrac{1}{2} \int_{t_o}^{t_o+\delta t} \left[x^{\mathrm{T}}(t) Q x(t) + u^{\mathrm{T}}(t) R u(t) \right] dt + J^*\left[x(t_o + \delta t), t_o + \delta t \right] \right\} \tag{11.31}$$

That is, the optimal cost in going from t_o to t_f is the cost of going from t_o to $t_o + \delta t$ plus the *optimal* cost in going from $t_o + \delta t$ to t_f, which itself includes the terminal cost. In this expression, we will expand the term $J^*\left[x(t_o + \delta t), t_o + \delta t \right]$ into a Taylor series about the point $(x(t_o), t_o)$ as follows:

$$\begin{aligned} J^*(x(t_o + \delta t), t_o + \delta t) = {} & J^*(x(t_o), t_o) + \left. \frac{\partial J^*}{\partial t} \right|_{(x(t_o),t_o)} (t_o + \delta t - t_o) \\ & + \left. \frac{\partial J^*}{\partial x} \right|_{(x(t_o),t_o)} \left[x(t_o + \delta t) - x(t_o) \right] + \text{h.o.t.} \end{aligned} \tag{11.32}$$

where "h.o.t." stands for "higher-order terms." Optimal cost, Equation (11.31), can then be written as:

$$J^*(x,t)=\min_{u(t)}\left\{\tfrac{1}{2}\int_{t_o}^{t_o+\delta t}\left[x^T(t)Qx(t)+u^T(t)Ru(t)\right]dt+J^*(x(t_o),t_o)+\left.\frac{\partial J^*}{\partial t}\right|_{(x(t_o),t_o)}\delta t\right.$$
$$\left.+\left.\frac{\partial J^*}{\partial x}\right|_{(x(t_o),t_o)}\left[x(t_o+\delta t)-x(t_o)\right]\right\} \tag{11.33}$$

Within this expression, the terms $J^*(x(t_o),t_o)$ and $\left(\partial J^*/\partial t\right)\delta t$ do not depend on $u(t)$ and can therefore be moved outside the *min* operator. Furthermore, assuming that δt is very small, we can make the approximations:

$$\tfrac{1}{2}\int_{t_0}^{t_0+\delta t}\left[x^T(t)Qx(t)+u^T(t)Ru(t)\right]dt\approx\tfrac{1}{2}\left[x^T(t)Qx(t)+u^T(t)Ru(t)\right]\delta t \tag{11.34}$$

and

$$\begin{aligned} x(t_o+\delta t)-x(t_o)&=\delta x\\ &=\frac{\delta x}{\delta t}\delta t\\ &\approx\frac{dx}{dt}\delta t\\ &=(Ax+Bu)\delta t \end{aligned} \tag{11.35}$$

making (11.33) appear as

$$J^*(x,t)=J^*(x,t)+\frac{\partial J^*}{\partial t}\delta t$$
$$+\min_{u(t)}\left\{\tfrac{1}{2}\left[x^T(t)Qx(t)+u^T(t)Ru(t)\right]\delta t+\frac{\partial J^*}{\partial x}(Ax+Bu)\delta t\right\}$$

or

$$0 = \frac{\partial J^*}{\partial t}\delta t + \min_{u(t)}\left\{\tfrac{1}{2}\left[x^{\mathrm{T}}(t)Qx(t) + u^{\mathrm{T}}(t)Ru(t)\right]\delta t + \frac{\partial J^*}{\partial x}(Ax + Bu)\delta t\right\} \tag{11.36}$$

Dividing out the δt term from both sides:

$$0 = \frac{\partial J^*}{\partial t} + \min_{u(t)}\left\{\tfrac{1}{2}\left[x^{\mathrm{T}}(t)Qx(t) + u^{\mathrm{T}}(t)Ru(t)\right] + \frac{\partial J^*}{\partial x}(Ax + Bu)\right\} \tag{11.37}$$

This equation is known as the *Hamilton-Jacobi-Bellman* equation. It is useful in a wide range of optimal control problems, even for nonlinear systems and nonquadratic cost functions. (Note that we have not yet exploited the linear system equations and the quadratic terms in *J*.) It is a partial differential equation that can be numerically solved in the general case using a boundary condition derived by setting $t_o = t_f$ in (11.27):

$$J^*(x(t_f), t_f) = \tfrac{1}{2}x^{\mathrm{T}}(t_f)Sx(t_f) \tag{11.38}$$

The term in braces in (11.37) is known as the *hamiltonian*. Recall that we used the same terminology in Section 3.3.1, when we discussed solving underdetermined systems of equations. In that chapter, the hamiltonian had the same structure: the optimization criterion plus a multiple of the constraint expression. Thus, we define

$$H(x, u, J^*, t) \triangleq \tfrac{1}{2}\left[x^{\mathrm{T}}(t)Qx(t) + u^{\mathrm{T}}(t)Ru(t)\right] + \frac{\partial J^*}{\partial x}(Ax + Bu) \tag{11.39}$$

If we can find the *optimal* control u^*, then (11.39) will become (dropping the t arguments):

$$\begin{aligned} H(x, u^*, J^*, t) &\triangleq \tfrac{1}{2}\left(x^{\mathrm{T}}Qx + u^{*\mathrm{T}}Ru^*\right) + \frac{\partial J^*}{\partial x}\left(Ax + Bu^*\right) \\ &= \min_u\left[\tfrac{1}{2}\left(x^{\mathrm{T}}Qx + u^{\mathrm{T}}Ru\right) + \frac{\partial J^*}{\partial x}(Ax + Bu)\right] \end{aligned} \tag{11.40}$$

making the Hamilton-Jacobi-Bellman equation appear as

$$\begin{aligned} 0 &= \frac{\partial J^*}{\partial t} + \min_u \left[\tfrac{1}{2}\left(x^{\mathrm{T}} Q x + u^{\mathrm{T}} R u\right) + \frac{\partial J^*}{\partial x}(Ax + Bu) \right] \\ &= \frac{\partial J^*}{\partial t} + H(x, u^*, J^*, t) \end{aligned} \tag{11.41}$$

In order to minimize the hamiltonian itself, over u, we take the derivative:

$$\begin{aligned} \frac{\partial^{\mathrm{T}} H}{\partial u} &= 0 \\ 0 &= \frac{\partial^{\mathrm{T}}}{\partial u}\left[\tfrac{1}{2}\left(x^{\mathrm{T}} Q x + u^{\mathrm{T}} R u\right) + \frac{\partial J^*}{\partial x}(Ax + Bu) \right] \\ 0 &= Ru + B^{\mathrm{T}} \frac{\partial J^{*\mathrm{T}}}{\partial x} \end{aligned} \tag{11.42}$$

giving the solution

$$u^* = -R^{-1} B^{\mathrm{T}} \frac{\partial J^{*\mathrm{T}}}{\partial x} \tag{11.43}$$

(hence the need for R to be positive definite). That this is a minimizing solution rather than a maximizing solution may be checked with a second derivative, similar to (11.8). This expression is still not suitable for use as an optimal control because of the partial derivative of J^*. To solve for this term, substitute (11.43) into (11.40):

$$\begin{aligned} H(x, u^*, J^*, t) &= \tfrac{1}{2}\left(x^{\mathrm{T}} Q x + u^{*\mathrm{T}} R u^* \right) + \frac{\partial J^*}{\partial x}\left(Ax + Bu^*\right) \\ &= \tfrac{1}{2} x^{\mathrm{T}} Q x + \tfrac{1}{2} \frac{\partial J^*}{\partial x} B R^{-1} R R^{-1} B^{\mathrm{T}} \frac{\partial J^{*\mathrm{T}}}{\partial x} + \frac{\partial J^*}{\partial x}\left(Ax - B R^{-1} B^{\mathrm{T}} \frac{\partial J^{*\mathrm{T}}}{\partial x} \right) \\ &= \tfrac{1}{2} x^{\mathrm{T}} Q x + \tfrac{1}{2} \frac{\partial J^*}{\partial x} B R^{-1} B^{\mathrm{T}} \frac{\partial J^{*\mathrm{T}}}{\partial x} + \frac{\partial J^*}{\partial x} A x - \frac{\partial J^*}{\partial x} B R^{-1} B^{\mathrm{T}} \frac{\partial J^{*\mathrm{T}}}{\partial x} \\ &= \tfrac{1}{2} x^{\mathrm{T}} Q x - \tfrac{1}{2} \frac{\partial J^*}{\partial x} B R^{-1} B^{\mathrm{T}} \frac{\partial J^{*\mathrm{T}}}{\partial x} + \frac{\partial J^*}{\partial x} A x \end{aligned} \tag{11.44}$$

Substituting this into the Hamilton-Jacobi-Bellman equation in (11.41):

$$0 = \frac{\partial J^*}{\partial t} + \tfrac{1}{2} x^{\mathrm{T}} Q x - \tfrac{1}{2} \frac{\partial J^*}{\partial x} B R^{-1} B^{\mathrm{T}} \frac{\partial J^{*\mathrm{T}}}{\partial x} + \frac{\partial J^*}{\partial x} A x \tag{11.45}$$

The remaining task is to solve this equation. The solution is facilitated by the assumption that the optimal cost J^* from *any* point until the final time can be represented as a quadratic form in terms of the state vector. We have already seen that this is true for the discrete-time system, Equation (11.24), and we know it is true for the end state, Equation (11.38), so it is a reasonable assumption here as well. We therefore denote

$$J^*(x,t) = \tfrac{1}{2} x^{\mathrm{T}}(t) P(t) x(t) \tag{11.46}$$

where the matrix $P(t)$ is presumed to be symmetric but as yet, unknown. Using this assumption,

$$\frac{\partial J^*(x,t)}{\partial x} = x^{\mathrm{T}}(t) P^{\mathrm{T}}(t) = x^{\mathrm{T}}(t) P(t) \tag{11.47}$$

and

$$\frac{\partial J^*(x,t)}{\partial t} = \tfrac{1}{2} x^{\mathrm{T}}(t) \dot{P}(t) x(t) \tag{11.48}$$

Substituting these relations into (11.45) gives

$$0 = \tfrac{1}{2} x^{\mathrm{T}} \dot{P} x + \tfrac{1}{2} x^{\mathrm{T}} Q x - \tfrac{1}{2} x^{\mathrm{T}} P B R^{-1} B^{\mathrm{T}} P x + x^{\mathrm{T}} P A x \tag{11.49}$$

In this expression, each term, except for the last, is a symmetric quadratic form. However, appealing to Equation (5.17), which allows us to treat *any* quadratic form as a symmetric one, we can express

$$\begin{aligned} x^{\mathrm{T}} P A x &= x^{\mathrm{T}} \frac{\left(PA + A^{\mathrm{T}} P\right)}{2} x \\ &= \tfrac{1}{2} x^{\mathrm{T}} P A x + \tfrac{1}{2} x^{\mathrm{T}} A^{\mathrm{T}} P x \end{aligned} \tag{11.50}$$

Using (11.50) in (11.49) gives

$$\begin{aligned} 0 &= \tfrac{1}{2} x^{\mathrm{T}} \dot{P} x + \tfrac{1}{2} x^{\mathrm{T}} Q x - \tfrac{1}{2} x^{\mathrm{T}} P B R^{-1} B^{\mathrm{T}} P x + \tfrac{1}{2} x^{\mathrm{T}} P A x + \tfrac{1}{2} x^{\mathrm{T}} A^{\mathrm{T}} P x \\ &= \tfrac{1}{2} x^{\mathrm{T}} \left[\dot{P} + Q - P B R^{-1} B^{\mathrm{T}} P + P A + A^{\mathrm{T}} P \right] x \end{aligned} \tag{11.51}$$

This equation must hold for all x, so

$$0 = \dot{P} + Q - PBR^{-1}B^{\mathrm{T}}P + PA + A^{\mathrm{T}}P \tag{11.52}$$

or equivalently,

$$\dot{P} = PBR^{-1}B^{\mathrm{T}}P - Q - PA - A^{\mathrm{T}}P \tag{11.53}$$

This is a famous equation known as the *differential matrix Riccati equation*. It is a nonlinear differential equation whose boundary condition is given by (11.38):

$$P(t_f) = S \tag{11.54}$$

where S is the end-state weighting matrix from the cost criterion in (11.27). Like the matrix equation in (11.22) for the discrete-time case, it is a backward-time equation whose solution can be obtained numerically. If we solve for $P(t)$ from (11.53), then we have from (11.43) and (11.47):

$$\begin{aligned} u^*(t) &= -R^{-1}B^{\mathrm{T}}P(t)x(t) \\ &\triangleq K(t)x \end{aligned} \tag{11.55}$$

which is clearly a state variable feedback solution, albeit one with time-varying feedback gain. As in continuous-time, the backward-time nature of the matrix $P(t)$ will give the feedback matrix a transient near t_f rather than t_o, as is usually the case. Together, (11.53) and (11.55) represent the complete specification of the continuous-time LQ controller. [M]

```
lqr(A,B,Q,R)
reg(sys,K,L)
```

It should be noted that there are practical reasons for *not* solving the differential matrix Riccati equation, (11.53). Recall that in the discrete-time case, the corresponding equation in (11.22) was computed backward and the results at each step stored for application in forward time. If the solution of the differential Riccati equation is solved numerically, how can it be stored for use in the continuous-time feedback control (11.55)? If only discrete points of the solution can be stored, then it is actually the discrete-time equivalent of the system that is being controlled, and we are back to the discrete-time LQ controller. In scalar systems, the Riccati equation can be solved in closed form so that it can be used in (11.55) to generate the control signal $u(t)$. If continuous-time optimal controllers are required for higher-order systems, they are most practical in the *infinite-horizon* case discussed in the next section.

11.1.3 Infinite-Horizon Control

We may wonder how to determine the final *time* as well as the penalty term for the final state. Usually in classical control systems, the time-domain characteristics such as rise time and settling time are important, but often a finite final time is not a factor in the design. Rather, it is intended that the regulating system operate continually after it is turned on. Yet in the LQ controllers discussed above, the final time is a critical piece of information. It is the starting point for the solution of the Riccati equations, for both discrete- and continuous-time, which then determine the values of the time-varying state feedback gains.

If we examine the plots in Figure 11.4, we might guess that if the final time were to approach infinity, the values for the elements of the S-matrix would remain constant for most of the operating interval (or at least until the state vector decays to a very small value). Then, the contribution to the total cost of the portion occurring during the end-time transient would be comparatively small, and we might justify using the constant values throughout. Figures 11.3 and 11.4 show Riccati equation solutions that hold steady at their initial horizontal values for a considerable simulation time.

For continuous-time systems, this might suggest solving the Riccati equation (11.53) by assuming that $\dot{P}=0$; this gives the *algebraic Riccati equation*^{M}:

`care(A,B,Q,R,S,E)`

$$0 = PBR^{-1}B^{\mathrm{T}}P - Q - PA - A^{\mathrm{T}}P \tag{11.56}$$

and the corresponding feedback function

$$u = Kx = -R^{-1}B^{\mathrm{T}}Px \tag{11.57}$$

In discrete-time, the presumption of steady state would imply that $S_k = S_{k+1} = S$, giving

$$S = (A+BK)^{\mathrm{T}}S(A+BK) + Q + K^{\mathrm{T}}RK \tag{11.58}$$

and

$$K = -\left[R + B^{\mathrm{T}}SB\right]^{-1}B^{\mathrm{T}}SA \tag{11.59}$$

which together can be combined into a decoupled equation:

$$S = A^{\mathrm{T}}SA - A^{\mathrm{T}}SB\left[R + B^{\mathrm{T}}SB\right]^{-1}B^{\mathrm{T}}SA + Q \tag{11.60}$$

This would then imply a constant state feedback gain K in each case, and, instead of the difference or differential equations in (11.22) or (11.53), we care(A,B,Q,R,S,E)

dare(A,B,Q,R,S,E)

would have instead *algebraic* Riccati equation[M] (11.56) or (11.60). In essence, this is exactly what happens, but some complications may arise. In particular, we have not yet established that all optimal control solutions do reach a bounded steady state as $t \to -\infty$ and that these solutions stabilize the system over *infinite* time. Nor have we considered whether these steady-state solutions are unique.

In order to investigate these issues briefly, we will consider the continuous-time situation in more detail. Consider the problem of controlling the time-invariant system

$$\dot{x} = Ax + Bu$$

with static state feedback

$$u = Kx \tag{11.61}$$

such that the cost function

$$J = \int_0^\infty \left[x^T(t)Qx(t) + u^T(t)Ru(t) \right] dt \tag{11.62}$$

is minimized. Using the feedback equation, (11.61), we obtain the closed-loop system

$$\dot{x} = (A + BK)x \tag{11.63}$$

making the cost function, (11.62), appear as

$$J = \int_0^\infty \left[x^T(t)Qx(t) + x^T(t)K^T RKx(t) \right] dt \tag{11.64}$$

Knowing that without an exogenous input, (11.63) has the solution

$$x(t) = e^{(A+BK)t} x_0 \tag{11.65}$$

then (11.64) can be written as

$$\begin{aligned} J &= x_0^T \left[\int_0^\infty e^{(A+BK)^T t} (Q + K^T RK) e^{(A+BK)t} \, dt \right] x_0 \\ &= x_0^T P x_0 \end{aligned} \tag{11.66}$$

We know from our study of Lyapunov stability in Chapter 7 (Theorem 7.32) that the system in (11.63) is asymptotically stable if and only if the matrix P defined in (11.66) is the unique positive-definite solution of the Lyapunov equation

$$(A+BK)^{\mathrm{T}}P+P(A+BK)+Q+K^{\mathrm{T}}RK=0 \tag{11.67}$$

Therefore, for *any* constant feedback gain K that asymptotically stabilizes the system, the solution P of (11.67) will give a convergent cost integral computed by (11.66). [Also recall that if $Q+K^{\mathrm{T}}RK$ is merely positive *semi*definite, then we must also have observability of the pair $(A, Q+K^{\mathrm{T}}RK)$.]

Now suppose that we hypothesize an *optimal* control input of

$$u=Kx=-R^{-1}B^{\mathrm{T}}P_{ss}x \tag{11.68}$$

where P_{ss} is the steady-state solution of the differential matrix Riccati equation solved for the same Q and R. Substituting $K=-R^{-1}B^{\mathrm{T}}P_{ss}$, (11.67) becomes

$$\begin{aligned} 0 &= (A-BR^{-1}B^{\mathrm{T}}P_{ss})^{\mathrm{T}}P_{ss}+P_{ss}(A-BR^{-1}B^{\mathrm{T}}P_{ss})+Q+P_{ss}BR^{-1}B^{\mathrm{T}}P_{ss} \\ &= A^{\mathrm{T}}P_{ss}-P_{ss}BR^{-1}B^{\mathrm{T}}P_{ss}+P_{ss}A-P_{ss}BR^{-1}B^{\mathrm{T}}P_{ss}+Q+P_{ss}BR^{-1}B^{\mathrm{T}}P_{ss} \\ &= A^{\mathrm{T}}P_{ss}+P_{ss}A+Q-P_{ss}BR^{-1}B^{\mathrm{T}}P_{ss} \end{aligned} \tag{11.69}$$

which we can see is exactly the differential matrix Riccati equation, (11.53), subject to the constraint that $\dot{P}_{ss}=0$. Therefore, the optimal solution of the infinite-time LQ problem is given by a solution to the *algebraic* Riccati equation, (11.56), when it exists. This does not necessarily mean that solutions to (11.56) are unique. If there is a unique stabilizing solution, it is optimal.

For discrete-time cases, exactly the same result holds: *if* a unique stabilizing solution to the steady-state algebraic Riccati equation, (11.60), exists, it solves the infinite-horizon LQ control problem via the feedback function in (11.21).

The question that remains in both cases is: When does such a suitable solution exist, and is it unique? The answer is that two theorems are required:

> THEOREM: If (A,B) is stabilizable, then regardless of the final state weighting matrix S, a finite steady-state solution P_{ss} (S_{ss}) exists for the differential (difference) Riccati equation, (11.53) or (11.22). This solution will also be a solution to the corresponding *algebraic* Riccati equation, and

will be the *optimizing* solution of the infinite-time case; i.e., it will result in $J = J^*$. (11.70)

THEOREM: Let state weighting matrix Q be factored into its square roots as $Q = T^{\mathrm{T}} T$. Then the steady-state solutions P_{ss} (S_{ss}) of the differential (difference) Riccati equations are unique positive-definite solutions of their corresponding *algebraic* Riccati equations if and only if (A, T) is *detectable* (see Section 8.3.3). (11.71)

In summary, if both of these conditions hold, i.e., if (A, B) is stabilizable *and* (A, T) is detectable, the solution to (11.56) or (11.60) exists, is unique, and provides the optimal static state feedback gain through (11.55) or (11.21). The reason for the first condition is fairly obvious given our infinite-length control interval. It guarantees that a static feedback gain will result in a convergent cost integral. When cost minimization over finite time was desired, there was actually no guarantee or assumption that the system was controllable or even stabilizable. If over finite time the integrand of cost function J remains bounded, then the optimal feedback control will minimize it, whether it stabilizes the system or not. If optimization of J over *infinite* time occurs, then the integrand of (11.64) must asymptotically approach zero in order for the costs to converge to a finite value and for a unique optimal solution to exist. By asking that the system be stabilizable, we are guaranteeing that a feedback K does exist that drives the state $x(t)$ asymptotically to zero and gives a finite cost. This fact does not alone guarantee that the system is stable if J converges, or consequently, that the optimal feedback will be found by solving (11.56). For that guarantee we need the other condition, i.e., detectability.

Understanding the reason for detectability of the pair (A, T) is slightly less obvious but just as reasonable. To justify this condition, consider the expression for cost as in (11.62):

$$
\begin{aligned}
J &= \int_0^\infty \left[x^{\mathrm{T}}(t) Q x(t) + u^{\mathrm{T}}(t) R u(t) \right] dt \\
&= \int_0^\infty x^{\mathrm{T}}(t) T^{\mathrm{T}} T x(t)\, dt + \int_0^\infty u^{\mathrm{T}}(t) R u(t)\, dt
\end{aligned}
\tag{11.72}
$$

Note that the first term might converge even if the system is not stable. If, for example, $x(t)$ is an eigenvector of A corresponding to an unstable eigenvalue and it is in the null space of T ($Tx = 0$), it would not contribute to the cost J. By the results of Section 8.2.2, this is equivalent to (A,T) being unobservable. Even if such modes were stable, two identical costs might result from two different feedback gains and, hence, two different P-matrices (S-matrices). If the system (A,T) is observable, we can guarantee that *no* modes escape the integration of (11.72), and if the system (A,T) is merely *detectable*, we guarantee that no *unstable* modes escape integration.

It should be remembered that multiple stabilizing solutions to (11.56) might exist if (A,B) is stabilizable but (A,T) is *not* detectable. The Riccati equation is, after all, a quadratic equation, which is easily seen if one considers scalar systems. Its solution would therefore not be unique, requiring us to find the "optimal" one, i.e., the one that is the asymptotic solution (as $t \to -\infty$) of the *differential* matrix Riccati equation. Rigorous proof of the theorems and methods for finding alternate solutions require an alternative formulation of the optimal control problem and are not given here (but can be found in [4]).

Example 11.2: Continuous-Time LQ Control

Consider the continuous-time system given by the equations:

$$\begin{aligned} \dot{x} &= \begin{bmatrix} 1 & 0 \\ 2 & 0 \end{bmatrix} x + \begin{bmatrix} 1 \\ 0 \end{bmatrix} u \\ y &= cx \end{aligned} \tag{11.73}$$

Solve the differential matrix Riccati equation that results in the control signal that minimizes the cost function

$$J = x^{\mathrm{T}}(5)Sx(5) + \int_0^5 \left[y^{\mathrm{T}}(t)y(t) + u^{\mathrm{T}}(t)u(t) \right] dt \tag{11.74}$$

Use the two different final-state weighting matrices

$$S_1 = \begin{bmatrix} 0 & 0 \\ 0 & 0 \end{bmatrix} \qquad S_2 = \begin{bmatrix} 2 & 0 \\ 0 & 2 \end{bmatrix} \tag{11.75}$$

Then find the solution to the algebraic Riccati equation that gives the optimal control for the cost function

$$\int_0^\infty \left[y^{\mathrm{T}}(t)y(t) + u^{\mathrm{T}}(t)u(t) \right] dt \tag{11.76}$$

Do this for both of the c-matrices, i.e., $c_1 = \begin{bmatrix} 0 & 1 \end{bmatrix}$ and $c_2 = \begin{bmatrix} 1 & 0 \end{bmatrix}$. Compare the results. Is the system asymptotically stabilized in each case?

Solution:

We should first analyze the problem to anticipate what results are expected. The system is clearly open-loop unstable, with eigenvalues of 0 and 1. It is easy to check that it is controllable (and hence, stabilizable). It is observable through output matrix c_1 but not through c_2. Because at least one of the modes is unobservable through c_2 but neither is asymptotically stable, we will expect that the infinite-time control of the system will then give an algebraic Riccati equation whose solution is unable to stabilize the system.

The second feature to notice from the problem statement is that the cost function is written in terms of the output signal $y(t)$ instead of the state vector $x(t)$. However because there is no feedthrough term, we can substitute $y(t) = cx(t)$ to get the new integral

$$J = x^{\mathrm{T}}(5)Sx(5) + \int_0^5 \left[x^{\mathrm{T}}(t)c^{\mathrm{T}}cx(t) + u^{\mathrm{T}}(t)u(t) \right] dt \qquad (11.77)$$

(and similarly for the infinite-time cost). This implies that the weighting matrices are $Q = c^{\mathrm{T}}c$ and $R = r = 1$ (scalar). This provides a ready-made factorization for Q.

For the case of output matrix c_1 and the finite-horizon cost function, numerical simulation of the differential matrix Riccati equation, (11.53) (see Problem 11.5) gives the solutions for matrix P as depicted in Figure 11.5. Note that in each case, the limiting solution (as $t \to 0$) is the same, but that the simulation with the nonzero final state weighting matrix converges faster (as we consider traveling backward in time).

If we now seek the solution to the *algebraic* Riccati equation, we can guess that it will be equal to the limiting values of the curves in Figure 11.5, because the system is stabilizable and detectable when using c_1. This solution can be obtained using several numerical techniques,[M] which are usually necessary because of the nonlinearity of the equations involved. However, for this particular system, the Riccati equation may be solved analytically:

```
care(A,B,Q,
  R,S,E)
```

$$0 = A^{\mathrm{T}}P + PA + c^{\mathrm{T}}c - Pbr^{-1}b^{\mathrm{T}}P$$

$$\begin{aligned}
\begin{bmatrix} 0 & 0 \\ 0 & 0 \end{bmatrix} &= \begin{bmatrix} 1 & 2 \\ 0 & 0 \end{bmatrix}\begin{bmatrix} p_{11} & p_{12} \\ p_{12} & p_{22} \end{bmatrix} + \begin{bmatrix} p_{11} & p_{12} \\ p_{12} & p_{22} \end{bmatrix}\begin{bmatrix} 1 & 0 \\ 2 & 0 \end{bmatrix} + \begin{bmatrix} 0 & 0 \\ 0 & 1 \end{bmatrix} \\
&\quad - \begin{bmatrix} p_{11} & p_{12} \\ p_{12} & p_{22} \end{bmatrix}\begin{bmatrix} 1 & 0 \\ 0 & 0 \end{bmatrix}\begin{bmatrix} p_{11} & p_{12} \\ p_{12} & p_{22} \end{bmatrix} \\
&= \begin{bmatrix} p_{11}+2p_{12} & p_{12}+2P_{22} \\ 0 & 0 \end{bmatrix} + \begin{bmatrix} p_{11}+2p_{12} & 0 \\ p_{12}+2P_{22} & 0 \end{bmatrix} \\
&\quad + \begin{bmatrix} 0 & 0 \\ 0 & 1 \end{bmatrix} - \begin{bmatrix} p_{11}^2 & p_{11}p_{12} \\ p_{11}p_{12} & p_{12}^2 \end{bmatrix}
\end{aligned} \tag{11.78}$$

This matrix equality represents a set of three simultaneous equations [because the (1,2) and (2,1) terms are the same]. They are:

$$2p_{11} - p_{11}^2 + 4p_{12} = 0 \tag{11.79}$$

$$p_{12} + 2p_{22} - p_{11}p_{12} = 0 \tag{11.80}$$

$$1 - p_{12}^2 = 0 \tag{11.81}$$

Solving (11.81) results in $p_{12} = \pm 1$. If we select $p_{12} = -1$, then Equation (11.79) will have complex roots, which is clearly undesirable. So instead we select $p_{12} = +1$, giving a quadratic equation for (11.79) that provides the two solutions $p_{11} = 1 \pm \sqrt{5}$. If $p_{11} = 1 - \sqrt{5}$, then the *P*-matrix cannot be positive definite. So instead take $p_{11} = 1 + \sqrt{5} = 3.236$. Finally, these two results substituted into (11.80) give the final element $p_{22} = \sqrt{5}/2 = 1.118$. Notice that these values are the steady-state solutions to the differential Riccati equations shown in Figure 11.5. Thus, the static state feedback result is

$$u = -r^{-1}b^{\mathrm{T}}Px = Kx = \begin{bmatrix} -3.236 & -1 \end{bmatrix}x \tag{11.82}$$

When this feedback is used to compute the closed loop eigenvalues, we find that

$$\sigma(A + bK) = -\tfrac{\sqrt{5}}{2} \pm \tfrac{\sqrt{3}}{2}$$

which indicates an asymptotically stable system.

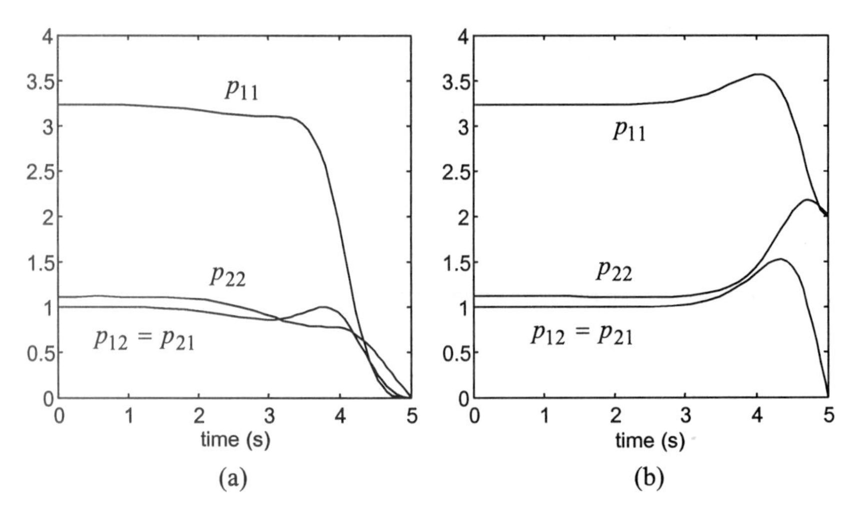

Figure 11.5 Numerical solutions for $P(t)$ for two different final-state weighting matrices using $c_1 = \begin{bmatrix} 0 & 1 \end{bmatrix}$. (a) $S_1 = 0_{2\times 2}$ (b) $S_2 = 2 \cdot I_{2\times 2}$.

For the second output matrix $c_2 = \begin{bmatrix} 1 & 0 \end{bmatrix}$, the numerical simulations for the two cases are shown in Figure 11.6. Note that in this case, the simulation with the nonzero final-state matrix does not reach a steady state (again, backward) as fast as the $S = 0_{2\times 2}$ case does. In fact, the curves approach their steady-state very slowly and are always poor approximations for constants. Note also that the limiting solutions are different from those of Figure 11.5.

To find the limiting solutions in Figure 11.6, we can solve the algebraic Riccati equations exactly as we did in (11.79) through (11.81), which results in

$$P = \begin{bmatrix} 1+\sqrt{2} & 0 \\ 0 & 0 \end{bmatrix} \tag{11.83}$$

This positive-semidefinite solution provides steady-state (i.e., static) feedback of

$$u = -r^{-1} b^{\mathrm{T}} P x = K x = \begin{bmatrix} -2.414 & 0 \end{bmatrix} x \tag{11.84}$$

which results in closed-loop eigenvalues of $\sigma(A + bK) = \{0, 1\}$. The closed-loop system is therefore still not asymptotically stable. In fact, its eigenvalues have not changed at all.

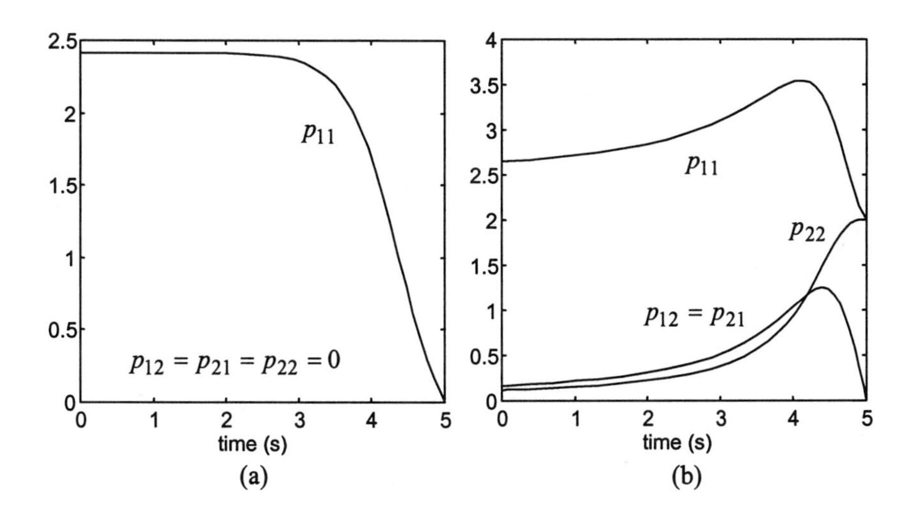

Figure 11.6 Numerical simulation of the differential Riccati equation using the output matrix $c_2 = [1 \quad 0]$. (a) $S_1 = 0_{2\times 2}$ (b) $S_2 = 2 \cdot I_{2\times 2}$.

11.2 Optimal Estimators

One of the most important but most often neglected factors in the performance of linear control systems is the influence of noise and disturbances. All systems are subject to noise, be it in the form of unmodeled input effects, unmodeled dynamics, or undesired signals that act as inputs to the system at different points. We have, in fact, ignored noise up to this point in our analysis and design of state space control systems. Yet, noise is known to have significant effects on the convergence of regulators and numerical design algorithms, excitation of unmodeled dynamics, and stability. A growing subdiscipline of control systems, *robust control,* is dedicated to the investigation of such influences and their *rejection* from the desired outputs of the system.

We will treat some simple noise models in this section and determine the ways we may deal with it in the context of our knowledge of observers and state feedback. First, we consider noise as being a corrupting influence in the process of observing the state of a system. With a state space model that includes noise, we will generate the best possible observer, i.e., the observer that most effectively rejects the effects of the noise. We call such observers *estimators,* and the particular estimator we develop is known as the *Kalman filter.*

11.2.1 Models of Systems with Noise

Consider the discrete-time linear system: *

$$\begin{aligned} x(k+1) &= Ax(k) + Bu(k) + Gv(k) \\ y(k) &= Cx(k) + w(k) \end{aligned} \tag{11.85}$$

In this expression, two new terms appear, $v(k)$ and $w(k)$. These signals are considered to be random processes consisting of white, stationary noise, with zero mean, and uncorrelated with each other. Thus, they have the following properties:

$$\begin{aligned} E[v(k)v^{\mathrm{T}}(k)] &= V \\ E[v(k)v^{\mathrm{T}}(j)] &= 0 \text{ if } k \neq j \\ E[v(k)] &= 0 \end{aligned} \tag{11.86}$$

and

$$\begin{aligned} E[w(k)w^{\mathrm{T}}(k)] &= W \\ E[w(k)w^{\mathrm{T}}(j)] &= 0 \text{ if } k \neq j \\ E[w(k)] &= 0 \end{aligned} \tag{11.87}$$

as well as the joint property that $E[v(j)w^{\mathrm{T}}(k)] = 0$ for all j and k. Furthermore, we will consider the initial condition on the state to be a random variable x_0, because we seldom know it exactly. We also consider x_0 to be white noise, with

$$E\left[((x_0 - E(x_0))((x_0 - E(x_0))^{\mathrm{T}}\right] \stackrel{\Delta}{=} S_0 \tag{11.88}$$

Furthermore, we assume that x_0 is uncorrelated with $v(k)$ and $w(k)$.

The plant input term $v(k)$ is called the *plant noise* (sometimes called the *process* noise), and the noise term in the measurement equation $w(k)$ is called the *measurement noise*. The plant noise models the effects of noise inputs on the state space variables themselves, and the measurement noise models such

* Time-varying systems are easily included in the derivations of this section. However, the time argument on the system matrices, i.e., the k in $A_d(k)$, as well as the d subscript we have been using to denote discrete-time quantities are omitted for brevity of notation.

factors as noisy sensors. The other input, $u(k)$, is the conventional input term and is considered noise free, i.e., deterministic.

11.2.2 The Discrete-Time Kalman Filter

When the state space system is modeled as in (11.85), we must reconsider our derivation of the observer. In particular, we must consider the effects of the noise on the selection of the observer gain L (from Chapter 10). This gain should ideally be chosen to provide the best estimate of the system's state while simultaneously rejecting any influence due to the two noise inputs. Such an estimator can therefore be thought of as a filter. For the model we have given, the most commonly used estimator for the state is known as the *Kalman filter*. We will present one of several possible approaches to the Kalman filter, one that follows our previous understanding of observers.

Our derivation of the Kalman filter begins with the formulation for the current estimator as seen in (10.98). Recall that in the current estimator, the observation process is separated into a time update and a measurement update. This is convenient for us now because it initially allows us to separate the effects of the two different noise terms.

For the time update of the current estimator, we will alter (10.97) by finding the best estimate for $x(k+1)$ given only the current best estimate $\hat{x}(k)$ and the plant matrices (but not the output that is received at time $k+1$). The best estimate is the *expected value*

$$\begin{aligned} \bar{x}(k+1) &= E[A\hat{x}(k) + Bu(k) + Gv(k)] \\ &= A\hat{x}(k) + Bu(k) \end{aligned} \tag{11.89}$$

Now, we will propose an observer following (10.98) to update this estimate after the output at time $k+1$ becomes available:

$$\hat{x}(k+1) = \bar{x}(k+1) + L(k+1)[y(k+1) - C\bar{x}(k+1)] \tag{11.90}$$

Note that we have written the estimator gain L as a time-varying quantity. We will see that this is indeed necessary.

The Kalman filter is an *optimal* estimator in the sense that it provides the best estimate for the state while rejecting the noise. Therefore, it is derived through the minimization of an error criterion, which can be defined in two ways. The error in the a priori estimate [i.e., (11.89), before the measurement is considered] is

$$\bar{e}(k) = x(k) - \bar{x}(k) \tag{11.91}$$

while the error in the a posteriori estimate (i.e., after the measurement is considered) is:

$$e(k) = x(k) - \hat{x}(k) \tag{11.92}$$

These errors are of course measures of the accuracy of the estimates. However because they now depend on noise and are therefore random variables, it is necessary to also consider their noise statistical properties, for their sample values at any given time may not be indicative of their accuracy. Therefore we define

$$\bar{S}(k) \stackrel{\Delta}{=} E[\bar{e}(k)\bar{e}^{\mathrm{T}}(k)] \tag{11.93}$$

and

$$S(k) \stackrel{\Delta}{=} E[e(k)e^{\mathrm{T}}(k)] \tag{11.94}$$

These quantities are known as the a priori and a posteriori error covariances, respectively. Given that our guess at the initial condition of the system is a random one, we will therefore say that $S(k) = S_0$.

With the current estimator, we simply formed an error system, Equation (10.100), and selected L such that the error dynamics were asymptotically stable. With the inclusion of the noise terms, this process is insufficient. Instead, we will find a statistical error criterion and use it to minimize the effects of the noise. If we start by finding the error dynamics, we will later see how the error covariance is minimized:

$$\begin{aligned}
e(k+1) &= x(k+1) - \hat{x}(k+1) \\
&= Ax(k) + Bu(k) + Gv(k) - A\hat{x}(k) - Bu(k) \\
&\quad - L(k+1)\left[Cx(k+1) + w(k+1) - C\bar{x}(k+1)\right] \\
&= Ax(k) - A\hat{x}(k) + Gv(k) \\
&\quad - L(k+1)\left\{C\left[Ax(k) + Bu(k) + Gv(k)\right] + w(k+1) - C\left[A\hat{x}(k) + Bu(k)\right]\right\} \\
&= \left[A - L(k+1)CA\right]e(k) + \left[G - L(k+1)CG\right]v(k) - L(k+1)w(k+1) \\
&= \left[I - L(k+1)C\right]\left[Ae(k) + Gv(k)\right] - L(k+1)w(k+1)
\end{aligned} \tag{11.95}$$

Using this measure, it is reasonable to optimize the estimator by finding

$$\min_{L(k+1)} \left\| E\left[e(k+1)\right] \right\|^2 = \min_{L(k+1)} E\left[e^{\mathrm{T}}(k+1)e(k+1)\right] \tag{11.96}$$

For such a minimization process, the scalar measure $E\left\|\left[e(k+1)\right]\right\|^2$ must then be differentiated with respect to the *matrix* $L(k+1)$. Such differentiation is undefined. Instead, we recognize that for any vector x, $x^{\mathrm{T}}x = tr(xx^{\mathrm{T}})$, where $tr(\cdot)$ is the *trace* operator.[M] Therefore, (11.96) becomes

trace(A)

$$\begin{aligned}\min_{L(k+1)} \left\| E[e(k+1)] \right\|^2 &= \min_{L(k+1)} E\left[e^{\mathrm{T}}(k+1)e(k+1)\right] \\ &= \min_{L(k+1)} tr\left(E\left[e(k+1)e^{\mathrm{T}}(k+1)\right]\right) \\ &= \min_{L(k+1)} tr\left(S(k+1)\right)\end{aligned} \tag{11.97}$$

where we have assumed the commutativity of the trace operator and the expectation operator.

This presents us with the need to compute $S(k+1)$ from (11.95):

$$\begin{aligned}S(k+1) &= E\left[e(k+1)e^{\mathrm{T}}(k+1)\right] \\ &= E\Big\{\left[(I - L(k+1)C)(Ae(k) + Gv(k)) - L(k+1)w(k+1)\right] \cdot \\ &\qquad \left[(I - L(k+1)C)(Ae(k) + Gv(k)) - L(k+1)w(k+1)\right]^{\mathrm{T}}\Big\} \\ &= E\Big\{(I - L(k+1)C)(Ae(k) + Gv(k))(Ae(k) + Gv(k))^{\mathrm{T}}(I - L(k+1)C)^{\mathrm{T}} \\ &\qquad - 2L(k+1)w(k+1)\left[e^{\mathrm{T}}(k)A^{\mathrm{T}} + v^{\mathrm{T}}(k)G^{\mathrm{T}}\right](I - L(k+1)C)^{\mathrm{T}} \\ &\qquad + L(k+1)w(k+1)w^{\mathrm{T}}(k+1)L^{\mathrm{T}}(k+1)\Big\}\end{aligned} \tag{11.98}$$

Because of our assumptions on the statistics of the noise terms, we have that

$$\begin{aligned}E\left[w(k+1)w^{\mathrm{T}}(k+1)\right] &= W \\ E\left[v(k)v^{\mathrm{T}}(k)\right] &= V \\ E\left[w(k+1)v^{\mathrm{T}}(k)\right] &= 0\end{aligned} \tag{11.99}$$

It is also easy to compute that

$$\begin{aligned}E\left[w(k+1)e^{\mathrm{T}}(k)\right] &= E\left[w(k+1)(x(k) - \hat{x}(k))^{\mathrm{T}}\right] \\ &= E\left[w(k+1)x^{\mathrm{T}}(k)\right] - E\left[w(k+1)\hat{x}^{\mathrm{T}}(k)\right] \\ &= 0\end{aligned}$$

and

$$\begin{aligned} E\left[v(k)e^{\mathrm{T}}(k)\right] &= E\left[v(k)(x(k)-\hat{x}(k))^{\mathrm{T}}\right] \\ &= E\left[v(k)x^{\mathrm{T}}(k)\right] - E\left[v(k)\hat{x}^{\mathrm{T}}(k)\right] \\ &= 0 \end{aligned} \tag{11.100}$$

[We should be careful to remember that $x(k+1)$ and $v(k)$ *are* correlated through Equation (11.85).] Using these results and the fact that $E\left[e(k)e^{\mathrm{T}}(k)\right] = S(k)$, we may reduce (11.98) to

$$\begin{aligned} S(k+1) = {} & [I - L(k+1)C]\left[AS(k)A^{\mathrm{T}} + GVG^{\mathrm{T}}\right][I - L(k+1)C]^{\mathrm{T}} \\ & + L(k+1)WL^{\mathrm{T}}(k+1) \end{aligned} \tag{11.101}$$

Note that this equation represents the "dynamics" of the error covariance $S(k)$ in the sense that it is a recursion relationship for $S(k+1)$ in terms of $S(k)$. However, at this point the equation is not entirely useful, as we do not yet know $L(k+1)$. To determine $L(k+1)$, we must find the trace of $S(k+1)$ as given in (11.101), then differentiate it with respect to $L(k+1)$:

$$\begin{aligned} tr\left[S(k+1)\right] = tr\Big\{ & AS(k)A^{\mathrm{T}} + GVG^{\mathrm{T}} \\ & + L(k+1)C\left[AS(k)A^{\mathrm{T}} + GVG^{\mathrm{T}}\right]C^{\mathrm{T}}L^{\mathrm{T}}(k+1) \\ & -2L(k+1)C\left[AS(k)A^{\mathrm{T}} + GVG^{\mathrm{T}}\right] + L(k+1)WL^{\mathrm{T}}(k+1)\Big\} \end{aligned} \tag{11.102}$$

When we take the derivative with respect to $L(k+1)$, the first two terms will give zero because they do not depend on $L(k+1)$. The derivatives of the other terms are computed with the help of some interesting identities for derivatives of the trace of matrix quantities, see Appendix A, Equations (A.78) and (A.79):

$$\frac{\partial}{\partial X} tr\left(XYX^{\mathrm{T}}\right) = 2XY \qquad \frac{\partial}{\partial X} tr(YXZ) = Y^{\mathrm{T}}Z^{\mathrm{T}}$$

Using these,

$$\frac{\partial}{\partial L(k+1)} tr[S(k+1)] = 2L(k+1)C\left[AS(k)A^{\mathrm{T}} + GVG^{\mathrm{T}}\right]C^{\mathrm{T}}$$
$$-2\left[AS(k)A^{\mathrm{T}} + GVG^{\mathrm{T}}\right]C^{\mathrm{T}} + 2L(k+1)W$$
$$= 0$$

which has the solution

$$L(k+1) = \left[AS(k)A^{\mathrm{T}} + GVG^{\mathrm{T}}\right]C^{\mathrm{T}}\left\{C\left[AS(k)A^{\mathrm{T}} + GVG^{\mathrm{T}}\right]C^{\mathrm{T}} + W\right\}^{-1} \quad (11.103)$$

This is the best gain for the estimator given in Equation (11.90). It is known as the *Kalman gain*. Note that even if all the system matrices are time-invariant, this gain will vary with time because of its dependence on $S(k)$, which evolves according to (11.101).

Together with (11.89), (11.90) and (11.101), the Kalman gain given in (11.103) completes the specification of the Kalman filter.[M] To summarize, the following sequence of computations may be performed:

```
lqgreg(Kest,K)
estim(sys,L)
```

1. The "time" update can be computed with (11.89):

$$\bar{x}(k+1) = A\hat{x}(k) + Bu(k) \quad (11.104)$$

which is initialized with the initial guess $x(k_0)$.

2. The Kalman gain is computed according to (11.103):

$$L(k+1) = \left[AS(k)A^{\mathrm{T}} + GVG^{\mathrm{T}}\right]C^{\mathrm{T}}\left\{C\left[AS(k)A^{\mathrm{T}} + GVG^{\mathrm{T}}\right]C^{\mathrm{T}} + W\right\}^{-1} \quad (11.105)$$

which may be initialized with the original covariance $S(k_0) = S_0$.

3. The measurement update can then be applied to the time update according to (11.90):

$$\hat{x}(k+1) = \bar{x}(k+1) + L(k+1)\left[y(k+1) - C\bar{x}(k+1)\right] \quad (11.106)$$

where $y(k+1)$ is the measurement from the plant. This can be further simplified as

$$
\begin{aligned}
\hat{x}(k+1) &= A\hat{x}(k) + Bu(k) + L(k+1)\{y(k+1) - C[A\hat{x}(k) + Bu(k)]\} \\
&= (I - L(k+1)C)A\hat{x}(k) + [I - L(k+1)C]Bu(k) + L(k+1)y(k+1) \\
&= (I - L(k+1)C)[A\hat{x}(k) + Bu(k)] + L(k+1)y(k+1)
\end{aligned} \tag{11.107}
$$

4. The accuracy of the estimate is gauged and the next iteration of (11.103) is prepared for by calculating the error covariance update according to (11.101):

$$
\begin{aligned}
S(k+1) = [I - L(k+1)C][AS(k)A^{\mathrm{T}} + GVG^{\mathrm{T}}][I - L(k+1)C]^{\mathrm{T}} \\
+ L(k+1)WL^{\mathrm{T}}(k+1)
\end{aligned} \tag{11.108}
$$

The process then repeats itself at the next time step.

Simplifications of the Kalman Filter

The Kalman filter can be written in a number of different ways, some for simplicity, some for numerical properties, and some for physical insight. For example, the calculation of the Kalman gain in (11.103) can be simplified if we investigate the significance of the term that appears twice in that relationship, $AS(k)A^{\mathrm{T}} + GVG^{\mathrm{T}}$. Consider taking the "time" update of the error covariance S, just as we did with the state estimate. Then from (11.93),

$$
\begin{aligned}
\bar{S}(k+1) &= E\left[\bar{e}(k+1)\bar{e}^{\mathrm{T}}(k+1)\right] \\
&= E\left\{[x(k+1) - \bar{x}(k+1)][x(k+1) - \bar{x}(k+1)]^{\mathrm{T}}\right\} \\
&= E\{[Ax(k) + Bu(k) + Gv(k) - [A\hat{x}(k) + Bu(k)]] \cdot \\
&\qquad [Ax(k) + Bu(k) + Gv(k) - [A\hat{x}(k) + Bu(k)]]^{\mathrm{T}}\} \\
&= E\left\{[Ae(k) + Gv(k)][Ae(k) + Gv(k)]^{\mathrm{T}}\right\} \\
&= AS(k)A^{\mathrm{T}} + GVG^{\mathrm{T}}
\end{aligned} \tag{11.109}
$$

Thus, the Kalman gain can be more efficiently expressed as

$$
L(k+1) = \bar{S}(k+1)C^{\mathrm{T}}\left[C\bar{S}(k+1)C^{\mathrm{T}} + W\right]^{-1} \tag{11.110}
$$

if Equation (11.109) is first used to compute $\bar{S}(k+1)$.

Likewise, (11.101) could then be written

$$S(k+1) = [I - L(k+1)C]\bar{S}(k+1)[I - L(k+1)C]^{\mathrm{T}} + L(k+1)WL^{\mathrm{T}}(k+1) \tag{11.111}$$

This, in fact, can be further simplified:

$$\begin{aligned} S(k+1) &= [I - L(k+1)C]\bar{S}(k+1)[I - L(k+1)C]^{\mathrm{T}} + L(k+1)WL^{\mathrm{T}}(k+1) \\ &= [I - L(k+1)C]\bar{S}(k+1) + L(k+1)C\bar{S}(k+1)C^{\mathrm{T}}L^{\mathrm{T}}(k+1) \\ &\qquad - \bar{S}(k+1)C^{\mathrm{T}}L^{\mathrm{T}}(k+1) + L(k+1)WL^{\mathrm{T}}(k+1) \\ &= [I - L(k+1)C]\bar{S}(k+1) - \bar{S}(k+1)C^{\mathrm{T}}L^{\mathrm{T}}(k+1) \\ &\qquad + L(k+1)\left[C\bar{S}(k+1)C^{\mathrm{T}} + W\right]L^{\mathrm{T}}(k+1) \\ &= [I - L(k+1)C]\bar{S}(k+1) - \bar{S}(k+1)C^{\mathrm{T}}L^{\mathrm{T}}(k+1) \\ &\qquad + \bar{S}(k+1)C^{\mathrm{T}}\left[C\bar{S}(k+1)C^{\mathrm{T}} + W\right]^{-1}\left[C\bar{S}(k+1)C^{\mathrm{T}} + W\right]L^{\mathrm{T}}(k+1) \\ &= [I - L(k+1)C]\bar{S}(k+1) \end{aligned} \tag{11.112}$$

which, if (11.109) is used to find $\bar{S}(k+1)$, can be readily substituted for (11.101).

Now, combining (11.112) and (11.110),

$$\begin{aligned} S(k+1) &= [I - L(k+1)C]\bar{S}(k+1) \\ &= \left\{I - \bar{S}(k+1)C^{\mathrm{T}}\left[C\bar{S}(k+1)C^{\mathrm{T}} + W\right]^{-1}C\right\}\bar{S}(k+1) \\ &= \bar{S}(k+1) - \bar{S}(k+1)C^{\mathrm{T}}\left[C\bar{S}(k+1)C^{\mathrm{T}} + W\right]^{-1}C\bar{S}(k+1) \end{aligned}$$

or

$$S(k) = \bar{S}(k) - \bar{S}(k)C^{\mathrm{T}}\left[C\bar{S}(k)C^{\mathrm{T}} + W\right]^{-1}C\bar{S}(k) \tag{11.113}$$

Equation (11.113) represents the effect of the measurement on the a priori error covariance. Comparing this formula to the matrix inversion lemma given in Appendix A results in

$$
\begin{aligned}
S(k) &= \bar{S}(k) - \bar{S}(k)C^{\mathrm{T}}\left[C\bar{S}(k)C^{\mathrm{T}} + W\right]^{-1} C\bar{S}(k) \\
&= \left[\bar{S}^{-1}(k) + C^{\mathrm{T}}W^{-1}C\right]^{-1}
\end{aligned}
\tag{11.114}
$$

This equation can also be used as an update for the error covariance. Inverting it gives the backward relationship:

$$
\bar{S}(k) = \left[S^{-1}(k) - C^{\mathrm{T}}W^{-1}C\right]^{-1}
$$

or

$$
\bar{S}(k+1) = \left[S^{-1}(k+1) - C^{\mathrm{T}}W^{-1}C\right]^{-1} \tag{11.115}
$$

To further simplify the expression for the Kalman gain, take (11.110) and again use the matrix inversion lemma:

$$
\begin{aligned}
L(k+1) &= \bar{S}(k+1)C^{\mathrm{T}}\left[C\bar{S}(k+1)C^{\mathrm{T}} + W\right]^{-1} \\
&= \bar{S}(k+1)C^{\mathrm{T}}\left\{W^{-1} - W^{-1}C\left[C^{\mathrm{T}}W^{-1}C + \bar{S}^{-1}(k+1)\right]^{-1} C^{\mathrm{T}}W^{-1}\right\} \\
&= \bar{S}(k+1)\left\{I - C^{\mathrm{T}}W^{-1}C\left[C^{\mathrm{T}}W^{-1}C + \bar{S}^{-1}(k+1)\right]^{-1}\right\}C^{\mathrm{T}}W^{-1}
\end{aligned}
\tag{11.116}
$$

Now substitute (11.115) into (11.116):

$$
\begin{aligned}
L(k+1) &= \left[S^{-1}(k+1) - C^{\mathrm{T}}W^{-1}C\right]^{-1} \cdot \\
&\qquad \left\{I - C^{\mathrm{T}}W^{-1}C\left[C^{\mathrm{T}}W^{-1}C + S^{-1}(k+1) - C^{\mathrm{T}}W^{-1}C\right]^{-1}\right\}C^{\mathrm{T}}W^{-1} \\
&= \left[S^{-1}(k+1) - C^{\mathrm{T}}W^{-1}C\right]^{-1}\left[I - C^{\mathrm{T}}W^{-1}CS(k+1)\right]C^{\mathrm{T}}W^{-1} \\
&= \left[S^{-1}(k+1) - C^{\mathrm{T}}W^{-1}C\right]^{-1}\left[S^{-1}(k+1) - C^{\mathrm{T}}W^{-1}C\right]S(k+1)C^{\mathrm{T}}W^{-1} \\
&= S(k+1)C^{\mathrm{T}}W^{-1}
\end{aligned}
\tag{11.117}
$$

This will further reduce (11.90) to the form:

$$
\hat{x}(k+1) = \bar{x}(k+1) + S(k+1)C^{\mathrm{T}}W^{-1}\left[y(k+1) - C\bar{x}(k+1)\right] \tag{11.118}
$$

Note the similarity between (11.117) and the formula for state feedback gain K from Equation (11.55), particularly its dependence on the matrix W and the time-varying matrix quantity $S(k+1)$. To develop this similarity further, one additional manipulation is in order. Inserting (11.113) into (11.109) gives

$$\begin{aligned}\bar{S}(k+1) &= AS(k)A^{\mathrm{T}} + GVG^{\mathrm{T}} \\ &= A\left\{\bar{S}(k) - \bar{S}(k)C^{\mathrm{T}}\left[C\bar{S}(k)C^{\mathrm{T}} + W\right]^{-1} C\bar{S}(k)\right\}A^{\mathrm{T}} + GVG^{\mathrm{T}} \\ &= A\bar{S}(k)A^{\mathrm{T}} - A\bar{S}(k)C^{\mathrm{T}}\left[C\bar{S}(k)C^{\mathrm{T}} + W\right]^{-1} C\bar{S}(k)A^{\mathrm{T}} + GVG^{\mathrm{T}}\end{aligned} \tag{11.119}$$

This can be compared to (11.22) to realize that it is a difference (discrete-time) Riccati equation that governs the time update of the error covariance. This is a remarkable and important fact and again illustrates the inherent duality seen so often in linear systems. More will be said concerning this duality following the next section, in which we derive the continuous-time counterpart to this Kalman filter [8].

11.2.3 The Continuous-Time Kalman Filter

The discrete-time Kalman filter was derived from the formulation of the current estimator, as presented in Section 10.3.2. We will begin the derivation of the continuous-time Kalman filter in the same way, but we will then proceed somewhat differently. In practice, the continuous-time Kalman filter is rarely used because it is almost always implemented on a computer, in which case the discrete-time version is more natural. However, it has some interesting parallels to the continuous-time LQ controller and has some steady-state properties that are important to the theory of analog signal processing.

Assume that the continuous-time system equations are of the form

$$\begin{aligned}\dot{x}(t) &= Ax(t) + Bu(t) + Gv(t) \\ y(t) &= Cx(t) + w(t)\end{aligned} \tag{11.120}$$

where, as with the discrete-time case, the system *could* be time varying, although we will drop the t arguments for simplicity. Analogously to (11.85), we have included in this model the two white, zero-mean, mutually uncorrelated noise signals $v(t)$ and $w(t)$. Following (11.86) and (11.87), we assume that

$$\begin{aligned}E[v(t)v^{\mathrm{T}}(\tau)] &= V\delta(t-\tau) \\ E[v(t)] &= 0\end{aligned} \tag{11.121}$$

and

$$\begin{aligned} E[w(t)w^{\mathrm{T}}(\tau)] &= W\delta(t-\tau) \\ E[w(t)] &= 0 \end{aligned} \tag{11.122}$$

where $\delta(t)$ is the Dirac delta and $E[w(t)v^{\mathrm{T}}(\tau)] = 0$. Furthermore, we assume the initial guess at the system state to be denoted by $x(t_0) = x_0$ and let it be uncorrelated with the plant and measurement noise:

$$E[x_0 v^{\mathrm{T}}(\tau)] = 0 \qquad E[x_0 w^{\mathrm{T}}(\tau)] = 0$$

The covariance of the initial guess is defined as

$$E\left\{[x_0 - E(x_0)][x_0 - E(x_0)]^{\mathrm{T}}\right\} \stackrel{\Delta}{=} P_0 \tag{11.123}$$

Again, we need to determine the estimator that best estimates the state of (11.120) while rejecting the influence of the noisy inputs and random initial condition. As before, we will pose this problem as an observer design, where the observer gain will be chosen to minimize an error criterion. For the observer, we cannot use the separate "time" and "measurement" updates, because time is continuous. The estimator is therefore proposed in the structure

$$\dot{\hat{x}}(t) = A\hat{x}(t) + Bu(t) + L(t)[y(t) - C\hat{x}(t)] \tag{11.124}$$

Having discovered that the discrete-time Kalman gain $L(k+1)$ was time-varying even for time-invariant plants, we should be prepared for the gain $L(t)$ to be a function of time as well.

First, we construct the estimation error signal $e(t) = x(t) - \hat{x}(t)$ and then find its derivative to determine the error dynamics:

$$\begin{aligned} \dot{e}(t) &= \dot{x}(t) - \dot{\hat{x}}(t) \\ &= Ax + Bu + Gv - A\hat{x} - Bu - L(Cx + w - C\hat{x}) \\ &= (A - LC)e(t) + Gv(t) - Lw(t) \end{aligned} \tag{11.125}$$

Because we are not interested in the covariance of $\dot{e}$, we will solve (11.125) according to the methods of Chapter 6. First, let $\Phi(t, t_0)$ denote the state-transition matrix for the error system in (11.125). Then the complete solution of (11.125) can be written as

$$e(t) = \Phi(t,t_0)e(t_0) + \int_{t_0}^{t} \Phi(t,\tau)[Gv(\tau) - Lw(\tau)]\,d\tau \tag{11.126}$$

Now finding the error covariance from this expression:

$$\begin{aligned} P(t) &= E\left[e(t)e^{\mathrm{T}}(\tau)\right] \\ &= E\left\{\left[\Phi(t,t_0)e(t_0) + \int_{t_0}^{t} \Phi(t,\tau)[Gv(\tau) - Lw(\tau)]\,d\tau\right]\cdot\right. \\ &\qquad \left.\left[\Phi(t,t_0)e(t_0) + \int_{t_0}^{t} \Phi(t,s)[Gv(s) - Lw(s)]\,ds\right]^{\mathrm{T}}\right\} \end{aligned} \tag{11.127}$$

In simplifying this expression, we immediately recognize that because of the mutual uncorrelatedness of the signals $x(t)$, $v(t)$, and $w(t)$, the cross-terms within the braces of (11.127) will have and expected value of zero. Therefore, (11.127) reduces to

$$\begin{aligned} P(t) &= E\Big[\Phi(t,t_0)e(t_0)e^{\mathrm{T}}(t_0)\Phi^{\mathrm{T}}(t,t_0) \\ &\qquad + \int_{t_0}^{t}\int_{t_0}^{t} \Phi(t,\tau)\left[Gv(\tau)v^{\mathrm{T}}(s)G^{\mathrm{T}} + Lw(\tau)w^{\mathrm{T}}(s)L^{\mathrm{T}}\right]\Phi^{\mathrm{T}}(t,s)\,d\tau\,ds\Big] \\ &= \Phi(t,t_0)P_0\Phi^{\mathrm{T}}(t,t_0) \\ &\qquad + \int_{t_0}^{t}\int_{t_0}^{t} \Phi(t,\tau)\left[GV\delta(\tau-s)G^{\mathrm{T}} + LW\delta(\tau-s)L^{\mathrm{T}}\right]\Phi^{\mathrm{T}}(t,s)\,d\tau\,ds \\ &= \Phi(t,t_0)P_0\Phi^{\mathrm{T}}(t,t_0) + \int_{t_0}^{t} \Phi(t,s)\left[GVG^{\mathrm{T}} + LWL^{\mathrm{T}}\right]\Phi^{\mathrm{T}}(t,s)\,ds \end{aligned} \tag{11.128}$$

Next we take the derivative of (11.128) with respect to time t in order to determine the dynamics of the error covariance.*

* See Appendix A, Equation (A.94), for the formula for finding the derivative of an integral with respect to one of its arguments.

$$
\begin{aligned}
\dot{P}(t) &= \dot{\Phi}(t,t_0)P_0\Phi^{\mathrm{T}}(t,t_0) + \Phi(t,t_0)P_0\dot{\Phi}^{\mathrm{T}}(t,t_0) \\
&\quad + \int_{t_0}^{t}\left[\dot{\Phi}(t,s)\left(GVG^{\mathrm{T}} + LWL^{\mathrm{T}}\right)\Phi^{\mathrm{T}}(t,s) + \Phi(t,s)\left(GVG^{\mathrm{T}} + LWL^{\mathrm{T}}\right)\dot{\Phi}^{\mathrm{T}}(t,s)\right]ds \\
&\quad + GVG^{\mathrm{T}} + LWL^{\mathrm{T}} \\
&= (A - LC)\Phi(t,t_0)P_0\Phi^{\mathrm{T}}(t,t_0) + \Phi(t,t_0)P_0\Phi^{\mathrm{T}}(t,t_0)(A - LC)^{\mathrm{T}} \\
&\quad + (A - LC)\int_{t_0}^{t}\left[\Phi(t,s)\left(GVG^{\mathrm{T}} + LWL^{\mathrm{T}}\right)\Phi^{\mathrm{T}}(t,s)\right]ds \\
&\quad + \int_{t_0}^{t}\left[\Phi(t,s)\left(GVG^{\mathrm{T}} + LWL^{\mathrm{T}}\right)\Phi^{\mathrm{T}}(t,s)\right]ds\,(A - LC)^{\mathrm{T}} \\
&\quad + GVG^{\mathrm{T}} + LWL^{\mathrm{T}} \\
&= (A - LC)P(t) + P(t)(A - LC)^{\mathrm{T}} + GVG^{\mathrm{T}} + LWL^{\mathrm{T}}
\end{aligned}
\tag{11.129}
$$

We can immediately recognize this as a differential matrix Riccati equation for the error covariance $P(t)$ whose initial condition is $P(t_0) = P_0$. However, we have not yet optimized the norm of the error over all possible gains $L(t)$.

To perform the optimization, we will attempt to minimize the squared error at any time *t*. This squared error may be expressed as

$$
\begin{aligned}
E[e^{\mathrm{T}}(t)e(t)] &= tr\left\{E[e(t)e^{\mathrm{T}}(t)]\right\} \\
&= tr\left[P(t)\right]
\end{aligned}
\tag{11.130}
$$

To do this, we ask that the *negative change* in *P* be as large as possible at any given instant, i.e., if the derivative of *P* is as large (and negative) as possible at every instant, then *P* is decreasing at a maximal rate, and *P* itself will be minimized over time. Therefore we will seek to find

$$
\max_{L(t)}\left\{tr\left[\dot{P}(t)\right]\right\} = \max_{L(t)}\left\{tr\left[(A - LC)P + P(A - LC)^{\mathrm{T}} + GVG^{\mathrm{T}} + LWL^{\mathrm{T}}\right]\right\} \tag{11.131}
$$

Proceeding with this maximization,

$$\begin{aligned}\frac{\partial}{\partial L}\left[tr\left(\dot{P}\right)\right] &= \frac{\partial}{\partial L}\left\{tr\left[(A-LC)P+P(A-LC)^{\mathrm{T}}+GVG^{\mathrm{T}}+LWL^{\mathrm{T}}\right]\right\} \\ &= \frac{\partial}{\partial L}tr(-LCP)+\frac{\partial}{\partial L}tr\left(-PC^{\mathrm{T}}L^{\mathrm{T}}\right)+\frac{\partial}{\partial L}tr\left(LWL^{\mathrm{T}}\right) \\ &= -2P(t)C^{\mathrm{T}}+2L(t)W \\ &= 0\end{aligned} \tag{11.132}$$

giving

$$L(t)=P(t)C^{T}W^{-1} \tag{11.133}$$

This is the Kalman gain for the continuous-time Kalman filter.[M] Note the similarity between it and the discrete-time Kalman gain in (11.117).

```
estim(sys,L)
kalman(sys,Q,
  R,N)
```

Using (11.133) for the gain, the error covariance dynamics simplify as well:

$$\begin{aligned}\dot{P}(t) &= (A-LC)P(t)+P(t)(A-LC)^{\mathrm{T}}+GVG^{\mathrm{T}}+LWL^{\mathrm{T}} \\ &= AP(t)-P(t)C^{\mathrm{T}}W^{-1}CP(t)+P(t)A-P(t)C^{\mathrm{T}}W^{-1}CP(t) \\ &\quad +GVG^{\mathrm{T}}+P(t)C^{\mathrm{T}}W^{-1}WW^{-1}CP(t) \\ &= AP(t)+P(t)A^{\mathrm{T}}-P(t)C^{\mathrm{T}}W^{-1}CP(t)+GVG^{\mathrm{T}}\end{aligned} \tag{11.134}$$

This is recognizable as a differential matrix Riccati equation, independent of the gain L.

11.2.4 Properties of Kalman Filters

Optimality of Kalman Filters

In each of the derivations above, an error criterion was minimized in order to arrive at an "optimal" estimator gain. The error criterion was chosen as the squared error, and it was used to optimize the behavior of the observer structures that were proposed, Equations (11.90) and (11.124). However, these observers were chosen merely because of their familiarity. There is, in most cases, no guarantee that the estimator equations, (11.90) or (11.124), are the best possible estimator structures for a given linear system. The exception is when the noise signals v and w are gaussian. If the plant and measurement noise are both gaussian, then the Kalman filters we have presented are indeed the best possible estimators of the plant's state. If the noise signals follow a different probability distribution, then the Kalman filters are the best *linear* estimators of the state. To show this would require a derivation from statistical principles that we will not explore.

Steady-State Behavior

It may be noticed that the discrete and continuous-time Kalman filters above have strong similarities to the LQ controllers derived in Section 11.1. In particular, each system required the solution of a Riccati equation in parallel with the computation of the optimal gain. In the LQ controller, this was a backward time Riccati equation, and in the Kalman filter, it is a forward-time equation. When discussing the Riccati equations for the LQ controllers, we took some time to investigate their steady-state behavior in the hopes that a constant solution would lead to an *approximately* optimal system that is easier to implement numerically. Clearly, the same questions can now be asked regarding the Kalman filter. Thus, when does the Riccati equation that governs the Kalman filter have a stabilizing steady state that is also a solution of the corresponding *algebraic* Riccati equation? We will briefly answer this question by appealing to the duality of the equations.

Consider the continuous-time case, i.e., the LQ controller's Riccati equation, (11.53), and the Kalman filter's Riccati equation, (11.134). For convenience, they are repeated here:

LQ controller:

$$\begin{aligned} \dot{P} &= PBR^{-1}B^{\mathrm{T}}P - Q - PA - A^{\mathrm{T}}P \\ K &= -R^{-1}B^{\mathrm{T}}P \end{aligned} \tag{11.135}$$

Kalman filter:

$$\begin{aligned} \dot{P} &= -PC^{\mathrm{T}}W^{-1}CP + GVG^{\mathrm{T}} + AP + PA^{\mathrm{T}} \\ L &= PC^{T}W^{-1} \end{aligned} \tag{11.136}$$

In the case of the LQ controller, the algebraic version of (11.135) needed the following conditions to provide a unique positive-definite stabilizing solution that corresponds to the steady-state value of the differential equations: (A, B) must be stabilizable, and (A, T) must be detectable, where $Q = T^{\mathrm{T}}T$. By simply comparing Equations (11.135) and (11.136) above, the following can be said about the steady-state behavior of the Kalman filter:

> THEOREM: The algebraic Riccati equation has a unique, stabilizing, positive-definite solution P that is the steady-state solution of the differential equation in (11.136) if and only if $(A^{\mathrm{T}}, C^{\mathrm{T}})$ is stabilizable, and $(A^{\mathrm{T}}, T^{\mathrm{T}})$ is detectable, where $GVG^{\mathrm{T}} = TT^{\mathrm{T}}$. (11.137)

If these conditions are met, then the error dynamics, (11.125), will be stable and the Kalman filter will converge. The meanings of these conditions can be better understood by their stabilizability and detectability duality. First, stabilizability of $(A^{\mathrm{T}}, C^{\mathrm{T}})$ is equivalent to detectability of (A, C). This is the expected necessity for the unstable modes of the system to be observable. This condition is fairly obvious. Second, detectability of $(A^{\mathrm{T}}, T^{\mathrm{T}})$ is equivalent to stabilizability of (A, T). This condition is less obvious. In essence it dictates that the noise term $v(t)$ must somehow affect each unstable mode of the system. The system is therefore not stabilizable unless it is sufficiently corrupted by noise! This will prevent the error covariance for this mode from approaching zero, which would thereby keep such a mode from affecting the error dynamics. Naturally, similar analogies can be drawn for the discrete-time LQ controller and Kalman filters.

Example 11.3: A Continuous-Time Kalman Filter

Generate and simulate a continuous-time Kalman filter to estimate the state variables of the system:

$$\begin{aligned} \dot{x} &= \begin{bmatrix} -4 & 2 \\ -2 & -4 \end{bmatrix} x + \begin{bmatrix} 0 \\ 1 \end{bmatrix} u + \begin{bmatrix} 1 \\ -1 \end{bmatrix} v \\ y &= \begin{bmatrix} 1 & 0 \end{bmatrix} x + w \end{aligned} \tag{11.138}$$

where the noise term $v(t)$ has zero mean and covariance $V = 0.09$. The measurement noise term is assumed to have zero mean and covariance $W = 0.25$. Use as an input $u(t) = \sin t$ over a period of $t \in (0, 10)$ s. Guess that the initial state of the plant is $x(0) = [0.5 \quad -0.5]^{\mathrm{T}}$, with a covariance of this initial estimate of $P_0 = I_{2\times 2}$.

Solution:

Because Equations (11.124), (11.133), and (11.134) completely describe the Kalman filter, there is very little analysis necessary before simulating the system. The result is shown in Figure 11.7. In the plot, the noisy state variables are clearly seen to be filtered by the estimator, giving smoothed versions. In Figure 11.7, the time-varying Kalman filter is used with the Riccati equation solved explicitly using MATLAB. The resulting error covariance is plotted in Figure 11.8. Note that the elements of $P(t)$ reach a steady state value relatively quickly. Note also that the time at which the error covariance reaches its approximate steady-state (~ 0.5 s) is also approximately the time at which the estimates approach the true states.

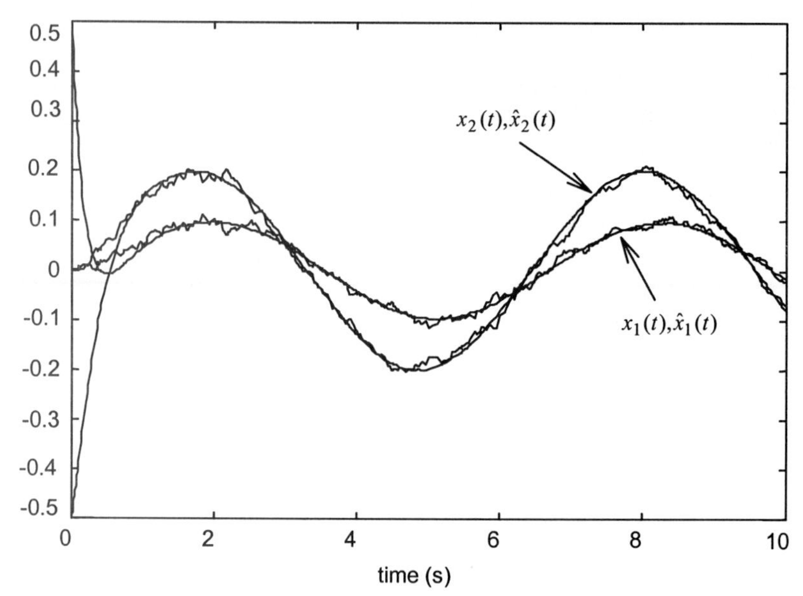

Figure 11.7 True and estimated state variables using the continuous-time Kalman filter for the system of Example 11.3. Shown are the filtered estimates using time-varying gains.

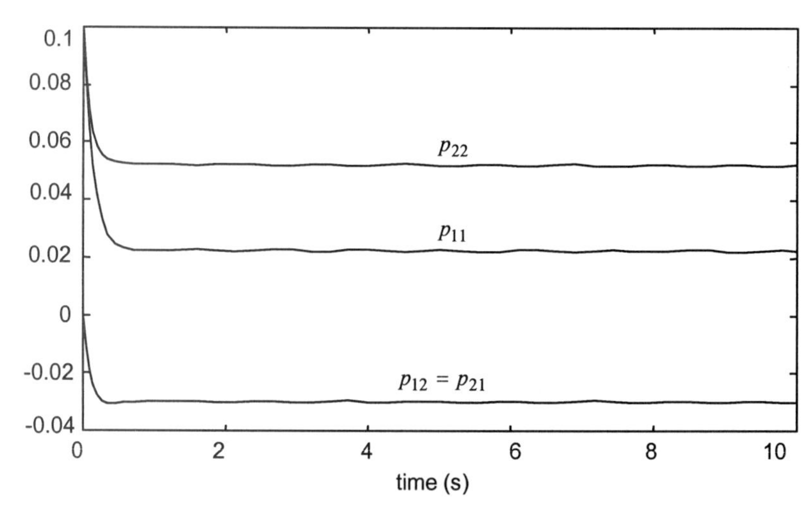

Figure 11.8 Elements of the error covariance matrix $P(t)$ from numerical solution of the differential matrix Riccati equation.

Because of the difficulty of solving the Riccati equation (11.134), we will also consider the steady-state Kalman gain formulation. The relatively flat values for $P(t)$ in Figure 11.8 suggest that this will be a good approximation. To do this, we must first verify that the conditions on (A^T, C^T) and (A^T, T^T) are true, according to the theorem above (they are). When the corresponding algebraic Riccati equation is solved, we get

$$P(t) = \begin{bmatrix} 0.0224 & -0.0299 \\ -0.0299 & 0.0522 \end{bmatrix} \tag{11.139}$$

giving a constant Kalman gain of

$$L = \begin{bmatrix} 0.0449 \\ -0.0598 \end{bmatrix} \tag{11.140}$$

The result of applying this constant gain is shown in Figure 11.9. Comparison of this result with that of Figure 11.7 indicates that the approximation of the Kalman gain as constant is a reasonable one.

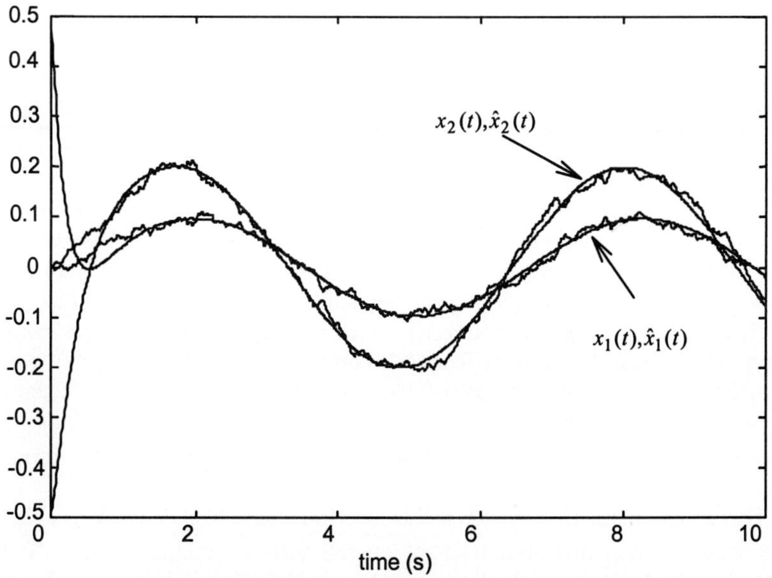

Figure 11.9 True and estimated state variables when the algebraic Riccati equation is used to compute a constant Kalman gain. Shown are the filtered estimates with steady-state gains.

The open-loop system given in (11.138) happens to be asymptotically stable. Appropriate choice of the estimator gain L ensures that the estimator itself is stable, so we have the estimate tracking seen in the figures. Note however that we are still simulating an open-loop system, i.e., one without a controller. If the plant itself were unstable, we could still design a stable Kalman filter that would track the unstable plant, even while the plant state variables diverge to infinity. To stabilize the plant itself, the Kalman filter must be combined with a controller, such as the LQ regulator. This is our next topic.

11.3 LQG Control

The next logical step in controller design is to combine the LQ regulator with the Kalman filter so that a system with noise can be controlled to minimize a cost criterion. Of course if the system contains plant and/or measurement noise, then the cost to be minimized must in fact be an expected value. For continuous time, we can therefore state the LQG control problem for the state space equations as follows:

$$\begin{aligned} \dot{x} &= Ax + Bu + Gv \\ y &= Cx + w \end{aligned} \tag{11.141}$$

find the control input $u^*(t)$ that minimizes the cost criterion

$$\bar{J}(x(t_0),t_0) = E\left\{\tfrac{1}{2}x^{\mathrm{T}}(t_0)Sx(t_0) + \tfrac{1}{2}\int_{t_0}^{t}\left[x^{\mathrm{T}}(\tau)Qx(\tau) + u^{\mathrm{T}}(t)Ru(\tau)\right]d\tau\right\} \tag{11.142}$$

The complete derivation of the solution to this problem is beyond the scope of this book. However, we are already prepared to *use* the solution based on our investigations of the deterministic LQ control problem and the Kalman filter. This is because *the solution of the LQ problem for stochastic systems* (*i.e., with the noise models we have introduced*) *is exactly the same as the deterministic LQ problem*. Thus, the optimal feedback control is of the form (11.55), where the matrix $P(t)$ is the solution of the differential matrix Riccati equation, (11.53). This solution will be truly optimal when the noise is gaussian. The steady-state solution with constant P may of course be used when the infinite-horizon approximation is made.

There is a distinction between the deterministic and stochastic problems, but it occurs in the computation of the overall cost of a control, i.e., $\bar{J}$. If the feedback in (11.55) is applied to a system with plant noise in the form of (11.120), the cost $\bar{J}$ will include a term for the effect of the noise. This is unavoidable as the noise affects the plant states. However, in practice the cost is not usually computed. The cost criterion is merely the optimization guideline, and its numerical value is not very meaningful.

The feedback solution (11.55) presumes that the state $x(t)$ is available for computation of $u(t)$. This is sometimes referred to as the "complete information" LQG problem. As we are well aware, though, the state is not always available, and must often be estimated. This is the "incomplete information" LQG problem. To solve it, we simply *employ the Kalman filter in order to generate the "optimal" state estimate, then use this estimate to compute the state feedback.* Of course, the Kalman filter estimates the state variables from noisy measurements, and as a result, this measurement noise appears (though filtered) in the estimates themselves, which are then included in the state feedback. Therefore, some additional noise will appear in the overall closed-loop cost, just as the plant noise resulted in a term in this cost. However, the extra cost due to the plant noise is independent of the extra cost due to the measurement noise. This fact implies the separation principle for LQG control: *The best linear controller/estimator for the stochastic system, (11.120), consists of the LQ controller, (11.135), and the Kalman filter, (11.136).* The gains of each can be computed independently. The optimal control will therefore be

$$u^*(t) = -R^{-1}B^{\mathrm{T}}P(t)\hat{x}(t) \tag{11.143}$$

where $P(t)$ is the solution to (11.135) and $\hat{x}(t)$ is the state vector estimate provided by the Kalman filter. All of these results apply, with only a change to the corresponding discrete-time governing equations, to discrete-time systems as well as continuous-time systems.

Example 11.4: LQG Controller for Disturbance Rejection

We will end this chapter with an example illustrating the LQG controller. Consider a plant with state equations:

$$\begin{aligned} \dot{x} &= Ax + bu = \begin{bmatrix} 0 & 1 \\ -50 & -10 \end{bmatrix} x + \begin{bmatrix} 0 \\ 1 \end{bmatrix} u \\ y &= cx + w = \begin{bmatrix} 20 & 0 \end{bmatrix} x + w \end{aligned} \tag{11.144}$$

where $w(t)$ is zero-mean white noise with covariance $W = 0.01$. It is observed that, in operation, the state vector of this plant is corrupted by a noisy signal of approximately 4 Hz, i.e., a narrow-band noise signal. The goal of the control problem is to reject this 4-Hz disturbance.

To model the disturbance, we will assume that there is a dynamic system of the form

$$\begin{aligned} \dot{x}_n &= A_n x_n + g_n v = \begin{bmatrix} 0 & 4(2\pi) \\ -4(2\pi) & 0 \end{bmatrix} x_n + \begin{bmatrix} 0 \\ 1 \end{bmatrix} v \\ \zeta &= c_n x_n = \begin{bmatrix} 100 & 0 \end{bmatrix} x_n \end{aligned} \tag{11.145}$$

where $v(t)$ is a zero-mean white noise with covariance $V = 0.01$. This noise therefore excites the system to produce a signal of 4 Hz, scaled by the factor of 100. We will then treat the noisy output of (11.145), i.e., ζ, as a noise-source disturbance for (11.144):

$$\begin{aligned} \dot{x} &= Ax + bu + g\zeta \\ y &= cx + w \end{aligned} \tag{11.146}$$

where $g = \begin{bmatrix} 0 & 1 \end{bmatrix}^{\mathrm{T}}$. Combining the two systems, (11.145) and (11.146), together gives the coupled ("augmented") system

$$\begin{aligned} \dot{\xi} &= A_{\text{aug}}\xi + b_{\text{aug}}u + g_{\text{aug}}v = \begin{bmatrix} A & gc_n \\ 0 & A_n \end{bmatrix}\xi + \begin{bmatrix} b \\ 0 \end{bmatrix}u + \begin{bmatrix} 0 \\ g_n \end{bmatrix}v \\ y &= c_{\text{aug}}\xi + w = \begin{bmatrix} c & 0 \end{bmatrix}\xi + w \end{aligned} \tag{11.147}$$

where $\xi \stackrel{\Delta}{=} \begin{bmatrix} x & x_n \end{bmatrix}^{\mathrm{T}}$.

To control this system, we will compute state feedback of the form $u = K\hat{\xi}$, where $\hat{\xi}$ is an optimal estimate of the true state ξ (which is clearly not available, because we have modeled the disturbance ζ as the output of a hypothetical plant). First, we will generate an optimal feedback gain K by minimizing the cost criterion

$$J = \int_0^{\infty} \left(x^{\mathrm{T}} Q x + u^{\mathrm{T}} R u \right) dt \tag{11.148}$$

This is the infinite-time LQ criterion for which we must solve the algebraic Riccati equation, (11.135). We will choose weighting matrices as

$$Q = \begin{bmatrix} 1 & 0 & 0 & 0 \\ 0 & 0 & 0 & 0 \\ 0 & 0 & 0 & 0 \\ 0 & 0 & 0 & 0 \end{bmatrix} \qquad R \text{ variable} \tag{11.149}$$

This form of Q penalizes the "position" coordinate of the state vector (note that the original plant is in "phase-variable" form) but not the velocity coordinate. It does not penalize the "noise" states, which are not controllable anyway. We will use two different values for r to see the effect of each. Finding the solution of the Riccati equation, and using (11.55) to form state-feedback matrix K, we now set $u = K\hat{\xi}$, where $\hat{\xi}$ is the output of an estimator.

To construct the estimator, assume the form of the Kalman filter,

$$\dot{\hat{\xi}} = A_{\text{aug}}\hat{\xi} + b_{\text{aug}}u + L(y - c_{\text{aug}}\hat{\xi}) \tag{11.150}$$

and, with the known noise covariances, solve the algebraic Riccati equation necessary to compute the optimal Kalman gain L from (11.133). With this gain, an overall composite system can be constructed as

$$\begin{bmatrix} \dot{\xi} \\ \dot{\hat{\xi}} \end{bmatrix} = \begin{bmatrix} A_{\text{aug}} & b_{\text{aug}}K \\ Lc_{\text{aug}} & A_{\text{aug}} - Lc_{\text{aug}} + b_{\text{aug}}K \end{bmatrix} \begin{bmatrix} \xi \\ \hat{\xi} \end{bmatrix} + \begin{bmatrix} 0 & 0 \\ g_n & 0 \\ \hline 0 & L \end{bmatrix} \begin{bmatrix} v \\ w \end{bmatrix}$$

$$y = \begin{bmatrix} c_{\text{aug}} & 0 \end{bmatrix} \begin{bmatrix} \xi \\ \hat{\xi} \end{bmatrix} + \begin{bmatrix} 0 & 1 \end{bmatrix} \begin{bmatrix} v \\ w \end{bmatrix} \tag{11.151}$$

This system is in a form suitable for simulating in MATLAB, where the noisy input signal is created with the random number generator. The resulting output signal $y(t)$ is shown in Figure 11.10 below for the open-loop case (i.e., setting $K = 0$) and for $R = 10^{-6}$ and 10^{-8}.

`reg(sys,K,L)`

The important feature to note is that as R gets smaller, the rejection of the noisy disturbance improves. This is because by penalizing the inputs *less*, the feedback signal is allowed to grow larger, thus allowing more effective, higher gains [as seen by the inverse effect of R on feedback in (11.55)]. With $R = 10^{-6}$, the controller results in moderate improvement, while with $R = 10^{-8}$, the disturbance has all but disappeared. With a further decrease in the input weighting R, not much further improvement will result because of the "noise floor" of the system.

To get a comparative illustration of the effects of the weighting factor R, we can plot the magnitude of the frequency response function from the disturbance input ζ (Equation 11.145) to the output y. These are shown in Figure 11.11, and the block diagram representing such a system is shown in Figure 11.12. Note that the dynamic compensator that we have implicitly created becomes a notch filter at 4 Hz as R grows larger.

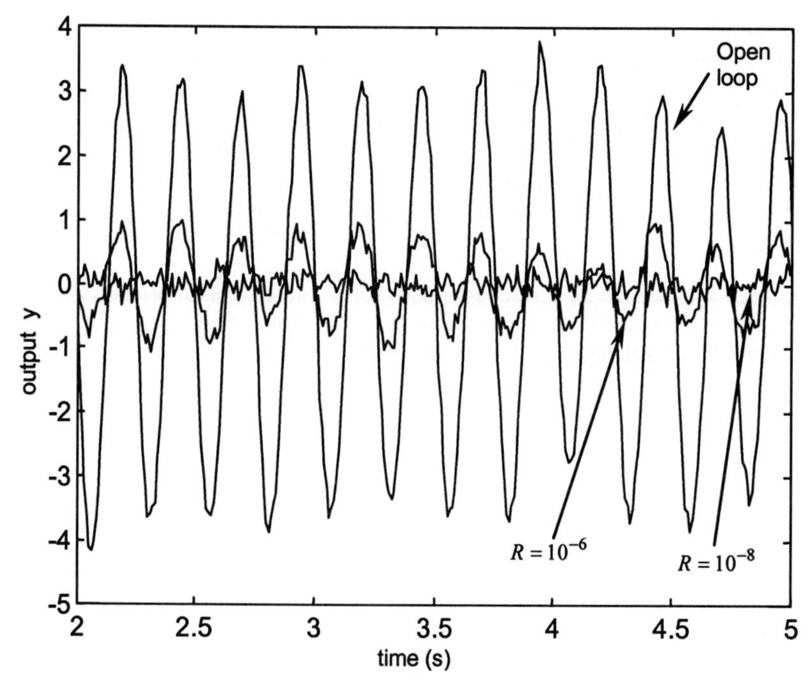

Figure 11.10 Output signal of the system in (11.146), with no feedback (open loop), and with optimal feedback computed using $R = 10^{-6}$ and 10^{-8}.

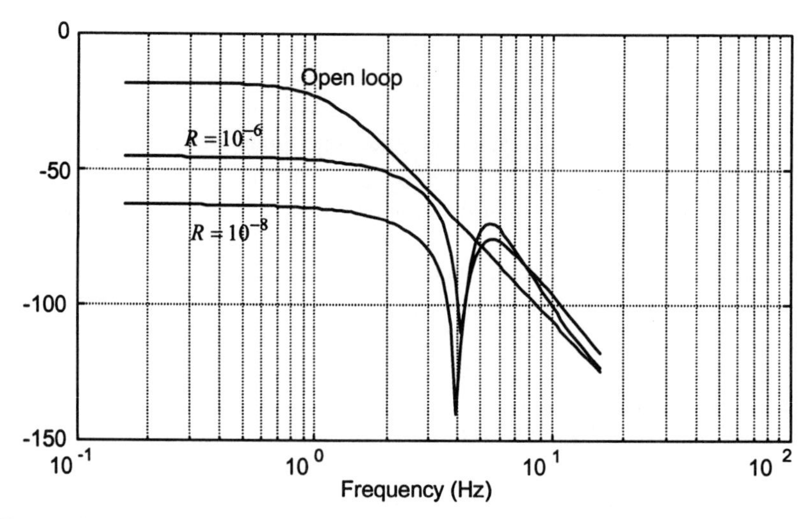

Figure 11.11 Frequency responses for the signal path from disturbance ζ, [Equation (11.145)], to the output y, for the open-loop case (i.e., $K = 0$) and for two different values of R.

This example presents not only an illustration of the use of the LQG controller, but also a method by which "colored" noise may be handled by the Kalman filter. Although we have assumed in the development of the filter that the plant noise was white, colored noise may be modeled with a state space system that is driven by white noise. The "noise system" can then be appended onto the plant, and the Kalman filter may be designed based on the augmented model.

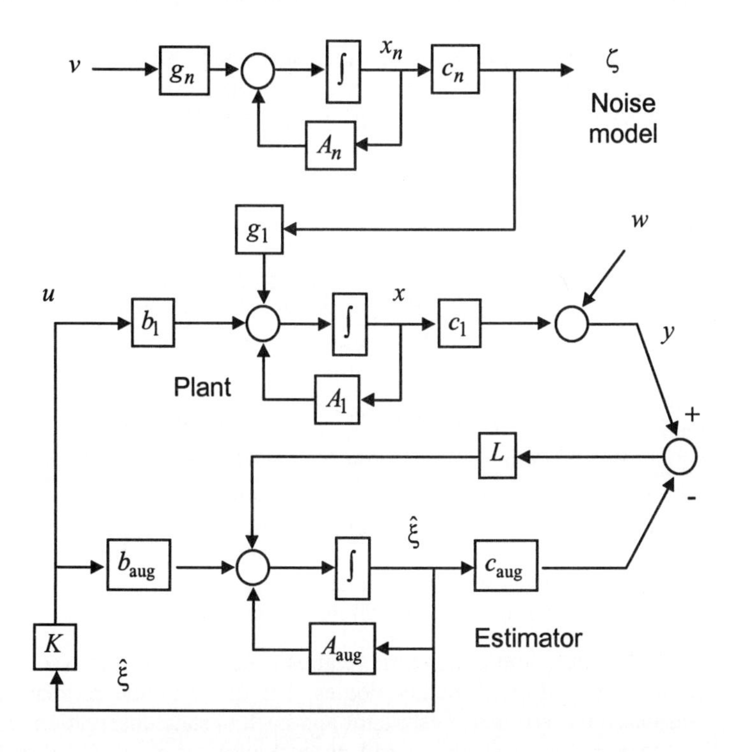

Figure 11.12 Block diagram for the plant and observer, with the noise model included. This diagram can be used to find the frequency response functions shown in Figure 11.11.

11.4 Summary

As we indicated at the beginning of the chapter, the LQG controller is one of the most basic and important tools in the control engineer's toolbox. For state space models of systems, it is the first choice control approach, and in most cases it will provide acceptable performance. Certainly, it can be criticized on the basis of various criteria, such as robustness, but for an accurately known plant it almost always works well.

We have chosen to present the LQG controller here because it also culminates the book well by tying together the concepts of stability, controllability, observability, state feedback, and observers (estimators), as well as introducing practical applications of matrix calculus in the optimization of the various cost and error criteria. While drawing all these skills into a single methodology for control design, the LQG controller illustrates the duality of a linear system even more strongly than in previous chapters. It is quite remarkable that the LQ regulator and the Kalman filter are so similar, both being governed by an optimal gain matrix that depends on the solution of a (difference or differential) matrix Riccati equation. The Riccati equation itself is an interesting mathematical relationship that is still the subject of active research.

To summarize the important topics covered throughout the course of this chapter:

- Discrete- and continuous-time LQ regulators were derived with similar procedures, each based on Bellman's optimality criterion. This criterion (that an optimal journey consists of a first step plus an optimal journey thereafter) is both elegant and powerful. It was used here to generate optimal gains but is also the basis for other optimization techniques, such as dynamic programming [2] and modern machine learning [12]. We should point out at this point, though, that while the equations for optimal control that we derived suffice for many cases, alternative formulations based on variational calculus can provide results with somewhat different, though ultimately equivalent, formats. Different formulations can provide better insight into a broader class of problems than the simple regulators discussed here. Furthermore, there are many kinds of optimal control that we have not discussed.

- Discrete- and continuous-time Kalman filters were derived using significantly different methodologies. The discrete-time Kalman filter was based on the current estimator approach to state observation, while the continuous-time version had no such starting point and therefore had to be derived differently. In fact, although the Kalman filters we present here are indeed accurate and equivalent to the filters derived from any other method, we have completely ignored many of the statistical technicalities that must be considered to prove the filter "optimal." Like the LQ regulator, Kalman filtering is a discipline in itself, and we urge caution in the use of this chapter. We have presented only the most basic formulation of the filter, with the most common model from which to start. Kalman filter behavior is complex and deserves a more thorough treatment if the filters are to be thoroughly appreciated.

- The LQ controller and Kalman filter both depend on a matrix Riccati equation, which is either a difference equation (discrete-time) or a differential equation (continuous-time). These equations will reach an optimizing steady state under certain stabilizability and detectability conditions. Such a steady-state result will provide a suboptimal but often acceptable approximation to the optimal controller (estimator). In practice, it is usually the steady-state solution that is used because of the difficulty in solving the dynamic Riccati equations.
- We have unapologetically given the definition of the LQG controller as the simple sum of an LQ controller and a Kalman filter. This is the practical definition of an LQG controller. However, the justification for this claim was not given here as it requires too lengthy a derivation and too much statistical analysis to justify in a one-chapter treatment. Nevertheless, for a state space model of a plant, if a reader designs an LQ controller and a Kalman filter from the basic techniques presented here, a good LQG controller design will result.

It would seem that this chapter raises as many unresolved questions about controller and observer design as it answers. A good chapter summary always points toward the next unsolved problem, and this one is no exception. However, as this concludes the basic information needed for competency in linear control systems, we simply invite the reader to pursue additional reading from the many excellent references in linear systems, control systems, and optimization theory.

11.5 Problems

11.1 Write the MATLAB program necessary to solve the discrete-time LQ problem given in Example 11.1.

11.2 For the discrete-time system

$$x(k+1)=\begin{bmatrix}0.6 & 0.3\\ 0.1 & 0.8\end{bmatrix}x(k)+\begin{bmatrix}0.03\\ 0.1\end{bmatrix}u(k)$$

Find the control sequence that minimizes the cost function

$$J=\tfrac{1}{2}\sum_{k=1}^{19}\left(x^{\mathrm{T}}(k)\begin{bmatrix}5 & 0\\ 0 & 1\end{bmatrix}x(k)+u^2(k)\right)$$

11.3 For the discrete-time system given as

$$x(k+1) = \begin{bmatrix} 0 & 1 \\ -1 & 1 \end{bmatrix} x(k) + \begin{bmatrix} 0 \\ 1 \end{bmatrix} u(k) \qquad x(0) = \begin{bmatrix} 1 \\ 1 \end{bmatrix}$$

determine the feedback control that minimizes the cost criterion

$$J = \tfrac{1}{2} \sum_{i=1}^{\infty} \left[x_1^2(i) + u^2(i) \right]$$

11.4 Show how Equations (11.58) and (11.59) can be combined to form (11.60).

11.5 Consider the definition of the Kronecker product $\otimes$ for two matrices A and B, see Equation (A.10):

$$A \otimes B = \left[a_{ij} B \right]$$

[This means that if A is $m \times n$, and B is $p \times q$, then $A \otimes B$ is $mp \times nq$ and the $(i, j)^{\text{th}}$ $p \times q$ block of $A \otimes B$ is $a_{ij} B$]. Use the stacking operator also defined in Appendix A, Equation (A.10), and devise a method by which the differential matrix Riccati equation, (11.53), can be transformed into a *vector* differential Riccati equation and thereby numerically solved with MATLAB's ODE routines [9].

11.6 For the system

$$\dot{x} = \begin{bmatrix} 0 & 1 \\ -10 & -7 \end{bmatrix} x + \begin{bmatrix} 0 \\ 1 \end{bmatrix} u$$

Find a control $u(t)$ that minimizes the performance measure

$$J[x(t), t] = 10x^2(5) + \tfrac{1}{2} \int_0^5 \left[5x_1^2(t) + x_2^2(t) + .25u^2(t) \right] dt$$

Then discretize the system using a step size of $T = 0.05$ and find a control that minimizes the performance measure

$$J[x(k), k] = 10x^2(100) + \tfrac{1}{2} \sum_{k=1}^{99} 5x_1^2(k) + x_2^2(k) + 0.25u^2(k)$$

11.7 For a given single-input continuous-time system, show how the modified LQ cost criterion

$$J = \int_0^{\infty} e^{2\lambda}\left[x^{\mathrm{T}}(t)Qx(t) + ru^2(t)\right]dt$$

can be minimized. Show that the resulting closed-loop poles will be to the left of $s = -\lambda$.

11.8 A scalar system has the equations

$$\dot{x} = -x + u \qquad x(0) = 1$$

Find the feedback control that minimizes the *minimum energy* criterion

$$J = x(1) + \int_0^1 u^2(t)\, dt$$

11.9 Point-masses under the influence of a force are described by Newton's law $F = m\ddot{x}$, and are known as "double integrators." Consider the double-integrator system given by:

$$\dot{x} = \begin{bmatrix} 0 & 1 \\ 0 & 0 \end{bmatrix} x + \begin{bmatrix} 0 \\ 1 \end{bmatrix} u$$

$$y = cx$$

Find and simulate the optimal control system that minimizes the cost functional

$$J = \int_0^{\infty} \left(y^{\mathrm{T}} y + u^{\mathrm{T}} u\right)\, dt$$

given the two *c*-matrices $c_1 = \begin{bmatrix} 1 & 0 \end{bmatrix}$. Determine all solutions that exist if $c_2 = \begin{bmatrix} 0 & 1 \end{bmatrix}$.

11.10 Consider the system given by

$$\dot{x} = \begin{bmatrix} -1 & 0 \\ 2 & 0 \end{bmatrix} x + \begin{bmatrix} 1 \\ 0 \end{bmatrix} u$$
$$y = \begin{bmatrix} 0 & 1 \end{bmatrix} x$$

Suppose we desire to use output feedback of the form $u = ky$ to control the system such that the cost criterion

$$J = \int_0^\infty \left(y^2 + u^2\right) dt$$

is minimized. Find the cost J in terms of feedback gain k by solving a Lyapunov equation similar to Equation (11.67). Plot this cost versus k and determine the minimum J and the k at which it occurs. Then plot the root locus of the system and show where the chosen k places the poles.

11.11 A two-link robot arm shown in Figure P11.11 is *underactuated* in the sense that it has a motor at the elbow joint but not at the base [11].

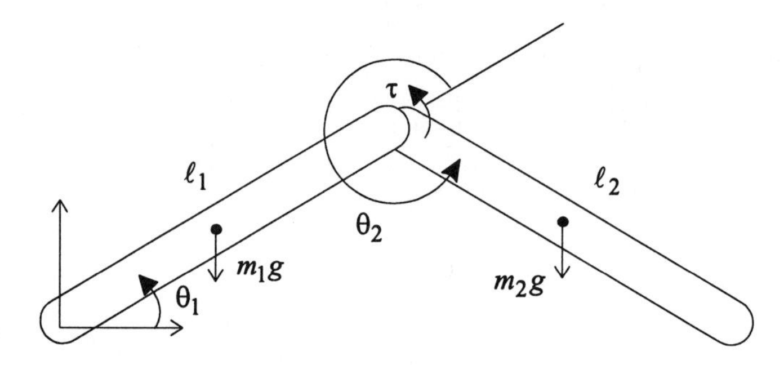

P11.11

The equations of motion for the robot in terms of the two angles θ_1 and θ_2 and the input torque τ are

$$(2.66 + 2\cos\theta_2)\ddot{\theta}_1 + (1.33 + \cos\theta_2)\ddot{\theta}_2$$
$$-(\sin\theta_2)\dot{\theta}_2^2 - 2(\sin\theta_2)\dot{\theta}_1\dot{\theta}_2 + 24.5\cos\theta_1 + 9.8\cos(\theta_1 + \theta_2) = 0$$

$$(1.33 + \cos\theta_2)\ddot{\theta}_1 + 1.33\ddot{\theta}_2 + (\sin\theta_2)\dot{\theta}_1^2 + 9.8\cos(\theta_1 + \theta_2) = \tau$$

Let the state variables be

$$x = [x_1 \quad x_2 \quad x_3 \quad x_4]^T = [\theta_1 - \pi/2 \quad \theta_2 \quad \dot{\theta}_1 \quad \dot{\theta}_2]^T$$

and linearize the equations about the equilibrium point $x_0 = [0 \quad 0 \quad 0 \quad 0]^T$. Then design an LQ controller that minimizes the cost function

$$J = \int_0^\infty \left(x^T Q x + u^T R u\right) dt$$

where $Q = I_{4\times4}$ and $R = 1$. Simulate the controller using initial conditions $x(0) = \begin{bmatrix}0.01 & -0.01 & 0 & 0\end{bmatrix}^T$. The system is judged to have exceeded its linear region if any of the state variables exceed $0.5\,\text{rad}$ (or $0.5\,\text{rad/s}$ for the velocities). How far off can the initial angles be before the transients in the controller exceed the linear bounds?

11.12 Consider the problem of fitting a straight line $y = mt + b$ to a set of points given by the data pairs $\left[t(i), y(i)\right]$ for $i = 1, 2, 3, \ldots,$ obtained sequentially. Model the problem as one of estimating a constant plant

$$x(k+1) = x(k) = \begin{bmatrix} m \\ b \end{bmatrix}$$

with the noisy measurements

$$y(k) = \begin{bmatrix} t(k) & 1 \end{bmatrix} x(k)$$

Formulate a discrete-time Kalman filter to recursively provide the best fit for the line parameters m and b. Simulate the filter with some corrupted test data, and plot the error covariance and the squared errors $\left\|m - \hat{m}\right\|^2$ and $\left\|b - \hat{b}\right\|^2$.

11.13 Consider the scalar Kalman filtering problem presented by the plant

$$\dot{x} = -2x + u + v$$
$$y = x + w$$

where $v(t) \sim (0,1)$ and $w(t) \sim (0, \frac{1}{2})$. Solve the differential matrix Riccati equation *analytically* and use it to simulate the Kalman filter. Use zero input and plot the state variable $x(t)$, estimate $\hat{x}(t)$, and Kalman gain $L(t)$ over the interval $t \in [0,5]$ s.

11.14 In the underactuated robot problem, Problem 11.11, suppose that only the joint angles θ_1 and θ_2 are measured, but that their measurements are given by sensors that also add zero-mean white noise w with covariance $W = 0.01$ to each reading. Suppose further that the actuator at the elbow provides noise v of covariance $V = 0.25$. Design and simulate an LQG controller using the same weighting matrices as in Problem 11.11.

11.15 The continuous-time system

$$\dot{x} = x + u - 2d$$
$$y = 2x + d$$

represents a first-order linear system that is subject to a deterministic disturbance signal d. If $d(t) = d_0 e^{0.1t}$, find a feedback controller and observer system that minimizes the effect of the disturbance on the output.

11.6 References and Further Reading

While the material presented in this chapter is considerably more advanced than the state space fundamentals in previous chapters, it is still only a taste of linear state-feedback control theory. Our derivations may not be the same as the original papers, the most elegant, nor the most theoretically general. Rather, all the presentations given here are simply the most direct results from the viewpoints presented in previous chapters. Alternative derivations can be enlightening and instructive. For example, our use of the current estimator as the motivation for the Kalman filter is quite dissimilar from the optimal projection development given in [6]. Other derivations, discussions, and applications of Kalman filters may be found in [1], [3], and [10].

Likewise, the optimal LQR can be developed from different viewpoints, such as variational calculus, as opposed to the dynamic programming approach [2] we use. Our use of dynamic programming was simply expedient given the contents of the rest of the book. See [4], [7], [8], and [9] for other presentations. The steady-state behavior of the Riccati equations is particularly well addressed in [4] and [5]. The LQG controller, i.e., the LQR in conjunction with the Kalman filter, is specifically addressed in [4] and [10].

[1] Anderson, Brian D. O., and John B. Moore, *Optimal Filtering*, Prentice Hall, 1979.

[2] Bertsekas, Dimitri P., *Dynamic Programming and Stochastic Control*, Academic Press, 1976.

[3] Brown, Robert Grover, and Patrick Y. C. Hwang, *Introduction to Random Signals and Applied Kalman Filtering*, 3rd edition, John Wiley & Sons, 1997.

[4] Casti, John L., *Linear Dynamical Systems*, Academic Press, 1987.

[5] Kailath, Thomas, *Linear Systems*, Prentice Hall, 1980.

[6] Kalman, Rudolph E., "A New Approach to Linear Filtering and Prediction Problems," *Transactions of the ASME, Journal of Basic Engineering*, vol. 82, 1960, pp. 35-45.

[7] Kirk, Donald E., *Optimal Control Theory*, Prentice Hall, 1970.

[8] Kwakernaak, Huibert, and Raphael Sivan, *Linear Optimal Control Systems*, John Wiley & Sons, 1972.

[9] Lewis, Frank L., *Optimal Control*, John Wiley & Sons, 1986.

[10] Lewis, Frank L., *Optimal Estimation: with an Introduction to Stochastic Control Theory*, Wiley-Interscience, 1986.

[11] Spong, Mark W., "The Swing Up Control Problem for the Acrobat," *IEEE Control Systems Magazine*, vol. 15, no. 1, February 1995, pp. 49-55.

[12] Watkins, C.J.C.H., *Learning from Delayed Rewards*, Ph.D. Thesis, University of Cambridge, England, 1989.

A

Mathematical Tables and Identities

Throughout this book, references have been made to certain matrix identities, definitions, and properties. These, as well as other formulae such as selected integrals, are gathered here as a reference for the reader.

Basic Matrix Operations

An $n \times m$ matrix A consists of the nm elements a_{ij}, $i = 1,\ldots,n$ and $j = 1,\ldots,m$. To represent a matrix by the array of its entries we use the notation:

$$A = \left[a_{ij}\right] \tag{A.1}$$

The sum of two matrices gives a third matrix of the same dimension: $A + B = C$, where

$$c_{ij} = a_{ij} + b_{ij} \tag{A.2}$$

The product of two matrices AB is only allowed if the number of columns of A is the same as the number of rows of B. In that case, $A_{n\times m}B_{m\times p} = C_{n\times p}$, where

$$c_{ij} = \sum_{k=1}^{m} a_{ik} b_{kj} \tag{A.3}$$

This multiplication operation may be performed block-by-block such that if matrices A and B are partitioned into compatibly dimensioned blocks, i.e., if

$$A = \begin{bmatrix} A_{11} & A_{12} \\ A_{21} & A_{22} \end{bmatrix} \quad \text{and} \quad B = \begin{bmatrix} B_{11} & B_{12} \\ B_{21} & B_{22} \end{bmatrix}$$

then

$$AB = \begin{bmatrix} A_{11}B_{11} + A_{12}B_{21} & A_{11}B_{12} + A_{12}B_{22} \\ A_{21}B_{11} + A_{22}B_{21} & A_{21}B_{12} + A_{22}B_{22} \end{bmatrix} \tag{A.4}$$

Except for special examples, matrix multiplication is not commutative, i.e.,

$$AB \neq BA \tag{A.5}$$

but other elementary properties of matrix algebra do hold:

1. Associativity over addition and multiplication:

$$\begin{aligned} (AB)C &= A(BC) \\ (A+B)+C &= A+(B+C) \end{aligned} \tag{A.6}$$

2. Distributivity of addition and multiplication:

$$\begin{aligned} A(B+C) &= AB+AC \\ (A+B)C &= AC+BC \end{aligned} \tag{A.7}$$

3. Multiplication of a matrix by a scalar is performed in an element-wise fashion: if $B = \alpha A$, then

$$b_{ij} = \alpha a_{ij} \tag{A.8}$$

The *Kronecker* product of two matrices *A* and *B*:

$$A \otimes B = \left[a_{ij} B \right] \tag{A.9}$$

This means that if A is $m \times n$ and B is $p \times q$, then $A \otimes B$ is $mp \times nq$ and the $(i, j)^{\text{th}}$ $p \times q$ block of $A \otimes B$ is $a_{ij}B$.

The *stacking* of a matrix is reshaping it in order to make it a single column. If we denote the jth column of matrix A as $a_{.j}$, then the stacking operator can be written as:

$$s(A) \triangleq \begin{bmatrix} a_{\cdot 1} \\ a_{\cdot 2} \\ \vdots \\ a_{\cdot n} \end{bmatrix} \qquad \text{(A.10)}$$

Transposes

The *transpose* of a matrix A is denoted A^{T}. If $B = A^{\mathrm{T}}$ and $A = \left[a_{ij}\right]$, then

$$B = \left[a_{ji}\right] \quad \text{or} \quad b_{ij} = a_{ji} \qquad \text{(A.11)}$$

The complex-conjugate transpose of a matrix is denoted A^* or sometimes A^{H}. If $B = A^*$, then

$$b_{ij} = \bar{a}_{ji} \qquad \text{(A.12)}$$

where the overbar indicates complex-conjugation.

For matrix products and sums,

$$(A_1 + A_2 + \cdots + A_n)^{\mathrm{T}} = (A_1^{\mathrm{T}} + A_2^{\mathrm{T}} + \cdots + A_n^{\mathrm{T}}) \qquad \text{(A.13)}$$

$$(A_1 A_2 \cdots A_n)^{\mathrm{T}} = (A_n^{\mathrm{T}} \cdots A_2^{\mathrm{T}} A_1^{\mathrm{T}}) \qquad \text{(A.14)}$$

A real-valued matrix A is said to be *symmetric* (sometimes called *self-adjoint*) if $A = A^{\mathrm{T}}$ and *skew-symmetric* if $A = -A^{\mathrm{T}}$. A complex-valued matrix A is said to be *hermitian* if $A = A^*$. It is *skew-hermitian* if $A = -A^*$. Regardless of the symmetry of matrix A, the products AA^{T} and $A^{\mathrm{T}}A$ are both hermitian.

Determinants

The determinant of a matrix A is defined only if it is square and is denoted $|A|$ or $\det(A)$. For any square matrix A,

$$\left|A^{\mathrm{T}}\right| = |A| \qquad \text{(A.15)}$$

and for products of square matrices,

$$|AB| = |A||B| \tag{A.16}$$

The elementary properties of determinants are:

1. If any two rows or columns of a matrix are interchanged, its determinant changes sign.
2. If any single row or column of a matrix is multiplied by a scalar α, the determinant of the result is also multiplied by α.
3. If the entire matrix is multiplied by α, then

$$|\alpha A| = \alpha|A| \tag{A.17}$$

4. If A is an $n \times m$ matrix and B is an $m \times n$ matrix, then

$$|I_{n\times n} - AB| = |I_{m\times m} - BA| \tag{A.18}$$

The $(i, j)^{\text{th}}$ *cofactor* of matrix A is denoted $\text{cof}_{ij}(A)$ and is the determinant obtained by eliminating the ith row and j^{th} column from A. It is the *negative* of such a determinant if $i + j$ is an odd number. The determinant of $n \times n$ matrix A can be defined in terms of cofactors:

$$|A| = a_{i1}\text{cof}_{i1}(A) + a_{i2}\text{cof}_{i2}(A) + \cdots + a_{in}\text{cof}_{in}(A) \tag{A.19}$$

or

$$|A| = a_{1j}\text{cof}_{1j}(A) + a_{2j}\text{cof}_{2j}(A) + \cdots + a_{nj}\text{cof}_{nj}(A) \tag{A.20}$$

for *any* row i or column j.

Inverses

The *inverse* of square matrix A is denoted A^{-1} and is a matrix such that

$$AA^{-1} = A^{-1}A = I \tag{A.21}$$

where I is the identity matrix. The inverse of A exists if and only if $|A| \neq 0$. The inverse may be computed from the cofactors and determinant. If $B = A^{-1}$, then

$$b_{ij} = \frac{\text{cof}_{ij}^{\text{T}}(A)}{|A|} \tag{A.22}$$

Some properties of matrix inverses are:

$$
\begin{aligned}
(A_1 A_2 \cdots A_n)^{-1} &= (A_n^{-1} \cdots A_2^{-1} A_1^{-1}) \\
(A^{-1})^{\mathrm{T}} &= (A^{-\mathrm{T}})^{-1} \triangleq A^{-\mathrm{T}} \\
(\alpha A)^{-1} &= \frac{1}{\alpha} A^{-1} \\
\left|A^{-1}\right| &= \frac{1}{|A|}
\end{aligned}
\tag{A.23}
$$

When matrices are written in block form, further identities result. Let

$$
A = \begin{bmatrix} A_{11} & A_{12} \\ A_{21} & A_{22} \end{bmatrix} \quad \text{and} \quad A^{-1} = B = \begin{bmatrix} B_{11} & B_{12} \\ B_{21} & B_{22} \end{bmatrix} \tag{A.24}
$$

Then $AB = I$, giving

$$
A_{11} B_{11} + A_{12} B_{21} = I \tag{A.25}
$$

$$
A_{11} B_{12} + A_{12} B_{22} = 0 \tag{A.26}
$$

$$
A_{21} B_{11} + A_{22} B_{21} = 0 \tag{A.27}
$$

$$
A_{21} B_{12} + A_{22} B_{22} = I \tag{A.28}
$$

If it is known that $|A_{11}| \neq 0$ and $|A_{22}| \neq 0$, then these four equations can be solved to give

$$
B_{11} = \left(A_{11} - A_{12} A_{22}^{-1} A_{21}\right)^{-1} \tag{A.29}
$$

$$
B_{22} = \left(A_{22} - A_{21} A_{11}^{-1} A_{12}\right)^{-1} \tag{A.30}
$$

$$
B_{12} = -A_{11}^{-1} A_{12} \left(A_{22} - A_{21} A_{11}^{-1} A_{12}\right)^{-1} \tag{A.31}
$$

$$
B_{21} = -A_{22}^{-1} A_{21} \left(A_{11} - A_{12} A_{22}^{-1} A_{21}\right)^{-1} \tag{A.32}
$$

The first two of these, (A.29) and (A.30), are known as the *Schur* complements of A_{11} and A_{22}, respectively. If $|A_{22}| = 0$, then the solutions of (A.29) through (A.32) change to

$$B_{11} = A_{11}^{-1} + A_{11}^{-1}A_{12}\left(A_{22} - A_{21}A_{11}^{-1}A_{12}\right)^{-1}A_{21}A_{11}^{-1} \tag{A.33}$$

$$B_{22} = \left(A_{22} - A_{21}A_{11}^{-1}A_{12}\right)^{-1} \tag{A.34}$$

$$B_{12} = -A_{11}^{-1}A_{12}\left(A_{22} - A_{21}A_{11}^{-1}A_{12}\right)^{-1} \tag{A.35}$$

$$B_{21} = -\left(A_{22} - A_{21}A_{11}^{-1}A_{12}\right)^{-1}A_{21}A_{11}^{-1} \tag{A.36}$$

A similar result can be obtained if $|A_{22}| \neq 0$ and $|A_{11}| = 0$.

Comparing (A.29) and (A.33) gives

$$\left(A_{11} - A_{12}A_{22}^{-1}A_{21}\right)^{-1} = A_{11}^{-1} + A_{11}^{-1}A_{12}\left(A_{22} - A_{21}A_{11}^{-1}A_{12}\right)^{-1}A_{21}A_{11}^{-1} \tag{A.37}$$

or, more generally,

$$(A + BCD)^{-1} = A^{-1} - A^{-1}B\left(C^{-1} + DA^{-1}B\right)^{-1}DA^{-1} \tag{A.38}$$

This is known as the *matrix inversion lemma*. It can be used for any set of dimensionally compatible matrices, and is also useful for reducing the size of certain matrix inversion operations:

$$(I_{n\times n} + BD)^{-1} = I_{p\times p} - B\left(I_{p\times p} + DB\right)^{-1}D \tag{A.39}$$

where B is $n \times p$ and D is $p \times n$. Alternatively, it is used for deriving a recursive matrix inversion formula useful in recursive estimation:

$$(A + xy^{\mathrm{T}})^{-1} = A^{-1} - \frac{\left(A^{-1}x\right)\left(y^{\mathrm{T}}A^{-1}\right)}{1 + y^{\mathrm{T}}A^{-1}x} \tag{A.40}$$

A similar formula applies to determinants:

$$|A| = \begin{vmatrix} A_{11} & A_{12} \\ A_{21} & A_{22} \end{vmatrix} = |A_{11}|\left|A_{22} - A_{21}A_{11}^{-1}A_{12}\right| = |A_{22}|\left|A_{11} - A_{12}A_{22}^{-1}A_{21}\right| \tag{A.41}$$

The *resolvent* of a matrix A is $(sI - A)^{-1}$. The *resolvent identity* used in the derivation of the Bass-Gura formula of Chapter 10 is

$$(sI - A)^{-1} = \frac{\left[s^{n-1}I + (A + a_{n-1}I)s^{n-2} + \cdots + (A^{n-1} + a_{n-1}A^{n-2} + \cdots + a_0 I)\right]}{\left|(sI - A)\right|} \tag{A.42}$$

where the coefficients a_i come from the characteristic polynomial:

$$|sI - A| \stackrel{\Delta}{=} s^n + a_{n-1}s^{n-1} + \cdots + a_1 s + a_0 \tag{A.43}$$

The *generalized inverse* (or *pseudoinverse*) of a matrix A is a matrix denoted A^+ such that

$$AA^+A = A^+ \tag{A.44}$$

The *Moore-Penrose* generalized inverse of matrix A is a generalized inverse with the additional properties

$$A^+AA^+ = A \tag{A.45}$$

$$\left(A^+A\right)^* = A^+A \tag{A.46}$$

$$\left(AA^+\right)^* = AA^+ \tag{A.47}$$

Trace

The *trace* of an $n \times n$ matrix A is defined as the sum of all elements on the diagonal:

$$tr(A) = \sum_{i=1}^{n} a_{ii} \tag{A.48}$$

For compatible nonsquare matrices A and B,

$$tr(AB) = tr(BA) \tag{A.49}$$

or for three compatible matrices A, B, and C,

$$tr(ABC) = tr(CAB) = tr(BCA) \tag{A.50}$$

Special Matrices and Matrix Forms

A square matrix A is *idempotent* if

$$A^2 = AA = I \tag{A.51}$$

A square matrix A is *nilpotent* if there exists a positive integer r, called the *index of* A, such that

$$A^r = 0 \tag{A.52}$$

A *Vandermonde* matrix is a matrix of the form

$$A = \begin{bmatrix} 1 & a_1 & a_1^2 & \cdots & a_1^{n-1} \\ 1 & a_2 & a_2^2 & \cdots & a_2^{n-1} \\ 1 & a_3 & a_3^2 & \cdots & a_3^{n-1} \\ \vdots & \vdots & \vdots & \ddots & \vdots \\ 1 & a_n & a_n^2 & \cdots & a_n^{n-1} \end{bmatrix} \tag{A.53}$$

Matrices whose transposes are of this form are also called Vandermonde.

A *Toeplitz* matrix is a matrix of the form

$$A = \begin{bmatrix} a & b & c & \cdots & x \\ \beta & a & b & c & \vdots \\ \gamma & \beta & a & b & \ddots \\ \vdots & \gamma & \beta & a & \ddots \\ \chi & \cdots & \ddots & \ddots & \ddots \end{bmatrix} \tag{A.54}$$

One way to describe such a matrix is that entry a_{ij} is the same for every element for which $i - j$ is the same.

A matrix is said to be in *Hankel* form if

$$A = \begin{bmatrix} a & b & c & \cdots \\ b & c & \cdot^{\cdot^{\cdot}} & \\ c & \cdot^{\cdot^{\cdot}} & \ddots & \\ \vdots & & & \end{bmatrix} \tag{A.55}$$

A matrix A is said to be upper (or lower) *triangular* if it has the form

$$A = \begin{bmatrix} a_{11} & a_{12} & \cdots & a_{1n} \\ 0 & a_{22} & \ddots & a_{2n} \\ \vdots & \ddots & \ddots & \vdots \\ 0 & \cdots & 0 & a_{nn} \end{bmatrix} \quad \text{or} \quad A = \begin{bmatrix} a_{11} & 0 & \cdots & 0 \\ a_{21} & a_{22} & \ddots & \vdots \\ \vdots & \ddots & \ddots & 0 \\ a_{n1} & a_{n2} & \cdots & a_{nn} \end{bmatrix} \tag{A.56}$$

A matrix A is said to be *diagonal* if it has the form

$$A = \begin{bmatrix} a_{11} & 0 & \cdots & 0 \\ 0 & a_{22} & \ddots & \vdots \\ \vdots & \ddots & \ddots & 0 \\ 0 & \cdots & 0 & a_{nn} \end{bmatrix} \tag{A.57}$$

A matrix A is said to be *orthogonal* if $A^T A$ is diagonal. In that case, the inner product of any two different columns is zero, i.e., the columns are pair-wise orthogonal.

A matrix A is said to be *orthonormal* if $A^{-1} = A^T$.

Matrices in *companion form* are discussed in the text.

Matrix Decompositions

Matrices may be decomposed into factors in several ways, each convenient for different purposes. We will give the notation here but will not present the methods by which such decompositions may be computed.

A *Cholesky* decomposition of matrix A is a factorization

$$A = B^* B \tag{A.58}$$

where B is a nonsingular upper triangular matrix.

A *QR factorization* of matrix A is a factorization

$$A = QR \tag{A.59}$$

such that Q is orthonormal and R is upper triangular. If matrix A is any $n \times m$ and has rank m, then the decomposition will be of the form

$$A = Q \begin{bmatrix} R \\ \hline 0 \end{bmatrix} \tag{A.60}$$

where R is again upper triangular. QR factorizations may be used to determine the eigenvalues or rank of a matrix.

An *LU decomposition* of matrix A is a factorization

$$A = LU \tag{A.61}$$

such that matrix L is lower triangular and matrix U is upper triangular.

Singular value decompositions are discussed in the text.

Matrix Calculus

For matrices whose elements are functions of a scalar parameter t, matrix differentiation and integration is performed in an element-wise fashion:

$$\frac{d}{dt}A(t) = \left[\frac{d}{dt}a_{ij}(t)\right] \tag{A.62}$$

and

$$\int A(t) = \left[\int a_{ij}(t)\,dt\right] \tag{A.63}$$

In the following definitions, $x = \begin{bmatrix} x_1 & x_2 & \cdots & x_n \end{bmatrix}^{\mathrm{T}}$ and $y = \begin{bmatrix} y_1 & y_2 & \cdots & y_n \end{bmatrix}^{\mathrm{T}}$ are vectors of length n, $g(x)$ is a scalar function of such a vector, and

$$f(x) = \begin{bmatrix} f_1(x) \\ f_2(x) \\ \vdots \\ f_m(x) \end{bmatrix}$$

is a vector function of length m.

The differential of g is

$$
\begin{aligned}
dg &= \frac{\partial g}{\partial x_1}dx_1 + \frac{\partial g}{\partial x_2}dx_2 + \cdots + \frac{\partial g}{\partial x_n}dx_n \\
&= \begin{bmatrix} \frac{\partial g}{\partial x_1} & \frac{\partial g}{\partial x_2} & \cdots & \frac{\partial g}{\partial x_n} \end{bmatrix} \begin{bmatrix} dx_1 \\ dx_2 \\ \vdots \\ dx_n \end{bmatrix}
\end{aligned} \tag{A.64}
$$

This implies the vector derivative

$$
\frac{dg}{dx} = \begin{bmatrix} \frac{\partial g}{\partial x_1} & \frac{\partial g}{\partial x_2} & \cdots & \frac{\partial g}{\partial x_n} \end{bmatrix} \tag{A.65}
$$

Note that this is properly written as a row vector rather than as a column vector as often presented. We define the *gradient* of g as the column-vector counterpart:

$$
\mathrm{grad}(g) = \nabla g = \frac{d^{\mathrm{T}} g}{dx} = \begin{bmatrix} \frac{\partial g}{\partial x_1} \\ \frac{\partial g}{\partial x_2} \\ \vdots \\ \frac{\partial g}{\partial x_n} \end{bmatrix} \tag{A.66}
$$

The derivative of a vector with respect to another vector is then consistently defined as the *jacobian* matrix

$$
\frac{df}{dx} = \left[\frac{\partial f_i}{\partial x_j} \right] = \begin{bmatrix} \frac{\partial f_1}{\partial x_1} & \frac{\partial f_1}{\partial x_2} & \cdots & \frac{\partial f_1}{\partial x_n} \\ \frac{\partial f_2}{\partial x_1} & \frac{\partial f_2}{\partial x_2} & \cdots & \frac{\partial f_2}{\partial x_n} \\ \vdots & \vdots & \ddots & \vdots \\ \frac{\partial f_m}{\partial x_1} & \frac{\partial f_m}{\partial x_2} & \cdots & \frac{\partial f_m}{\partial x_n} \end{bmatrix} \tag{A.67}
$$

which of course is an $m \times n$ matrix. Note that this is consistent with the definition of the derivative of a scalar function with respect to a vector, Equation (A.65).

If A is an $n \times n$ matrix

$$\frac{\partial}{\partial y}(Ay) = A \tag{A.68}$$

$$\frac{\partial}{\partial y}\left(x^{\mathrm{T}} Ay\right) = x^{\mathrm{T}} A \tag{A.69}$$

$$\frac{\partial}{\partial x}\left(x^{\mathrm{T}} Ay\right) = (Ay)^{\mathrm{T}} = y^{\mathrm{T}} A^{\mathrm{T}} \tag{A.70}$$

$$\frac{\partial}{\partial x^{\mathrm{T}}}\left(x^{\mathrm{T}} Ay\right) = \frac{\partial^{\mathrm{T}}}{\partial x}\left(x^{\mathrm{T}} Ay\right) = A^{\mathrm{T}} y \tag{A.71}$$

As a result of these,

$$\frac{\partial}{\partial x}\left(x^{\mathrm{T}} Ax\right) = x^{\mathrm{T}} A + x^{\mathrm{T}} A^{\mathrm{T}} = x^{\mathrm{T}}(A + A^{\mathrm{T}}) \tag{A.72}$$

or

$$\frac{\partial^{\mathrm{T}}}{\partial x}\left(x^{\mathrm{T}} Ax\right) = \nabla\left(x^{\mathrm{T}} Ax\right) = (A + A^{\mathrm{T}})x \tag{A.73}$$

A result of this is that if A is symmetric,

$$\frac{\partial^{\mathrm{T}}}{\partial x}\left(x^{\mathrm{T}} Ax\right) = \nabla\left(x^{\mathrm{T}} Ax\right) = 2Ax \tag{A.74}$$

When performing matrix differentiation, the chain rule and product rule apply, but we must always be careful to preserve the proper order of matrix multiplication. If A and B are compatible matrices and x, y, and z are vectors, then

$$\begin{aligned} \frac{d}{dt}(A(t)B(t)) &= \frac{dA(t)}{dt}B(t) + A(t)\frac{dB(t)}{dt} \\ &\neq \frac{dA(t)}{dt}B(t) + \frac{dB(t)}{dt}A(t) \end{aligned} \tag{A.75}$$

$$\begin{aligned} \frac{d[x(y(z)]}{dt} &= \frac{\partial x}{\partial y}\frac{\partial y}{\partial z} \\ &\neq \frac{\partial y}{\partial z}\frac{\partial x}{\partial y} \end{aligned} \tag{A.76}$$

The differentiation of trace operators provides several useful results for optimizing the squared-error criterion and error covariances:

$$\frac{\partial}{\partial A} tr(A) = I \tag{A.77}$$

$$\frac{\partial}{\partial A} tr(CAB) = (CB)^{\mathrm{T}} = B^{\mathrm{T}} C^{\mathrm{T}} \tag{A.78}$$

$$\frac{\partial}{\partial A} tr\left(ABA^{\mathrm{T}}\right) = AB + AB^{\mathrm{T}} \tag{A.79}$$

Integrals

Sometimes integral tables are helpful in the computation of vector inner products and norms, and in the projection operations such as those performed during the Gram-Schmidt orthogonalization process of Chapter 3.

$$\int \sin ax \, dx = -\frac{1}{a} \cos ax \tag{A.80}$$

$$\int \sin^2 ax \, dx = \frac{1}{2} x - \frac{1}{4a} \sin 2ax \tag{A.81}$$

$$\int \sin^n ax \, dx = -\frac{\sin^{n-1} ax \cos ax}{na} + \frac{n-1}{n} \int \sin^{n-2} ax \, dx \tag{A.82}$$

$$\int \cos ax \, dx = \frac{1}{a} \sin ax \tag{A.83}$$

$$\int \cos^2 ax \, dx = \frac{1}{2} x + \frac{1}{4a} \sin 2ax \tag{A.84}$$

$$\int \cos^n ax \, dx = \frac{\sin ax \cos^{n-1} ax}{na} + \frac{n-1}{n} \int \cos^{n-2} ax \, dx \tag{A.85}$$

$$\int (\sin ax)(\cos ax) \, dx = \frac{1}{2a} \sin^2 ax \tag{A.86}$$

$$\int x e^{ax} \, dx = \frac{e^{ax}}{a^2} (ax - 1) \tag{A.87}$$

$$\int x^n e^{ax}\,dx = \frac{x^n e^{ax}}{a} - \frac{n}{a}\int x^{n-1} e^{ax}\,dx \tag{A.88}$$

$$\int_0^\infty e^{-x^2}\,dx = \frac{\sqrt{\pi}}{2} \tag{A.89}$$

$$\int_0^\infty x e^{-x^2}\,dx = \frac{1}{2} \tag{A.90}$$

$$\int_0^\infty x^2 e^{-x^2}\,dx = \frac{\sqrt{\pi}}{4} \tag{A.91}$$

$$\int_0^\infty x^{2n} e^{-x^2}\,dx = \frac{1\cdot 3\cdot 5\cdots(2n-1)}{2^{n+1}}\sqrt{\pi} \tag{A.92}$$

$$\int_0^\infty x^{2n+1} e^{-x^2}\,dx = \frac{n!}{2} \tag{A.93}$$

$$\frac{d}{dt}\left[\int_{g(t)}^{h(t)} f(x,t)\,dx\right] = \int_{g(t)}^{h(t)} \frac{\partial}{\partial t} f(x,t)\,dx + f(h(t),t)\frac{dh(t)}{dt} - f(g(t),t)\frac{dg(t)}{dt} \tag{A.94}$$

B

MATLAB® Command Summaries

MATLAB is a software package that is nearly a de facto standard for linear systems analysis. It was written *by* linear systems researchers and practitioners *for* linear systems researchers and practitioners. It therefore has included in it many specialized commands that general-purpose mathematical packages do not contain.

Throughout this book, certain terms have been marked with an *M* superscript (such as *rank*^{M}). These are terms that describe linear algebraic or control systems operations and procedures that are easily performed with the aid of MATLAB. Given in this appendix are the "help" texts for the commands that are indexed in the book. These descriptions are the responses that a user will receive upon typing `help <command>` at the command prompt.

MATLAB is a registered trademark of The MathWorks, Inc. For further information, users can contact

The Mathworks, Inc.
24 Prime Park Way
Natick, MA 01760-1500
Phone: (508) 647-7000
Fax: (508) 647-7001
Email: info@mathworks.com

There are also a great many textbooks and other references on the use of MATLAB in linear systems analysis and control system design (as well as in other specialties in science and engineering). Some of these references are included at the end of this appendix.

Command Reference

```
ACKER  Pole placement gain selection using Ackermann's formula.
   K = ACKER(A,B,P)  calculates the feedback gain matrix K such that
   the single input system
           .
           x = Ax + Bu

   with a feedback law of  u = -Kx  has closed loop poles at the
   values specified in vector P, i.e.,  P = eig(A-B*K).

   Note: This algorithm uses Ackermann's formula.  This method
   is NOT numerically reliable and starts to break down rapidly
   for problems of order greater than 10, or for weakly controllable
   systems.  A warning message is printed if the nonzero closed-loop
   poles are greater than 10 from the desired locations specified
   in P.

   See also  PLACE.

   Algorithm is from page 201 of:
   Kailath, T.  "Linear Systems", Prentice-Hall, 1980.

   Copyright (c) 1986-96 by The MathWorks, Inc.


BALREAL  Gramian-based balancing of state-space realizations.
   SYSb = BALREAL(SYS) returns a balanced state-space
   realization of the reachable, observable, stable system SYS.

   [SYSb,G,T,Ti] = BALREAL(SYS) also returns a vector G containing
   the diagonal of the Gramian of the balanced realization.  The
   matrices T is the state transformation xb = Tx used to convert SYS
   to SYSb, and Ti is its inverse.

   If the system is normalized properly, small elements in the balanced
   Gramian G indicate states that can be removed to reduce the model
   to lower order.

   See also  MODRED, GRAM, SSBAL.
   J.N. Little 3-6-86
   Revised 12-30-88
       Alan J. Laub 10-30-94
       P. Gahinet 6-27-96
   Copyright (c) 1986-96 by The MathWorks, Inc.
   $Revision: 1.2 $  $Date: 1996/10/18 15:16:05 $

   Reference:
   [1] Laub, A.J., M.T. Heath, C.C. Paige, and R.C. Ward,
       ``Computation of System Balancing Transformations and Other
       Applications of Simultaneous Diagonalization Algorithms,''
       IEEE Trans. Automatic Control, AC-32(1987), 115--122.
```

```
C2D  Conversion of continuous-time systems to discrete time.
   SYSD = C2D(SYSC,TS,METHOD)  converts the continuous system SYSC
   to a discrete-time system SYSD with sample time TS.  The string
   METHOD selects the discretization method among the following:
      'zoh'        Zero-order hold on the inputs.
      'foh'        Linear interpolation of inputs (triangle appx.)
      'tustin'     Bilinear (Tustin) approximation.
      'prewarp'    Tustin approximation with frequency prewarping.
                   The critical frequency Wc is specified last as in
                   C2D(SysC,Ts,'prewarp',Wc)
      'matched'    Matched pole-zero method (for SISO systems only).
   The default is 'zoh' when METHOD is omitted.

   If SYS is a delay-free state-space model with initial state x0, the
   matching initial state for its FOH discretization is  x0-G*u(1,:)
   where the matrix G is given by    [SYSD,G] = C2D(SYSC,TS,'foh').

   See also D2C, D2D.

   Other syntax
   C2D  Conversion of state space models from continuous to discrete time.
   [Phi, Gamma] = C2D(A,B,T)  converts the continuous-time system:
          .
          x = Ax + Bu

   to the discrete-time state-space system:

          x[n+1] = Phi * x[n] + Gamma * u[n]

   assuming a zero-order hold on the inputs and sample time T.

   See also D2C.

   J.N. Little 4-21-85
   Copyright (c) 1986-96 by The MathWorks, Inc.
   $Revision: 1.4 $  $Date: 1996/08/28 21:49:08 $


CANON  Canonical state-space realizations.
   CSYS = CANON(SYS,TYPE)  computes a canonical state-space
   realization CSYS of the LTI model SYS.  The string TYPE
   selects the type of canonical form:
     'modal'     :  Modal canonical form where the system
                    eigenvalues appear on the diagonal.
                    The state matrix A must be diagonalizable.
     'companion':   Companion canonical form where the characteristic
                    polynomial appears in the right column.

   [CSYS,T] = CANON(SYS,TYPE)  also returns the state transformation
   matrix T such that z = Tx where z is the new state.  This syntax
   is only meaningful when SYS is a state-space model.

   The modal form is useful for determining the relative controll-
   ability of the system modes.  Note: the companion form is ill-
   conditioned and should be avoided if possible.

   See also  SS2SS, CTRB, and CTRBF.
```

```
   Clay M. Thompson  7-3-90
   Copyright (c) 1986-96 by The MathWorks, Inc.
   $Revision: 1.2 $  $Date: 1996/10/18 15:16:11 $


CARE  Solve continuous-time algebraic Riccati equations.
   [X,L,G,RR] = CARE(A,B,Q,R,S,E)  computes the unique symmetric
   stabilizing solution X of the continuous-time algebraic Riccati
   equation
                                -1
      A'XE + E'XA - (E'XB + S)R  (B'XE + S') + Q = 0

   or, equivalently,
                       -1            -1                   -1
      F'XE + E'XF - E'XBR  B'XE + Q - SR  S' = 0  with  F:=A-BR  S'.

   When omitted, R,S and E are set to the default values R=I, S=0,
   and E=I.  Additional optional outputs include the gain matrix
                -1
          G = R   (B'XE + S') ,

   the vector L of closed-loop eigenvalues (i.e., EIG(A-B*G,E)),
   and the Frobenius norm RR of the relative residual matrix.

   [X,L,G,REPORT] = CARE(A,B,Q,...,'report')  turns off error
   messages and returns a success/failure diagnosis REPORT instead.
   The value of REPORT is
     * -1 if Hamiltonian matrix has eigenvalues too close to jw axis
     * -2 if X=X2/X1 with X1 singular
     * the relative residual RR when CARE succeeds.

   [X1,X2,L,REPORT] = CARE(A,B,Q,...,'implicit')  also turns off
   error messages, but now returns matrices X1,X2 such that X=X2/X1
   and [X1;X2] has orthonormal columns.  REPORT=0 indicates success.

   See also  DARE.

   Author(s): Alan J. Laub (1993)  (laub@ece.ucsb.edu)
              with key contributions by Pascal Gahinet, Cleve Moler,
              and Andy Potvin
   Revised: 94-10-29, 95-07-20, 96-01-09, 8-21-96
   Copyright (c) 1986-96 by Alan J. Laub and The MathWorks, Inc.
   $Revision: 1.9 $

   Assumptions: E is nonsingular, Q=Q', R=R' with R nonsingular, and
                the associated Hamiltonian pencil has no eigenvalues
                on the imaginary axis.
   Sufficient conditions to guarantee the above are stabilizability,
   detectability, and [Q S;S' R] >= 0, with R > 0.

   Reference: W.F. Arnold, III and A.J. Laub, ``Generalized Eigenproblem
          Algorithms and Software for Algebraic Riccati Equations,''
          Proc. IEEE, 72(1984), 1746--1754.
```

```
CDF2RDF Complex diagonal form to real block diagonal form.
   [V,D] = CDF2RDF(V,D) transforms the outputs of EIG(X) (where X is
   real) from complex diagonal form to a real diagonal form.  In
   complex diagonal form, D has complex eigenvalues down the
   diagonal.  In real diagonal form, the complex eigenvalues are in
   2-by-2 blocks on the diagonal.  Complex conjugate eigenvalue pairs
   are assumed to be next to one another.

   See also EIG, RSF2CSF.

   J.N. Little 4-27-87
   Based upon M-file from M. Steinbuch, N.V.KEMA & Delft Univ. of Tech.
   Copyright (c) 1984-96 by The MathWorks, Inc.
   $Revision: 5.3 $  $Date: 1996/05/10 19:15:48 $


CHOL   Cholesky factorization.
   CHOL(X) uses only the diagonal and upper triangle of X.
   The lower triangular is assumed to be the (complex conjugate)
   transpose of the upper.  If X is positive definite, then
   R = CHOL(X) produces an upper triangular R so that R'*R = X.
   If X is not positive definite, an error message is printed.

   [R,p] = CHOL(X), with two output arguments, never produces an
   error message.  If X is positive definite, then p is 0 and R
   is the same as above.   But if X is not positive definite, then
   p is a positive integer.
   When X is full, R is an upper triangular matrix of order q = p-1
   so that R'*R = X(1:q,1:q).
   When X is sparse, R is an upper triangular matrix of size q-by-n
   so that the L-shaped region of the first q rows and first q
   columns of R'*R agree with those of X.

   See also CHOLINC.

   Copyright (c) 1984-96 by The MathWorks, Inc.
   $Revision: 5.4 $  $Date: 1996/09/04 19:20:15 $
   Built-in function.


COMPAN Companion matrix.
   COMPAN(P) is a companion matrix of the polynomial
   with coefficients P.

   Copyright (c) 1984-96 by The MathWorks, Inc.
   $Revision: 5.1 $  $Date: 1996/01/01 23:53:56 $


COND   Condition number with respect to inversion.
   COND(X) returns the 2-norm condition number (the ratio of the
   largest singular value of X to the smallest).  Large condition
   numbers indicate a nearly singular matrix.

   COND(X,P) returns the condition number of X in P-norm:

      NORM(X,P) * NORM(INV(X),P).
```

```
   where P = 1, 2, inf, or 'fro.'

   See also CONDEST, CONDEIG, NORM, NORMEST.

   Copyright (c) 1984-96 by The MathWorks, Inc.
   $Revision: 5.7 $  $Date: 1996/04/04 16:49:41 $
```

CONV Convolution and polynomial multiplication.

```
   C = CONV(A, B) convolves vectors A and B.  The resulting
   vector is length LENGTH(A)+LENGTH(B)-1.
   If A and B are vectors of polynomial coefficients, convolving
   them is equivalent to multiplying the two polynomials.

   See also XCORR, DECONV, CONV2, FILTER, and CONVMTX in the Signal
            Processing Toolbox.

   J.N. Little 4-21-85
   Revised 9-3-87 JNL
   Copyright (c) 1984-96 by The MathWorks, Inc.
   $Revision: 5.4 $  $Date: 1996/10/28 22:46:30 $
```

CROSS Vector cross product.

```
   C = CROSS(A,B) returns the cross product of the vectors
   A and B.  That is, C = A x B.  A and B must be 3 element
   vectors.

   C = CROSS(A,B) returns the cross product of A and B along the
   first dimension of length 3.

   C = CROSS(A,B,DIM), where A and B are N-D arrays, returns the cross
   product of vectors in the dimension DIM of A and B. A and B must
   have the same size, and both SIZE(A,DIM) and SIZE(B,DIM) must be 3.

   Note: Use sum(A.*B) to compute the dot product.

   Clay M. Thompson
   updated 12-21-94, Denise Chen
   Copyright © 1984-96 by The MathWorks, Inc.
   $Revision: 5.9 $  $Date: 1996/04/03 02:53:39 $
```

CTRB   Form the controllability matrix.

```
   CO = CTRB(A,B)  returns the controllability matrix [B AB A^2B ...].

   CO = CTRB(SYS)  returns the controllability matrix of the
   state-space system SYS with realization (A,B,C,D).  This is
   equivalent to  CTRB(sys.a,sys.b)

   See also  CTRBF.

   Copyright (c) 1986-96 by The MathWorks, Inc.
   $Revision: 1.1 $  $Date: 1996/06/28 18:21:21 $
```

CTRBF Controllability staircase form.

```
   [ABAR,BBAR,CBAR,T,K] = CTRBF(A,B,C) returns a decomposition
   into the controllable/uncontrollable subspaces.

   [ABAR,BBAR,CBAR,T,K] = CTRBF(A,B,C,TOL) uses tolerance TOL.

   If Co=CTRB(A,B) has rank r <= n, then there is a similarity
   transformation T such that

   Abar = T * A * T' ,  Bbar = T * B ,  Cbar = C * T'
   and the transformed system has the form

          | Anc    0 |           | 0 |
   Abar =  ----------  ,  Bbar =  ---  ,  Cbar = [Cnc| Cc].
          | A21   Ac |           |Bc |
                                                    -1          -1
   where (Ac,Bc) is controllable, and Cc(sI-Ac)Bc = C(sI-A)B.

   See also  CTRB, OBSVF.

   Author : R.Y. Chiang  3-21-86
   Revised 5-27-86 JNL
   Copyright (c) 1986-96 by The MathWorks, Inc.
   $Revision: 1.2 $  $Date: 1996/08/09 14:23:31 $

   This M-file implements the Staircase Algorithm of Rosenbrock, 1968.
```

DARE Solve discrete-time algebraic Riccati equations.

```
   [X,L,G,RR] = DARE(A,B,Q,R,S,E)  computes the unique symmetric
   stabilizing solution X of the discrete-time algebraic Riccati
   equation
                                             -1
    E'XE = A'XA - (A'XB + S)(B'XB + R)   (A'XB + S)' + Q

   or, equivalently (if R is nonsingular)
                                 -1              -1               -1
    E'XE = F'XF - F'XB(B'XB + R)   B'XF + Q - SR   S'  with  F:=A-BR   S'.

   When omitted, R,S and E are set to the default values R=I, S=0,
   and E=I.  Additional optional outputs include the gain matrix
                              -1
          G = (B'XB + R)   (B'XA + S'),

   the vector L of closed-loop eigenvalues (i.e., EIG(A-B*G,E)),
   and the Frobenius norm RR of the relative residual matrix.

   [X,L,G,REPORT] = DARE(A,B,Q,...,'report')  turns off error
   messages and returns a success/failure diagnosis REPORT instead.
   The value of REPORT is
     * -1 if symplectic pencil has eigenvalues too close to unit circle,
     * -2 if X=X2/X1 with X1 singular
     * the relative residual RR when DARE succeeds.

   [X1,X2,L,REPORT] = DARE(A,B,Q,...,'implicit')  also turns off
   error messages, but now returns matrices X1,X2 such that X=X2/X1
   and [X1;X2] has orthonormal columns.  REPORT=0 indicates success.
```

```
   See also  CARE.

      Author(s): Alan J. Laub (1993)  (laub@ece.ucsb.edu)
           with key contributions by Pascal Gahinet, Cleve Moler,
           and Andy Potvin
   Revised: 94-10-29, 95-07-20, 95-07-24, 96-01-09
       Copyright (c) 1986-96 by Alan J. Laub and The MathWorks, Inc.
       $Revision: 1.8 $

   Assumptions: E is nonsingular, Q=Q', R=R', and the associated
                symplectic pencil has no eigenvalues on the unit circle.
   Sufficient conditions to guarantee the above are stabilizability,
   detectability, and [Q S;S' R] >= 0.

   Reference: W.F. Arnold, III and A.J. Laub, ``Generalized Eigenproblem
              Algorithms and Software for Algebraic Riccati Equations,''
              Proc. IEEE, 72(1984), 1746--1754.
```

DECONV Deconvolution and polynomial division.

```
   [Q,R] = DECONV(B,A) deconvolves vector A out of vector B.  The result
   is returned in vector Q and the remainder in vector R such that
   B = conv(A,Q) + R.

   If A and B are vectors of polynomial coefficients, deconvolution
   is equivalent to polynomial division.  The result of dividing B by
   A is quotient Q and remainder R.

   See also CONV.

   J.N. Little 2-6-86
   Copyright (c) 1984-96 by The MathWorks, Inc.
   $Revision: 5.3 $  $Date: 1996/01/01 21:44:52 $
```

DET Determinant.

```
   DET(X) is the determinant of the square matrix X.

   Use COND instead of DET to test for matrix singularity.

   See also COND.

   Copyright (c) 1984-96 by The MathWorks, Inc.
   $Revision: 5.2 $  $Date: 1996/05/10 16:00:06 $
   Built-in function.
```

DIAG Diagonal matrices and diagonals of a matrix.

```
   DIAG(V,K) when V is a vector with N components is a square matrix
   of order N+ABS(K) with the elements of V on the K-th diagonal. K = 0
   is the main diagonal, K > 0 is above the main diagonal and K < 0
   is below the main diagonal.

   DIAG(V) is the same as DIAG(V,0) and puts V on the main diagonal.

   DIAG(X,K) when X is a matrix is a column vector formed from
```

```
   the elements of the K-th diagonal of X.

   DIAG(X) is the main diagonal of X. DIAG(DIAG(X)) is a diagonal matrix.

   Example
      m = 5;
      diag(-m:m) + diag(ones(2*m,1),1) + diag(ones(2*m,1),-1)
   produces a tridiagonal matrix of order 2*m+1.

   See also SPDIAGS, TRIU, TRIL.

   Copyright (c) 1984-96 by The MathWorks, Inc.
   $Revision: 5.2 $  $Date: 1996/03/29 20:02:29 $
   Built-in function.
```

DLQR Linear-quadratic regulator design for discrete-time systems.

```
   [K,S,E] = DLQR(A,B,Q,R,N)  calculates the optimal gain matrix K
   such that the state-feedback law  u[n] = -Kx[n]  minimizes the
   cost function

         J = Sum {x'Qx + u'Ru + 2*x'Nu}

   subject to the state dynamics   x[n+1] = Ax[n] + Bu[n].

   The matrix N is set to zero when omitted.  Also returned are the
   Riccati equation solution S and the closed-loop eigenvalues E:
                                          -1
    A'SA - S - (A'SB+N)(R+B'SB) (B'SA+N') + Q = 0,    E = EIG(A-B*K).


   See also  DLQRY, LQRD, LQGREG, and DARE.

   Author(s): J.N. Little 4-21-85
   Revised     P. Gahinet  7-24-96
   Copyright (c) 1986-96 by The MathWorks, Inc.
   $Revision: 1.4 $  $Date: 1996/11/05 16:13:12 $
```

DLYAP Discrete Lyapunov equation solver.

```
    X = DLYAP(A,Q) solves the discrete Lyapunov equation:

        A*X*A' - X = Q

    See also  LYAP.

    J.N. Little 2-1-86, AFP 7-28-94
    Copyright (c) 1986-96 by The MathWorks, Inc.
    $Revision: 1.2 $  $Date: 1996/10/18 15:16:26 $

 How to prove the following conversion is true.  Re: show that if
         (1) Ad X Ad' + Cd = X                 Discrete lyaponuv eqn
         (2) Ac = inv(Ad + I) (Ad - I)         From dlyap
         (3) Cc = (I - Ac) Cd (I - Ac')/2      From dlyap
 Then
         (4) Ac X + X Ac' + Cc = 0             Continuous lyapunov
```

```
Step 1) Substitute (2) into (3)
        Use identity 2*inv(M+I) = I - inv(M+I)*(M-I)
                                = I - (M-I)*inv(M-I) to show
        (5) Cc = 4*inv(Ad + I)*Cd*inv(Ad' + I)
Step 2) Substitute (2) and (5) into (4)
Step 3) Replace (Ad - I) with (Ad + I -2I)
        Replace (Ad' - I) with (Ad' + I -2I)
Step 4) Multiply through and simplify to get
        X -inv(Ad+I)*X -X*inv(Ad'+I) +inv(Ad+I)*Cd*inv(Ad'+I) = 0
Step 5) Left multiply by (Ad + I) and right multiply by (Ad' + I)
Step 6) Simplify to (1)
```

DOT Vector dot product.

```
C = DOT(A,B) returns the scalar product of the vectors A and B.
A and B must be vectors of the same length.  When A and B are both
column vectors, DOT(A,B) is the same as A'*B.

DOT(A,B), for N-D arrays A and B, returns the scalar product
along the first non-singleton dimension of A and B. A and B must
have the same size.

DOT(A,B,DIM) returns the scalar product of A and B in the
dimension DIM.

See also CROSS.

Clay M. Thompson
Copyright (c) 1984-96 by The MathWorks, Inc.
$Revision: 5.6 $
```

EIG Eigenvalues and eigenvectors.

```
E = EIG(X) is a vector containing the eigenvalues of a square
matrix X.

[V,D] = EIG(X) produces a diagonal matrix D of eigenvalues and a
full matrix V whose columns are the corresponding eigenvectors so
that X*V = V*D.

[V,D] = EIG(X,'nobalance') performs the computation with balancing
disabled, which sometimes gives more accurate results for certain
problems with unusual scaling.

E = EIG(A,B) is a vector containing the generalized eigenvalues
of square matrices A and B.

[V,D] = EIG(A,B) produces a diagonal matrix D of generalized
eigenvalues and a full matrix V whose columns are the
corresponding eigenvectors so that A*V = B*V*D.

See also CONDEIG.

Copyright (c) 1984-96 by The MathWorks, Inc.
$Revision: 5.3 $  $Date: 1996/05/10 16:01:07 $
Built-in function.
```

```
ESTIM  Form estimator given estimator gain.
   EST = ESTIM(SYS,L)  produces an estimator EST with gain L for
   the outputs and states of the state-space model SYS, assuming
   all inputs of SYS are stochastic and all outputs are measured.
   For a continuous system
                 .
      SYS:      x = Ax + Bw ,    y = Cx + Dw   (with w stochastic),

   the resulting estimator
          .
       x_e  = [A-LC] x_e + Ly

      |y_e| = |C| x_e
      |x_e|   |I|

   generates estimates x_e and y_e of x and y.  ESTIM behaves
   similarly when applied to discrete-time systems.

   EST = ESTIM(SYS,L,SENSORS,KNOWN)  handles more general plants
   SYS with both deterministic and stochastic inputs, and both
   measured and non-measured outputs.  The index vectors SENSORS
   and KNOWN specify which outputs y are measured and which inputs
   u are known, respectively.  The resulting estimator EST uses
   [u;y] as input to produce the estimates [y_e;x_e].

   You can use pole placement techniques (see PLACE) to design
   the estimator (observer) gain L, or use the Kalman filter gain
   returned by KALMAN or KALMD.

   See also  REG, PLACE, KALMAN, KALMD, LQGREG.

   Clay M. Thompson 7-2-90
   Copyright (c) 1986-96 by The MathWorks, Inc.
   $Revision: 1.2 $  $Date: 1996/10/18 15:15:51 $


EXPM   Matrix exponential.
   EXPM(X) is the matrix exponential of X.  EXPM is computed using
   a scaling and squaring algorithm with a Pade approximation.

   Although it is not computed this way, if X has a full set
   of eigenvectors V with corresponding eigenvalues D, then
   [V,D] = EIG(X) and EXPM(X) = V*diag(exp(diag(D)))/V.

   EXPM1, EXPM2 and EXPM3 are alternative methods.

   EXP(X) (that's without the M) does it element-by-element.

   See also EXPM1, EXPM2, EXPM3, LOGM, SQRTM, FUNM.

   Copyright (c) 1984-96 by The MathWorks, Inc.
   $Revision: 5.3 $  $Date: 1996/05/10 16:01:45 $
   Built-in function.
```

```
EYE Identity matrix.
   EYE(N) is the N-by-N identity matrix.

   EYE(M,N) or EYE([M,N]) is an M-by-N matrix with 1's on
   the diagonal and zeros elsewhere.

   EYE(SIZE(A)) is the same size as A.

   See also ONES, ZEROS, RAND, RANDN.

   Copyright (c) 1984-96 by The MathWorks, Inc.
   $Revision: 5.2 $  $Date: 1996/03/29 20:03:29 $
   Built-in function.


FUNM Evaluate general matrix function.
   F = FUNM(A,'fun') for a square matrix argument A, evaluates the
   matrix version of the function specified by 'fun'.  For example,
   FUNM(A,'sin') is the matrix sine.  For matrix exponentials,
   logarithms and square roots, use EXPM(A), LOGM(A) and SQRTM(A)
   instead.

   FUNM uses a potentially unstable algorithm.  If A is close to a
   matrix with multiple eigenvalues and poorly conditioned eigenvectors,
   FUNM may produce inaccurate results.  An attempt is made to detect
   this situation and print a warning message.  The error detector is
   sometimes too sensitive and a message is printed even though the
   the computed result is accurate.

   [F,ESTERR] = FUNM(A,'fun') does not print any message, but returns
   a very rough estimate of the relative error in the computed result.

   If A is symmetric or Hermitian, then its Schur form is diagonal and
   FUNM is able to produce an accurate result.

   S = SQRTM(A) and L = LOGM(A) use FUNM to do their computations,
   but they can get more reliable error estimates by comparing S*S
   and EXPM(L) with A.  E = EXPM(A) uses a completely different
   algorithm.

   See also EXPM, SQRTM, LOGM.

   C.B. Moler 12-2-85, 7-21-86, 7-11-92, 5-2-95.
   Copyright (c) 1984-96 by The MathWorks, Inc.
   $Revision: 5.7 $  $Date: 1996/05/10 16:02:21 $

 Parlett's method.  See Golub and VanLoan (1983), p. 384.


GRAM   Controllability and observability gramians.
   Wc = GRAM(SYS,'c')  returns the controllability gramian of
   the state-space system SYS.

   Wo = GRAM(SYS,'o')  returns its observability gramian.

   In both cases, the state-space model SYS should be stable.
   The gramians are computed by solving the Lyapunov equations:
```

```
    *  A*Wc + Wc*A' + BB' = 0  and   A'*Wo + Wo*A + C'C = 0
       for continuous-time systems
              dx/dt = A x + B u  ,   y = C x + D u

    *  A*Wc*A' - Wc + BB' = 0  and   A'*Wo*A - Wo + C'C = 0
       for discrete-time systems
          x[n+1] = A x[n] + B u[n] ,  y[n] = C x[n] + D u[n].

 See also  BALREAL, CTRB, OBSV.

 J.N. Little 3-6-86
 P. Gahinet  6-27-96
 Copyright (c) 1986-96 by The MathWorks, Inc.
 $Revision: 1.2 $  $Date: 1996/10/18 15:17:53 $

 Laub, A., "Computation of Balancing Transformations", Proc. JACC
   Vol.1, paper FA8-E, 1980.
```

HANKEL Hankel matrix.

```
 HANKEL(C) is a square Hankel matrix whose first column is C and
 whose elements are zero below the first anti-diagonal.

 HANKEL(C,R) is a Hankel matrix whose first column is C and whose
 last row is R.

 Hankel matrices are symmetric, constant across the anti-diagonals,
 and have elements H(i,j) = P(i+j-1) where P = [C R(2:END)]
 completely determines the Hankel matrix.

 See also TOEPLITZ.

 J.N. Little 4-22-87
 Revised 1-28-88 JNL
 Revised 2-25-95 Jim McClellan
 Copyright (c) 1984-96 by The MathWorks, Inc.
 $Revision: 5.5 $  $Date: 1996/08/08 21:13:27 $
```

INITIAL Initial condition response of state-space models.

```
 INITIAL(SYS,X0) plots the undriven response of the state-space
 system SYS with initial condition X0 on the states.  This
 response is characterized by the equations
                                 .
   Continuous time:    x = A x ,  y = C x ,  x(0) = x0

   Discrete time:  x[k+1] = A x[k],  y[k] = C x[k],  x[0] = x0 .

 The time range and number of points are chosen automatically.

 INITIAL(SYS,X0,TFINAL)  simulates the time response from t = 0
 to the final time t = TFINAL.  For discrete-time systems with
 unspecified sample time, TFINAL should be the number of samples.

 INITIAL(SYS,X0,T)  specifies a time vector T to be used for
 simulation.  For discrete systems, T should be of the form
```

```
0:Ts:Tf where Ts is the sample time of the system.  For continuous
systems, T should be of the form 0:dt:Tf where dt will become the
sample time of a discrete approximation of the continuous system.

INITIAL(SYS1,SYS2,...,X0,T)  plots the response of multiple LTI
systems SYS1,SYS2,... on a single plot.  The time vector T is
optional.  You can also specify a color, line style, and marker
for each system, as in  initial(sys1,'r',sys2,'y--',sys3,'gx',x0).

When invoked with left hand arguments,
    [Y,T,X] = INITIAL(SYS,X0,...)
returns the output response Y, the time vector T used for simulation,
and the state trajectories X.  No plot is drawn on the screen.  The
matrix Y has LENGTH(T) rows and as many columns as outputs in SYS.
Similarly, X has LENGTH(T) rows and as many columns as states.

See also  IMPULSE, STEP, LSIM.

 Clay M. Thompson  7-6-90
 Revised ACWG 6-21-92
 Revised AFP 9-21-94,  PG 4-25-96
 Copyright (c) 1986-94 by The MathWorks, Inc.
 $Revision: 1.4 $  $Date: 1996/10/18 15:16:41 $
```

INV   Matrix inverse.

```
INV(X) is the inverse of the square matrix X.
A warning message is printed if X is badly scaled or
nearly singular.

See also SLASH, PINV, COND, CONDEST, NNLS, LSCOV.

Copyright (c) 1984-96 by The MathWorks, Inc.
$Revision: 5.3 $  $Date: 1996/05/10 16:03:33 $
Built-in function.
```

KALMAN Continuous or discrete Kalman estimator

```
[KEST,L,P] = KALMAN(SYS,Qn,Rn,Nn)  designs a Kalman estimator
KEST for the continuous or discrete LTI plant SYS.  For a
continuous plant
     .
     x = Ax + Bu + Gw                {State equation}
     y = Cx + Du + Hw + v            {Measurements}

with known inputs u, process noise w, measurement noise v, and
noise covariances

     E{ww'} = Qn,     E{vv'} = Rn,     E{wv'} = Nn,

the estimator KEST has input [u;y] and generates optimal
estimates y_e,x_e of y,x by:
      .
      x_e  = Ax_e + Bu + L(y - Cx_e - Du)

     |y_e| = | C | x_e + | D | u
     |x_e|   | I |       | 0 |
```

```
Type HELP DKALMAN for details on the discrete-time counterpart.

The LTI system SYS contains the plant data (A,[B G],C,[D H]),
and Nn is set to zero when omitted.  The Kalman estimator KEST
is continuous when SYS is continuous, discrete otherwise.
Also returned are the estimator gain L and the steady-state
error covariance P.  In continuous time with H=0, P solves the
Riccati equation
                                  -1
     AP + PA' - (PC'+G*N)R  (CP+N'*G') + G*Q*G' = 0 .


[KEST,L,P] = KALMAN(SYS,Qn,Rn,Nn,SENSORS,KNOWN)  handles more
general plants SYS where the known and stochastic inputs u,w
are mixed together, and not all outputs are measured.  The
index vectors SENSORS and KNOWN then specify which outputs y
of SYS are measured and which inputs u are known.  All other
inputs are assumed stochastic.

See also  KALMD, ESTIM, LQGREG, CARE, DARE.

Author(s): P. Gahinet  8-1-96
Copyright (c) 1986-96 by The MathWorks, Inc.
$Revision: 1.5 $  $Date: 1996/11/05 16:13:17 $
```

KRON   Kronecker tensor product.

```
KRON(X,Y) is the Kronecker tensor product of X and Y.
The result is a large matrix formed by taking all possible
products between the elements of X and those of Y.   For
example, if X is 2 by 3, then KRON(X,Y) is

   [ X(1,1)*Y  X(1,2)*Y  X(1,3)*Y
     X(2,1)*Y  X(2,2)*Y  X(2,3)*Y ]

If either X or Y is sparse, only nonzero elements are multiplied
in the computation, and the result is sparse.

Paul L. Fackler, North Carolina State, 9-23-96
Copyright (c) 1984-96 by The MathWorks, Inc.
$Revision: 5.4 $
```

LQE Linear quadratic estimator design.

```
For the continuous-time system:
        .
        x = Ax + Bu + Gw              {State equation}
        z = Cx + Du + v               {Measurements}

with process noise and measurement noise covariances:

        E{w} = E{v} = 0,  E{ww'} = Q,  E{vv'} = R, E{wv'} = 0

L = LQE(A,G,C,Q,R) returns the gain matrix L such that the
stationary Kalman filter:
```

```
        .
    x = Ax + Bu + L(z - Cx - Du)

produces an LQG optimal estimate of x. The estimator can be formed
with ESTIM.

[L,P,E] = LQE(A,G,C,Q,R) returns the gain matrix L, the Riccati
equation solution P which is the estimate error covariance, and
the closed loop eigenvalues of the estimator: E = EIG(A-L*C).

[L,P,E] = LQE(A,G,C,Q,R,N) solves the estimator problem when the
process and sensor noise is correlated: E{wv'} = N.

J.N. Little 4-21-85
Revised Clay M. Thompson  7-16-90
Copyright (c) 1986-93 by the MathWorks, Inc.
```

LQR Linear-quadratic regulator design for continuous-time systems.

```
[K,S,E] = LQR(A,B,Q,R,N)  calculates the optimal gain matrix K
such that the state-feedback law  u = -Kx  minimizes the cost
function

    J = Integral {x'Qx + u'Ru + 2*x'Nu} dt
                                      .
subject to the state dynamics  x = Ax + Bu.

The matrix N is set to zero when omitted.  Also returned are the
Riccati equation solution S and the closed-loop eigenvalues E:
                      -1
  SA + A'S - (SB+N)R   (B'S+N') + Q = 0 ,    E = EIG(A-B*K) .


See also  LQRY, DLQR, LQGREG, CARE, and REG.

Author(s): J.N. Little 4-21-85
Revised    P. Gahinet  7-24-96
Copyright (c) 1986-96 by The MathWorks, Inc.
$Revision: 1.4 $  $Date: 1996/11/05 16:13:08 $
```

LSIM Simulation of the time response of LTI systems to arbitrary inputs.

```
LSIM(SYS,U,T)  plots the time response of the LTI system SYS to the
input signal described by U and T.  The time vector T consists of
regularly spaced time samples and U is a matrix with as many columns
as inputs and whose i-th row specifies the input value at time T(i).
For instance,
        t = 0:0.01:5;   u = sin(t);   lsim(sys,u,t)
simulates the response of SYS to u(t) = sin(t) during 5 seconds.

In discrete time, U should be sampled at the same rate as the system
(T is then redundant and can be omitted or set to the empty matrix).
In continuous time, the sampling period  T(2)-T(1)  should be chosen
small enough to capture the details of the input signal.  The time
vector T is resampled when intersample oscillations may occur.

LSIM(SYS,U,T,X0)  specifies an additional nonzero initial state X0
(for state-space systems only).
```

```
   LSIM(SYS1,SYS2,...,U,T,X0)  simulates the response of multiple LTI
   systems SYS1,SYS2,... on a single plot.  The initial condition X0
   is optional.  You can also specify a color, line style, and marker
   for each system, as in  lsim(sys1,'r',sys2,'y--',sys3,'gx',u,t).

   When invoked with left hand arguments,
        [Y,T] = LSIM(SYS,U,...)
   returns the output history Y and time vector T used for simulation.
   No plot is drawn on the screen.  The matrix Y has LENGTH(T) rows
   and as many columns as outputs in SYS.

   For state-space systems,
        [Y,T,X] = LSIM(SYS,U,...)
   also returns the state trajectory X, a matrix with LENGTH(T) rows
   and as many columns as states.

   See also  GENSIG, STEP, IMPULSE, INITIAL.

  LSIM normally linearly interpolates the input (using a first order hold)
  which is more accurate for continuous inputs. For discrete inputs such
  as square waves LSIM tries to detect these and uses a more accurate
  zero-order hold method. LSIM can be confused and for accurate results
  a small time interval should be used.

   J.N. Little 4-21-85
   Revised 7-31-90  Clay M. Thompson
       Revised A.C.W.Grace 8-27-89 (added first order hold)
                    1-21-91 (test to see whether to use foh or zoh)
   Revised 12-5-95 Andy Potvin
   Revised 5-8-96  P. Gahinet
   Copyright (c) 1986-96 by The MathWorks, Inc.
   $Revision: 1.5 $  $Date: 1996/10/06 12:59:31 $
```

LYAP Solve continuous-time Lyapunov equations.

```
   X = LYAP(A,C) solves the special form of the Lyapunov matrix
   equation:

           A*X + X*A' = -C

   X = LYAP(A,B,C) solves the general form of the Lyapunov matrix
   equation (also called Sylvester equation):

           A*X + X*B = -C

   See also  DLYAP.

    S.N. Bangert 1-10-86
    Copyright (c) 1986-96 by The MathWorks, Inc.
    $Revision: 1.2 $  $Date: 1996/10/18 15:16:28 $
    Last revised JNL 3-24-88, AFP 9-3-95
```

NORM   Matrix or vector norm.

```
   For matrices...
     NORM(X) is the largest singular value of X, max(svd(X)).
```

```
    NORM(X,2) is the same as NORM(X).
    NORM(X,1) is the 1-norm of X, the largest column sum,
                    = max(sum(abs((X)))).
    NORM(X,inf) is the infinity norm of X, the largest row sum,
                    = max(sum(abs((X')))).
    NORM(X,'fro') is the Frobenius norm, sqrt(sum(diag(X'*X))).
    NORM(X,P) is available for matrix X only if P is 1, 2, inf or 'fro'.

  For vectors...
    NORM(V,P) = sum(abs(V).^P)^(1/P).
    NORM(V) = norm(V,2).
    NORM(V,inf) = max(abs(V)).
    NORM(V,-inf) = min(abs(V)).

  See also COND, CONDEST, NORMEST.
```



```
  $Revision: 5.5 $  $Date: 1996/04/05 15:20:20 $
  Built-in function.
```

NULL Null space.

```
  Z = NULL(A) is an orthonormal basis for the null space of A obtained
  from the singular value decomposition.  That is,  A*Z has negligible
  elements, size(Z,2) is the nullity of A, and Z'*Z = I.

  Z = NULL(A,'r') is a "rational" basis for the null space obtained
  from the reduced row echelon form.  A*Z is zero, size(Z,2) is an
  estimate for the nullity of A, and, if A is a small matrix with
  integer elements, the elements of R are ratios of small integers.

  The orthonormal basis is preferable numerically, while the rational
  basis may be preferable pedagogically.

  Example:

      A =
          1     2     3
          1     2     3
          1     2     3

     null(A) =

         -0.1690   -0.9487
          0.8452    0.0000
         -0.5071    0.3162

     null(A,'r') =

          -2    -3
           1     0
           0     1

  See also SVD, ORTH, RANK, RREF.
```



```
  $Revision: 5.6 $  $Date: 1996/10/04 16:12:51 $
```

```
OBSV   Form the observability matrix.
   OB = OBSV(A,C)  returns the observability matrix [C; CA; CA^2 ...]

   CO = OBSV(SYS)  returns the observability matrix of the state-space
   system SYS with realization (A,B,C,D).  This is equivalent to
   OBSV(sys.a,sys.c)

   See also  OBSVF.

   Copyright (c) 1986-96 by The MathWorks, Inc.
   $Revision: 1.1 $  $Date: 1996/06/28 18:26:28 $
```

```
OBSVF  Observability staircase form.
   [ABAR,BBAR,CBAR,T,K] = OBSVF(A,B,C) returns a decomposition
   into the observable/unobservable subspaces.

   [ABAR,BBAR,CBAR,T,K] = OBSVF(A,B,C,TOL) uses tolerance TOL.

   If Ob=OBSV(A,C) has rank r <= n, then there is a similarity
   transformation T such that

      Abar = T * A * T' ,  Bbar = T * B  ,  Cbar = C * T' .

   and the transformed system has the form

          | Ano   A12|           |Bno|
   Abar =  ----------  ,  Bbar =  ---  ,  Cbar = [ 0 | Co].
          |  0    Ao |           |Bo |
                                              -1          -1
   where (Ao,Bo) is controllable, and Co(sI-Ao) Bo = C(sI-A) B.

   See also  OBSV, CTRBF.

   Author : R.Y. Chiang  3-21-86
   Revised 5-27-86 JNL
   Copyright (c) 1986-96 by The MathWorks, Inc.
   $Revision: 1.1 $  $Date: 1996/08/02 15:11:23 $
```

```
ODE23  Solve non-stiff differential equations, low order method.
  [T,Y] = ODE23('F',TSPAN,Y0) with TSPAN = [T0 TFINAL] integrates the
  system of differential equations y' = F(t,y) from time T0 to TFINAL with
  initial conditions Y0.  'F' is a string containing the name of an ODE
  file.  Function F(T,Y) must return a column vector.  Each row in
  solution array Y corresponds to a time returned in column vector T.  To
  obtain solutions at specific times T0, T1, ..., TFINAL (all increasing
  or all decreasing), use TSPAN = [T0 T1 ... TFINAL].

  [T,Y] = ODE23('F',TSPAN,Y0,OPTIONS) solves as above with default
  integration parameters replaced by values in OPTIONS, an argument
  created with the ODESET function.  See ODESET for details.  Commonly
  used options are scalar relative error tolerance 'RelTol' (1e-3 by
  default) and vector of absolute error tolerances 'AbsTol' (all
  components 1e-6 by default).
```

```
[T,Y] = ODE23('F',TSPAN,Y0,OPTIONS,P1,P2,...) passes the additional
parameters P1,P2,... to the ODE file as F(T,Y,FLAG,P1,P2,...) (see
ODEFILE).  Use OPTIONS = [] as a place holder if no options are set.

It is possible to specify TSPAN, Y0 and OPTIONS in the ODE file (see
ODEFILE).  If TSPAN or Y0 is empty, then ODE23 calls the ODE file
[TSPAN,Y0,OPTIONS] = F([],[],'init') to obtain any values not supplied
in the ODE23 argument list.  Empty arguments at the end of the call list
may be omitted, e.g. ODE23('F').

As an example, the commands

    options = odeset('RelTol',1e-4,'AbsTol',[1e-4 1e-4 1e-5]);
    ode23('rigidode',[0 12],[0 1 1],options);

solve the system y' = rigidode(t,y) with relative error tolerance 1e-4
and absolute tolerances of 1e-4 for the first two components and 1e-5
for the third.  When called with no output arguments, as in this
example, ODE23 calls the default output function ODEPLOT to plot the
solution as it is computed.

[T,Y,TE,YE,IE] = ODE23('F',TSPAN,Y0,OPTIONS) with the Events property in
OPTIONS set to 'on', solves as above while also locating zero crossings
of an event function defined in the ODE file.  The ODE file must be
coded so that F(T,Y,'events') returns appropriate information.  See
ODEFILE for details.  Output TE is a column vector of times at which
events occur, rows of YE are the corresponding solutions, and indices in
vector IE specify which event occurred.

[T,X,Y] = ODE23('MODEL',TSPAN,Y0,OPTIONS,UT,P1,P2,...) simulates a
SIMULINK model using the 'ODE23' integrator.  The OPTIONS argument is
created with the ODESET function.  See also SIM.

See also ODEFILE and
    other ODE solvers:  ODE45, ODE113, ODE15S, ODE23S
    options handling:   ODESET, ODEGET
    output functions:   ODEPLOT, ODEPHAS2, ODEPHAS3, ODEPRINT
    odefile examples:   ORBITODE, ORBT2ODE, RIGIDODE, VDPODE

ODE23 is an implementation of the explicit Runge-Kutta (2,3) pair of
Bogacki and Shampine called BS23.  It uses a "free" interpolant of order
3.  Local extrapolation is done.

Details are to be found in The MATLAB ODE Suite, L. F. Shampine and
M. W. Reichelt, SIAM Journal on Scientific Computing, 18-1, 1997.

Mark W. Reichelt and Lawrence F. Shampine, 6-14-94
Copyright (c) 1984-96 by The MathWorks, Inc.
$Revision: 5.43 $  $Date: 1996/11/10 17:46:29 $
```

ORTH Orthogonalization.

```
Q = ORTH(A) is an orthonormal basis for the range of A.
That is, Q'*Q = I, the columns of Q span the same space as
the columns of A, and the number of columns of Q is the
rank of A.
```

```
   See also SVD, RANK, NULL.

   Major revision, 4-13-93, C. Moler.
   Copyright (c) 1984-96 by The MathWorks, Inc.
   $Revision: 5.5 $  $Date: 1996/05/10 16:06:37 $
```

Beginning with MATLAB 4.1, the algorithms for NULL and ORTH use singular value decomposition, SVD, instead of orthogonal factorization, QR. This doubles the computation time, but provides more reliable and consistent rank determination.

PINV Pseudoinverse.

```
   X = PINV(A) produces a matrix X of the same dimensions
   as A' so that A*X*A = A, X*A*X = X and A*X and X*A
   are Hermitian. The computation is based on SVD(A) and any
   singular values less than a tolerance are treated as zero.
   The default tolerance is MAX(SIZE(A)) * NORM(A) * EPS.

   PINV(A,TOL) uses the tolerance TOL instead of the default.

   See also RANK.

   Copyright (c) 1984-96 by The MathWorks, Inc.
   $Revision: 5.5 $  $Date: 1996/05/10 16:07:14 $
```

PLACE Pole placement technique

```
   K = PLACE(A,B,P)  computes a state-feedback matrix K such that
   the eigenvalues of  A-B*K  are those specified in vector P.
   No eigenvalue should have a multiplicity greater than the
   number of inputs.

   [K,PREC,MESSAGE] = PLACE(A,B,P)  returns PREC, an estimate of how
   closely the eigenvalues of A-B*K match the specified locations P
   (PREC measures the number of accurate decimal digits in the actual
   closed-loop poles).  If some nonzero closed-loop pole is more than
   10 off from the desired location, MESSAGE contains a warning
   message.

   See also  ACKER.

   M. Wette 10-1-86
   Revised 9-25-87 JNL
   Revised 8-4-92 Wes Wang
   Revised 10-5-93, 6-1-94 Andy Potvin

   Ref:  Kautsky, Nichols, Van Dooren, "Robust Pole Assignment in Linear
         State Feedback," Intl. J. Control, 41(1985)5, pp 1129-1155


   Copyright (c) 1986-96 by The MathWorks, Inc.
   $Revision: 1.2 $  $Date: 1996/09/27 21:43:23 $
```

```
POLYVALM Evaluate polynomial with matrix argument.
   POLYVALM(V,X) where V is a vector whose elements are the
   coefficients of a polynomial, is the value of the polynomial
   evaluated with matrix argument X.  X must be a square matrix.

   See also POLYVAL, POLYFIT.

   J.N.Little 4-20-86
   Copyright (c) 1984-96 by The MathWorks, Inc.
   $Revision: 5.4 $  $Date: 1996/10/27 14:01:11 $

 Polynomial evaluation c(x) using Horner's method


QR  Orthogonal-triangular decomposition.
   [Q,R] = QR(A) produces an upper triangular matrix R of the same
   dimension as A and a unitary matrix Q so that A = Q*R.

   [Q,R,E] = QR(A) produces a permutation matrix E, an upper
   triangular R and a unitary Q so that A*E = Q*R.  The column
   permutation E is chosen so that abs(diag(R)) is decreasing.

   [Q,R] = QR(A,0) produces the "economy size" decomposition.
   If A is m-by-n with m > n, then only the first n columns of Q
   are computed.

   [Q,R,E] = QR(A,0) produces an "economy size" decomposition in
   which E is a permutation vector, so that Q*R = A(:,E).  The column
   permutation E is chosen so that abs(diag(R)) is decreasing.

   By itself, QR(A) returns the output of LINPACK'S ZQRDC routine.
   TRIU(QR(A)) is R.

   For sparse matrices, QR can compute a "Q-less QR decomposition",
   which has the following slightly different behavior.

   R = QR(A) returns only R.  Note that R = chol(A'*A).
   [Q,R] = QR(A) returns both Q and R, but Q is often nearly full.
   [C,R] = QR(A,B), where B has as many rows as A, returns C = Q'*B.
   R = QR(A,0) and [C,R] = QR(A,B,0) produce economy size results.

   The sparse version of QR does not do column permutations.
   The full version of QR does not return C.

   The least squares approximate solution to A*x = b can be found
   with the Q-less QR decomposition and one step of iterative refinement:

       x = R\(R'\(A'*b))
       r = b - A*x
       e = R\(R'\(A'*r))
       x = x + e;

   See also LU, NULL, ORTH, QRDELETE, QRINSERT.

   Copyright (c) 1984-96 by The MathWorks, Inc.
   $Revision: 5.5 $  $Date: 1996/08/15 21:52:13 $
   Built-in function.
```

```
RANDN  Normally distributed random numbers.
  RANDN(N) is an N-by-N matrix with random entries, chosen from
  a normal distribution with mean zero and variance one.
  RANDN(M,N) and RANDN([M,N]) are M-by-N matrices with random entries.
  RANDN(M,N,P,...) or RANDN([M,N,P...]) generate random arrays.
  RANDN with no arguments is a scalar whose value changes each time it
  is referenced.  RANDN(SIZE(A)) is the same size as A.

  S = RANDN('state') is a 2-element vector containing the current state
  of the normal generator.  RANDN('state',S) resets the state to S.
  RANDN('state',0) resets the generator to its initial state.
  RANDN('state',J), for integer J, resets the generator to its J-th state.
  RANDN('state',sum(100*clock)) resets it to a different state each time.

  MATLAB Version 4.x used random number generators with a single seed.
  RANDN('seed',0) and RANDN('seed',J) cause the MATLAB 4 generator
  to be used.
  RANDN('seed') returns the current seed of the MATLAB 4 normal generator.
  RANDN('state',J) and RANDN('state',S) cause the MATLAB 5 generator
  to be used.

  See also RAND, SPRAND, SPRANDN, RANDPERM.

  Copyright (c) 1984-96 by The MathWorks, Inc.
  $Revision: 5.6 $  $Date: 1996/04/19 22:11:11 $
  Built-in function.


RANK   Matrix rank.
   RANK(A) provides an estimate of the number of linearly
   independent rows or columns of a matrix A.
   RANK(A,tol) is the number of singular values of A
   that are larger than tol.
   RANK(A) uses the default tol = max(size(A)) * norm(A) * eps.

   Copyright (c) 1984-96 by The MathWorks, Inc.
   $Revision: 5.5 $  $Date: 1996/08/29 17:43:14 $


REG  Form regulator given state-feedback and estimator gains.
   RSYS = REG(SYS,K,L) produces an observer-based regulator RSYS
   for the state-space system SYS, assuming all inputs of SYS are
   controls and all outputs are measured.  The matrices K and L
   specify the state-feedback and observer gains.  For
                .
       SYS:    x = Ax + Bu ,    y = Cx + Du

   the resulting regulator is
        .
       x_e = [A-BK-LC+LDK] x_e + Ly
         u = -K x_e

   This regulator should be connected to the plant using positive
   feedback.  REG behaves similarly when applied to discrete-time
   systems.
```

```
RSYS = REG(SYS,K,L,SENSORS,KNOWN,CONTROLS)  handles more
general regulation problems where
  * the plant inputs consist of controls u, known inputs Ud,
    and stochastic inputs w,
  * only a subset y of the plant outputs are measured.
The I/O subsets y, Ud, and u are specified by the index vectors
SENSORS, KNOWN, and CONTROLS.  The resulting regulator RSYS
uses [Ud;y] as input to generate the commands u.

You can use pole placement techniques (see PLACE) to design the
gains K and L, or alternatively use the LQ and Kalman gains
produced by LQR/DLQR and KALMAN.

See also  ESTIM, PLACE, LQR, DLQR, LQGREG, KALMAN.

Clay M. Thompson 6-29-90
Copyright (c) 1986-96 by The MathWorks, Inc.
$Revision: 1.2 $  $Date: 1996/10/18 15:15:58 $
```

RESIDUE Partial-fraction expansion (residues).

```
[R,P,K] = RESIDUE(B,A) finds the residues, poles and direct term of
a partial fraction expansion of the ratio of two polynomials B(s)/A(s).
If there are no multiple roots,
   B(s)       R(1)       R(2)             R(n)
   ----  =  -------- + -------- + ... + -------- + K(s)
   A(s)     s - P(1)   s - P(2)         s - P(n)
Vectors B and A specify the coefficients of the numerator and
denominator polynomials in descending powers of s.  The residues
are returned in the column vector R, the pole locations in column
vector P, and the direct terms in row vector K.  The number of
poles is n = length(A)-1 = length(R) = length(P). The direct term
coefficient vector is empty if length(B) < length(A), otherwise
length(K) = length(B)-length(A)+1.

If P(j) = ... = P(j+m-1) is a pole of multplicity m, then the
expansion includes terms of the form

             R(j)        R(j+1)                R(j+m-1)
           -------- + ------------   + ... + ------------
           s - P(j)   (s - P(j))^2           (s - P(j))^m

[B,A] = RESIDUE(R,P,K), with 3 input arguments and 2 output arguments,
converts the partial fraction expansion back to the polynomials with
coefficients in B and A.

Warning: Numerically, the partial fraction expansion of a ratio of
polynomials represents an ill-posed problem.  If the denominator
polynomial, A(s), is near a polynomial with multiple roots, then
small changes in the data, including roundoff errors, can make
arbitrarily large changes in the resulting poles and residues.
Problem formulations making use of state-space or zero-pole
representations are preferable.

See also POLY, ROOTS, DECONV.
```

```
   Reference:  A.V. Oppenheim and R.W. Schafer, Digital
   Signal Processing, Prentice-Hall, 1975, p. 56-58.

   C.R. Denham and J.N. Little, MathWorks, 1986, 1989.
   Copyright (c) 1984-96 by The MathWorks, Inc.
   $Revision: 5.4 $  $Date: 1996/10/27 14:01:22 $
```

ROOTS Find polynomial roots.

```
   ROOTS(C) computes the roots of the polynomial whose coefficients
   are the elements of the vector C. If C has N+1 components,
   the polynomial is C(1)*X^N + ... + C(N)*X + C(N+1).

   See also POLY, RESIDUE, FZERO.

   J.N. Little 3-17-86
   Copyright (c) 1984-96 by The MathWorks, Inc.
   $Revision: 5.4 $  $Date: 1996/10/05 14:22:34 $

 ROOTS finds the eigenvalues of the associated companion matrix.
```

RREF   Reduced row echelon form.

```
   R = RREF(A) produces the reduced row echelon form of A.

   [R,jb] = RREF(A) also returns a vector, jb, so that:
       r = length(jb) is this algorithm's idea of the rank of A,
       x(jb) are the bound variables in a linear system, Ax = b,
       A(:,jb) is a basis for the range of A,
       R(1:r,jb) is the r-by-r identity matrix.

   [R,jb] = RREF(A,TOL) uses the given tolerance in the rank tests.

   Roundoff errors may cause this algorithm to compute a different
   value for the rank than RANK, ORTH and NULL.

   See also RREFMOVIE, RANK, ORTH, NULL, QR, SVD.

   CBM, 11/24/85, 1/29/90, 7/12/92.
   Copyright (c) 1984-96 by The MathWorks, Inc.
   $Revision: 5.3 $  $Date: 1996/03/26 02:31:24 $
```

Slash

```
Matrix division.
 \   Backslash or left division.
     A\B is the matrix division of A into B, which is roughly the
     same as INV(A)*B , except it is computed in a different way.
     If A is an N-by-N matrix and B is a column vector with N
     components, or a matrix with several such columns, then
     X = A\B is the solution to the equation A*X = B computed by
     Gaussian elimination. A warning message is printed if A is
     badly scaled or nearly singular.  A\EYE(SIZE(A)) produces the
     inverse of A.
     If A is an M-by-N matrix with M < or > N and B is a column
     vector with M components, or a matrix with several such columns,
     then X = A\B is the solution in the least squares sense to the
```

```
     under- or overdetermined system of equations A*X = B. The
     effective rank, K, of A is determined from the QR decomposition
     with pivoting. A solution X is computed which has at most K
     nonzero components per column. If K < N this will usually not
     be the same solution as PINV(A)*B.  A\EYE(SIZE(A)) produces a
     generalized inverse of A.

 /   Slash or right division.
     B/A is the matrix division of A into B, which is roughly the
     same as B*INV(A) , except it is computed in a different way.
     More precisely, B/A = (A'\B')'. See \.

 ./  Array right division.
     B./A denotes element-by-element division.  A and B
     must have the same dimensions unless one is a scalar.
     A scalar can be divided with anything.

 .\  Array left division.
     A.\B. denotes element-by-element division.  A and B
     must have the same dimensions unless one is a scalar.
     A scalar can be divided with anything.

   Copyright (c) 1984-96 by The MathWorks, Inc.
   $Revision: 5.2 $  $Date: 1996/04/16 22:02:49 $
```

SS Create state-space models or convert LTI model to state space.

```
   You can create a state-space model by:
    SYS = SS(A,B,C,D)        Continuous-time model
    SYS = SS(A,B,C,D,T)      Discrete-time model with sampling time T
                             (Set T=-1 if undetermined)
    SYS = SS                 Default empty SS object
    SYS = SS(D)              Static gain matrix
    SYS = SS(A,B,C,D,LTI)    State-space model with LTI properties
                             inherited from the system LTI (can be
                             SS, TF, or ZPK)
   All the above syntaxes may be followed by Property/Value pairs.
   (Type "help ltiprops" for details on assignable properties).
   Setting D=0 is interpreted as the zero matrix of adequate
   dimensions.  The output SYS is an SS object.

   SYS_SS = SS(SYS)  converts an arbitrary LTI model SYS to state space,
   i.e., computes a state-space realization SYS_SS of SYS.

   See also   SET, SSDATA, DSS, TF, ZPK.

       Author(s): A. Potvin, 3-1-94
       Revised: P. Gahinet, 4-1-96
       Copyright (c) 1986-96 by The MathWorks, Inc.
       $Revision: 1.6 $  $Date: 1996/10/31 21:47:15 $
```

SSDATA Quick access to state-space data.

```
    [A,B,C,D] = SSDATA(SYS)  returns the values of the A,B,C,D
    matrices.  If SYS is not a state-space model, it is first
    converted to state space.
```

```
[A,B,C,D,TS,TD] = SSDATA(SYS)  also returns the sample time
TS and input delays TD.  For continuous systems, TD is a vector
with one entry per input channel.  For discrete systems, TD is
the empty matrix [].

Other properties of SYS can be accessed with GET or by direct
structure-like referencing (e.g., SYS.Ts)

See also  GET, DSSDATA.

   Author(s): P. Gahinet, 4-1-96
   Copyright (c) 1994 by The MathWorks, Inc.
   $Revision: 1.4 $
```

SS2SS Change of state coordinates for state-space systems.

```
 SYS = SS2SS(SYS,T) performs the similarity transformation z = Tx
 on the state vector x of the system SYS.  The resulting state-space
 system is described by:

                .       -1
                z = [TAT   ] z + [TB] u
                       -1
                y = [CT    ] z + D u

 (Respectively,

          -1   .        -1
      [TET   ] z = [TAT   ] z + [TB] u
                       -1
                y = [CT    ] z + D u

 in the descriptor case).

 See also   CANON, AUGSTATE, SSBAL, BALREAL.

Clay M. Thompson  7-3-90
Copyright (c) 1986-96 by The MathWorks, Inc.
$Revision: 1.2 $  $Date: 1996/06/28 18:28:36 $
```

SS2TF State-space to transfer function conversion.

```
[NUM,DEN] = SS2TF(A,B,C,D,iu)  calculates the transfer function:

            NUM(s)          -1
    H(s) = -------- = C(sI-A) B + D
            DEN(s)
of the system:
    .
    x = Ax + Bu
    y = Cx + Du

from the iu'th input.  Vector DEN contains the coefficients of the
denominator in descending powers of s.  The numerator coefficients
are returned in matrix NUM with as many rows as there are
outputs y.
```

```
   See also TF2SS.

   J.N. Little 4-21-85
   Revised 7-25-90 Clay M. Thompson, 10-11-90 A.Grace
   Copyright (c) 1984-96 by The MathWorks, Inc.
   $Revision: 1.16 $  $Date: 1996/03/18 21:06:12 $
```

STEP Step response of LTI systems.

```
   STEP(SYS)  plots the step response of each input channel of
   the LTI system SYS.  The time range and number of points are
   chosen automatically.

   STEP(SYS,TFINAL)  simulates the step response from t = 0 to the
   final time t = TFINAL.  For discrete-time systems with unspecified
   sampling time, TFINAL is interpreted as the number of samples.

   STEP(SYS,T)  uses the user-supplied time vector T for simulation.
   For discrete-time systems, T should be of the form  Ti:Ts:Tf
   where Ts is the sample time of the system.  For continuous systems,
   T should be of the form  Ti:dt:Tf  where dt will become the sample
   time of a discrete approximation to the continuous system.

   STEP(SYS1,SYS2,...,T)  plots the step response of multiple LTI
   systems SYS1,SYS2,... on a single plot.  The time vector T is
   optional.  You can also specify a color, line style, and marker
   for each system, as in  step(sys1,'r',sys2,'y--',sys3,'gx').

   When invoked with left-hand arguments,
        [Y,T] = STEP(SYS,...)
   returns the output response Y and the time vector T used for
   simulation.  No plot is drawn on the screen.  If SYS has
   NU inputs and NY outputs, and LT=length(T), the array Y is
   LT-by-NY-by-NU  and Y(:,:,j) gives the step response of the
   j-th input channel.

   For state-space systems,
        [Y,T,X] = STEP(SYS,...)
   also returns the state trajectory X which is an LT-by-NX-by-NU
   array if SYS has NX states.

   See also  INITIAL, IMPULSE, LSIM.

Extra notes on user-supplied T: For continuous-time systems, the system is
converted to discrete time with a sample time of dt=t(2)-t(1).  The time
vector plotted is then t=t(1):dt:t(end).

    J.N. Little 4-21-85
    Revised A.C.W.Grace 9-7-89, 5-21-92
    Revised A. Potvin 12-1-95
    Copyright (c) 1986-96 by The MathWorks, Inc.
    $Revision: 1.8 $  $Date: 1996/10/06 12:59:17 $
```

SUBSPACE Angle between subspaces.

```
   SUBSPACE(A,B) finds the angle between two subspaces specified
   by the columns of A and B.  If A and B are vectors of unit
```

```
   length, this is the same as ACOS(A'*B).

   If the angle is small, the two spaces are nearly linearly
   dependent.  In a physical experiment described by some
   observations A, and a second realization of the experiment
   described by B, SUBSPACE(A,B) gives a measure of the amount
   of new information afforded by the second experiment not
   associated with statistical errors of fluctuations.

   L. Shure 11-03-88, CBM 5-3-93, 4-18-94.
```


```
   $Revision: 5.3 $  $Date: 1996/10/28 22:49:14 $
```

SVD   Singular value decomposition.

```
   [U,S,V] = SVD(X) produces a diagonal matrix S, of the same
   dimension as X and with nonnegative diagonal elements in
   decreasing order, and unitary matrices U and V so that
   X = U*S*V'.

   S = SVD(X) returns a vector containing the singular values.

   [U,S,V] = SVD(X,0) produces the "economy size"
   decomposition. If X is m-by-n with m > n, then only the
   first n columns of U are computed and S is n-by-n.
```


```
   $Revision: 5.2 $  $Date: 1996/05/10 19:19:43 $
   Built-in function.
```

TF2SS Transfer function to state-space conversion.

```
   [A,B,C,D] = TF2SS(NUM,DEN)  calculates the state-space
   representation:
       .
       x = Ax + Bu
       y = Cx + Du

   of the system:
             NUM(s)
       H(s) = --------
             DEN(s)

   from a single input.  Vector DEN must contain the coefficients of
   the denominator in descending powers of s.  Matrix NUM must
   contain the numerator coefficients with as many rows as there are
   outputs y.  The A,B,C,D matrices are returned in controller
   canonical form.  This calculation also works for discrete systems.
   To avoid confusion when using this function with discrete systems,
   always use a numerator polynomial that has been padded with zeros
   to make it the same length as the denominator.  See the User's
   guide for more details.

   See also SS2TF.

   J.N. Little 3-24-85
```

```
   $Revision: 1.13 $  $Date: 1996/07/29 19:45:19 $
   Latest revision 4-29-89 JNL, 7-29-96 PG
```

```
TOEPLITZ Toeplitz matrix.
   TOEPLITZ(C,R) is a non-symmetric Toeplitz matrix having C as its
   first column and R as its first row.

   TOEPLITZ(C) is a symmetric (or Hermitian) Toeplitz matrix.

   See also HANKEL.

   Revised 10-8-92, LS - code from A.K. Booer.
   Copyright (c) 1984-96 by The MathWorks, Inc.
   $Revision: 5.2 $  $Date: 1996/08/08 21:13:31 $
```

```
TRACE  Sum of diagonal elements.
   TRACE(A) is the sum of the diagonal elements of A, which is
   also the sum of the eigenvalues of A.

   Copyright (c) 1984-96 by The MathWorks, Inc.
   $Revision: 5.2 $  $Date: 1996/01/01 23:16:22 $
```

References

[B1] Cavallo, Alberto, Roberto Setola, and Francesco Vasca, *Using MATLAB, SIMULINK and Control System Toolbox: A Practical Approach*, Prentice Hall, 1996. (ISBN 0-13-261058-2)

[B2] Chen, Chi-Tsong, *Analog and Digital Control System Design: Transfer-Function, State-Space, and Algebraic Methods*, Oxford University Press, 1993. (ISBN 0-03-094070-2)

[B3] Close, Charles and Dean Frederick, *Modeling and Analysis of Dynamic Systems*, 2nd edition, John Wiley and Sons, Inc., 1993. (ISBN 0-471-12517-2)

[B4] D'Azzo, John J., and Constantine H. Houpis, *Linear Control System Analysis and Design: Conventional and Modern*, 4th edition, McGraw-Hill, 1995. (ISBN 0-07-016321-9)

[B5] Dorato, Peter, Chaouki Abdallah, and Vito Cerone, *Linear-Quadratic Control: An Introduction*, Prentice Hall, 1995. 9ISBN 0-02-329962-2)

[B6] Dorf, Richard C., and Robert H. Bishop, *Modern Control Systems*, 7th edition, Addison-Wesley, 1995. (ISBN 0-201-50174-0)

[B7] Franklin, Gene F., J. David Powell, and Abbas Emami-Naeini, *Feedback Control of Dynamic Systems*, 3rd edition, Addison-Wesley, 1994. (ISBN 0-201-52747-2)

[B8] Franklin Gene F., J. David Powell, and Michael L. Workman, *Digital Control of Dynamic Systems*, 2nd edition, Addison-Wesley, 1990. (ISBN 0-201-11938-2)

[B9] Frederick, Dean K., and Joe H. Chow, *Feedback Control Problems Using MATLAB and the Control System Toolbox*, PWS Publishing Company, 1995. (ISBN 0-534-93798-5)

[B10] Hanselman, Duane C., and Benjamin C. Kuo, *MATLAB Tools for Control System Analysis and Design*, 2nd edition, Prentice Hall, 1995. (PC version ISBN 0-13-202293-1 - Macintosh version ISBN 0-13-202574-4)

[B11] Kuo, Benjamin C., *Automatic Control Systems*, 7th edition, Prentice Hall, 1995. (ISBN 0-13-304759-8)

[B12] Leonard, Naomi Ehrich, and William S. Levine, *Using MATLAB to Analyze and Design Control Systems*, 2nd edition, Addison-Wesley, 1995. (ISBN 0-8053-2193-4)

[B13] Moscinski, Jerzy, and Zbigniew Ogonowski, editors, *Advanced Control with MATLAB and SIMULINK*, Prentice Hall, 1996. (ISBN 0-13-309667X)

[B14] Nise, Norman S., *Control Systems Engineering*, 2nd edition, Addison-Wesley, 1995. (ISBN 0-8053-5424-7)

[B15] Ogata, Katsuhiko, *Designing Linear Control Systems with MATLAB*, Prentice Hall, 1994. (ISBN 0-13-293226-1)

[B16] Ogata, Katsuhiko, *Solving Control Engineering Problems with MATLAB*, Prentice Hall, 1994. (ISBN 0-13-045907-0)

[B17] Phillips, Charles L., and Royce D. Harbor, *Feedback Control Systems*, 3rd edition, Prentice Hall, 1996. (ISBN 0-13-371691-0)

[B18] Phillips, Charles L., and H. Troy Nagle, *Digital Control System Analysis and Design*, 3rd edition, Prentice Hall, 1995. (ISBN 0-13-309832-X)

[B19] Rohrs, Charles E., James L. Melsa, and Donald G. Schultz, *Linear Control Systems*, McGraw Hill, 1993. (ISBN 0-07-041525-0)

[B20] Sciavicco, Lorenzo, and Bruno Siciliano, *Modeling and Control of Robot Manipulators* McGraw-Hill, 1996. (ISBN 0-07-057217-8)

[B21] Strum, Robert, and Donald Kirk, *Contemporary Linear Systems Using MATLAB 4.0*, PWS Publishing Company, 1994. (ISBN 0-534-947107)

[B22] Vaccaro, Richard J., D*igital Control: A State-Space Approach*, McGraw-Hill, 1995. (ISBN 0-07-066781-0)

Index

D

E

F

G

H

I

J

K

L

M

P

Q

R

S

T

U

V

W

Z